Sensory Evaluation of Food

FOOD SCIENCE AND TECHNOLOGY

A Series of Monographs, Textbooks, and Reference Books

1. Flavor Research: Principles and Techniques, *R. Teranishi, I. Hornstein, P. Issenberg, and E. L. Wick*
2. Principles of Enzymology for the Food Sciences, *John R. Whitaker*
3. Low-Temperature Preservation of Foods and Living Matter, *Owen R. Fennema, William D. Powrie, and Elmer H. Marth*
4. Principles of Food Science
   Part I: Food Chemistry, *edited by Owen R. Fennema*
   Part II: Physical Methods of Food Preservation, *Marcus Karel, Owen R. Fennema, and Daryl B. Lund*
5. Food Emulsions, *edited by Stig E. Friberg*
6. Nutritional and Safety Aspects of Food Processing, *edited by Steven R. Tannenbaum*
7. Flavor Research: Recent Advances, *edited by R. Teranishi, Robert A. Flath, and Hiroshi Sugisawa*
8. Computer-Aided Techniques in Food Technology, *edited by Israel Saguy*
9. Handbook of Tropical Foods, *edited by Harvey T. Chan*
10. Antimicrobials in Foods, *edited by Alfred Larry Branen and P. Michael Davidson*
11. Food Constituents and Food Residues: Their Chromatographic Determination, *edited by James F. Lawrence*
12. Aspartame: Physiology and Biochemistry, *edited by Lewis D. Stegink and L. J. Filer, Jr.*
13. Handbook of Vitamins: Nutritional, Biochemical, and Clinical Aspects, *edited by Lawrence J. Machlin*
14. Starch Conversion Technology, *edited by G. M. A. van Beynum and J. A. Roels*
15. Food Chemistry: Second Edition, Revised and Expanded, *edited by Owen R. Fennema*

16. Sensory Evaluation of Food: Statistical Methods and Procedures, *Michael O'Mahony*
17. Alternative Sweetners, *edited by Lyn O'Brien Nabors and Robert C. Gelardi*
18. Citrus Fruits and Their Products: Analysis and Technology, *S. V. Ting and Russell L. Rouseff*
19. Engineering Properties of Foods, *edited by M. A. Rao and S. S. H. Rizvi*
20. Umami: A Basic Taste, *edited by Yojiro Kawamura and Morley R. Kare*
21. Food Biotechnology, *edited by Dietrich Knorr*
22. Food Texture: Instrumental and Sensory Measurement, *edited by Howard R. Moskowitz*
23. Seafoods and Fish Oils in Human Health and Disease, *John E. Kinsella*
24. Postharvest Physiology of Vegetables, *edited by J. Weichmann*
25. Handbook of Dietary Fiber: An Applied Approach, *Mark L. Dreher*
26. Food Toxicology, Parts A and B, *Jose M. Concon*
27. Modern Carbohydrate Chemistry, *Roger W. Binkley*
28. Trace Minerals in Foods, *edited by Kenneth T. Smith*
29. Protein Quality and the Effects of Processing, *edited by R. Dixon Phillips and John W. Finley*
30. Adulteration of Fruit Juice Beverages, *edited by Steven Nagy, John A. Attaway, and Martha E. Rhodes*
31. Foodborne Bacterial Pathogens, *edited by Michael P. Doyle*
32. Legumes: Chemistry, Technology, and Human Nutrition, *edited by Ruth H. Matthews*
33. Industrialization of Indigenous Fermented Foods, *edited by Keith H. Steinkraus*
34. International Food Regulation Handbook: Policy · Science · Law, *edited by Roger D. Middlekauff and Philippe Shubik*
35. Food Additives, *edited by A. Larry Branen, P. Michael Davidson, and Seppo Salminen*
36. Safety of Irradiated Foods, *J. F. Diehl*
37. Omega-3 Fatty Acids in Health and Disease, *edited by Robert S. Lees and Marcus Karel*
38. Food Emulsions: Second Edition, Revised and Expanded, *edited by Kåre Larsson and Stig E. Friberg*
39. Seafood: Effects of Technology on Nutrition, *George M. Pigott and Barbee W. Tucker*
40. Handbook of Vitamins: Second Edition, Revised and Expanded, *edited by Lawrence J. Machlin*
41. Handbook of Cereal Science and Technology, *Klaus J. Lorenz and Karel Kulp*
42. Food Processing Operations and Scale-Up, *Kenneth J. Valentas, Leon Levine, and J. Peter Clark*
43. Fish Quality Control by Computer Vision, *edited by L. F. Pau and R. Olafsson*
44. Volatile Compounds in Foods and Beverages, *edited by Henk Maarse*

45. Instrumental Methods for Quality Assurance in Foods, *edited by Daniel Y. C. Fung and Richard F. Matthews*
46. Listeria, Listeriosis, and Food Safety, *Elliot T. Ryser and Elmer H. Marth*
47. Acesulfame-K, *edited by D. G. Mayer and F. H. Kemper*
48. Alternative Sweeteners: Second Edition, Revised and Expanded, *edited by Lyn O'Brien Nabors and Robert C. Gelardi*
49. Food Extrusion Science and Technology, *edited by Jozef L. Kokini, Chi-Tang Ho, and Mukund V. Karwe*
50. Surimi Technology, *edited by Tyre C. Lanier and Chong M. Lee*
51. Handbook of Food Engineering, *edited by Dennis R. Heldman and Daryl B. Lund*
52. Food Analysis by HPLC, *edited by Leo M. L. Nollet*
53. Fatty Acids in Foods and Their Health Implications, *edited by Ching Kuang Chow*
54. *Clostridium botulinum*: Ecology and Control in Foods, *edited by Andreas H. W. Hauschild and Karen L. Dodds*
55. Cereals in Breadmaking: A Molecular Colloidal Approach, *Anne-Charlotte Eliasson and Kåre Larsson*
56. Low-Calorie Foods Handbook, *edited by Aaron M. Altschul*
57. Antimicrobials in Foods: Second Edition, Revised and Expanded, *edited by P. Michael Davidson and Alfred Larry Branen*
58. Lactic Acid Bacteria, *edited by Seppo Salminen and Atte von Wright*
59. Rice Science and Technology, *edited by Wayne E. Marshall and James I. Wadsworth*
60. Food Biosensor Analysis, *edited by Gabriele Wagner and George G. Guilbault*
61. Principles of Enzymology for the Food Sciences: Second Edition, *John R. Whitaker*
62. Carbohydrate Polyesters as Fat Substitutes, *edited by Casimir C. Akoh and Barry G. Swanson*

Additional Volumes in Preparation

Engineering Properties of Foods: Second Edition, Revised and Expanded, *edited by M. A. Rao and Syed S. H. Rizvi*

Handbook of Brewing, *edited by William A. Hardwick*

Sensory Evaluation of Food

Statistical Methods and Procedures

Michael O'Mahony

Taylor & Francis
Taylor & Francis Group
Boca Raton London New York Singapore

A CRC title, part of the Taylor & Francis imprint, a member of the Taylor & Francis Group, the academic division of T&F Informa plc.

Library of Congress Cataloging-in-Publication Data

O'Mahony, Michael, [date]
Sensory evaluation of food.

(Food science and technology ; 16)
Includes index.
1. Food–Sensory evaluation–Statistical methods.
I. Title. II. Series: Food science and technology
(Marcel Dekker, Inc.) ; 16.
TX546.043 1985 664'.07 85-16090
ISBN 0-8247-7337-3

MARCEL DEKKER, INC.
270 Madison Avenue, New York, New York 10016

To Eileen

Preface

The aim of this book is to provide basic knowledge of the logic and computation of statistics for the sensory evaluation of food, or for other forms of sensory measurement encountered in, say, psychophysics. All that is required is that the reader be able to add, subtract, multiply, and divide.

The book is aimed at three types of reader: first, the undergraduate or graduate student who needs some statistics for a research project or who is simply trying to fulfill the requirements for a degree before escaping academia; second, the unwitting food scientist who has just been told that he or she is now in charge of sensory evaluation, despite having had no training, and needs to catch up quickly; and third, the established sensory professional who wants a ready reference, some guidance, and an instant introduction to some new tricks.

The book will take you from "ground zero" into the traditional realms of statistics and experimental design. It will lead you there gently, while concentrating on the logic behind the tests. It will look at some of the special difficulties involved in applying statistics to sensory data and will also give you a good, healthy dose of nonparametric statistics.

The book is hard-headed, but it is also very simple to understand. This is because it is written by a very simple nonstatistician. If I can understand it, you can too!

Michael O'Mahony

Contents

Preface *v*

Key to Statistical Tables *xiv*

1 BEFORE YOU BEGIN 1

1.1 How to Use This Book 1
1.2 The Use of Statistics in Sensory Evaluation Versus Its Use in Psychophysics, Experimental Psychology, and Consumer Testing 2
1.3 Why Use Statistics Anyway? 6
1.4 Some Perils of Sampling 7

2 AVERAGES, RANGES, AND THE NATURE OF NUMBERS 9

2.1 What Is the Average? Means, Medians, and Modes 9
2.2 Which Measure of Central Tendency Should You Use? 11
2.3 How are Numbers Spread? Range, Interquartile Range, Mean Deviation 12
2.4 Variance and Standard Deviation 14

2.5 Worked Example: Find the Mean, Standard Deviation, and Variance 17
2.6 The Nature of Numbers: Types of Scale 19
2.7 A First Look at Parametric and Nonparametric Statistical Tests 22

3 PROBABILITY, SORTING, AND ARRANGING 27

3.1 What Is Probability? 27
3.2 How to Compute a Probability: The Total Number of Events 29
3.3 The Multiplication Law of Probability: Getting Two Heads and One Tail 30
3.4 The Addition Law of Probability: Getting a Red or a White Ball 33
3.5 Worked Examples: Lots of Probability Calculations 35
3.6 Sorting 40
3.7 Permutations and Combinations 40
3.8 Worked Examples: Sorting, Permutations, and Combinations 42

4 NORMAL (GAUSSIAN) DISTRIBUTION: *z* TESTS 45

4.1 What Is a Normal Distribution? 45
4.2 *z* Scores 47
4.3 *z* Tests: Computing Probabilities 48
4.4 When Are Normal (Gaussian) Distributions Used? 51
4.5 How Normal (Gaussian) Distributions Are Used in Statistical Theory: Sampling Distributions, Standard Errors, and the Central Limit Theorem 52
4.6 Worked Example: *z* Tests—Computing Probabilities from Normal Distributions 55

5 THE BINOMIAL TEST: APPLICATIONS IN SENSORY DIFFERENCE AND PREFERENCE TESTING 57

5.1 How to Compute Probabilities Using the Binomial Expansion 57
5.2 The Binomial Test and the Null Hypothesis 61
5.3 Levels of Significance: Type I and Type II Errors 66
5.4 One- and Two-Tailed Tests 68
5.5 Why the Binomial Test Is Nonparametric 70
5.6 Large Samples: When the Binomial Expansion Approximates to a Normal Distribution 71
5.7 The Poisson Distribution 72

5.8 The Sign Test. One- and Two-Sample Difference Tests: Related- and Independent-Samples Design 73
5.9 Using the Binomial Test to Compare Two Proportions: A Two-Sample Binomial Test 78
5.10 Brief Recapitulation on Design for the Sign Test and the Binomial Comparison of Proportions 79
5.11 Application of Binomial Statistics to Sensory Difference and Preference Testing 79
5.12 Assumptions: Use and Abuse of Binomial Statistics in Sensory Testing 84
5.13 Worked Examples: Some Binomial Tests 85
5.14 Final Example: What is a Flavor Difference? Some Thoughts on the Philosophy of Difference Testing 89

6 CHI-SQUARE 91

6.1 What Is Chi-Square? 91
6.2 Chi-Square: Single-Sample Test—One-Way Classification 92
6.3 Chi-Square: Two-Way Classification 96
6.4 Assumptions and When to Use Chi-Square 99
6.5 Special Case of Chi-Square: The 2 × 2 Table 101
6.6 Further Note on the Chi-Square Distribution 104
6.7 The McNemar Test 104
6.8 The Cochran Q Test 105
6.9 The Contingency Coefficient 107
6.10 Worked Examples: Chi-Square Tests 108

7 STUDENT'S *t* TEST 111

7.1 What Is a *t* Test? 111
7.2 The One-Sample *t* Test: Computation 112
7.3 Worked Example: One-Sample *t* Test 114
7.4 The Two-Sample *t* Test, Related Samples: Computation 115
7.5 Worked Example: Two-Sample *t* Test, Related Samples 116
7.6 The Two-Sample *t* Test, Independent Samples: Computation 117
7.7 Worked Example: Two-Sample *t* Test, Independent Samples 119
7.8 Assumptions and When to Use the *t* Test 121
7.9 The Independent-Samples *t* Test with Unequal Variances 123
7.10 Further Statistical Theory: The Relationship Between *t* Distributions and Normal Distributions, *t* Tests and *z* Tests 124
7.11 Can You Use *t* Tests to Analyze Scaling Data? 127
7.12 Additional Worked Examples: *t* Tests 129

8 INTRODUCTION TO ANALYSIS OF VARIANCE AND THE ONE-FACTOR COMPLETELY RANDOMIZED DESIGN 135

8.1 Why You Don't Use Multiple t Tests 135
8.2 Logic Behind Analysis of Variance and the F Ratio 136
8.3 How to Compute F 140
8.4 Computational Formulas for Sums of Squares and Degrees of Freedom 143
8.5 Computational Summary for the One-Factor, Completely Randomized Design 145
8.6 Worked Example: One-Factor, Completely Randomized Design 147
8.7 Computation When Sample Sizes Are Unequal 149
8.8 Is Analysis of Variance a One- or a Two-Tailed Test? 150
8.9 How F Is Related to t 151
8.10 Assumptions and When to Use Analysis of Variance 151
8.11 Can You Use Analysis of Variance for Scaling Data? 152

9 MULTIPLE COMPARISONS 153

9.1 Logic Behind Multiple Comparisons 153
9.2 Various Multiple-Comparison Tests: Formulas for Calculating Ranges 156
9.3 Which Multiple-Comparison Test Should You Use? 158
9.4 How to Compute Fisher's LSD Test 161
9.5 Worked Example: Fisher's LSD Test 162
9.6 How to Compute Duncan's Multiple-Range Test 164
9.7 Worked Example: Duncan's Multiple-Range Test 165
9.8 Additional Worked Example: A Completely Randomized Design with Multiple Comparisons 166

10 ANALYSIS OF VARIANCE: TWO-FACTOR DESIGN WITHOUT INTERACTION, REPEATED MEASURES 171

10.1 Logic Behind the Two-Factor Design Without Interaction, Repeated Measures 171
10.2 Computational Summary for the Two-Factor Design Without Interaction, Repeated Measures 173
10.3 Worked Example: Two-Factor Design Without Interaction, Repeated Measures 176

10.4 Assumptions and When to Use the Two-Factor Analysis of Variance Without Interaction 178
10.5 A Further Look at the Two-Factor ANOVA: The Completely Randomized Design 179
10.6 Additional Worked Example: Two-Factor Design Without Interaction, Repeated Measures 180

11 ANALYSIS OF VARIANCE: TWO-FACTOR DESIGN WITH INTERACTION 183

11.1 What Is Interaction? 183
11.2 Interpretation of Interaction and Between-Treatments *F* Values 189
11.3 How to Compute a Two-Factor ANOVA with Interaction 193
11.4 Worked Example: Two-Factor ANOVA with Interaction 195
11.5 Assumptions and When to Use the Two-Factor Design with Interaction 200
11.6 What if the Interaction Is Not Significant? 200
11.7 Sometimes You Cannot Calculate Interaction Mean Squares 201
11.8 A Cautionary Tale 203
11.9 Additional Worked Example: Two-Factor Analysis of Variance with Interaction 204

12 ANALYSIS OF VARIANCE: THREE- AND FOUR-FACTOR DESIGNS 211

12.1 How to Compute a Three-Factor ANOVA 211
12.2 How to Interpret Two- and Three-Way Interactions 218
12.3 Sometimes You Cannot Compute Higher-Order Interactions 221
12.4 How to Compute a Four-Factor ANOVA 222
12.5 Worked Examples: Computation and Interpretation of ANOVA with Two or More Factors 226

13 FIXED- AND RANDOM-EFFECTS MODELS 247

13.1 What Are Fixed- and Random-Effects Models? 247
13.2 Theory Behind Fixed- and Random-Effects Models 250
13.3 Summary of the Denominators for *F* Ratios for Fixed and Random Effects 256

14 SPLIT-PLOT DESIGN 259

14.1 How the Split-Plot Design Is Related to Other Designs 259
14.2 Logic of the Split-Plot Design 261
14.3 Split-Plot Computation Scheme 266
14.4 Worked Example: Split-Plot Analysis of Variance 272

15 CORRELATION AND REGRESSION 279

15.1 Some Preliminaries 279
15.2 What Are Correlation Coefficients? 283
15.3 How to Compute Pearson's Product-Moment Correlation Coefficient 286
15.4 Worked Example: Pearson's Product-Moment Correlation 287
15.5 Is the Correlation Significant? 287
15.6 The Coefficient of Determination 288
15.7 Assumptions and When to Use the Correlation Coefficient 289
15.8 Correlation Does Not Imply Causality 293
15.9 Further Topics for Correlation: Means, Differences, and Significance 293
15.10 Logic of Linear Regression 295
15.11 Computation for Linear Regression 297
15.12 Worked Example: Linear Regression 298
15.13 Assumptions and When to Use Linear Regression 299
15.14 Worked Example: Correlation and Linear Regression 300

16 ADDITIONAL NONPARAMETRIC TESTS 303

16.1 When to Use Nonparametric Tests 303
16.2 The Wilcoxon Test: A Two-Sample, Related-Samples Difference Test 304
16.3 Worked Example: The Wilcoxon Test 308
16.4 The Mann-Whitney U Test: A Two-Sample, Independent-Samples Difference Test 309
16.5 Worked Example: The Mann-Whitney U Test 314
16.6 The Wilcoxon-Mann-Whitney Rank Sums Test 315
16.7 Worked Example: The Wilcoxon-Mann-Whitney Rank Sums Test 325
16.8 Spearman's Ranked Correlation 327
16.9 Worked Example: Spearman's Ranked Correlation 331
16.10 The Friedman Two-Factor Ranked Analysis of Variance (Two-Factor Repeated-Measures Design) 332
16.11 Worked Example: The Friedman Test 336

16.12 The Kramer Two-Factor Ranked Analysis of Variance (Two-Factor Repeated-Measures Design) 337
16.13 Worked Example: The Kramer Test 340
16.14 The Kruskal-Wallis One-Factor Ranked Analysis of Variance (One-Factor Completely Randomized Design) 341
16.15 Worked Example: The Kruskal-Wallis Test 345
16.16 The Page Test: Testing for a Ranked Tendency 346
16.17 Worked Example: The Page Test 349
16.18 The Jonckheere Test: Testing for a Ranked Tendency 350
16.19 Worked Example: The Jonckheere Test 362
16.20 Final Theoretical Note Regarding Nonparametric Tests 365

APPENDIX A PROOF THAT $\Sigma(X - \bar{X})^2 = \Sigma X^2 - (\Sigma X)^2/N$ 371

APPENDIX B BINOMIAL EXPANSION 373

B.1 How to Calculate the Expansion: First Principles 373
B.2 How to Calculate the Expansion Without Effort 376

APPENDIX C PROOF THAT $SS_T = SS_B + SS_E$ 379

C.1 Total Variance: Sum of Squares 380
C.2 Between-Treatments Estimate of the Total Variance: Sum of Squares 380
C.3 Within-Treatments Estimate of the Total Variance: Sum of Squares 381

APPENDIX D A NOTE ON CONFIDENCE INTERVALS 385

APPENDIX E SENSORY MULTIPLE-DIFFERENCE TESTING, USING THE R-INDEX TO ASCERTAIN DEGREES OF DIFFERENCE 389

E.1 Degree of Difference Between Two Products Using Rating 390
E.2 Degrees of Difference Between Several Products Using Rating 393

APPENDIX F DO WE HAVE THE WRONG TOOLS? 399

F.1 Related Versus Independent Samples Designs and Storage Studies 399
F.2 Rejecting H_0 and Quality Assurance 399

APPENDIX G STATISTICAL TABLES 403

Index *475*

Key to Statistical Tables

			Two-Sample Difference Tests	
		One-Sample Tests	Independent Samples	Related Samples
Nonparametric Tests	Nominal Data (Categories)	Binomial test, page 57 Chi-square one-sample test, page 92 Test a sample to investigate distribution of categories in population	Binomial comparison of proportions, page 78 Chi-square comparison of distributions over two independent samples, page 96	McNemar test (from chi-square), page 104
Nonparametric Tests	Ordinal Data (Ranks)		Mann-Whitney U test, page 309 Wilcoxon-Mann-Whitney rank sums test, page 315	Sign test (from binomial test), page 73 Wilcoxon test, page 304
Parametric Tests	Interval/Ratio Data (Numbers) and Data from Normally Distributed Populations	One-sample t test, page 112 Test whether sample mean is significantly different from population mean	Independent-samples t test, page 117 Completely randomized ANOVA, page 135	Related-samples t test, page 115 Repeated-measures ANOVA, page 171

More-Than-Two-Sample Difference Tests			
Independent Samples (Completely Randomized)	Related Samples (Repeated Measures)	Two-or-More-Factor Difference Tests	Correlation
Chi-square comparison of distributions over more than two independent samples, page 96	Cochran Q test (from chi-square), page 105		Contingency coefficient (from chi-square), page 107
Kruskal-Wallis one-way ANOVA, page 341 Jonckheere test: for ranked tendency, page 350	Friedman two-way ANOVA, page 332 Kramer test, page 337 Page test: for ranked tendency, page 346	TWO-FACTOR TESTS ONLY Friedman two-way ANOVA, page 332 Kramer test, page 337 Page test: for ranked tendency, page 346	Spearman's rank correlation, ρ, page 327
Completely randomized ANOVA with multiple comparisons, page 135, 153	Repeated-measures ANOVA with multiple comparisons, page 153, 171	Two-or-more-factor ANOVA with multiple comparisons, page 153, 171, 211	Pearson's product-moment correlation, r, page 279

Sensory Evaluation of Food

1
Before You Begin

1.1 HOW TO USE THIS BOOK

This book presents a simplified look at statistics, as do all introductory texts. In the interests of clarity and simplicity some shortcuts will be taken, but wherever possible, these will be indicated. Although the study of mathematical statistics is clear enough, its application to real-life situations is not always so clear. Statistical tests involve, in their derivation, certain assumptions about the data and it is sometimes doubtful whether these assumptions are always fulfilled in behavioral measurement. If they are not fulfilled, it is not always clear how important this may be. For instance, in some experimental situations, should a one- or a two-tailed test be employed? How much do data from intensity scales violate the assumptions for analysis of variance, and how important are such violations? In what situations are nonparametric tests more desirable? These matters will be discussed in more detail throughout the text; suffice it to say that in the area of sensory and behavioral measurement such doubts should be taken seriously. As far as possible, we will give recommendations and try to indicate where there is disagreement. However, this text is not a statistical treatise; it is not intended to give an exhaustive account of these controversies—for that, go and get a degree or two in statistics! The aim of this text is to give a working knowledge of statistical analysis with as few frills as possible. Enough background will be given so that the reader can understand what a statistical test is doing and when and how it should be used. The arguments and logic will always be expressed as nonmathematically as possible.

The mathematical formulas will be expressed simply; formal mathematical notation with its welter of subscripts will not be used here. If it offends the mathematical purist, I apologize, but a simpler, albeit more casual approach makes teaching simpler at this level. So be warned that the notation used here will not always be the standard notation used by statisticians. We will not bother to put all algebra referring to populations in capital letters and that referring to samples in lowercase letters. We will not bother to identify each value with its own subscript; that only terrifies the reader and is bad teaching practice. Furthermore, although most of the statistical tests will be illustrated by examples from sensory measurement and the sensory analysis of food, this principle will not be adhered to slavishly. Sometimes a statistical principle is better illustrated by other examples, and in the interests of clarity and better teaching, these will be adopted. For example, some probability theory is more simply explained using familiar examples such as tossing of dice and coins.

Unfortunately, there is little agreement among the available texts on the best approach to explaining the application of various statistical tests to behavioral measurement; there is even disagreement in such texts over the names to be given to various tests. The babel is best seen by a consideration of all the different names given in behavioral texts to the various analysis-of-variance procedures (Table G.10 in Appendix G). In this text, we will give statistical tests the simplest yet most fully descriptive names possible; we will also keep the use of jargon to a minimum—we are not trying to impress anyone.

Finally, this text is brief, sparse, and not very conversational. It is hoped that the starkness will eliminate distraction and allow the reader to stick strictly to the point. It is hoped that this will speed the process of learning without leaving the reader with a sense of desolation.

1.2 THE USE OF STATISTICS IN SENSORY EVALUATION VERSUS ITS USE IN PSYCHOPHYSICS, EXPERIMENTAL PSYCHOLOGY, AND CONSUMER TESTING

The experimental and statistical skills used for the sensory analysis of foods or consumer testing are very similar to those used in experimental psychology. For further study, texts on psychological statistics are recommended. There are, however, some important differences between the application of statistics to sensory evaluation and its application to psychology or consumer testing. These are due to the different goals and approaches of the various disciplines. Often the differences are not important, but sometimes they are crucial because they can alter the actual mathematical operations involved in statistical tests; for example, analysis of variance. So, dear reader, if you are an experimental psychologist or a consumer researcher about to use this text as a simple reminder, or if you are a sensory analyst about to consult a psychology text, please read this section beforehand to avoid confusion.

First, let us look at differences between sensory evaluation and experimental psychology, of which sensory psychophysics is an important part. Generally, in experimental psychology the goal is to find out something about the behavior of people in general, or even of animals. In sensory psychophysics, specifically, the goal is to examine the functioning of the senses and the brain, and the focus is generally on people. In both these areas, the general strategy is to take a random sample from the population of people under consideration. The sample is examined in a suitable experiment, the appropriate statistical analysis is performed on the sample, and using this analysis, conclusions are drawn about the population. A sample is used because there is generally not enough time or money to examine the whole population. The sample is generally picked as randomly as possible in the hope that it will match the population as closely as possible.

In the sensory evaluation of foods or other products, there can be more than one goal. First, in some situations, people may be used for a preliminary analysis to measure the flavor of the food and provide clues for later chemical or physical analysis. They may be used to measure, say, flavor changes due to a change in the processing of the food. The question here is: Does the new processing procedure alter the sensory characteristics of the food? Changes in sensory characteristics can provide clues about changes in physical characteristics. Should judges detect a change in the odor, it would provide a clue to chemists to investigate changes in the volatiles. Should judges detect a change in sourness it would indicate that chemists might first want to investigate changes in acid content or even sugar content. The goal is to seek information about the food, not the people who taste the food. A random sample of people is not selected. A panel comprising of a few people is selected and trained to become judges in the specific task required, whether it be detecting off-flavors in beverages or rating the texture of cakes. Generally, potential judges are screened to see whether they are suitable for the task at hand. This careful selection of judges is in no way a random sampling and is the first reason why the judge could not be regarded as representative of the population in general. The training then given to the selected judges will usually make them more skilled in their task than the general population would be; this is a second reason why they can no longer be regarded as representative of the population from which they were drawn. Our judges are not a sample to be examined so that conclusions can be drawn about a population; they are the population! They are a population of laboratory instruments used to study samples of food. On the other hand, the food samples tested are samples from which conclusions can be drawn about the populations of foods. They may be samples of a food processed in different ways from which conclusions may be drawn about the whole population of that food processed in these ways. So the focus of the study is very much on the food; the judges are merely instruments used for measuring samples of food. Logically, one good judge would be adequate for this purpose, as is one good gas chromatograph or atomic absorption spectrometer. More than one judge is used as a precaution because unlike a gas chromatograph, a judge can become distracted; the cautious use of second opinions can provide a useful fail-safe mechanism.

Having detected a change in flavor under such circumstances, the question becomes: Can this flavor change be detected by the consumers or people other than trained panelists? It is possible that the flavor change is so slight as to be detectable only by highly trained judges. One way of answering this question is to take a sample of untrained people and test them to see whether they can detect the flavor change. Then from this sample, inferences can be made about the population of untrained people in general. The experiment now becomes similar to one in sensory psychophysics. People are sampled, tested, and conclusions are then drawn about people in general. This is the second use of sensory evaluation.

It can thus be seen that in the sensory evaluation of food, depending on the aims of the study, conclusions may be drawn only for those judges actually tested or they may be drawn for the whole population of people from which the judges were sampled. It is reasonable to expect the probabilities and the statistical procedures to vary for these two cases. The difference becomes important for analysis of variance, where different denominators are used for calculating F values. Where the conclusions apply only to the judges tested, the judges are said to be a *fixed effect*. Where they apply to the whole population from which the judges were drawn, the judges are said to be a *random effect*. In this text, the analysis-of-variance procedures all use people as fixed effects because they are easier to understand and learn that way. To use people as random effects when the aim is to study people per se, the calculation is altered very slightly. The brief chapter (Chapter 13) on fixed- and random-effects models shows the slight differences required when people are "random effects." Be sure to read it before you perform an analysis of variance on data designed to give you facts about people in general rather than only those people tested in the experiment. It is worth noting that texts on psychological statistics differ from this text. In psychological texts people are generally treated only as random effects.

The traditions of psychophysics and sensory evaluation vary even more. In psychology or psychophysics, people are the subject of the investigation and thus tend to be called *subjects*. In sensory evaluation, or sensory analysis as it is also called, people are often specially selected and trained and tend to be referred to as *judges*. In sensory analysis, judges are tested while isolated in booths; experimenters and judges communicate in writing. In sensory psychophysics, the experimenter and subject often interact and communicate verbally; this requires special training for experimenters so that they do not influence or bias the subjects. The methods of sensory analysis are often derived from those of psychophysics, but care must be taken when adapting psychophysical tests to sensory evaluation. This is because the goals of the two disciplines are different and they can affect the appropriateness of various behavioral methods of measurement to be used, as well as the statistical analysis employed. In psychophysics, the aim is to measure the "natural" functioning of the senses and cognitive processes. Extensive training will not be given if it alters a subject's mode of responding or recalibrates the person, so obscuring his or her "natural" functioning.

Consumer testing must also be considered. Whereas judges are used in sensory analysis to assess the flavor of a food per se, consumer testing selects samples of consumers from the marketplace to find out whether they like or will buy the food. For consumer testing, the people must be as representative as possible of the potential buying population so that they do not give a misleading indication of potential sales. To perform consumer tests on members of a panel used for sensory analysis would be to use a sample that was not representative of the potential buying public. The panel would be comprised of a few people who had been specially selected for their skill in sensory evaluation and who may also have been made atypical as a result of their training. There is another important point: The judges on a sensory panel are usually recruited from the company; they are often laboratory or office workers who take time from their regular duties to participate in the panel. These judges may have a knowledge of the product which may bias their judgments of whether they like it; they may even select responses that they feel might impress their bosses. Thus, to use a sensory panel for consumer testing would seem unwise. Despite this, some food companies blindly adopt this research strategy.

The sample of consumers tested should be as representative as possible of the potential buying population. It is a sample from which conclusions are drawn about the population and so, as with sensory psychophysics, the people tested are considered to be random effects! Psychophysicists tend to pick a sample as randomly as possible in the hope that it will turn out to be representative of the population. The strategy of random sampling is used in consumer testing, but another strategy is also adopted. Here the demographics of the population are studied and a sample carefully chosen so as to match the population demographically as closely as possible. If the population has 80% males, so will the sample. If the population has 50% Hispanics, the sample will also. This approach to making the sample as representative as possible of the population is called *stratified sampling*. Sampling will be discussed further in the section on the perils of sampling (Section 1.4).

As you read through the text, it will become apparent that statistical tests are set up to show that differences exist between sets of data. However, should the reader be using statistics for quality assurance, trying to show that a reformulated product has the same flavor as the regular product, the aim will be to show that differences do not exist. This is using statistical analysis in a rather different manner, and readers are advised to consult Appendix F before analyzing their data.

1.3 WHY USE STATISTICS ANYWAY?

In Section 1.2 we mentioned the various disciplines that use statistics. We discussed their sampling techniques and whether inferences made from samples were carried to populations. To summarize, psychophysicists, experimental psychologists, and consumer testers are interested primarily in the behavior or likes and dislikes of people. They make their measurements on a sample of people, for reasons of time and economy, and use a statistical analysis to make inferences about the population from which the sample was taken. Naturally, the larger the sample, the more representative it is of the population. The more powerful the statistical test applied to the sample, the better the information obtained about the population. This is true, of course, as long as all the conditions required for the proper application of the test are fulfilled. There are procedures for estimating how large a sample should be taken for a given test of a given power, but these will not be dealt with here. Sensory analysts select and train a small panel to assess samples of foods. Conclusions resulting from the statistical analysis may or may not be generalized further beyond the actual panelists tested; the panelists may be random or fixed effects. In the same way, conclusions may or may not be generalized beyond the actual food servings tested. Replicate servings of a given food processed in a given way (often called "replicates" or "reps") can themselves be seen as a fixed or a random effect, depending on how far the conclusions are to be generalized.

Statistical tests may be inferential; inferences are made about populations from samples. Statistical procedures may also be merely descriptive; the data in the sample may merely be summarized in some way with no inferences being made about the population. Thus a mean is a summary of the data in a given sample; it gives a central summary value of the data. In this way, it is descriptive. However, the best estimate of the mean of a population is also the mean of the sample. In this way, it is inferential.

It is important when taking a sample or designing an experiment to remember that no matter how powerful the statistics used, the inferences made from a sample are only as good as the data in that sample. Furthermore, the inferences made from a sample apply only to the population from which the sample was taken. This can be crucial in medical or consumer-orientated research.

Some Definitions and Jargon

Descriptive statistics are used to describe the data obtained: graphs, tables, averages, ranges, etc.

Inferential statistics are used to infer, from the sample, facts about the population.
A *parameter* is a fact concerning a population.
A *statistic* is a fact concerning a sample.
Data is plural; the singular is *datum.*

1.4 SOME PERILS OF SAMPLING

Generally, statistical tests require that a sample be selected randomly from the population it is representing. This may be a complex matter but there are additional points of design which are worth mentioning here.

When sampling, you must select your sample in an unbiased manner. If you wish to take a sample of university students and examine the sample to see whether there are more men than women in the university, it is no use standing outside the men's toilet. You will only get the occasional brave, or lost, lady and so have a biased sample. This type of thing happened when in 1936 the *Literary Digest* polled 10 million *Digest* readers and phone subscribers as to who would win the U.S. presidential election. The poll indicated Landon, but in fact Roosevelt was elected. The *Literary Digest* had taken a biased sample. Only the richer people read the *Literary Digest* or had phones at that time. The poorer Americans tended not to be included in the poll; it was their vote which elected Roosevelt.

So take care how you collect your data. If the sample is chosen so as to be representative of the population, the conclusions drawn from the statistical analysis apply only to that population. The sample may be selected by carefully matching its demographics to that of the population (*stratified sampling*), or a representative sample may be obtained by randomly sampling from the population. In practice, a so-called *random sample* is rarely completely random. Subjects are usually volunteers and volunteers are not necessarily representative of the population from which they were drawn. Also, an experimenter's method of choosing subjects is hardly likely to be random; he or she will tend to select those who are friendly and enthusiastic, as well as those who are conveniently near the testing location. Because much psychological research is conducted in psychology departments at universities, many of the subjects will be students. Thus, many of the findings in psychophysics were established using samples of willing, perhaps friendly, and enthusiastic students who work in or near psychology departments; such samples are hardly representative of the population at large. The importance of such difficulties is more a matter of philosophy than mathematics; it is certainly hard to gauge. Unfortunately, such questions are rarely considered; experimenters seem to hope that they are unimportant.

Members of the sample may also tell lies. This can occur when questions of a personal nature (money, sex, etc.) are asked. In fact, if a subject guesses the correct answer or feels that a certain response is right, he or she may give it to you, regardless of whether or not it is true for him or her. The tendency to please the experimenter and give the "right" answer is strong among human subjects.

Control groups are often used to control for any other effects that might be happening. With drug research, some patients are put in a *control* group and are not given the drug being tested. Instead, they are given a harmless placebo, which resembles the drug but has no action. Often these control patients are cured as well—the "placebo effect." By comparing those who took the drug with those who thought they took the drug, a measure can be obtained for the efficacy of the drug itself, independent of any suggestion effects. Control groups are often used in behavioral research and could sometimes be useful in the sensory analysis of foods.

It is not the intention here to go into detail about sampling procedures; there are plenty of texts that do this. Suffice it to say that one of the common errors in sensory work with foods is to sample from the wrong population. If you wish to sample the likes and dislikes of consumers, do not ask people who are not representative of consumers, such as expert panelists or scientists working on the food product itself.

A final point. No amount of sophisticated statistical analysis will make good data out of bad data. There are many scientists who try to disguise badly constructed experiments by blinding their readers with a complex statistical analysis. Be cautious about all scientific papers, especially in behavioral research (sensory analysis, psychology, etc.); some of them contain grotesque nonsense.

2

Averages, Ranges, and the Nature of Numbers

2.1 WHAT IS THE AVERAGE? MEANS, MEDIANS, AND MODES

A sample of data can be difficult to comprehend; it is often difficult to see a trend in a set of numbers. Thus, various procedures are used to aid the understanding of data. Graphs, histograms, and various diagrams can be of help. Furthermore, some middle or average value, a measure of the *central tendency* of the numbers, is useful. A measure of the spread or dispersion of the numbers is also useful. These measures will now be considered.

The Mean

The *mean* is what is commonly called an *average* in everyday language. The mean of a sample of numbers is given by the formula

$$\bar{X} = \frac{\Sigma X}{N}$$

where ΣX (Σ is the Greek capital letter sigma) denotes the sum of all the X scores, N is the number of scores present, and $\bar{X}$ is the common symbol for the mean of a given sample of numbers.

The mean of the whole population would be calculated in the same way. It is generally denoted by μ (the Greek lowercase letter mu). Because the mean of a population is usually not possible to obtain, it is estimated. The best estimate of the population mean, μ, is the sample mean $\bar{X}$. The mean is the common measure used when inferences are to be drawn from the sample about the population. Strictly, the mean we are discussing is called the *arithmetic mean*; there are other means, which are discussed in Section 2.2.

For the mean to be used, the spread of the numbers needs to be fairly symmetrical; one very large number can unduly influence the mean so that it ceases to be a central value. When this is the case, the median or the mode can be used.

The Median

The *median* is the middle number of a set of numbers arranged in order. To find a median of several numbers, first arrange them in order and then pick out the one in the middle.

Example 1

Consider

1 2 2 3 4 7 7

Here the number in the middle is 3. That is, the median = 3.

Example 2

Consider

1 2 3 : 3 3 4

Here there is no one middle number, there are two: 3 and 3. The median in this case is the number halfway between the two middle numbers. So the median = 3.

Example 3

Consider

1 2 2 : 3 8 90

Again, there are two middle numbers: 2 and 3. The number halfway between these is 2.5. So the median = 2.5. Note how the extreme value of 90 did not raise the median the way it would have the mean.

The Mode

The *mode* is the most frequently occurring value

Example 4

Find the mode of

1, 1, 2, 2, 3, 3, 3, 4, 5, 6, 7, 7, 7, 7, 8, 8

The most commonly occurring number is 7; it occurs four times. So the mode is 7.

Of course, it is possible for a set of numbers to have more than one mode, or if no number occurs more than any other, no mode at all. A distribution with two modes is said to be *bimodal*; three modes, *trimodal*; and so on.

2.2 WHICH MEASURE OF CENTRAL TENDENCY SHOULD YOU USE?

The mean is the most useful statistic when inferences are to be drawn about a population from a sample; the mean of a sample is the best estimate of the mean of the population. It is commonly used by statisticians but is affected unduly by extreme values. The mode is easily computed and interpreted (except where there are several possible modes) but is rarely employed; no common inferential procedure makes use of the mode. The median is a useful descriptive statistic which fits the requirements of clear and effective communication, but it is of only limited use where inferences about population parameters are to be made from a sample. It is less affected by a preponderance of extreme values on one side of a distribution (skew) than the mean, and is to be preferred to the mean in cases where there are extreme values or values of indeterminable magnitude (e.g., when values are too big to be measured on the scale being used). The mean uses more information than either of the other measures, shows less fluctuation over successive samples, and is employed in all classical (parametric: see Section 2.7) statistical procedures.

The mean that we have been discussing is the arithmetic mean; there are other means. Instead of adding all the N values in a sample and dividing the total by N, we could multiply all the numbers together and take the Nth root of the value obtained; this is called the *geometric mean*. It is equivalent to calculating a mean by converting all the numbers to logarithms, taking their arithmetic mean, and then converting the answer back to an antilogarithm. The geometric mean is the appropriate mean to use when plotting power functions of perceived versus physical intensities, obtained by the magnitude estimation intensity scaling procedure. Another type of mean is the *harmonic mean* which is obtained from reciprocal values. This is discussed in Section 9.6, dealing with Duncan's multiple range test.

There is one more important point. An average or mean of a sample of scores is a middle value or a measure of central tendency. Obviously, not all the scores are equal to the mean; it is the middle value where the high scores cancel out the low scores. However, a lot of researchers have made the mistake of assuming that mean values represent the values for each individual subject. An example will make this clear. It is of interest to food product developers to know how the liking for a food may increase as the sugar content of that food increases. So measurements are made to provide data so that a graph of degree of liking (measured on a hedonic scale) can be plotted against sugar content. The shape of the graph will provide clues about the optimum sugar content required. However, for some studies, only graphs from the mean scores of a group of judges have been constructed. A mean graph does not necessarily represent each individual. When graphs are drawn for each individual, they are often seen to have quite different shapes from the mean graph. Some judges may like the food more as the sugar content increases, others may dislike it more. Still other judges may like the food more as the sugar content increases to an optimum level but then dislike any higher levels of sugar as being too sweet. All these nuances would be lost by merely plotting a mean response function; a misleading picture would be obtained. When dealing with the interpretation of any form of average data, care must be taken so that the hidden assumption that it represents the data of each individual is not inadvertently made.

2.3 HOW ARE NUMBERS SPREAD? RANGE, INTERQUARTILE RANGE, AND MEAN DEVIATION

Not only are some measures of central tendency needed to describe the data in a sample, but some measure of the spread of the numbers is also required. 100 is the mean of 99, 100, and 101. It is also the mean of 25, 100, and 175. Clearly, these two samples are very different; the spread or *dispersion* of the latter is greater. As with central tendency, there are several measures of how the numbers are spread, which are now considered.

The Range

The *range* is the absolute difference between the largest and smallest value in the data. It is the best known everyday measure of dispersion.

Generally, we would like to know the range of the numbers in the population but can only find the range of numbers in a sample. Unlike the mean, the range of the sample is not a good estimate of the range of the population; it is too small. It is obvious that the spread of a whole large population of numbers will be greater than the spread of a small sample of numbers. Measures of dispersion called the variance or standard deviation have special modifications to get around this problem, so that the population value can be estimated from a sample (see Section 2.4). Naturally, the larger the sample, the nearer the sample range will be to the population range because the spread of the scores will increase as the sample gets larger. This means, incidentally, that because the range varies with sample size, comparing ranges from samples of different sizes should be done with caution.

One problem with the range is that it makes use of only a fraction of the data, the highest and lowest values. It may also misrepresent the bulk of the data if extreme values are present. For example, should the age of most of a group of consumers be between 25 and 30 years but one be 7 years and one 97 years, the range will be from 7 to 97 years. This could make it appear as though there were a lot of children, young adults, middle-aged, and older people in this group of consumers when, in fact, the bulk of the group fell in the narrow age range 25 to 30 years. Thus a measure of *middle range*, a measure that leaves out the extreme values, may also be desirable. The following are measures of middle range.

Interquartile Range

Just as the median is that value above (and below) which half the cases in the distribution lie, so the *lower quartile* is that point above which three-quarters of the cases lie, and the *upper quartile* that point above which one-quarter of the cases lie.

The *interquartile range* is the difference between the values of the upper and lower quartiles (i.e., the range of the middle 50% of the distribution).

1, 3, 5, <u>7</u>, 6, 8, 9, <u>10</u>, 12, 13, 15, <u>17</u>, 20, 20, 90

lower quartile (7) — median (10) — upper quartile (17)

interquartile range (lower quartile to upper quartile)

The range and interquartile range do not depend intimately on the mean. The following measures of central tendency do, in that they all involve the mean in their computation.

Mean Deviation

Another approach to establishing a middle-range value is to look at the deviation of each score from the mean value. The mean of all these values will give a useful measure of middle range. The mean deviation of a set of scores is given by the formula

$$\frac{\Sigma |X - \bar{X}|}{N}$$

$X - \bar{X}$ indicates that we subtract the mean ($\bar{X}$) from each score (X). The vertical lines on either side of $X - \bar{X}$ indicate that we should take the modulus, that is, we should arrange that all the values are positive. Σ indicates that all the positive $|X - \bar{X}|$ values are added before dividing by N, the number of X scores that are present.

Another way of ensuring that the deviations have positive values is to square them; this will tend to give greater weight to extreme values. Such an approach is generally used by statisticians to compute the most used measures of spread: the variance and the standard deviation.

2.4 VARIANCE AND STANDARD DEVIATION

The variance and standard deviation are the most commonly used middle-range values, the most common measures of dispersion.

Variance and Standard Deviation of Populations

The *variance* of a population of numbers is given by a formula similar to that for the mean deviation:

$$\sigma^2 = \frac{\Sigma(X - \mu)^2}{N}$$

Squaring all the deviations from the mean, $X - \mu$ values, ensures that they are all positive. It also makes σ^2 more sensitive than the mean deviation to large deviations. μ is the population mean.

Again, the variance of a sample could be calculated in the same way. It will, however, be smaller than the variance of the population. It is computed from the formula

$$\sigma^2_{\text{sample}} = \frac{\Sigma(X - \bar{X})^2}{N}$$

where $\bar{X}$ is now the mean of the sample.

The *standard deviation* is merely the square root of the variance. It is symbolized by σ rather than σ^2. Hence the standard deviation of a population is given by

$$\sigma = \sqrt{\frac{\Sigma(X - \mu)^2}{N}}$$

Again, the standard deviation of a sample smaller than the population is given by

$$\sigma_{\text{sample}} = \sqrt{\frac{\Sigma(X - \bar{X})^2}{N}}$$

Estimates of Population Variance and Standard Deviation from Samples

Generally, however, we wish to estimate the variance or standard deviation of the population from the data in the sample. This is done by adjusting the formulas so that the denominator N is replaced by $N - 1$. We are not going to go into why this is so mathematically; suffice it to say that if we divide by $N - 1$, a smaller number, the variance or standard deviation value will then be larger. Just as we expect the range of a population to be larger than that of a sample drawn from it, we expect the variance or standard deviation of the population to be larger, too.

To avoid confusion, we denote the estimate of the population variance or standard deviation, obtained from the sample, by S^2 or S (some texts use lowercase s). Thus we can write

$$S^2 = \frac{\Sigma(X - \bar{X})^2}{N - 1}$$

$$S = \sqrt{\frac{\Sigma(X - \bar{X})^2}{N - 1}}$$

Rearranging Variance and Standard Deviation Formulas

These formulas for standard deviation (S and σ) and variance (S^2 and σ^2) are inconvenient. They involve having to take the mean away from every value and squaring the difference, $\Sigma(X - \bar{X})^2$, a tedious operation. The formulas can be simplified to avoid this chore by using the relationship

$$\Sigma(X - \bar{X})^2 = \Sigma X^2 - \frac{(\Sigma X)^2}{N}$$

The proof of this relationship is given in Appendix A. Thus the formulas now become

$$S^2 = \frac{\Sigma(X - \bar{X})^2}{N - 1} = \frac{\Sigma X^2 - \frac{(\Sigma X)^2}{N}}{N - 1}$$

$$S = \sqrt{\frac{\Sigma(X - \bar{X})^2}{N - 1}} = \sqrt{\frac{\Sigma X^2 - \frac{(\Sigma X)^2}{N}}{N - 1}}$$

With these convenient formulas, we can more easily calculate S and S^2. For example, consider the following X scores:

	X	X^2
	7	49
	2	4
	3	9
	5	25
	2	4
	3	9
	4	16
	1	1
	3	9
	3	9
$\Sigma X =$	33	$\Sigma X^2 = 135$
$(\Sigma X)^2 =$	1089	

$$S^2 = \frac{\Sigma X^2 - \frac{(\Sigma X)^2}{N}}{N - 1}$$

$$= \frac{135 - \frac{1089}{10}}{10 - 1}$$

$$= \frac{135 - 108.9}{9} = \frac{26.1}{9}$$

$$= 2.9$$

The standard deviation is given by the square root of the variance. Hence

$$S = \sqrt{2.9} = 1.7$$

Had the denominator been N, we would have calculated the variance and standard deviation of the sample per se rather than estimated the population values. In this case, the variance (= 2.6) and standard deviation (= 1.6) would have been smaller.

Thus σ and σ^2 denote the standard deviation and variance of the population (population data, denominator N), while S and S^2 denote estimates of the standard deviation and variance of the population, from the data in a sample (sample data, denominator $N - 1$). If you use N as a denominator for the data in the sample, you will get a smaller value representing the sample per se; this is denoted in various ways. I have used σ_{sample}. We have been at pains to point out the differences. However, many texts do not stress this. They may not differentiate the two and may use the symbols interchangeably. Calculators that have a key to give standard deviations may not always indicate whether the denominator is N or $N - 1$.

Naturally, the actual symbols used are not important; what is important are the actual concepts. Although it is important to understand the different concepts, the actual values may not differ much in practice. Whether the denominator is 30 (N) or 29 ($N - 1$) will often not affect the overall value a great deal. Generally, however, S and S^2 are most commonly used.

Variances and standard deviations are generally used with means and, like the mean, can be misleading unless the numbers are spread symmetrically. When the distribution of numbers is skewed, medians and interquartile ranges may be more useful descriptions.

In the statistical technique called analysis of variance (Chapter 8) the variance formula is employed:

$$S^2 = \frac{\Sigma X^2 - \frac{(\Sigma X)^2}{N}}{N - 1}$$

Several variances are calculated and to simplify the computation, the formula is dismantled and the numerators and denominators are calculated separately. These parts of the formula are even given separate names. The numerator is called the *sum of squares* and the denominator is called the *degrees of freedom*. Even parts of the numerator can be given their own names. $(\Sigma X)^2/N$ is called the *correction term* or the *correction factor, C*. This introduction of a new dose of jargon can make beginners lose sight of the fact that all is being computed is a variance. But more of this in Chapter 8.

There is yet another range value called the *confidence interval*. This is a little too complicated to be dealt with here but it is described in Appendix C and should be noted when the reader has understood t tests and sampling distributions.

2.5 WORKED EXAMPLE: FIND THE MEAN, STANDARD DEVIATION, AND VARIANCE

Find the mean, standard deviation, and variance of the following sample of numbers: 21, 20, 25, 23, 23, 21, 25, 26, 24, 24. Also, what do you estimate are the mean, standard deviation, and variance of the population from which the sample was drawn?

The variance of the numbers in the sample is given by

$$\sigma^2_{\text{sample}} = \frac{\Sigma X^2 - \frac{(\Sigma X)^2}{N}}{N}$$

The standard deviation equals the square root, $\sqrt{\sigma_{\text{sample}}}$.

If the numbers came from a population, the estimate of the variance of that population is given by:

$$S^2 = \frac{\Sigma X^2 - \frac{(\Sigma X)^2}{N}}{N-1}$$

The standard deviation equals the square root, $\sqrt{S}$. The best estimate of the mean of the population is the mean of the sample. The calculation is as follows:

	X		X^2
	21		441
	20		400
	25		625
	23		529
	23		529
	21		441
	25		625
	26		676
	24		576
	24		576
$\Sigma X =$	232	$\Sigma X^2 =$	5418
$(\Sigma X)^2 =$	53,824	$N =$	10

$$\text{Mean, } \bar{X} = \frac{\Sigma X}{N} = \frac{232}{10} = 23.2$$

$$\text{Sample variance, } \sigma^2_{\text{sample}} = \frac{\Sigma X^2 - \frac{(\Sigma X)^2}{N}}{N}$$

$$= \frac{5418 - \frac{53{,}824}{10}}{10} = \frac{5418 - 5382.4}{10}$$

$$= \frac{35.6}{10} = 3.56$$

$$\sigma_{\text{sample}} = \sqrt{3.56} = 1.89$$

$$\text{Estimate of population variance, } S^2 = \frac{\Sigma X^2 - \frac{(\Sigma X)^2}{N}}{N-1}$$

We know the numerator from the above calculation, so

$$S^2 = \frac{35.6}{9} = 3.96$$

$$S = \sqrt{3.96} = 1.99$$

Considering only the sample of 10 numbers, the mean, standard deviation, and variance ($\bar{X}$, σ_{sample}, σ^2_{sample}) are 23.2, 1.89, and 3.56. We do not know the mean, standard deviation, and variance of the population (μ, σ, σ^2), but we can estimate them from the sample. The best estimate of the population mean is the sample mean ($\bar{X}$) = 23.2. The estimated standard deviation and variance of the population (S and S^2) are 1.99 and 3.96.

2.6 THE NATURE OF NUMBERS: TYPES OF SCALE

Before proceeding, it is now worth pausing to consider the nature of numbers. We have gotten used to the idea that numbers have several properties. For instance, the distance between 1 and 3 is the same as the distance between 11 and 13. 20 is twice as large as 10, which is itself twice as large as 5. Generally this is true, but sometimes it is not. When judges are making estimates of the intensity of various flavors of food or trying to rate the intensity of various sensory experiences, they will tend to use numbers inaccurately. They will use them in such a way that 10 is not necessarily twice 5 nor are the distances between 1 and 3 and 4 and 6 necessarily the same. People are just not very good at estimating their sensory experiences accurately.

Consider estimations of the sweetness of beverages being made on a 9-point scale, ranging from "1: very unsweet" to "9: very sweet." Some judges will be reluctant to use the ends of the scale (1 or 9); it is a psychological effect that they do not like to commit themselves to the extremes on the scale. So 1s and 9s will not occur very often; numbers nearer the center of the scale will be used more freqeuntly. It is as if it is more difficult to pass from 8 to 9 than from, say, 4 to 5. It is as if the psychological distance between 8 and 9 is greater than that between 4 and 5. Once the judge uses numbers in such a way that the distances between them are no longer the same, he breaks the basic rules for numbers. No longer is the distance between 7 and 9 or 4 and 6 the same; no longer is 6 twice as big as 3.

To allow discussion of these difficulties, psychologists have developed a language to denote the status of numbers. Numbers are said to be used as though they were on a nominal, ordinal, interval, or ratio scale. It is wise to be aware of this language because it helps in the discussion of the validity of scaling as well as in the choice of statistical test. On a *nominal scale*, numbers are used merely to denote categories; they are used as names. On an *ordinal scale*, numbers are used as ranks. On an *interval scale*, the numbers have equal distances between them. On a *ratio scale* the numbers have equal distances between them *and* the zero is a real zero. These will now be discussed in more detail.

Nominal Scale: Data Given in Terms of Categories

On a *nominal scale*, numbers represent nothing more than names: for example, numbers on football players' shirts, car registration numbers, numbers used rather than names to denote various flavor categories. These numbers have no numerical value; they are merely convenient labels. A football player with an 8 on his shirt is in no way greater than a player with a 6 on his shirt; he is merely different. It would be silly to take the mean of a set of numbers on a nominal scale. To say that the mean of the numbers on the shirts of Liverpool's soccer team is 6 or that their standard deviation (σ) is 3.162, is hardly giving truly numerical information; it is also rather useless unless you are an overenthusiastic Liverpool fan. But sometimes it is easier to make these mistakes when it is not so obvious that the numbers are really on a nominal scale. For instance, category scales used in the sensory evaluation of foods are not always as numerical as they might appear. Sometimes the numbers on a category scale denote categories that are only qualitatively rather than quantitatively different. For example, consider a category scale used for rating the color of bananas (1, dark green; 5, yellow with brown flecks; 9, dark brown) or the flavor of apples (1, green flavor; 5, weak flowerlike flavor; 9, weak old flavor); the numbers on these scales are no more than labels for qualitatively different sensations. "Yellow with brown flecks" is hardly the mean of "dark green" and "dark brown"; 5 is not the mean of 1 and 9. So to compute mean scores for these scales could be misleading. Certainly, not all scales are as nonnumerical as these but it is well to be cautious in analyzing and interpreting the numbers obtained from intensity scales.

Ordinal Scale: Data Given as Ranks

In an *ordinal scale*, numbers represent ranks: 1st, 2nd, 3rd, 4th, etc. 1 comes before 2, 2 before 3, and 3 before 4, but the distances between 1 and 2, 2 and 3, and 3 and 4 are not necessarily the same. So numbers on an ordinal scale are more like numbers than those on a nominal scale but they still do not have all the properties of numbers. At least 1st comes before 2nd, and so on. So we know the order of the numbers, if not their actual amounts. So, at least, numbers on an ordinal scale carry more information than numbers on a nomial scale.

Interval and Ratio Scales

In *interval and ratio scales*, numbers represent real quantities: for example, scores on a statistics exam or the weight of apples. Here not only is the order of the values known but also by how much they differ.

The difference between a ratio scale and an interval scale is a little more complex. With a *ratio scale*, the number zero represents a true zero, whereas with an interval scale it does not. What this means is that on a ratio scale, 8 is twice as big as 4, and 4 is twice as big as 2. On an *interval scale*, although the distance between 8 and 4 is the same as that between 4 and zero (the intervals are equal), 8 is not necessarily twice as big as 4, and 4 is not necessarily twice as big as 2, because zero is not a true zero. For example, the date A.D. is only an interval scale of world age. In the year A.D. 2000 the world will not be twice as old as it was in the year A.D. 1000 because the year A.D. zero was not the beginning of the world. However, the difference in time between zero and A.D. 1000 is the same as the difference between A.D. 1000 and 2000. So the scale has equal intervals but the lack of a true zero upsets the ratios on the scale. Most of the obviously numerical scales are ratio scales: weight, height, length, density, voltage, and so on. A weight of 10 kg is twice as heavy as one of 5 kg; 30 cm is three times as long as 10 cm. The zeros on the scale are real zeros. 0 kg means weightless, 0 cm means no length. In fact, interval scales are rather hard to find. Temperature in degrees Fahrenheit or Centigrade (Celsius) is another example. The Fahrenheit and Celsius scales are interval, not ratio, scales of the heat content of a body. The difference between 0 and 50 degrees is the same as that between 50 and 100 degrees, but a body at 100 degrees is not twice as hot as one at 50 degrees. This is because the heat content of a body is not zero at 0°C or 0°F.

Not all the scales encountered in real-life situations will fall neatly into these categories; they may fall in between. But at least we have now distinguished between them and have a language for discussing the nature of our data. For example, we may wish to investigate whether there is any relationship between a person's weight and how well he identifies odors. We may measure a person's weight on a ratio scale (actual weight in kilograms), on an ordinal scale (whether he is the heaviest, second most heavy, third most heavy, etc.) or on a nominal scale (categorize as "heavy," "medium," or "light"). We may measure how well he identifies odors on a ratio scale (how many odors he correctly identifies out of 200 tested), an ordinal scale (does he identify the most odors, second most, third most, etc.), or a nominal scale ("skilled" vs. "unskilled" at identifying odors). The status of the scale of measurement determines which statistical test should be used and so must be considered carefully. Snags arise when a scale may appear to be interval or ratio but is really ordinal or nominal; this may lead to an inappropriate choice of statistical analysis. We will be careful to keep this in mind as we proceed.

Having discussed the relationship between nominal, ordinal, interval, and ratio scales, there is one more point to consider: the difference between a continuous and a discrete variable.

Continuous variable: A continuous variable is one that can, in theory, have any value: 1, 20, 2, 2½, 2.7615, 2.499432. Such variables as length, chemical concentration, weight, density, and size are all continuous.

Discrete variable: A discrete variable is one that is not continuous; it can have only discrete values. For example, the number of children in a family is a discrete variable. You can have 1, 2, 3, 4, 5, etc., children; you cannot have 2½ children. There are no values between 1, 2, 3, 4, 5, etc.

A point of caution: It is important to note that a continuous variable may appear falsely to be discrete, because of the way it is measured. For instance, length is a continuous variable. However, if we only measure length to the nearest millimeter (1, 2, 3, 4, 5, 6 mm, etc.), it may appear falsely to be discreet because we have no fractional values. Thus the underlying variable, not just the scores obtained in an experiment, must be considered when deciding whether it is discrete or continuous.

2.7 A FIRST LOOK AT PARAMETRIC AND NONPARAMETRIC STATISTICAL TESTS

Statistical tests for inferential statistics are divided roughly into two types: parametric and nonparametric tests. *Parametric tests* are used for numbers on interval or ratio scales, while *nonparametric tests* are generally designed to handle ordinal data (ranks) and nominal data (categories). The parametric tests make use of more information than do nonparametric tests; actual numbers provide more exact information than mere ranks or categories. So parametric tests tend to be more discriminating or powerful than nonparametric tests. If there are differences between two sets of data, a parametric test will usually find that difference better than a nonparametric test (other things being equal). If there is a correlation, a parametric test will usually detect it better than a nonparametric test (other things, again, being equal). Thus parametric tests are the ones we choose to use as long as we can fulfill all the assumptions they require for their use.

The first prerequisite required for parametric tests is that the data must have been sampled from a population in which the data are spread or distributed in a very special way. They must be distributed according to what is called a *gaussian* or *normal distribution*. (The use of the word *normal* here in no way implies normality as opposed to abnormality.) The normal distribution has certain properties which are exploited in the design of parametric tests. We will examine the properties of normal distributions in more detail later (Chapter 4); suffice it

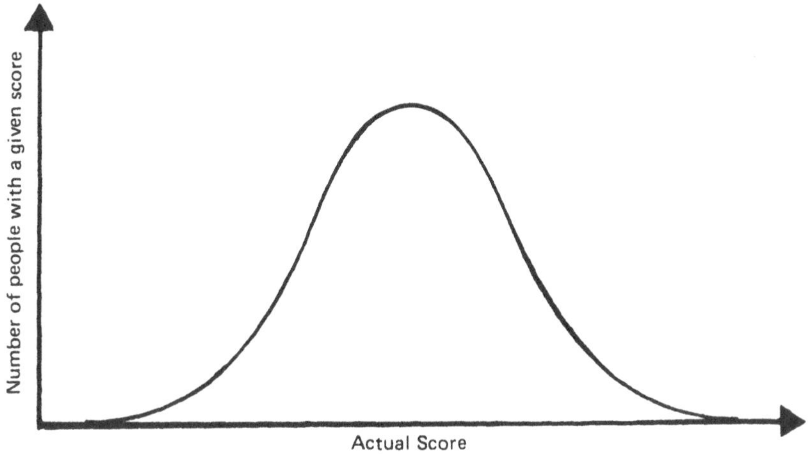

Figure 2.1 Bell-shaped normal or gaussian distribution.

to say here that if scores are distributed normally, the shape of a graph indicating the number of people who get a given score plotted against those actual scores will be bell-shaped (Figure 2.1).

For numbers to be distributed normally, they must be on an interval scale at least. Data on an ordinal scale (ranks) can hardly give a bell-shaped distribution. Each person will have a different rank; there will be one person in first place, one in second, one in third, etc. The distribution will have the shape of a horizontal line, not a normal distribution (Figure 2.2).

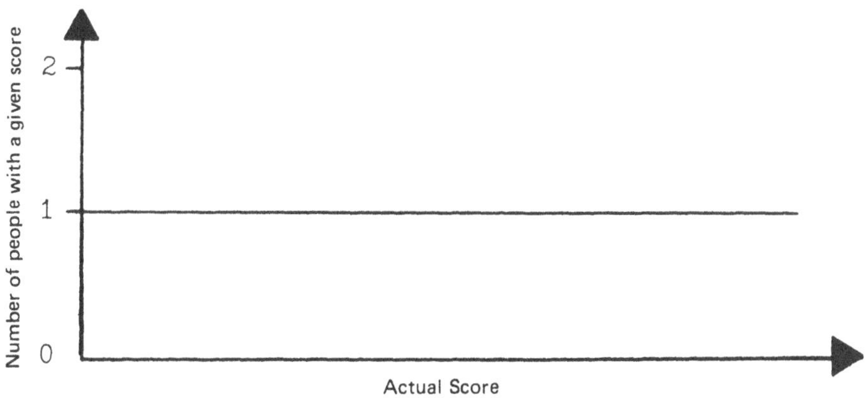

Figure 2.2 Distribution of ranked scores.

So, implicit in the assumption of normality is the assumption that the numbers we have are at least on an interval scale (better than ordinal scale). However, it makes things simpler if we deal with these two assumptions as though they were separate. For instance, we may not be able to tell, by inspection, whether our data came from a normally distributed population directly, but we may at least be able to inspect the data to see whether or not they are on an interval or ratio scale. If they are not, we know that they break the assumption of normality. On the other hand, data may be on a ratio scale but still not be normally distributed. So it helps to ask these two questions separately. First, are the data on at least an interval scale? If they are not they will not have come from a normally distributed population. If the data are on an interval or ratio scale, the second question should be asked: Are there any other reasons why the data may not have come from a normal distribution? For instance, our data may be a set of age scores. Age is not normally distributed. The younger the age, the more people there are; this is not a bell-shaped distribution with very few old and very few young people. There are ways of determining whether a sample of data come from a population of data that is normally distributed; these will be considered later (Chapter 4). There are other assumptions involved with parametric tests, such as different sets of data having to have the same variance, but these will be dealt with as we come to each test.

If the assumptions for a parametric test are not fulfilled by the data, the corresponding nonparametric test must then be used. Nonparametric tests are much simpler to use than parametric tests and are dealt with throughout this text (Chapter 5, Binomial Test; Chapter 6, Chi-Square; Chapter 16, Additional Nonparametric Tests). Sometimes researchers argue that parametric tests can still be used when the assumptions are violated slightly, that the tests are robust. Statisticians argue about this point and there seem to be no clear-cut rules as to how far the assumptions for the various tests can be violated. When in doubt, one strategy is to use both parametric and nonparametric analyses; if they agree, the answer is clear. If they disagree, the original data should be inspected carefully; more data may need to be collected. However, parametric and nonparametric tests rarely disagree if used intelligently (e.g., without a slavish adherence to the 5% level of significance; this is discussed in Section 5.4).

The Key to Statistical Tables indicates which test to use on which occasion. Across the top are given various categories of test. They may be tests that seek differences between two or more sets of data or tests that seek out correlations. Down the side of the chart are given sections for various categories of data: for nominal, ordinal, interval, or ratio data. The former two need nonparametric tests; the latter two can be analyzed by parametric tests as long as the data fulfill other prerequisites, such as coming from normal distributions. To use the chart, we first identify the type of statistical operation required. We examine the data to see their status (ranks, categories, ratio data, etc.) and then go to the appropriate section to see which test is required. Should we wish to use a correlation and our data are on an interval scale and come from a normally distributed population, we would use Pearson's product-moment correlation. Should the data be ranked, on an ordinal scale, we would use Spearman's rank correlation; should we only have categories (nominal data) we would use the contingency coefficient. As we proceed through the book, check the positions of the various tests on the chart.

3

Probability, Sorting, and Arranging

3.1 WHAT IS PROBABILITY?

Before dealing with the various statistical tests, it is worthwhile deviating for a while to examine the nature of probability and how to calculate various probability values. Basically, probability is a number used to express the likelihood of a given event happening. Instead of simply saying that something is likely or unlikely, we say that it has a 10% (or 50% or 90% or whatever) chance of happening.

Probability is described by a number which can range between 0 and 1. 0 indicates the absolute impossibility of the event concerned; 1 indicates absolute certainty; 0.5 indicates that there is an equal chance of the event occurring or not occurring. Similarly, probability can be described in percentages (i.e., 0 to 100%).

There are two approaches to the definition and calculation of the probability of an event:

1. If we know all the possible outcomes of a set of events, we can calculate the probability of one or more of these outcomes occurring. This value is known as *a priori probability* or *Laplacean* or *classical probability*. It is commonly used in problems about games involving dice, cards, etc. (this reflects something of the social history of statistics.). For example, the probability of a tossed coin coming down "heads" is 1/2. The key point here, however, is that this kind of probability can be computed before the event takes place; it can be calculated a priori.

2. The other approach rests on experience and is called *a posteriori* or *empirical probability*. Here the probability of an event happening is computed from experience rather than being predicted in advance. For instance, the probability of getting a "head" when tossing a coin is 1/2; it can be predicted in advance and is thus an a priori probability. However, if the coin were biased, "heads" would not be as equally likely as "tails"; it may come up more often than "tails." Because we do not know exactly how the coin was biased, we would have to throw the coin several times and see how much more often "heads" actually did come up. Probability would thus be computed a posteriori, after the event.

Further examples of a priori and a posteriori probabilities may be considered. What is the probability that a taste panel judge will pick by chance, from three samples of food, the one sample that has a slightly different flavor? Before doing any experiment we can say that the probability is one in three (1/3 or 33.3%), all other things being equal. This is an a priori probability. If, on eating the food, the judge can distinguish the food with the different flavor substantially more often than one-third of the time, we would conclude that the judge can distinguish the difference in flavors. On the other hand, the chance of the judge picking the odd sample without tasting anything may be rather higher than one in three. It could be that slight differences in appearance or odor of the food samples may sometimes be detected by the judges. The actual value of this probability of picking the odd sample without testing cannot be determined a priori; it would have to be determined empirically and thus will be an a posteriori probability. Probabilities for insurance premiums are computed from past experience and are thus a posteriori probabilities.

In general, if an event could occur on every one of N occasions, but in fact only occurs n times, the probability of occurrence of that event = n/N. Probability is usually symbolized by the letter p. Hence

$$p = \frac{n}{N} = \frac{\text{number of events that fulfill the required condition}}{\text{total number of events}}$$

For instance, the probability of getting "heads" on tossing a coin is 1/2. Only the coin landing "heads" fulfills the condition "heads," so $n = 1$. The total number of events that could happen are "heads" and "tails." Hence $N = 2$. So the probability of getting "heads" is one of two possibilities:

$$p = \frac{n}{N} = \frac{1}{2}$$

The probability of a judge picking from three samples of food, without tasting, the one with a different flavor is 1/3, other things being equal. Only picking the correct sample is the event that fulfills the required conditions;

there is only one correct sample, so $n = 1$. The total number of possible events is 3; there are three samples that could possibly be picked, so $N = 3$. So the probability of picking the odd sample, $p = n/N = 1/3$. In the following examples, we will concentrate on calculating a priori probabilities.

3.2 HOW TO COMPUTE A PROBABILITY: THE TOTAL NUMBER OF EVENTS

To calculate the probability, $p = n/N$, we have to be able to calculate the total number of events that can happen, N. We will confine ourselves at the moment to equally likely events, such as getting a "heads" or a "tails" on tossing a coin or throwing a "five" or a "four" on the toss of a die, or even picking by chance a different-tasting food sample from a set of three samples that are otherwise identical and thus equally likely to be chosen.

On tossing a coin, there are two possible equally likely events: "heads" and "tails," $N = 2$. On tossing a die, there are six equally likely events, because a die can come up six ways, hence $N = 6$. In trying to pick the odd food sample out of three, there are three possible choices, three equally likely events, and so $N = 3$. However, determining the total number of events is not always as easy as this, so the following procedures will help.

Consider the following. When one coin is tossed, there are two outcomes: H, heads; T, tails. So one coin

$$H, T = 2 = 2^1$$

With two coins, there are four possible outcomes:

$$HH, HT, TH, TT = 4 = 2 \times 2 = 2^2$$

Similarly, with three coins, we have

$$HHH, HHT, HTH, THH, TTH, THT, HTT, TTT = 8 = 2 \times 2 \times 2 = 2^3$$

For four coins, there are $16 = 2 \times 2 \times 2 \times 2 = 2^4$ outcomes; etc.

So for a given number of events (coin tossings), the total number of outcomes is the number of outcomes for one of these events raised to the power of the number of events. For example, for six coin tosses, the total number of outcomes is

(number of events: six coins)

$$2^6 = 64$$

(number of possibilities for one event: tossing one coin)

For 10 coins, the total number of outcomes that can occur is 2^{10}. Also, for a coin tossed 10 times, the total number of outcomes is 2^{10}. One coin tossed 10 times is equivalent to 10 coins each tossed once.

Using the same logic, our total number of outcomes from tossing one die is

$$6^1 = 6$$

The total number of outcomes that can occur on tossing three dice is

$$6^3 = 216$$

The total number of outcomes that can occur when a judge is required to pick an odd food sample from three food samples is

$$3^1 = 3$$

The total number of outcomes that can occur when this choice is made four successive times is

$$3^4 = 81$$

This rule of thumb can be a help in calculating N, the total possible number of events or outcomes.

Probability Calculations Using This Trick

If you toss a coin, the probability of getting "heads" is 1/2 and "tails" is 1/2. If you toss two coins, the probability of getting two heads is $1/2^2 = 1/4$ (where 2^2 is the total number of outcomes). The probability of tossing three coins and getting three "heads" is $1/2^3 = 1/8$. The probability of throwing a "one" when tossing a die is 1/6, of getting three "ones" on three separate dice is $1/6^3 = 1/216$ (where 6^3 is the total number of outcomes). The probability of picking by chance an odd food sample from three samples is 1/3 and of picking the odd sample four times in succession is $1/3^4 = 1/81$ (where 3^4 is the total number of outcomes).

3.3. THE MULTIPLICATION LAW OF PROBABILITY: GETTING TWO HEADS AND ONE TAIL

We can use the strategy outlined in Section 3.2 to calculate probabilities for various sets of events or we can take a shortcut and use the multiplication law of probability. The multiplication law is no more than another more formal way of expressing what we have already discussed.

The *multiplication law of probability* states that the probability of the joint occurrence of *independent* events is the product of the probabilities of the individual events. To illustrate this, take the last example, the probability of obtaining two "heads" when tossing two coins. By the multiplication law, this is the product of the probabilities of obtaining "heads" on each occasion— $1/2 \times 1/2 = 1/4$. We can see that this is correct if we go back to first principles. The total number of outcomes was 2^2 (= 4), of which getting two "heads" is one such outcome, so the probability of getting two "heads" = $1/2^2 = 1/4$, the same answer.

Example 1

With three coins, what is the probability of getting three "heads"? We could say that the total number of outcomes is $2^3 = 8$, and three "heads" is only one of these outcomes that fulfills our condition, so the probability of getting three "heads" is 1/8. Or the multiplication law could be used:

The probability of the first coin being $H = 1/2$.
The probability of the second coin being $H = 1/2$.
The probability of the third coin being $H = 1/2$.

Thus, by the multiplication law, the probability of getting three "heads" is $1/2 \times 1/2 \times 1/2 = 1/8$.

Example 2

I toss five coins; what is the probability that I get *H, H, H, H, H*?

By the multiplication law, it is

$$1/2 \times 1/2 \times 1/2 \times 1/2 \times 1/2 = 1/32$$

Similarly, for getting *T, T, T, T, T*, it is

$$1/2 \times 1/2 \times 1/2 \times 1/2 \times 1/2 = 1/32$$

Notice these values are all the same because the probability of getting *H* is the same as getting *T*, namely 1/2.

Example 3

Tossing three coins and a die, what is the probability of getting *H, H, H*, six?

It is

$$1/2 \times 1/2 \times 1/2 \times 1/6 = 1/48$$

Example 4

Tossing three coins and two dice, what is the probability of getting *T, T, T*, six, six?

It is

$$1/2 \times 1/2 \times 1/2 \times 1/6 \times 1/6 = 1/288$$

Example 5

Tossing three coins and two dice, what is the probability of getting *T, H, T*, five, three, in that order?

It is

$$1/2 \times 1/2 \times 1/2 \times 1/6 \times 1/6 = 1/288$$

Note there that we are specifying which coin is to be heads, which coin is to be tails, and which die is to have a given score. Only if we do this can we use the multiplication law. If we did not specify this, the calculation would be different, as we will see.

Example 6

If the probability of choosing a food with an off-flavor by chance from a pair of foods (one with an off-flavor, one without) is 1/2, and of choosing it from three samples of food (one with off-flavor, two without) is 1/3, and the probability of picking it from seven samples of food (two with the off-flavor, five without) is 2/7, then the probability of picking the food with an off-flavor by chance on all three tests in succession is

$$1/2 \times 1/3 \times 2/7 = 2/42 = 1/21$$

So far we have always specified exactly which event has to have a given result. With the tossing of coins we have stated specifically which coin had to be "heads" and which had to be "tails." However, we often do not want such specificity. We do not want to know the probability, when three coins are tossed, of having "tails" on the first coin, "heads" on the second, and "tails" on the third (*T, H, T*, in that order). Rather, we may want to know the probability of just getting two "tails" and one "head," no matter which coin shows which face. Here the first two coins may be "tails" or the last two or even the first and last ones. So the situation is different; we can get our specified pattern of "heads" and "tails" in more ways than one. We are satisfied as long as any two are "tails" and the other is "heads." But now we cannot blindly apply the multiplication law; we have to go back to first principles. Unless we specify which coin is to be "heads" or "tails," the events will not be independent. For the law to be applicable we must have *independent events*; it is a statement about the joint occurrence of independent events. If we only specify two "tails" and a "head" but not say which, the events are no longer independent. If the first two coins are "tails," the last must be "heads"; the way the third coin must land depends on how the first two landed. Thus to calculate probabilities in this case, when we do not have the independent events, we cannot use the multiplication law; we have to go back to first principles, as follows. There are $2^3 = 8$ outcomes:

HHH, HHT, HTH, THH, <u>HTT</u>, <u>THT</u>, <u>TTH</u>, TTT

The three underlined outcomes fulfill the condition of two "tails" and one "head." Thus the probability of getting two "tails" and one "head" = 3/8.

So beware of blindly using the multiplication law. Fools rush in where angels fear to multiply!

Example 7

We may be arranging food samples in groups of three, of which one food is sweeter than the other two. Let us say that we have five such groups of three food samples. What is the probability that the sweeter sample will be the first one to be tasted in each of the five groups of three foods?

The probability that it will be tasted first in any one group is 1/3. By the multiplication law, the probability that it will be tasted first in all five groups is

$$1/3 \times 1/3 \times 1/3 \times 1/3 \times 1/3 = 1/243$$

Example 8

What is the probability that the sweeter sample will be tasted first in the first two groups only and not in the other two?

The probability of it being tasted first is 1/3; the probability of it not being tasted first is 2/3. Thus, by the multiplication law, the probability of tasting the sweeter sample first in the first two groups and not in the other three is

$$1/3 \times 1/3 \times 2/3 \times 2/3 \times 2/3 = 8/243 = 0.03$$

In Examples 7 and 8, the outcomes in the five groups are independent of each other; whether our condition is fulfilled in one food group is not affected by whether it is fulfilled in another. On the other hand, if we were calculating the probability of the sweet food sample being tasted first in any two of the five groups, the outcomes for the groups would no longer be independent. Should the sweeter sample be tasted first in the last two groups it must not in the first three. Because the outcomes are no longer independent of each other, the multiplication law does not apply and we must go back to first principles. With five groups of three foods, there are $3^5 = 243$ possible outcomes, orders in which the various groups can have the sweeter samples tasted first or not. Of these, it can be seen from Figure 3.1 that only 80 outcomes satisfy the requirement that two (and only two) groups have the sweetest sample tasted first. Thus the probability of having the sweet food sample tasted first on any two (but only two) groups is 80/243 = 0.3.

3.4 THE ADDITION LAW OF PROBABILITY: GETTING A RED OR A WHITE BALL

A second law of probability, the *addition law of probability*, states that the probability of occurrence of one of a number of *mutually exclusive* events is the sum of the individual probabilities.

+	+	○	○	○
+	○	+	○	○
+	○	○	+	○
+	○	○	○	+
○	+	+	○	○
○	+	○	+	○
○	+	○	○	+
○	○	+	+	○
○	○	+	○	+
○	○	○	+	+

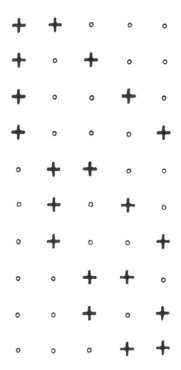

Figure 3.1 Number of arrangements of five groups of food samples in which the sweeter sample is tasted first in two groups only, where + denotes a group in which the sweeter sample is tasted first, and ○ denotes a group in which the sweeter sample is not tasted first.

Example 1

Imagine a bag full of four balls: one red, one white, one blue, one yellow. Imagine, also, that you put your hand in the bag and, without looking, pull out one of the balls. The probability that the ball you pull out is a red ball is 1/4. The probability of getting a white ball is also 1/4. The probability of getting a red *or* a white is 1/2 (because one-half of the set of balls is red + white, the other half is blue + yellow; i.e., 1/4 + 1/4 = 1/2). The probability of getting a red *or* a white *or* a blue ball is 1/4 + 1/4 + 1/4 = 3/4. The probability of getting a red *or* a white *or* a blue *or* a yellow ball is 1/4 + 1/4 + 1/4 + 1/4 = 1; it is a 100% certainty that you will get one ball with one of those colors.

Similarly, the probability of getting a "four" *or* a "six" in one throw of a die is 1/6 + 1/6 = 1/3.

Example 2

Imagine three solutions presented to a subject for tasting: one sweet, one sour, and one bitter. The probability that the subject will choose the sweet solution to taste first is 1/3. The probability of tasting the sour one first is also 1/3. The probability that the solution tasted first is sweet *or* is sour is 1/3 + 1/3 = 2/3.

Note: The events *must* be mutually exclusive for the addition law to be used. Picking a red ball from the bag means that you have not picked a white ball, and vice versa. They are mutually exclusive. If you toss a die and it gives you a "four," you cannot also have tossed a "six." You cannot get both at once; they are mutually exclusive. If you pick the sweet solution first, you cannot also pick the sour one first. A common mistake in probability calculations is blindly to add up probabilities when the events are not mutually exclusive.

Example 3

Above, we said that the probability of getting a "four" *or* a "six" in one throw of a die is 1/6 + 1/6 = 1/3, where getting a "four" or a "six" on the same die is mutually exclusive. But what if we had two dice and we wanted to know the probability of getting a "four" on one die and a "six" on the other. The probability is not 1/6 + 1/6, by the addition law, because getting a "four" on the first die does not exclude getting a "six" on the second die. So we have to go to first principles. The total number of outcomes is $6^2 = 36$. The outcomes that satisfy the condition are a "four" on the first die and a "six" on the second, or a "six" on the first die and a "four" on the second; two outcomes satisfy the condition. So the probability of getting a "six" and a "four" is 2/36 = 1/18. If we specified these conditions further by saying that we must have a "four" on the first die and a "six" on the second, the probability of getting this would then be 1/36, 1/6 × 1/6, by the multiplication law (for the joint occurrence of two independent events).

3.5 WORKED EXAMPLES: LOTS OF PROBABILITY CALCULATIONS

It is now worth getting some practice with worked examples.

Example 1

Imagine that you are presenting four taste solutions, one sweet, one sour, one salty, one bitter to a set of 10 subjects. What is the probability of the following events occurring? (We will leave the probabilities as fractions.)

a. The first subject chooses the sweet solution to taste first.

   There are four solutions so the probability of tasting a sweet solution first = 1/4.

b. The first subject chooses a sweet or a sour solution to taste first.

The probability of tasting a sweet solution first = 1/4. Similarly, the probability of tasting a sour solution first = 1/4. As these two events are mutually exclusive, the probability of tasting a sweet or a sour solution first, by the addition law, = 1/4 + 1/4 = 1/2.

c. The tenth subject chooses a sweet or a sour solution to taste first.

The probability that the first subject, or the tenth subject or any of the 10 subjects tasting a sweet or a sour solution first, is, from the last example, 1/4 + 1/4 = 1/2.

d. The tenth subject and the first subject chooses a sweet or sour solution to taste first.

From example (c) the probability of the tenth subject tasting a sweet or a sour solution first is 1/2. The same is true for the first subject. The probability that both these independent events will occur, that both subjects will taste a sweet or a sour solution first, is, by the multiplication law, $1/2 \times 1/2 = 1/4$.

e. The tenth subject or the first subject chooses a sweet or sour solution to taste first.

Don't use the addition law! The first subject picking sweet or sour does not exclude the tenth subject from doing so; the events are not mutually exclusive. So go back to first principles. Considering the two subjects, there are a total of 4^2 (= 16) possible outcomes. The 12 outcomes that fulfill our requirements are as follows:

First subject picking sweet, tenth subject picking sweet
First subject picking sweet, tenth subject picking sour
First subject picking sweet, tenth subject picking salty
First subject picking sweet, tenth subject picking bitter
First subject picking sour, tenth subject picking sweet
First subject picking sour, tenth subject picking sour
First subject picking sour, tenth subject picking salty
First subject picking sour, tenth subject picking bitter
First subject picking salty, tenth subject picking sweet
First subject picking salty, tenth subject picking sour
First subject picking bitter, tenth subject picking sweet
First subject picking bitter, tenth subject picking sour

Thus the probability that the tenth subject or the first subject will choose a sweet or a sour solution to taste first = 12/16 = 3/4.

f. All subjects taste the bitter solution first.

The probability that a subject will choose a bitter solution to taste first = 1/4. To find the probability that all 10 subjects will taste a bitter solution first, use the multiplication law, because the subjects are independent.

Thus the probability = $1/4 \times 1/4 \times 1/4 \times 1/4 \times 1/4 \times 1/4 \times 1/4 \times 1/4 \times 1/4 \times 1/4 = (1/4)^{10} = 1/1{,}048{,}576$.

g. The first subject will first taste a salty solution, the second a sweet, the third a sour, the fourth a bitter solution, the fifth a salty or sweet solution, the sixth a salty or bitter, the seventh a bitter or sour, the eighth a bitter or sweet, the ninth a sweet or sour or salty solution, and the tenth a sweet or bitter or salty solution.

For the first four subjects, we are looking at the probability of picking a specific one of four solutions, which is 1/4. By the multiplication law, the probability that all four subjects will taste a specified solution = $1/4 \times 1/4 \times 1/4 \times 1/4 = 1/256$.

For the next four subjects, we are looking at the probability of picking either one or another of two specified solutions. For each subject, by the addition law, this is $1/4 + 1/4 = 1/2$.

By the multiplication law, the probability that all four subjects will do this is $1/2 \times 1/2 \times 1/2 \times 1/2 = 1/16$.

For the last two subjects, we are looking at the probability of picking either one or another of three specified solutions. For each subject this is $1/4 + 1/4 + 1/4 = 3/4$.

By the multiplication law, the probability that both these subjects will do this is $3/4 \times 3/4 = 9/16$.

The probability that all these outcomes will occur is, by the multiplication law, $1/256 \times 1/16 \times 9/16 = 9/65536$.

Now a further example using coins and dice.

Example 2

You throw three pennies and two dice; this is called *one throw*. You make three such throws. What is the probability that:

a. The three coins are "heads" on the first throw.

The probability of getting "heads" for one coin = 1/2. The probability of getting a "head" for each of the three coins on the first throw (one throw), by the multiplication law, = $1/2 \times 1/2 \times 1/2 = 1/8$.

b. The three coins are "tails" on the second throw.

The probability of getting "tails" for one coin = 1/2. The probability of getting tails for each of the three coins on the second throw by the multiplication law = $1/2 \times 1/2 \times 1/2 = 1/8$, just as it would be for getting all "heads" on the first throw.

c. The three coins are tossed "heads," "heads," and "tails" (*H, H, T*) in that order as they fall on the first throw.

The probability that a first coin is "heads" = 1/2, second coin is "heads" = 1/2, third is "tails" = 1/2. So the probability of these three (independent) events happening, by the multiplication law = $1/2 \times 1/2 \times 1/2 = 1/8$. This is, of course, the same as the probability that all three are "heads" or all three are "tails."

d. On the first throw the three coins are arranged so that any two are "heads" while the remaining one is "tails."

Here it is not specified which coins are to be "heads" and which one is to be "tails," as it was in the last example. Thus, to get the required result, what happens to one coin affects what must happen to the other coins. The events are no longer independent as far as this problem is concerned, so the multiplication law is inappropriate. Two "heads" and one "tails" can occur in more than one way, not merely *HHT*. There are 2^3 (= 8) different ways in which the coins can fall: *HHH, TTT, HHT, HTH, THH, HTT, THT, TTH*. Of these, three (*HHT, HTH, THH*) fulfill the condition that two of the coins are "heads" and one "tails." Thus the probability that two of the coins will be "heads" while the other is "tails" on this first throw = 3/8.

e. The three coins are all "heads" on the first throw or they are all tails.

The probability that all coins are heads = $1/2 \times 1/2 \times 1/2 = 1/8$. The probability that all coins are tails = $1/2 \times 1/2 \times 1/2 = 1/8$. These two events are mutually exclusive (you cannot get all "heads" and get all "tails" simultaneously), so the addition law can apply. The probability that the coins on the first throw can be all "heads" or all "tails" is thus $1/8 + 1/8 = 1/4$.

f. The two dice both show "five" on the first throw.

The probability that a die will show "five" (or any given number, in fact) = 1/6. The probability that both dice will show "five" on the first throw, by the multiplication law, = $1/6 \times 1/6 = 1/36$.

g. The first die shows "three" and the second die shows "four" on the first throw.

The probability that the first die shows "three" = 1/6. The probability that the second die shows "four" = 1/6. The probability that the first die shows "three" and the second die shows "four" by the multiplication law = $1/6 \times 1/6 = 1/36$. This is, of course, the same as the probability that both show "five" or the probability of two specified dice showing any specified numbers.

h. Any one die shows "five" while the other shows "three" on the first throw.

Here, it is not specified which die is to show "five" and which is to show "three." Thus, in this example, what happens to one die affects what must happen to the other; they are no longer independent and the multiplication law is inappropriate. One die showing "five" and the other showing "three" can occur in more than one way. We go back to first principles. There are 6^2 (= 36) different ways in which the two dice can fall. There are only two ways which fulfill our conditions of one "five" and one "three." These are: first die shows "five," second shows "three" and first die shows "three," second shows "five." Thus the probability of getting any one die showing "five" while the other shows "three" = 2/36 = 1/18.

i. The first die shows "three" or "five" on the first throw.
The probability that this die shows "three" = 1/6. The probability that this die shows "five" = 1/6. These are mutually exclusive, if the die is showing "three" it cannot also show "five," so the addition law can apply. The probability that the first die shows "three" or "five" = 1/6 + 1/6 = 1/3.

j. All coins are "tails" and all dice show "four," on the first throw.
The probability that all coins are "tails" = 1/2 X 1/2 X 1/2 and all dice show "four" = 1/6 X 1/6. The probability of these both happening, by the multiplication law, = 1/2 X 1/2 X 1/2 X 1/6 X 1/6 = 1/288.

k. All coins are "tails" and all dice show "four" on all three throws.
The probability that all coins are "tails" and all dice show "four" on the first throw is 1/288. By the multiplication law, it is

$$\frac{1}{288} \times \frac{1}{288} \times \frac{1}{288} = \frac{1}{23{,}887{,}872}$$

for all three throws.

l. On the first throw, all coins are "heads" and both dice show "four" or all coins are "tails" and both dice show "six."
The probability that all coins are "heads" and both dice show "four" is

$$\frac{1}{2} \times \frac{1}{2} \times \frac{1}{2} \times \frac{1}{6} \times \frac{1}{6} = \frac{1}{288}$$

Similarly, the probability that all coins are "tails" and both dice show "six" = 1/288. These two events are mutually exclusive, so the addition law can apply. Thus the probability of getting all "heads" and "fours" *or* all "tails" and "sixes" on the first throw is

$$\frac{1}{288} + \frac{1}{288} = \frac{1}{144}$$

m. All coins are "heads" and both dice show "four" on the first or second throw only, but not on both throws.
Getting all "heads" and "fours" on the first throw does not prevent it from happening on the second throw, so the two events are not mutually exclusive. The addition law does not apply. So we go back to first principles. On the first throw there are $2^3 \times 6^2 = 288$ possible patterns for the coins and dice, and the same is true for the second throw. Thus there are a total number of 288^2 patterns over both throws.

We have specified that we want all "heads" and "fours" on one throw (the first or second) only. We do not allow it on both throws. We were careful to say "the first *or* second throw *only*" because if we had just said "first *or* second throw" it would have been ambiguous. "Or" has two meanings: "*A* or *B* only" (the exclusive or) and "*A* or *B* or both" (the inclusive or). Be careful when using this word.

If we get all "heads" and "fours" on the first throw, it must not happen on the second, and vice versa. If we get all "heads" and "fours" on the first throw, there are 288 possible patterns that can occur on the second throw, of which 287 are allowable (all "heads" and "fours" is not). In the same way, 287 patterns are allowable on the first throw if all "heads" and "fours" occurs on the second. So the total number of allowable patterns is $287 + 287 = 574$. The total number of patterns is $288^2 = 82{,}944$. Thus the probability of getting what we want (all "heads" and "fours" on the first or second throw only) = 574/82,944.

3.6 SORTING

Sometimes it is necessary to know how many different ways things can be sorted in order. To work this out, we must first consider factorials. The factorial of 6 (written 6! and called *six factorial* or "six shriek") is defined as $6 \times 5 \times 4 \times 3 \times 2 \times 1 = 720$. Similarly, $3! = 3 \times 2 \times 1 = 6$ and $7! = 7 \times 6 \times 5 \times 4 \times 3 \times 2 \times 1 = 5040$. Table G.2 gives factorial values up to 20!. Incidentally, by definition, $0! = 1$.

Now consider the problems of ordering or sorting. The number of ways three distinct objects, say food samples, can be sorted in order is $3! = 6$. This can be checked. If these objects are labeled *A*, *B*, and *C*, the possible six orders are *ABC, ACB, BAC, BCA, CAB, CBA*. There are 4! (= 24) ways of ordering four objects, 6! (= 720) different orders for six objects, and, in general, $n!$ ways of ordering n objects.

3.7 PERMUTATIONS AND COMBINATIONS

Permutations and combinations are used to calculate the number samples of a given size that can be chosen from a given population.

Permutations

Assume that one is drawing from a population of size N a sample of size n. Assume that each of these n people in the sample is assigned a particular place in the sample, like people being assigned positions in a football team. There are then ${}_NP_n$ possible *permutations* or ways of doing this, where

$$_NP_n = \frac{N!}{(N-n)!}$$

Permutation: position in sample specified

Thus, if we pick a four-man team from six players, with each man in a specified position (so that there are 4! possible ways of ordering each sample of four men), there are

$$_6P_4 = \frac{6!}{(6-4)!} = \frac{6!}{2!}$$

$$= \frac{6 \times 5 \times 4 \times 3 \times 2 \times 1}{2 \times 1} = 360 \text{ ways of doing it}$$

Combinations

On the other hand, we may not be concerned with each member of the sample having an assigned position (so that a group of four people cannot be ordered in 4! different ways; there is only one way they can be sampled). In this case, if we are drawing from a population of size N a sample of size n, where the positions in the sample are *not* specified, there are $_NC_n$ possible *combinations* or ways of doing this, where

$$_NC_n = \frac{N!}{(N-n)!\,n!}$$

Combination: position in sample unspecified

Thus, if we had six men, and we wish to pick samples of four, there are

$$_6C_4 = \frac{6!}{(6-4)!\,4!} = \frac{6!}{2!\,4!}$$

$$= \frac{6 \times 5 \times 4 \times 3 \times 2 \times 1}{2 \times 1 \times 4 \times 3 \times 2 \times 1} = \frac{6 \times 5}{2} = 15 \text{ ways of doing it}$$

Note: There are only 15 samples that we can choose if we do not specify the position of each of the four persons in the sample. If each sample could be arranged in 4! different ways, there would be more (360) samples.

$_NC_n$ is sometimes written $\binom{N}{n}$; hence $_6C_4$ can be written $\binom{6}{4}$. Note that

$_6P_4 = 360$ where the position of each member of the sample is specified

$_6C_4 = 15$ where the position of sample members is unspecified

So ${}_NP_n = n!\ {}_NC_n$ because there are $n!$ different ways of arranging in order a sample of size n.

Note that combinations are used as the coefficients for the binomial expansion (Appendix B); the binomial expansion is used in the calculation of probabilities (Chapter 5).

3.8 WORKED EXAMPLES: SORTING, PERMUTATIONS, AND COMBINATIONS

Example 1

You have four colored buttons on the control dial of an automatic processing plant and when an emergency bell rings, you are required to press three of them (i.e., three separate buttons, you cannot press the same one twice). How many distinguishable responses can you make if:

a. The buttons must be pressed at 0.5-second intervals?

If the buttons are pressed at 0.5-second intervals, different orders in which the three buttons are pressed give different responses. Thus we must specify the order. The problem becomes one of how many permutations of three can be drawn from a population of six.

$$ {}_5P_3 = \frac{N!}{(N-n)!} = \frac{5!}{(5-3)!} = \frac{5 \times 4 \times 3 \times 2 \times 1}{2 \times 1} = 60 $$

There are 60 distinguishable responses if the buttons are pressed at 0.5-second intervals.

b. The buttons are pressed simultaneously?

If the buttons are pressed simultaneously, there is only one response for a given set of three buttons; there is no distinguishable order in which they are pressed. Thus order is not distinguished and the problem is one of how many combinations of three can be drawn from a population of six.

$$ {}_5C_3 = \frac{N!}{(N-n)!\,n!} = \frac{5!}{(5-3)!\,3!} = \frac{5 \times 4 \times 3 \times 2 \times 1}{2 \times 1 \times 3 \times 2 \times 1} = 10 $$

There are 10 distinguishable responses if the buttons are pressed simultaneously.

Example 2

How many three-letter words can be formed from the letters of the word BACKGROUND? How many six-letter words can be formed by rearranging the letters of the word GROUND? Are they the same number or different? (A word is defined as a group of any letters, e.g., KGR, OAV, DNRGVO, etc.)

The number of three-letter words (order is important: KGR, KRG, GKR, GRK, etc.) that can be formed from the 10-letter word BACKGROUND is given by

$$_{10}P_3 = \frac{N!}{(N-n)!} = \frac{10!}{(10-3)!} = \frac{10 \times 9 \times 8 \times 7 \times 6 \times 5 \times 4 \times 3 \times 2 \times 1}{7 \times 6 \times 5 \times 4 \times 3 \times 2 \times 1}$$

$$= 720$$

The number of six-letter words that can be formed by rearranging the letters of the word GROUND is

$$6! = 6 \times 5 \times 4 \times 3 \times 2 \times 1 = 720$$

These two values are the same.

Example 3

Three types of cake and two types of beer are to be presented to judges who will judge their quality. The cakes and beer are to be arranged in a straight line; how many different ways can these be presented so that there is a cake at each end of the line and the beers are not put next to each other?

If there is to be a cake at each end of the line and the beers are not to be placed next to each other, the order must be

cake beer cake beer cake

There are two ways of arranging the beers in order. There are 3! (= 6) ways of arranging the cakes, for each way of arranging the beers. Thus there are $2 \times 6 = 12$ ways of arranging the cakes and beer.

Example 4

You are performing a psychophysical experiment on taste. You have six solutions which are different mixtures of chemicals and you wish your subjects to rank them for taste intensity. You also wish each subject to perform several replications. There are two difficulties. When the subject repeats his ranking task, you are worried that he may merely repeat his prior ranking by remembering the actual taste quality of the mixtures rather than making a second judment of the strength of the mixtures; he may remember that last time he placed the sweetest first and merely repeat this, regardless of its apparent taste intensity. The second difficulty is that it is not easy to rank six solutions; judges forget the taste and need a lot of retasting. However, four solutions can be ranked fairly easily and rapidly. Thus it was decided to have subjects rank different sets of four solutions from the group of six. This would mean that each replication would be a different task, subjects having a different set of solutions each time. How many sets of four solutions could you get from the six?

This is a simple combination problem. How many combinations of four (the order of the four does not matter) can you get from six? The answer is

$$_6C_4 = \frac{6!}{(6-4)!\,4!} = \frac{6 \times 5 \times 4 \times 3 \times 2 \times 1}{2 \times 1 \times 4 \times 3 \times 2 \times 1} = 15$$

Thus there are 15 different combinations of four taste solutions that can be drawn from six. To cover all these, a subject needs to perform 15 trials.

4

Normal (Gaussian) Distribution: z Tests

4.1 WHAT IS A NORMAL DISTRIBUTION?

Before proceeding further it is worth examining the normal or gaussian distribution in more detail. Remember that parametric statistical tests have as a basic assumption that samples should be taken from populations that are normally distributed. The easiest way of examining the normal distribution is by considering an example.

Imagine that we decide to test the whole population of the world for their liking of a new candy. Some people will like it extremely, some will dislike it extremely, and many will cluster around some central mean value of liking. We could plot the values on a graph. Such a graph is called a *frequency distribution*. There are many possible shapes the graph can take, but often the same shape of frequency distribution crops up. The particular distribution that occurs so often is called the *gaussian* or *normal distribution* and is bell-shaped, like the curve shown in Figure 4.1. It has many useful properties and because of this, much statistical theory is based on the normal distribution.

The normal distribution is a symmetrical bell-shaped curve which can vary in height and width. The mean (μ), mode, and median coincide. It is described by the function

$$Y = \frac{1}{\sqrt{2\pi\sigma^2}} \; e^{-(X-\mu)^2/2\sigma^2}$$

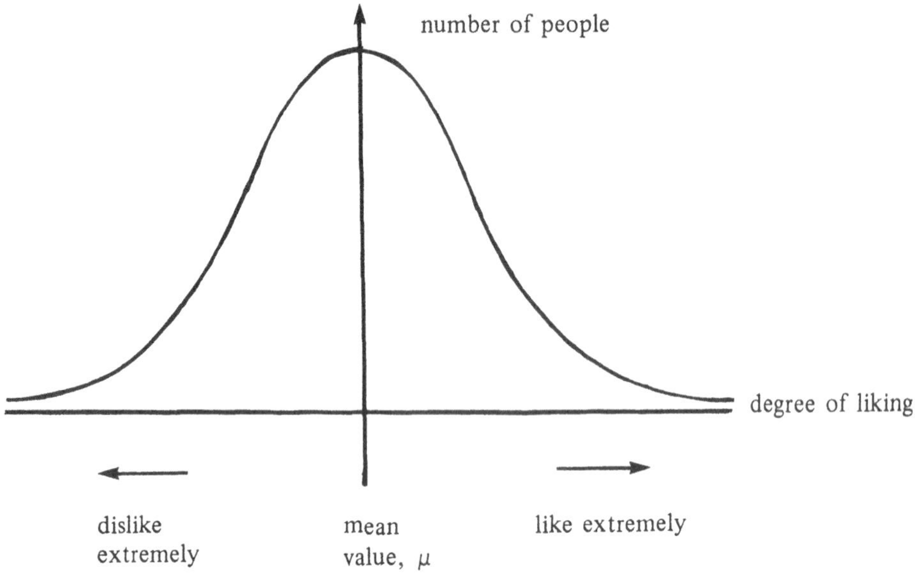

Figure 4.1 Normal distribution.

where

- X = any point along the X (horizontal) axis; degree of liking in the example above
- Y = height of the curve at any point X; the number of people who have that given degree of liking, in the example above
- μ = mean of the distribution
- σ = standard deviation of the distribution
- π = a constant, 3.1416
- e = a constant, 2.7183 (the base of napierian logarithms) (see Section 2.6)

We get a given curve for given values of μ (the mean) and σ (standard deviation). For each combination of μ and σ we get a different curve. So μ and σ are the parameters that must be specified for a normal distribution to be drawn.

Whatever the value of μ and σ, for a normal distribution 68.26% of all the cases fall within 1 standard deviation of the mean (see Figure 4.2). This is calculated from the function. Similarly, 95.44% of all readings occur within 2σ of the mean and 99.72% within 3σ of the mean. The rest of the readings do not fall within 3σ of the mean. 95.44% of the cases fall within 2σ of the mean. Or we can say that the probability of finding a case within 2σ of the mean is 95.44% (or 0.9544). The probability of a value being σ or less from the mean

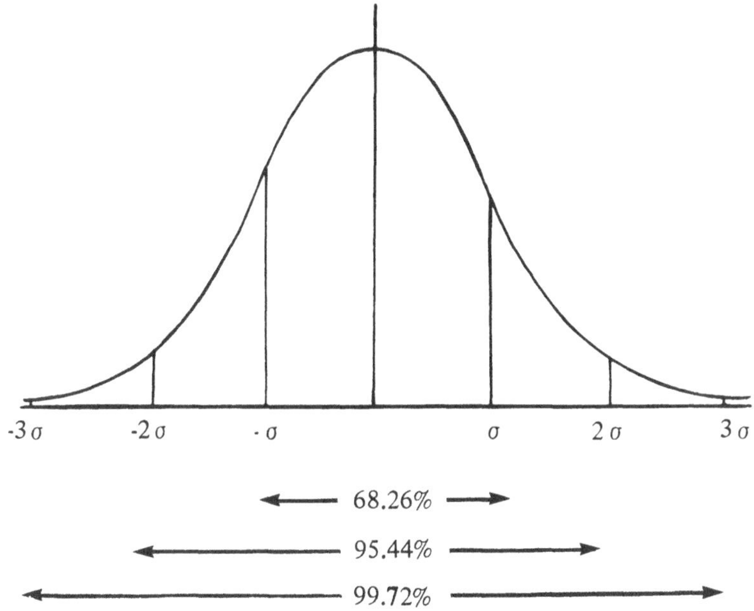

Figure 4.2 Percentage of readings that fall within 1, 2, and 3 standard deviations of the mean.

is 68.26% and 3σ or less from the mean is 99.72%. Similarly, we could calculate that the probability of a value being 0.675σ or less from the mean is 50%; 2.58σ or less from the mean is 99%. So we can see that the normal distribution is a probability distribution. Y gives the frequency (number of cases) for a given value of X, while the area under the curve gives the proportion of values falling between X and the mean (i.e., the probability of finding a value between X and μ). The area under the whole curve is thus 100% (i.e., p = 100% or 1.0).

4.2 *z* SCORES

If we know a given value of X (e.g., a score representing a degree of liking), we could calculate the probability of getting a score equal to that value or higher, or the probability of getting a score equal to that value or less, from the formula for normal distributions, if we knew μ and σ. This is a complicated operation, so tables can be constructed to give these probabilities. All you have to know is μ and σ to select the appropriate normal distribution and look up the probability for the given score. But such tables would be vast because there are millions of combinations of μ and σ and thus millions of possible normal distributions. So only one table is used and that is the one for which $\mu = 0$ and $\sigma = 1$. In this case, we know our mean score and the standard deviation and we transform our data so that the mean is zero and the standard deviation 1, to fit in with the table. Such scores are called *standard scores* or *z scores*. A z score is really the distance from the mean in terms of standard deviations, or in other words, the number of standard deviations above or below the mean. A z score of 1.5 represents a score 1.5 standard deviations above the mean. A z score of -2 represents a score 2 standard deviations below the mean.

For example, assume that we have a set of scores whose mean is 100 and whose standard deviation is 15. Then a score of 115 is 1σ above the mean; it will have a z score of value 1. Similarly, a score of 130 is equivalent to a z score of 2 and a score of 85 is equivalent to a z score of -1.

This can all be expressed by the formula

$$z = \frac{x - \mu}{\sigma} = \frac{\text{distance from the mean}}{\sigma}$$

4.3 z TESTS: COMPUTING PROBABILITIES

Having found the z score, we can now find the probability of obtaining a score equal to or less than this z score (or equal to or greater than the z score) from the tables. Basically, it is a matter of finding the area under a portion of the normal distribution curve.

Table G.1 gives areas in two different ways (Figure 4.3) for greater convenience. Some examples will make this clear.

Let us assume that we were scoring a questionnaire; it could be a consumer questionnaire measuring attitudes toward or liking for a new product. It could be an intelligence test or even an examination. We have measured a whole population and we want to know what proportion of the population falls above or below a given score. This may be needed for decisions about marketing the new product if the questionnaire were a consumer questionnaire, or decisions about whether to accept or reject a job applicant if the questionnaire were an intelligence test or an examination. Let us assume that the questionnaire scores turned out to be normally distributed with a mean of 100 and $\sigma = 15$.

Example 1

What is the probability of getting a score of 115 or more?

115 is 1 standard deviation above the mean; thus its z score is 1. Look up the area under the curve for a z score of 1 or more in Table G.1 (Figure 4.4), and the value is 0.1587. Thus the probability of getting a score of 115 or more is 0.1587, or 15.87%.

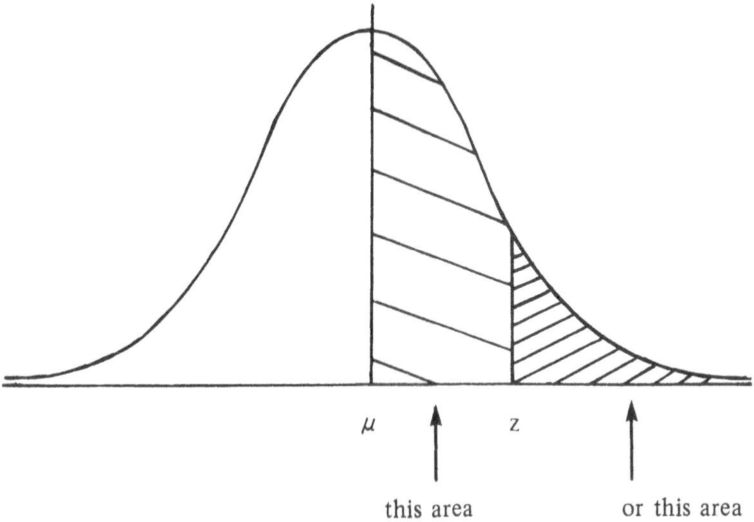

Figure 4.3 Areas under the normal curve, values from Table G.1.

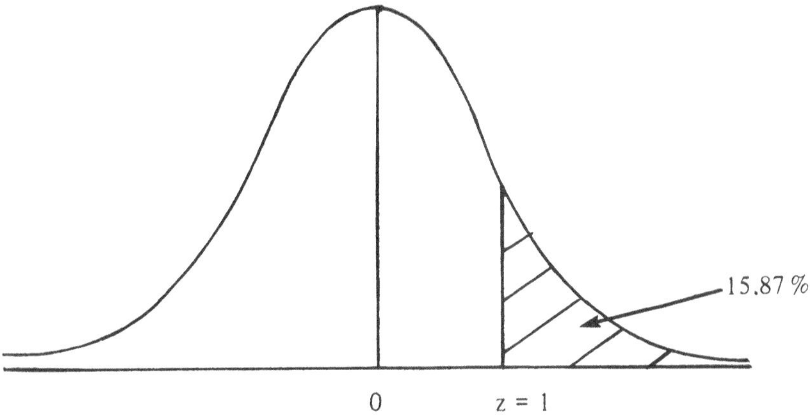

Figure 4.4 Curve for Example 1.

Example 2

What is the probability of getting a score of 115 or less?

Again, 115 is 1 standard deviation above the mean; thus the z score = 1. This time we want the shaded area for the probability of getting a z score of 1 or less (Figure 4.5). The probability of getting a score below the mean is 0.5 because 50% of the scores fall below the mean. The probability of getting a score between the mean and a z score of 1 is (from Table G.1) 0.3413 or 34.13%. Thus the probability of getting a score of 115 or less (z score of 1 or less) = 0.5 + 0.3413 = 0.8413, or 84.13%.

Example 3

What is the probability of getting a score of 70 or less?

70 is 2σ below the mean and negative z scores are not given in the table. But by symmetry, we can see that the probability of getting a score of 70 (-2σ) or less is the same as the probability of getting a score of 130 ($+2\sigma$) or more (Figure 4.6). This is found from Table G.1. For $z = 2$, the probability = 0.0228. Thus the probability of getting a score of 70 or less is 0.0228, or 2.28%.

It is important to note that these tests are called one-tailed tests and Table G.1 gives one-tailed probability values. For two-tailed tests, double the probability value. This will be discussed further in the section on binomial tests.

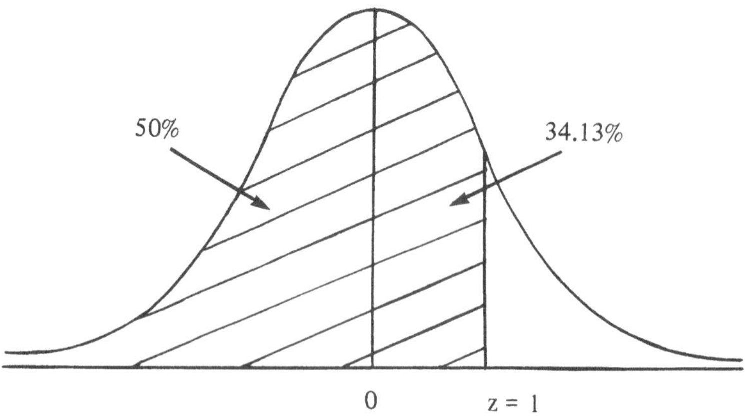

Figure 4.5 Curve for Example 2.

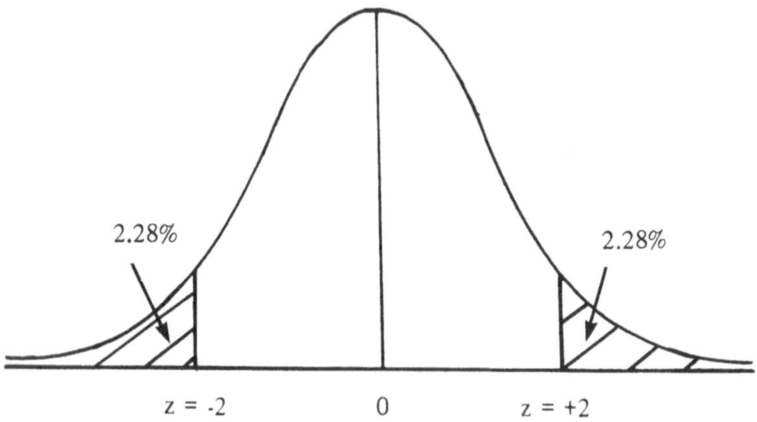

Figure 4.6 Curve for Example 3.

4.4 WHEN ARE NORMAL (GAUSSIAN) DISTRIBUTIONS USED?

Gaussian distributions are important because they often approximate distributions occurring in nature. Also, other distributions (Poisson distributions, binomial distributions) approximate to normal when the sample size is large. But there is a further reason for the importance of normal distributions. The parametric statistical tests assume that the population of scores from which a sample is drawn is normally distributed. How do we know whether this is true?

There are, in fact, statistical procedures for checking from the sample whether the distribution is normal, but we will not deal with these here. However, in practical behavioral situations one rule of thumb that can be used is that if the values are subject to a lot of randomly varying factors, the distribution will come out fairly near normal. This rule does not necessarily generalize to other disciplines; in health studies the populations are expected to be skewed by the many ill patients with atypical scores.

One of the best examples of normally distributed scores in behavioral studies is an intelligence test score. A person's score will depend on his or her motivation for the IQ test that day, family background, academic training, and interests; all these in turn are subject themselves to many variables. So the population of IQ scores will come out normally distributed. Other normally distributed scores would be scores of extraversion on a personality test, statistics exam scores, hedonic ratings for tomatoes, and number of calories consumed by a person per day.

There are notable cases when the distribution of scores in the population will not be normal. For instance, age is obviously not normally distributed in the population (see Section 2.7). There are many more young people than old people; the younger the age, the more people there are; thus a distribution of age scores will be skewed.

If there is only a small spread of scores possible, there will not be enough of them to be spread normally. On a three-point hedonic scale, scores could hardly be spread normally; there are only three possible values on the X axis. One needs at least around 10 possible values on the X axis before there are enough values to form a normal distribution. Ranked scores are not distributed normally. There is one of each (one 1st, one 2nd, one 3rd, one 4th, etc.) rank. So if a distribution were plotted, it would be a horizontal line of height = 1 above the X axis.

One final question. Can you compute z scores when the distribution is not normal? The answer is yes. However, you cannot use Table G.1 to compute probabilities for nonnormal distributions.

4.5 HOW NORMAL (GAUSSIAN) DISTRIBUTIONS ARE USED IN STATISTICAL THEORY: SAMPLING DISTRIBUTIONS, STANDARD ERRORS, AND THE CENTRAL LIMIT THEOREM

Before continuing, we will take a brief look at how normal distributions fit into statistical theory. This will give a wider theoretical background to the reader. However, if it is heavy going at first, skip this section and come back to it later.

Gaussian or normal distributions are important because of sampling distributions and the central limit theorem. The sampling distribution will be explained first.

Imagine a population with any distribution (skewed, uniformly distributed, etc.). Now imagine that a sample of say 20 persons is taken, tested, given a mean score, and returned to the population. Now imagine that another sample of 20 persons is taken and the mean again noted. Imagine that this is continued and a whole series of sample means taken; these means will themselves have a distribution. This is called the *sampling distribution*. The *central limit theorem* states that this sampling distribution will be normally distributed whatever the population distribution. The importance of the normal distribution is that generally, no matter what the population distribution, the sampling distribution will be normal as long as the samples are sufficiently large. Even skewed populations will have normal sampling distributions.

This can be seen in the following example. Imagine a population of which 10% has a score of 0, 10% has a score of 1, 10% has 2, 10% has 3, and so on up to a score of 9. This distribution of scores of this uniform population can be seen in the top diagram in Figure 4.7. The figure shows the distribution of means (sampling distributions) for samples of size 2, 3, and 4. It can be seen that the sampling distribution is beginning to look normal for means of samples as small as 4. Other population distributions would require larger samples.

The mean of the sampling distribution will be the same as the mean of the population. The scores in the sampling distribution will be less spread than in the total population; a little thought will indicate that a set of means will not be spread as widely as the original population. In fact, for a population of standard deviation of σ (it should be symmetrical to have a σ), the sampling distribution, for samples of size N, will have a standard deviation of $\sigma/\sqrt{N}$. This value is smaller, representing the smaller spread of the means.

$\dfrac{\sigma}{\sqrt{N}}$ is called the *standard error*

Now let us consider the calculations that we have been making using the normal distribution tables. Assume that we have a score, X, which is higher than the population mean, μ. We have used Table G.1 to find the probability of getting this score or higher, in a population that is normally distributed with a given mean μ and standard deviation σ. We could, in the same way, use Table G.1 to find the probability of getting a given sample mean ($\bar{X}$) or higher in the sampling distribution of means which has a mean μ and a standard deviation of $\sigma/\sqrt{N}$ (where σ is the standard deviation of the original population and N the sample size, $\sigma/\sqrt{N}$ being called the standard error). In the first case we would be seeing the probability of a given score being as big as or greater than the mean, so as to give some clue as to whether the score could have feasibly come from that population. In the second case we would be testing whether a sample mean was as big as or greater than the population mean, so as to decide whether that sample could have feasibly been drawn from the population. This is the most useful application of z tests. A sample mean is tested against the population mean and if it is significantly different, the chances are that the sample did not come from that population. That is, if the probability of getting that sample mean or higher in the distribution of sample means (sampling distribution) is low, it is likely that the sample did not come from that population. So if, say, a food was always tested under white light (so that a whole population of scores was available) and then a sample of the food was tested under red light, the mean of the red-light scores could be tested to see if it was sufficiently different as not to come from the population of white-light scores. If it did not, we could conclude that the presence of the red-colored light altered the test scores.

Two more points. We have been talking about testing whether sample means are significantly larger than the population mean. We could also repeat the

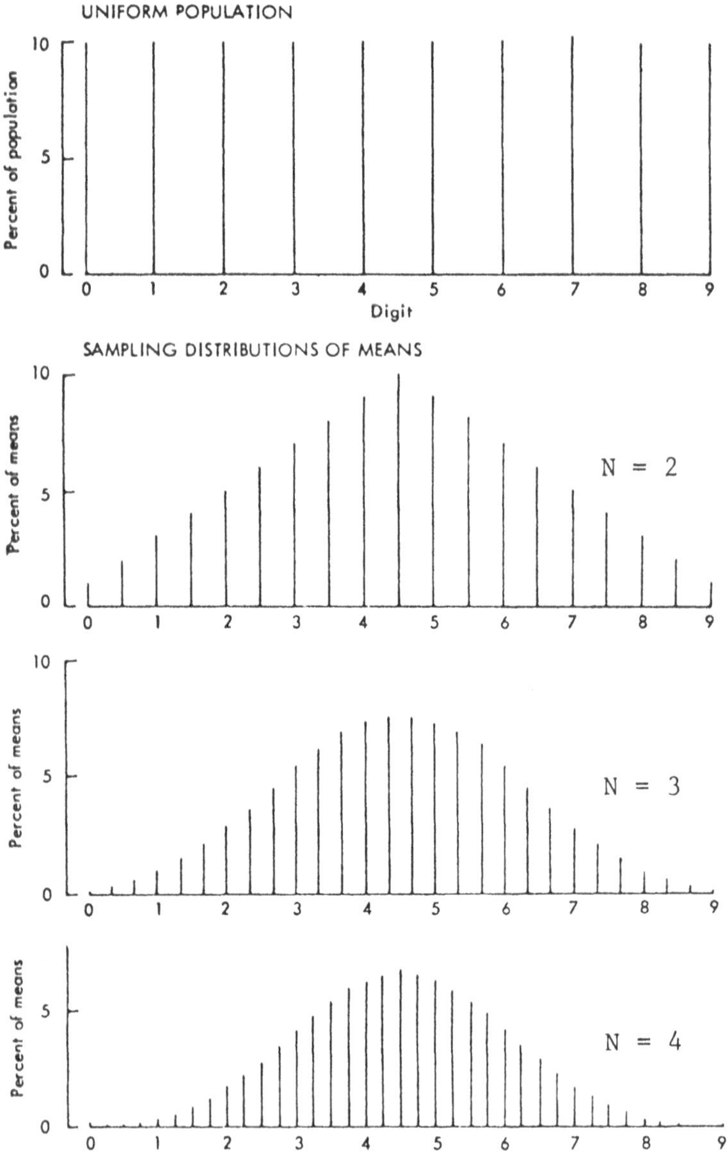

Figure 4.7 Uniform population and sampling distribution of means.

argument for whether the sample means are significantly smaller. Finally, this approach will work if we know the population mean and standard deviation (so as to be able to calculate the standard deviation of the sampling distribution $\sigma/\sqrt{N}$). Unfortunately, we rarely know σ for the population; we only have data from the sample. However, we can estimate the standard deviation of the population from the data in the sample by calculating S. S, however, is not an exact value for the population standard deviation, so an adjustment is made. The normal distribution is not used but a similar distribution, the t distribution, is used instead. This is discussed further in Chapter 7 and also in Appendix C. For large samples ($N > 30$ scores), the value S is pretty much the same as the population standard deviation, so a normal distribution can be used instead of the t distribution. In essence, the two distributions turn out to be pretty well the same for large samples.

4.6 WORKED EXAMPLE: z TESTS—COMPUTING PROBABILITIES FROM NORMAL DISTRIBUTIONS

The average output for a revolutionary, new, solar-powered fruit-juice bottling machine is 2000 bottles per day. The output varies slightly each day according to a gaussian distribution ($\sigma = 200$). What is the probability that the output per day will be:

a. 2300 bottles or more?
   2300 bottles is 300 above the mean, that is $300/200 = 1.5\sigma$ above the mean ($z = +1.5\sigma$). From Table G.1 for $z = 1.5$, $p = 0.0668$ (for $z = 1.5$ and above). Therefore the probability of producing 2300 bottles or more = 0.0668 (6.68%).
b. 2320 bottles or less?
   2320 bottles is 320 above the mean, that is, $320/200 = 1.6\sigma$ above the mean ($z = +1.6\sigma$). From Table G.1 for $z = 1.6$, $p = 0.4452$ (for scores between $z = +1.6\sigma$ and $z = 0$). By definition, $p = 0.5$ for scores equal to the mean and below. Thus $p = 0.5 + 0.4452 = 0.9452$ (for scores of $z = +1.6\sigma$ and below). Therefore, the probability of producing 2320 bottles or less is 0.9452 (94.52%).
c. 1300 bottles or less?
   1300 bottles is 700 below the mean, that is, $700/200 = 3.5\sigma$ below the mean ($z = -3.5\sigma$). By symmetry, probabilities associated with scores of -3.5σ or less are the same as those associated with scores of $+3.5\sigma$ or more. From Table G.1 for $z = 3.5$, $p = 0.0002$ (for scores of $+3.5\sigma$ or more and by symmetry for scores of -3.5σ or less). Therefore, the probability of producing 1300 bottles or less is 0.0002 (0.02%).

d. 1500 bottles or more?

1500 bottles is 500 below the mean, that is $500/200 = 2.5\sigma$ below the mean ($z = -2.5\sigma$). By symmetry, probabilities associated with scores of -2.5σ or more are the same as those associated with scores of $+2.5\sigma$ or less. From Table G.1 for $z = 2.5$, $p = 0.4938$ (for scores between $z = +2.5\sigma$ and the mean). By definition, $p = 0.5$ for scores equal to the mean and below. Thus $p = 0.5 + 0.4938 = 0.9938$ (for scores of $z = +2.5\sigma$ and below and by symmetry for scores of $z = -2.5\sigma$ and above). Therefore, the probability of producing 1500 bottles or more = 0.9938 (99.38%).

5

The Binomial Test: Applications in Sensory Difference and Preference Testing

5.1 HOW TO COMPUTE PROBABILITIES USING THE BINOMIAL EXPANSION

We will now examine the binomial test, one of the simpler statistical tests. It is generally used to determine whether more cases fall into one of two categories than into the other. For example, we may be interested in whether there are more men than women on a particular university campus. We would take a sample, note the number of men and women (one category is men, the other is women) in the sample, and use the binomial test to determine whether these numbers would indicate that there were more men than women in the whole population of the campus. A common use for the binomial test in sensory analysis is for the analysis of difference tests. Imagine samples of two types of cake, one with added sugar and one without. We are interested in whether a judge can taste the difference between these two types of cake. We ask him to taste the cakes and pick the sweeter of the two. He does it several times so that we can get a representative sample of his behavior. If he consistently picks the cake sample with added sugar, he can taste the difference between the two samples. If he sometimes picks the sample with added sugar and sometimes the sample without, it would be indicative that he cannot taste the difference. The binomial test is used to see whether this sample of his behavior indicates that he is a person (in his whole population of behavior) who can distinguish the two cake samples. Does he pick the cake sample with added sugar more than the cake sample without? Here, one category is cake with added sugar, the other is cake without added sugar.

In looking at the binomial test we will derive the test from first principles. We will go through the simple probability theory on which the test is based and show how the actual test ends up as merely a matter of looking up some tables. In doing this, we will demonstrate the logic behind the test and introduce the basic concepts of statistical testing. This book will not deal with the other statistical tests in such detail; essentially, the logic behind them is the same. We are more interested in applying the statistics than deriving them. But it is good to go through this long exercise at least once so as to understand how statistical tests work. The binomial test is ideal for this because of its simplicity.

We will derive the test right from first principles. First, we see how a mathematical exercise, called the *binomial expansion*, can be used as a shortcut for calculating probabilities. We will then see how this probability calculation can be used to analyze data from a sample so as to tell us about the population from which the sample was drawn. Then we will see how this whole procedure is streamlined by the use of ready-made tables.

We start this section assuming that you have the mathematical expertise to perform the binomial expansion, $(p + q)^n$. If you cannot do this, read Appendix B first, before proceeding. Now let us see how the binomial expansion can be a useful tool in the computation of probabilities. We will use the simple familiar example of tossing coins as an illustration.

If you toss one coin, you can get "heads" H or "tails" T. Let the probability of getting heads (H) be p and of getting "tails" (T) be q. Note $p = 1 - q$. Now consider the following examples. Let us consider tossing one coin; call it *one event* because there is only one coin. You can get H or T. The associated probabilities are p and q. The binomial expansion $(p + q)^1$ gives $p + q$.

Now consider tossing two coins; we will call this *two events*. The results you can get, with the associated probabilities of their happening are given below. You can get

All heads	HH with a probability of p^2 by the multiplication law
One head and one tail	HT with a probability of pq by the multiplication law
	TH with a probability of qp by the multiplication law
All tails	TT with a probability of q^2 by the multiplication law

These probabilities can be expressed more simply. $pq = qp$, so let us add these terms together to give $2pq$. It can be seen that all the probability terms required to calculate probabilities of getting various combinations of "heads" and "tails" (all "heads," one "head" and one "tail," all "tails") are given by the terms

$$p^2, 2pq, q^2$$

The binomial expansion $(p + q)^2$ also gives these terms: $p^2 + 2pq + q^2$.

If you toss three coins, the binomial expansion $(p + q)^3$ will give you all the terms you need and $(p + q)^4$ will do it for four coins.

The following makes this clear:

Toss one coin—one event:

Can get		Probabilities	
H		p	The required probability terms are:
T		q	$p + q = (p + q)^1$

Toss two coins—two events:

Can get		Probabilities	
HH		p^2	The required probability terms are:
HT, TH		$2pq$	$p^2 + 2pq + q^2 = (p + q)^2$
TT		q^2	

Toss three coins—three events:

Can get		Probabilities	
HHH		p^3	The required probability terms are:
HHT, HTH, THH		$3p^2q$	$p^3 + 3p^2q + 3pq^2 + q^3 = (p + q)^3$
HTT, THT, TTH		$3pq^2$	
TTT		q^3	

Toss four coins—four events:

Can get			
$HHHH$		p^4	The required probability terms are:
$HHHT$, $HTHH$	$HHTH$, $THHH$	$4p^3q$	$p^4 + 4p^3q + 6p^2q^2 + 4pq^3 + q^4 = (p + q)^4$
$THHT$, $TTHH$, $HTHT$	$THTH$, $HTTH$, $HHTT$	$6p^2q^2$	
$TTTH$, $THTT$	$TTHT$, $HTTT$	$4pq^3$	
$TTTT$		q^4	

Thus, in general, for n events, when p is the probability of an occurrence happening (e.g., getting "heads") and q the probability of it not happening $(q = 1 - p)$, the probabilities of the various combinations of occurrences happening are given by the terms in the binomial expansion $(p + q)^n$.

It is worth considering a few examples of p, q, and n values. Should you toss a coin, the probability of getting heads, $p = 1/2$; the probability of getting "not heads" (tails), $q = 1/2$. Should you throw a die, the probability of throwing a "five," $p = 1/6$. The probability of throwing "not a five" $q = 5/6$. Should you pick a playing card, the probability of picking "spades," $p = 1/4$; the probability of picking "not spades," $q = 3/4$. Should you be attempting to pick a food sample from a set of three food samples, the probability of getting the target sample by chance, $p = 1/3$; the probability of not getting the target sample, $q = 2/3$.

The number of events (n) is also important. If you toss a coin eight times, you have eight events. Toss a die nine times and you have nine events. Pick a card from each of six packs of cards and you have six events. If you were trying to pick one food sample from a set of three food samples on eight successive occasions, you have eight events.

Examples of Binomial Probability Computations

To understand the application of the binomial expansion to probability calculations, let us now examine a couple of examples.

Example 1

You draw one card from each of four packs. What is the probability that you draw:

a. No spades?
b. One spade?
c. One or less spades?

First, there are four packs of cards, four events, so we will be using $(p + q)^4$. Let the probability of drawing a spade, $p = 1/4$. Thus, the probability of drawing a card that is not a spade, $q = 3/4$. Now to begin. First expand $(p + q)^4$.

$$p^4 \quad + 4p^3q \quad + 6p^2q^2 \quad + \underline{4pq^3} \quad + \underline{q^4}$$

$$\left(\frac{1}{4}\right)^4 + 4\left(\frac{1}{4}\right)^3\left(\frac{3}{4}\right) + 6\left(\frac{1}{4}\right)^2\left(\frac{3}{4}\right)^2 + 4\left(\frac{1}{4}\right)\left(\frac{3}{4}\right)^3 + \left(\frac{3}{4}\right)^4$$

a. The probability of getting no spades is given by the term with no p in it, namely, $q^4 = (3/4)^4$. So the probability of getting no spades is $(3/4)^4$.
b. The probability of getting one spade is given by the term with p (raised to the power of 1) in it, namely, $4pq^3 = 4(1/4)(3/4)^3$. So the probability of getting one spade is $4(1/4)(3/4)^3$.
c. The probability of getting one or less spade is the probability of getting one spade or no spades, namely $4pq^3 + q^4 = 4(1/4)(3/4)^3 + (3/4)^4$. So the probability of getting one or less spade is $4(1/4)(3/4)^3 + (3/4)^4$.

It can be seen that it is a straightforward matter to be able to calculate all the probabilities of getting various combinations of cards by simply choosing the appropriate terms in the binomial distribution. Now, a second example.

Example 2

You toss a die three times. What is the probability that you will get at least two "sixes"?

There are three tosses of a die, three events, so we will use $(p + q)^3$. Let us make p = the probability of throwing a "six" = 1/6. Let us make q = the probability of not throwing a "six" = 5/6. Now let us begin. First expand $(p + q)^3$.

$$\underline{p^3} \quad + \underline{3p^2q} \quad + 3pq^2 \quad + \quad q^3$$

$$\left(\frac{1}{6}\right)^3 + 3\left(\frac{1}{6}\right)^2\left(\frac{5}{6}\right) + 3\left(\frac{1}{6}\right)\left(\frac{5}{6}\right)^2 + \left(\frac{5}{6}\right)^3$$

Getting at least two "sixes" can be obtained by getting two "sixes" or three "sixes." The probability of getting three "sixes" is given by $p^3 = (1/6)^3$. The probability of getting two "sixes" is given by $3p^2q = 3(1/6)^2(5/6)$. The probability of getting at least two "sixes" is thus

$$p^3 + 3p^2q = \left(\frac{1}{6}\right)^3 + 3\left(\frac{1}{6}\right)^2\left(\frac{5}{6}\right)$$

5.2 THE BINOMIAL TEST AND THE NULL HYPOTHESIS

Having used the binomial expansion to calculate probabilities for various problems such as tossing coins and drawing cards, we will now apply it to more of a statistical problem, one in which we are required to take a sample from a population.

Imagine that we want to know whether there are more men than women on a university campus. We take a random sample of 10 people. We would normally take a far larger sample, but we will choose 10 to make the mathematics simpler. Also, we make sure that we choose as unbiased a sample as possible. For example, it is more advisable to draw a sample from a campus restaurant, central quadrangle, or administration building. To sample from a ladies' toilet or a men's hall of residence would be foolish.

Forming a Null Hypothesis

We draw the sample of 10 and we find that 7 are men, 3 are women. Does this mean that there are more men than women in the university? Or could it be that there are equal numbers of men and women in the university and we just happened to pick more men by chance? Should there be equal numbers of men and women in the university, the most likely sample would be 5 men and 5 women. However, 6 men and 4 women is also quite likely as is 7 men and 3 women. Even a sample of 10 men is possible, although less likely. A sample of 10 men would be more likely to indicate that there were more men in the university, although it is still possible that there could be equal numbers of men and women. The point is: How possible is it? We use binomial probabilities to tackle the problem. So we approach the problem in this way. We say: Suppose that there were no difference in the number of men and women; what would be the probability of getting a sample with 7 or more men?

The hypothesis on which we are working, that the number of men is the same as the number of women in the university population (i.e., no difference in their numbers), is called the *null hypothesis*, "null" meaning "no effect," "no difference," "same," denoted by H_0. The *alternative hypothesis* (more men than women) is denoted by H_1. (The other alternative, of there being more women than men, will be considered in Section 5.4.)

In statistics, problems are approached by assuming that there is "no effect," and then the probability of getting our result, assuming no effect, is calculated. If that probability is low, it is unlikely that the hypothesis of "no effect" is true, so we reject it and say that there was an effect. We reject the hypothesis saying no effect, the null hypothesis (H_0), and accept the alternative hypothesis (H_1) that there was an effect.

The null hypothesis (H_0) may be a hypothesis stating that there is no difference between the numbers of people or things in various categories (here, men vs. women) or between the means of two sets of numbers, or rather, no difference between the mean of the two populations from which our samples came. We look at the difference between the sample means and how the data are spread. We calculate the probability of getting such differences should H_0 be true. If the probability of getting such difference if H_0 is true is low, we reject H_0. We reject the hypothesis that says there is no difference between the population means. We accept the alternative hypothesis (H_1) that there is a difference. If the probability of getting such differences is high, the null hypothesis could be true; we do not reject H_0. We say that the data are insufficient to reject H_0; it is insufficient to say that there is a difference between the means of the two populations of numbers from which our samples were drawn.

It is important to note that we do not actually accept H_0; we merely do not reject it. If our data are insufficient to show a difference, it may be that a small difference still exists in the population, but that our sample was not sufficiently large to pick it up. So we do not actually go as far as saying that H_0 is necessarily true; we merely do not reject it. So a statistical test will do one of two things. If the probability of getting our result is low, should H_0 be true, we reject H_0 and say there is a difference. If the probability of getting our result is high, should H_0 be true, we do not then reject H_0; we say that our data are insufficient to show a difference. We do not accept H_0; we merely do not reject it.

Consider another example. The null hypothesis may be a hypothesis that states that there is no correlation between two sets of numbers. We perform our statistical test to see whether the observed correspondence in our sample of numbers indicates a correlation in their respective populations. We calculate the probability of getting this degree of correlation should H_0 (no correlation) be true. If the probability of getting our result is low were H_0 true, we reject H_0 and accept the alternative hypothesis (H_1) that there is a correlation. Should the probability be high, we do not reject H_0. We say that our data are insufficient to indicate a correlation in the two respective populations. Note again that we do not accept H_0; we just do not reject it.

This covers the whole logic of statistics. Whatever the test, we set up a null hypothesis. Should the probability of getting our result on H_0 be low, we reject it; if not, we do not reject it. Statistical tests are designed simply to calculate such probabilities. You do not have to calculate these probabilities yourself; it has all been done—you merely use a set of tables.

Applying the Binomial Expansion

And now back to our example. H_0 states that there are no differences in the numbers of men and women in the university. In our sample we had 7 men and 3 women; what is the probability, if H_0 is true, of getting seven or more men? (We calculate the probability of getting 7 or more men, not the probability of getting 7 men exactly, because a sample of 8, 9, or 10 men would also be grounds for rejecting H_0.) The probability of getting 7 men or more can be calculated using the binomial expansion.

We first let the probability of picking a man be p and the probability of picking a woman (picking "not a man") be q. If the null hypothesis is true and there are equal numbers of men and women in the university, the probability of picking either a man or a woman is 50% (or 1/2). So on the null hypothesis, $p = q = 1/2$.

We have also sampled 10 people, 10 events, so we need to use the expansion $(p + q)^{10}$. This will tell us all we need to know:

$$p^{10} + 10p^9q + 45p^8q^2 + 120p^7q^3 + \cdots$$

The terms giving the probabilities that there are 7 or more men (10, 9, 8, or 7 men) are the terms involving p^{10}, p^9, p^8, and p^7. Filling in the values $p = q = 1/2$, the values we get by assuming the null hypothesis, we get

$$\left(\frac{1}{2}\right)^{10} + 10\left(\frac{1}{2}\right)^9\left(\frac{1}{2}\right) + 45\left(\frac{1}{2}\right)^8\left(\frac{1}{2}\right)^2 + 120\left(\frac{1}{2}\right)^7\left(\frac{1}{2}\right)^3 + \cdots$$

$$= 0.172 \text{ or } 17.2\%$$

Note: We did not even have to write out the whole binomial expansion; we took only the terms we needed.

Thus there is a 17.2% chance of getting 7 or more men in our sample of 10 people on the null hypothesis, the hypothesis that states there are equal numbers of men and women in the university. 17.2% is high. So it is quite likely that we would get such a sample on the null hypothesis. So we do not reject the null hypothesis. We do not reject the hypothesis that there are equal numbers of men and women in the university (in favor of a hypothesis saying that there are more men). We say that there are insufficient data to reject H_0. We have insufficient data to say that there are more men than women in the university.

Had the probability been not 17.2% but 1.0%, a low value, the probability of getting our result with the null hypothesis being true is low. Our result would have been unlikely had the null hypothesis been true. Thus we would have rejected the null hypothesis and accepted the alternative hypothesis (H_1) of there being more men than women. 17.2% being a high value and 1.0% a low value will be discussed in Section 5.3.

The rejection and nonrejection of H_0, depending on whether our result was likely on H_0 (17.2%) or unlikely (1.0%), is represented in the following diagram:

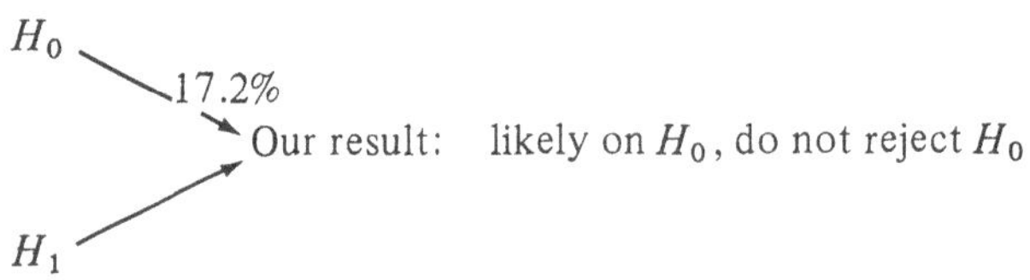

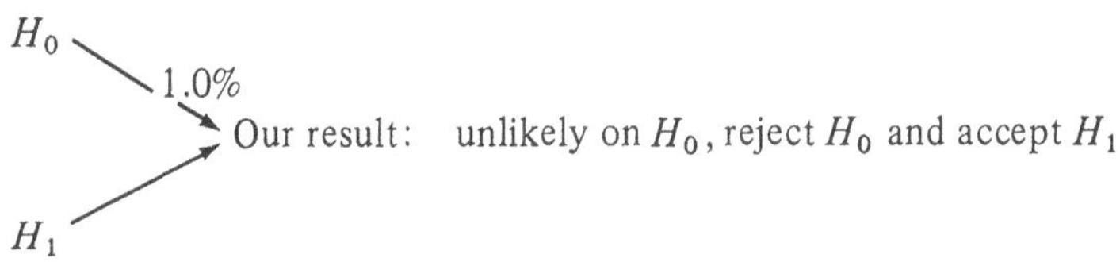

The Binomial Test Using Tables

In our example, our sample of 10 people (10 events) gave us 7 men and 3 women and we expanded $(p + q)^{10}$ to get the appropriate terms to calculate the probabilities of getting 7 or more men, should the null hypothesis be true:

$$p^{10} + 10p^9q + 45p^8q^2 + 120p^7q^3 + \cdots$$

To calculate the probability of getting a sample with 7 or more men (7, 8, 9, or 10 men), we used the first *four* terms of the expansion. Should our sample have had 8 men and 2 women we would have calculated the probability of getting 8 or more men, using the first *three* terms of the expansion. Had we obtained 9 men we would have added the first two terms; had we obtained 10 men, the first term only. Rather than adding the terms up ourselves, Table G.4.a will do it for us. The table gives the sums of various numbers of terms for a given

$(p + q)^n$, that is, for a given number of events (n) or given sample size. We simply look up the value in Table G.4.a for the appropriate size of sample (denoted in the table by N; $N = 10$ in our example) and the appropriate X value (X is the number of cases in the majority; 7 men in our example). In Table G.4.a for $N = 10$ and $X = 7$, the appropriate value is 172. This means that the probability of getting our sample with 7 or more men, should H_0 be true, is 0.172 or 17.2%. This is the answer we obtained above.

Note: Table G.4.a was calculated using the values $p = q = 1/2$. Substituting other values for p and q would result in the sums of the various terms in the binomial expansion being different. Only when the null hypothesis states that the chance probability of data falling into one category (p) is the same as the chance probability of data falling into the only other available category (q), that is, when $p = q = 1/2$, can Table G.4.a be used. So Table G.4.a can be used only when there are two categories available for the data to fall into (men vs. women; cats vs. dogs; picking a sweetened sample of food vs. not picking a sweetened sample of food).

Example 1

Assume that we are breeding shrimps in the warm water coming from a nuclear power station. We know that these particular shrimps mutate spontaneously and that generally there are equal numbers of mutants and normal shrimps. Here we wish to study their degree of mutation near the nuclear power station to see whether it increases. We wish to know whether there are more mutants than normal shrimps, so we sample them.

We sample 20 shrimps and obtain 15 mutants and 5 normal shrimps. Does this sample indicate that there are more mutants in the whole population of our breeding ground (H_1, the alternative hypothesis)? Or does it indicate that there are equal numbers of normal and mutant shrimps (H_0, the null hypothesis)? What is the probability of getting such a sample (or more extreme) on the null hypothesis?

We have a sample of 20 shrimps ($N = 20$). The majority category is "mutants" ($X = 15$). From Table G.4.a, for $N = 20$ and $X = 15$, $p = 0.021$ or 2.1%. This is a low value (we will discuss high vs. low later). The probability of getting our sample is low should the null hypothesis be true. Thus we reject the null hypothesis; we reject the hypothesis that there are equal numbers of normal shrimps and mutants, in favor of the alternative hypothesis (H_1), which states that there are more mutant shrimps.

What if we obtained 11 mutants and 9 normal shrimps? From Table G.4.a, when $N = 20$ and $X = 11$, the probability value given in the tables is 0.412 or 41.2%. This value is high. So it is likely that we could get such a sample (or more extreme) when the null hypothesis is true. We do not reject the null hypothesis. We have no evidence to lead us to believe that the population of shrimps has more mutants than normal shrimps.

5.3 LEVELS OF SIGNIFICANCE: TYPE I AND TYPE II ERRORS

In the preceding example, we called 41.2% a high probability and 2.1% a low probability. The question arises: What is "high" and what is "low"?

By convention, 5% (0.05) is the break point. Five percent or higher is considered high, less than that is low. Often in research papers, you will see written "$p < 0.05$," which means that the probability of getting this result on H_0 is less than 5%; we reject H_0 in favor of H_1 (the alternative hypothesis). The 5% break point is called the *level of significance*. A difference being significant at the 5% level means that the probability of obtaining such a result on the null hypothesis is less than 5%.

Now if the probability of getting our result on the null hypothesis was less than 5%, we would reject the null hypothesis. But there is a very small, but finite chance (< 5%) that it is still true. So we have a less than 5% chance of rejecting H_0 when it is, in fact, true. Rejecting H_0 when it is true is called making a *Type I error*. In our shrimp example, making a Type I error would be to deduce that there were more mutants than normal shrimps when, in fact, there were not.

There are also *Type II errors*—the acceptance of H_0 when it is untrue (thinking that there are equal numbers of normal shrimps and mutants when there are not). So:

Type I error: Reject H_0 when it is true.
Type II error: Accept H_0 when it is false.

The relationship of Type I and Type II errors is shown in the following diagram:

		The Truth: H_0	The Truth: H_1
Result of statistical test	Not reject H_0	√	Type II error
	Reject H_0 and accept H_1	Type I error	√

Five percent is chosen quite arbitrarily as the level of significance so that if the probability of getting our result on H_0 is less, we reject H_0. However, some statisticians are more cautious and choose 1% as their level of significance. The probability of getting our result on H_0 must be lower than 1% before we reject H_0.

In cases where finding a difference may be very critical (e.g., is a drug causing a cure or not?), we may choose the 1% level because we want to make quite sure that our results really indicate a cure (as opposed to the H_0, "no cure" hypothesis) before we deduce that the drug brings about the cure. Choosing a lower level of significance makes it tougher to reject H_0. The data must be more indicative to reject H_0 at the 1% level than at the 5% level. So we quite commonly will see written in papers:

$p < 0.05$	choosing 5% level
$p < 0.01$	choosing 1% level

What level you choose is a matter of opinion and depends on your research. Basically, it depends on how sure you wish to be when you reject H_0, that is, how prepared you are to risk making a Type I error. It depends on how afraid you are of rejecting H_0 should it be true. Sometimes, for the very cautious, the 0.001 level (0.1%) is chosen.

So if we took a sample to see if there were more men than women in the university and the probability of getting our result was 2% on H_0, then with $p < 0.05$, the probability is low, so you reject H_0. With $p < 0.01$, the probability is high, so you do not reject H_0. The conclusions are opposite for the same data. But there is no contradiction. Whatever you conclude, you conclude it at that level of significance and a person examining your results knows that it may not hold for another level of significance. So you always state your level of significance ($p < 0.05$ or $p < 0.01$) so that the reader knows exactly the status of your results.

In fact, it is far better, if possible, to quote the exact probability of getting your result on the null hypothesis. Unfortunately, few statistical tables are set up to do this. Care should be taken not to see the 5% and 1% levels usually given as anything more than mere tradition. A result with a probability on H_0 of 5.01% is not so different from one with a probability of 4.99%. So these levels should be used intelligently with due regard to their arbitrariness. If possible, it is far better to be precise and quote the exact probabilities, such as are given in Table G.4.a.

Statistics do not lie; they only appear to lie to the layman who does not know about null hypotheses and levels of significance. Politicians and advertisers are particularly fond of giving you conclusions drawn from statistical data without letting you know the significance levels. You can only fool the uninitiated with statistics.

Statistical tests are generally set up to seek out differences. We can be definite about rejecting a null hypothesis in favor of an alternative hypothesis. We worry about committing Type I errors, rejecting the null hypothesis when we shouldn't, in other words, deducing that there is a difference when no difference is there. Interestingly, in the sensory evaluation of a food we are often testing an original product against a reformulation of that product, to ensure that they are not different. We are more worried about committing a Type II error; we worry about not rejecting H_0 when it is false, not saying there is a difference when there is. Should we deduce that we do not have sufficient data to say that there is a difference between the products, when there was a difference, we would be in trouble. We would market the reformulated food when it had a different flavor from the original product. The consumers would notice this flavor change and stop buying the food. There is more discussion of this in Appendix F.

The probabilities of committing Type I and Type II errors are of concern to statisticians, who denote them by α (Greek lowercase letter alpha) and β (Greek lowercase letter beta).

α: the probability of making a Type I error (rejecting H_0 when it is true)
β: the probability of making a Type II error (not rejecting H_0 when it is false)

The power of a test is the probability of rejecting H_0 when it is false $= 1 - \beta$. (When it is false, you can either reject it or not reject it; if you do not reject it with a probability β, you will reject it with a probability $1 - \beta$.)

Obviously, we want our statistical tests to be powerful; we want tests that will reject the null hypothesis when it is false. We want tests designed to have a low β, a low chance of making a Type II error, especially if we are testing to ensure that an ingredient change in a food has not altered the flavor. As well as wanting a test designed to have a low β (high power, $1 - \beta$) we want to use a low α as well. We do this by picking a low level of significance as our cutoff point (0.05 or 0.01).

5.4 ONE- AND TWO-TAILED TESTS

Another question arises in the use of statistical tests. This is the problem of whether to use a one-tailed or a two-tailed test. One- and two-tailed tests are best explained by going back to our example concerning whether there were more men than women on a university campus. We took a sample of 10 people and found 7 men and 3 women.

Our H_0 was that there were an equal number of men and women. Our alternative hypothesis, H_1, was that there were more men than women. We used the binomial expression to calculate the probability of our sample being in the "7 or more men" region. We used $(p + q)^{10}$ with

p = probability of getting a male = 1/2 on H_0
q = probability of getting a female = 1/2 on H_0

$$\underbrace{\underline{\underline{p^{10}}} + \underline{\underline{10p^9q}} + \underline{\underline{45p^8q^2}} + \underline{\underline{120p^7q^3}}}_{7\text{ or more men } = 0.172} + 210p^6q^4$$

$$+$$

$$252p^5q^5$$

$$+$$

$$\underbrace{\underline{q^{10}} + \underline{10pq^9} + \underline{45p^2q^8} + \underline{120p^3q^7}}_{7\text{ or more women } = 0.172} + 210p^4q^6$$

The first four terms of the expansion (doubly underlined) gave us the probability, should the null hypothesis be true, of our sampling 7 or more men. This was 0.172 and was high. Thus we did not reject H_0 in favor of the alternative hypothesis (H_1) of there being more men in the population.

On the other hand, the last four terms give us the probability on H_0 of our sampling 7 or more women. Once again this is 0.172 and H_0 would not be rejected in favor of an alternative hypothesis of there being more women in the population.

So in both cases, whether the alternative hypothesis is "more men" or "more women" in the population, the probability value on the null hypothesis is 0.172. But consider this! What if the alternative hypothesis were neither of the above? What if it were that there were merely different numbers of men and women in the population?

If our alternative hypothesis (H_1) was that the numbers of men and women were different, this could occur either by having more men than women or by having more women than men. So to calculate this probability on H_0 we consider not only the tail of the distribution which deals with 7 or more men (underlined twice), but also the other tail of the distribution which deals with 7 or more women (underlined once), and the probability on H_0 is now 0.172 + 0.172 = 0.344. When one tail of the expansion is used, H_1 is called a *one-tailed alternative hypothesis* and the whole test is called a *one-tailed test*. When two tails of the expansion are used, the test is called a *two-tailed test.*

So for the binomial test, the two-tailed probability (0.344) is twice the value of the one-tailed probability (0.172). When using Table G.4 to look up the probability of getting your results for:

One-tailed test: Use probability values in Tables G.4a (one-tailed).
Two-tailed test: Use double the probability values; for convenience, these are given in Table G.4.b (two-tailed)

It is also worth noting that the normal distribution tables (Table G.1) are one-tailed; double the values for two-tailed tests.

When Do You Use One-Tailed and Two-Tailed Tests?

The test used depends on how we formulate the alternative hypothesis, H_1. (H_1 = there are more men than women—one-tailed; H_2 = there are different numbers of men and women—two-tailed). If you are absolutely sure before collecting the data that the only possible alternative to there being equal numbers of men and women (H_0) is more men than women, you would use a one-tailed test (H_1 would be the one-tailed alternative hypothesis). If you knew that the only alternative to equal numbers of men and women was more women than men, you would use a one-tailed test. If, on the other hand, you did not know if there were likely to be more men or women, there could be more of either, and you would use a two-tailed test.

At the beginning of this century when women did not advance to higher education, it was impossible to find more women than men on a university campus. You would not consider a two-tailed test; the only question would be whether there were significantly more men or not. Today, however, you may have no good reason for expecting either more men or women, so you would not risk predicting the direction of the difference; you would go for a two-tailed test. So, in general, if you know beforehand the direction of the difference, you use a one-tailed test; if not, you use a two-tailed test.

It is easier to get a difference using a one-tailed test. Should the one-tailed probability of getting a difference on the null hypothesis be 4% (this is sufficiently low to reject H_0), the two-tailed probability from the same data will be 8% (generally regarded as high enough not to reject H_0). These are not conflicting results. They are merely variations in the estimated probability that H_0 is true, which themselves vary because different alternatives to H_0 are being considered.

How do you decide beforehand whether you can use a one- or a two-tailed test? How can you know beforehand what trend the test will show (more men or more women?) should the null hypothesis be false? Can you ever know? Some statisticians would argue that you can never know and that you should always use two-tailed tests. Others say that sometimes it is reasonable to use one-tailed tests. It is a matter of controversy and you will have to judge for yourself. In this text we will use two-tailed tests unless we have a very good reason for using a one-tailed test. Such an occasion will be rare. Sensory analysts, however, will generally risk a one-tailed test to analyze the data from a directional paired-comparison sensory difference test (see Section 5.11).

5.5 WHY THE BINOMIAL TEST IS NONPARAMETRIC

The type of data that are used in a binomial test are merely categories, called *nominal data* or *data on a nominal scale*. In our examples in this chapter, we have dealt with nominal data (categories). We have used:

Men vs. women
Heads vs. tails
Drawing a spade vs. not drawing a spade
Throwing a "six" vs. not throwing a "six"
Normal shrimps vs. mutant shrimps
Cake with added sugar vs. cake without added sugar

A response can either be one category or the other (man or woman); this is nominal data. We did not rank our sample of students for manliness (the most hairy, muscular man coming first and the most beautiful, blonde, suntanned, curvy woman coming last), so it is not an ordinal scale. We did not give manliness scores to the people in our sample according to the amount of male hormone in their blood, so our data are not on an interval or ratio scale.

5.6 LARGE SAMPLES: WHEN THE BINOMIAL EXPANSION APPROXIMATES TO A NORMAL DISTRIBUTION

When the size of our sample is greater than 50, the relevant binomial distribution, $(p + q)^{50}$, approximates a normal distribution. This can be seen in Figure 5.1, where the frequency histograms are drawn for $(p + q)^{11}$ and $(p + q)^{26}$. It can be seen that the latter is taking on a shape very similar to the normal distribution.

When the binomial distribution approximates the normal distribution, a z value can be calculated and the probability (if H_0 is true) of getting a value as great as this or greater can be found using Table G.1 (as seen in Section 4.3). Normally for Table G.4, for the appropriate p and q and for a given sample size n, we select the number in the majority category, X. In the case where the

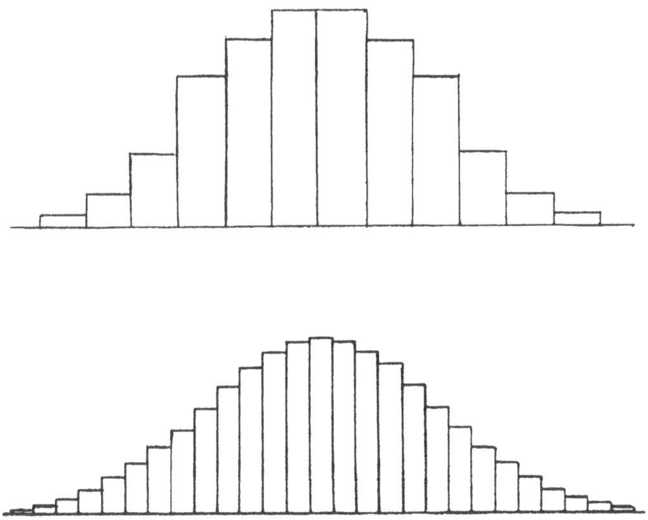

Figure 5.1 Binomial distribution for $(p + q)^{11}$ (top) and $(p + q)^{26}$ (bottom).

sample size is greater than 50, we use p, q, n, and X to calculate a z score for use with Table G.1. The z score is given by

$$z = \frac{\left(X - \frac{1}{2}\right) - np}{\sqrt{npq}}$$

Note that Table G.1 gives one-tailed probabilities; double the values for a two-tailed test.

Examining the formula above, it is worth noting that the value np is the mean of the binomial distribution. This can be seen to be true in the following way. If samples of 10 people are taken ($n = 10$) and the probability on H_0 of getting a male is one-half ($p = 1/2$), the sample that occurs most frequently—the middle or mean of the distribution—will be the one with five men. The mean = $np = 10 \times 1/2 = 5$. The denominator, $\sqrt{npq}$, in the formula is the standard deviation of the binomial distribution. Hence the formula resembles the formula for z in Chapter 4.

5.7 THE POISSON DISTRIBUTION

When p is a very small value (smaller than 0.1), representing a rare event such as catching a disease, a very large sample will be needed so that the event will actually occur a finite number of times. For instance, the probabilities of certain types of accidents happening are often of the order of 0.0001 (p small), while data would be available for samples of people such as 40,000 (n large), so that in the sample, one would expect about four cases (the mean, $np = 40{,}000 \times 0.0001 = 4$). To calculate the probability of getting various occurrences of the disease would require the expansion of $(p + q)^{40{,}000}$, with $p = 0.0001$ and $q = 0.9999$; this is a daunting task indeed. For instance, the probability of getting exactly two cases of the disease by chance is given by the expression

$$_{40{,}000}C_2 \times p^2 q^{39{,}998} = {}_{40{,}000}C_2 \times (0.0001)^2 (0.9999)^{39{,}998}$$

(see Appendix B)

This becomes an inconvenient calculation, so the Poisson distribution is used instead of the binomial distribution. The Poisson distribution is often used for calculations such as death rates or accident rates for insurance companies or disease rates in medical research. In the field of senory or behavioral measurement, the events studied are not so rare, so the Poisson distribution is not often used and we will not concern ourselves with it here. As with the binomial distribution, tables of probabilities are also available for the Poisson distribution.

5.8 THE SIGN TEST. ONE- AND TWO-SAMPLE DIFFERENCE TESTS: RELATED- AND INDEPENDENT-SAMPLES DESIGN

The sign test is a variation of the binomial test, and as we introduce it we will take the opportunity to introduce some additional statistical concepts.

One-Sample and Two-Sample Tests

The binomial test is a *one-sample test*. A sample is taken from the population, examined, and from it facts are inferred about the population. In the example we used, we were asking whether there were more men than women in a given population. We took a sample from that population and examined it to see whether there were more men than women. We then made relevant deductions from the single sample about the population.

A modified application of the binomial test is to compare two samples. We can compare the two samples to see whether they are different. Strictly, what we do is to compare the two samples of data to determine whether they were drawn from the same population (no difference between the samples) or from separate populations (the samples are different). Generally, we have two samples of actual scores (interval or ratio data rather than nominal data). We compare the means of the two samples of data to see whether they are sufficiently different to come from different populations (e.g., *t* tests, Chapter 7; analysis of variance, Chapter 8). Such a statistical test is called a *two-sample difference test*.

Even if two samples do not differ (they come from the same population) it is unlikely that they would have exactly the same mean; they would differ slightly by chance. We need a test to see whether the difference in mean scores for the two samples indicates that the samples are significantly different (the samples came from different populations, with different means) or whether it is merely an insignificant chance difference (the samples came from the same population, with only one mean). Again, we have a null hypothesis (H_0) which states that there is no difference between the two means (they come from the same population). We calculate the probability of getting the observed difference between the two samples if H_0 were true. If this probability is low, we reject H_0 and conclude that there is a significant difference between the two samples. If the probability is high, we do not reject H_0. The logic is similar to that of the one-sample binomial test.

Related- and Independent-Samples Design: Counterbalancing and Control Groups

A two-sample difference test can be set up in two ways. We could arrange for completely different people to be in each sample; this is called an *independent-* or *unrelated-samples design*. Alternatively, we could test the same people twice, in different conditions; this is called a *related-* or *paired-samples design* because each person contributes a pair of scores, one for each condition.

Assuming that we were examining the effect of two training protocols on panelists, we could test two panels, one trained with one and one trained with the other protocol to see which one was better (independent-samples design). This should indicate which protocol was superior. But there is a snag. One group could be better than the other, not because of training but just because they happened to be better tasters anyway. The differences between two groups for an unrelated-samples design may not be because of the experimental conditions but because one group of judges was superior, regardless of training.

One way around this is to use the related-samples design. In this, the same judges are tested under each experimental condition. The differences between the two sets of scores are not now due to individual differences, they are due to the differences in experimental procedure.

There is, however, a difficulty with the related-samples design. Judges may perform better in the second condition, merely because of practice. So, another strategy is used, that of *counterbalancing*. If two experimental conditions are being tested in a related samples design, counterbalancing is essential. If condition A is tested before condition B, judges may be better in condition B merely because of the practice received during condition A. So half the judges receive condition A before B and the other half receive B before A; the results are then pooled so that effects of improvement due to practice cancel out or counterbalance each other. In fact, counterbalancing applies to all experimental design. The order of presentation of food samples should be counterbalanced over judges or even for each judge. Any possible source of bias in the experiment should be anticipated and either eliminated or neutralized by counterbalancing.

With counterbalancing the related-samples design is a good experimental procedure. But what if conditions A and B were drug treatments? Perhaps the effects of drug A may not have worn off by the time drug B was tested; the second condition would then be examining the effects of drugs A + B, not drug B alone. Such *carryover effects* can negate the advantage of the related-samples design. So, it is not always possible to use a related-samples design.

There are further difficulties with this experiment. It could be that it was familiarity with the testing procedure and not training that was the factor that really improved performance. So if we tested untrained judges, trained them, and then retested them, they could be better not because of the training but because of the experience they gained while being tested as untrained judges. One way to control for this is to use a *control group*. Here a group of untrained judges would be tested, wait while the others were trained, and then be tested again. Should this control group improve as much, it would be evidence that judges improved anyway, without the training. It would then not be the training protocol but some other factor that improved performance.

Control groups are an essential part of many behavioral studies, especially studies on drugs. Patients treated with drugs are compared with patients treated with a placebo, a pill that looks the same as the drug but has no effect. If the patients on the placebo improve as much as those treated with the drug, it shows that the improvement is due to something other than the effect of the drug. Suggestion alone or belief that they will recover can sometimes be enough to initiate recovery. Control groups are extremely useful in behavioral research.

Going back to our original example, we could take half the judges, give them an initial test, train them on protocol A, retest them to measure the improvement, give further training on protocol B, and test them again to measure their improvement. The other half of the judges would experience the same except with the order of A and B reversed. This would work as long as the judges continued to improve in both conditions; there were no plateau effects. We could then pool the results from both groups of judges, so counterbalancing for order effects (practice or fatigue). But there are still problems. We have to give the same or equivalent tests to a judge, to test his or her improvement after protocol A and after protocol B. This is not such a problem with flavor testing, but it can be a very serious problem for other types of test, such as written examinations, where judges could improve merely because they could remember the correct answers from the first exam. Where it is a problem, equivalent tests must be devised, and this itself involves a lot of experimentation and is not always possible.

It is possible that condition A is the tasting of one food and condition B the tasting of a second food. A related-samples design would seem preferable here. However, it may be that each food is available on only one day, making a counterbalanced design impossible. The alternatives are to use a related-samples design and hope that temporal effects are small or use an unrelated-samples design or use a combination of both.

In general, it could be said that a related-samples design is superior to the independent-samples design, other things being equal. However, other things are generally not equal. Thus the first choice is a related-samples design, but if there are too many problems with this, an unrelated-samples design is used. There are problems with both approaches and each case must be decided on its merits. Sometimes a combination of both designs is best.

In sensory evaluation and other behavioral work it is relatively easy to see whether an experiment has a related-samples or independent-samples design. If different judges are tested in different experimental conditions, it is an independent-samples design; if the same judges are tested in different experimental conditions, the design is a related-samples design. It is related because the numbers in one condition can be matched to the numbers from the same judges in the other condition. Because of this, the design is often called a *matched design* or a *paired design*. It could also conceivably be a matched design should there be different judges in different experimental conditions who could be matched perfectly for all the important variables of the experiment. Perhaps clones would be considered similar enough to be matched pairs. Luckily, with human beings the problem is not too great; either there are the same judges or different judges, in different experimental conditions. But when other units of measurement are used, the problem can be more complex. How similar do two animals have to be before they can be considered matched? Is it enough that they came from the same litter, or should they have had the same food, same rearing conditions, and same life experiences? How similar do two samples of wine need to be before they can be considered matched? Do they need to come from the same vineyard, the same vine, the same branch on a vine in the same year; do the grapes need to have been picked on the same day? The answer to these questions will depend on the experiment at hand but it can be seen that the answer is not always straightforward. Such issues can cause problems in some disciplines. Luckily, in behavioral measurement the simple rule: "same judges–related samples" vs. "different judges–independent samples" suffices.

Refer to the Key to Statistical Tables to find the appropriate columns for one-sample tests and two-sample difference tests, both related and independent samples. See Section 16.20 for further discussion.

Computation for the Sign Test

The sign test is a modification of the binomial test. It is a related-samples difference test; it tests whether scores in one condition are higher than in another condition. The best way to examine this test is by using an example. For a change, we will take an example using plants, although it could equally well apply to food samples or behavioral responses.

Let us imagine that we wish to know whether talking to plants makes them grow more quickly. We take a plant and measure its growth rate while not talking to it. We then measure its growth rate for the following month while talking to it. Growth rates are measured in terms of "very fast," "fast," "medium," "slow," or "very slow." This is a related samples test because we have the *same* plants in both conditions.

Again, we are careful to control and counterbalance. We would take half of the plants and not talk to them in the first month and talk to them in the second month. The other half of the plants we would talk to in the first month and not in the second. We would then pool the results from both conditions. This counterbalancing controls for any other variables that may interfere with the experiment, such as weather conditions, over the two months. Also, we would take precautions to ensure that when we were speaking to the plants, the plants that were not being spoken to did not overhear or receive the message in any other way. Even with this lighthearted example, there are quite difficult experimental problems.

Let us now imagine a table of results.

Plant number	Growth rate: One month, no talking	Growth rate: One month, talking	Direction of difference
1	Medium	Fast	+
2	Very slow	Slow	+
3	Medium	Fast	+
4	Fast	Very fast	+
5	Slow	Slow	0
6	Very fast	Very slow	−
7	Fast	Very fast	+
8	Slow	Medium	+
9	Slow	Medium	+
10	Medium	Fast	+
11	Slow	Medium	+

Should the growth rate be greater in the "talking month" for a given plant, we score that plant "+." If the growth rate is less during the "talking month," the plant is scored "−." If there is no change, the plant is scored with a zero.

If talking to plants causes a higher growth rate than not talking to plants, there should be significantly more +'s than −'s. If not, growth rate changes will be random and subject to chance and there would be a roughly equal number of +'s and −'s. We can use the binomial test to see whether there are significantly more + values than − values. The zero values are ignored and dropped from the calculation. So in this case, $N = 10$ (not 11 because plant 5 is dropped from the calculation); and $X = 9$ (the number of "+" values).

Do we use a one- or two-tailed test? Do we know the direction of the difference should a difference occur? Do we know beforehand whether talking to plants would speed their growth or whether it would slow their growth? Let us reckon that we do not know this (after all, plants may hate being spoken to and slow their growth in protest), and so let us use a two-tailed test.

From Table G.4.b, for $N = 10$, $X = 9$, the probability of getting this on H_0 is 0.021 (two-tailed test) = 2.1%. This value is low (if we work at the 5% level). Thus we reject H_0; thus talking to plants increased their growth rate ($p < 0.05$).

The sign test is basically the binomial test and so is also a nonparametric test. The data must be on at least an ordinal scale so as to be able to tell whether there has been an increase (+) or a decrease (−) from one condition to another. The test is quick and easy to use but has two major disadvantages:

1. All ties (zero scores) are discounted. This does not matter if the number of ties is low, but if it is large (say over 25% of the scores), many cases would show no change and we could say that there was no difference between the two conditions, even if the sign test said there was a difference, by considering only the + and - values. An alternative statistical analysis would need to be used.
2. No account is taken of the size of the increase or decrease, so some information is lost. An increase of "very slow" to "very fast" is treated the same as an increase of "fast" to "very fast," both merely being given a +. Other tests take account of the degree of difference (e.g., Wilcoxon test, Section 16.2).

5.9 USING THE BINOMIAL TEST TO COMPARE TWO PROPORTIONS: A TWO-SAMPLE BINOMIAL TEST

A binomial test can also be used for another purpose: to compare two proportions, say the proportion of men on the U.C.L.A. campus and on the U.C.D. campus. Here a sample is taken from each campus and the proportion of men in each noted. These could be different because U.C.L.A. and U.C.D. really have different porportions of male students or may differ merely by chance, the proportion of men being the same at U.C.L.A. and U.C.D. (the null hypothesis). In this case, the approximation of the binomial distribution to the normal distribution is used, a z-value is calculated, and the normal probability tables (Table G.1) used to determine the probability of getting two proportions that differ this widely by chance (by H_0). Should this probability be low, the proportions will be surmised to be significantly different (reject H_0). In other words, the two proportions came from different populations. If P_A is the proportion in the sample with N_A persons (at U.C.L.A.) and P_B the proportion in the sample of N_B persons (at U.C.D.) and if p and q are the probabilities of finding males or females in each sample on the null hypothesis ($p = q = 1/2$), the z value is given by

$$z = \frac{P_A - P_B}{\sqrt{pq\left(\dfrac{1}{N_A} + \dfrac{1}{N_B}\right)}}$$

Note: Table G.1 gives one-tailed probability values; double probabilities for two-tailed values.

Because two different samples were taken (one from U.C.L.A. and one from U.C.D.) the test comes under the heading of an unrelated samples design. See the Key to Statistical Tables for the relationship of this test to others.

5.10 BRIEF RECAPITULATION ON DESIGN FOR THE SIGN TEST AND THE BINOMIAL COMPARISON OF PROPORTIONS

It is worth pausing to consolidate our ideas on one-sample and two-sample (related vs. unrelated samples) tests and see how they are placed in the Key to Statistical Tables. These points of design are important and bear repetition.

The original binomial test we examined is a one-sample test because we take one sample, examine it, and from it deduce something about the population (e.g., whether there are more men than women in the university population).

The sign test was a two-sample related-samples difference test. We took our plants and spoke to them for one month, measuring growth rates, and then did not speak to them for one month measuring growth rates. We also counterbalanced for order. We obtained two sets of growth rates, the scores from each set coming from the same plants. In this way the scores were related.

Another approach would have been to take one set of plants and talk to them and another separate set of plants and not speak to them. Again, we would have had two sets of growth rate scores but this time the scores in each sample would have come from quite different plants. In this way, the samples would have been unrelated or independent and the design called an independent-samples or unrelated-samples design. However, a sign test being a related-samples test, could not be used to analyze such data.

Of course, should we have measured growth rates in terms of increase in length of the plants' branches or increase in weight or volume, we would have had numerical data (on a ratio scale) rather than our rankings of "very fast," "fast," "medium," "slow," "very slow." With data on a ratio scale, a glance at the chart on the inside cover tells us that the data can be analyzed for a related- or an independent-samples design using *t* tests (see Chapter 7).

The binomial test that was used for comparing proportions from two populations was an unrelated-samples test. The proportion of men in a sample from U.C.L.A. was compared to the proportion in a sample from U.C.D. Each sample contained totally different people, so this version of the binomial test is a two-sample unrelated-samples test.

5.11 APPLICATION OF BINOMIAL STATISTICS TO SENSORY DIFFERENCE AND PREFERENCE TESTING

Because binomial tests are used so extensively for the analysis of difference and preference testing, in the sensory analysis of foods, it is worth examining this application in detail. A sensory difference test is used to determine whether two or more samples of food (or any other product) differ from each other. Such testing is generally performed using a trained taste panel, often members of the company concerned. For example, the samples of food may be products with and without an ingredient change. Testing is performed to see whether reformulation has altered the flavor or other sensory characteristics of the food. Alternatively, the food samples may be market leaders manufactured by a rival company and imitations marketed by your own company; testing may be performed, before marketing, to determine whether the imitations can be distinguished from the originals. The food samples may even have come from a food processed in several different ways, so as to determine the effects of processing variables. There are many uses for sensory difference tests.

Should the foods be distinguished from each other by sensory testing, it may then be necessary to perform some consumer testing to determine which product is preferred. Naturally, for consumer testing, it would be wise to test consumers rather than the sensory panelists from your own company; the latter having undergone training, are no longer representative of the consumers. This is discussed further in Section 5.14.

The most common sensory difference tests are

The paired-comparison test
The duo-trio test
The triangle test

The *paired-comparison difference test* presents two samples to a judge who has to determine which one has more or a given attribute (sweeter, spicier, saltier, etc.). Should he consistently pick the right sample over replicate tests, it would indicate that he can distinguish between the two; should he not do so, it would indicate that he cannot distinguish between the two. Should he pick the right sample more often than the wrong one, a binomial test will determine the probability of picking the right sample this often by chance, so as to be able to make a decision about whether the judge really could tell the difference.

The disadvantage of the paired-comparison test is that the experimenter has to specify which attribute the judge has to look for. It is relatively easy to ask a judge which sample is sweeter because we generally understand what is meant by "sweet." It is not easy, however, to explain exactly what is meant by "spicy," "having more off-flavor" or "young tasting." Our language for flavor is so underdeveloped that we often cannot describe exactly the flavor under consideration. To get around these problems, the duo-trio and triangle tests are used.

The *duo-trio test* does not require the judge to determine which of two samples has more of a given attribute; it is used to determine which of two samples are the same as a standard sample presented before the test. This saves the difficulty of having to describe the attribute under consideration. The test is analyzed statistically in the same way as the paired-comparison test, to see whether a judge demonstrated that he could distinguish the two samples, by consistently picking the sample that matched the standard.

For the paired-comparison and duo-trio tests, the approach to the statistical analysis is the same. Does the judge pick the correct sample enough times over replications for us to deduce that he can distinguish between the two? The binomial test can be used to see if from his replications of the test, the judge picks the correct sample so often that we can reject the null hypothesis (the null hypothesis states that the judge cannot distinguish between the two samples and will pick each an equal number of times).

Do we use a one- or a two-tailed test? Generally in sensory evaluation, two-tailed tests are used, but in this case a good argument can sometimes be made for a one-tailed test. Should the judge be able to distinguish between the two samples, we can often predict which one he will choose. For example, consider pair comparisons between samples of a food with and without an extra ingredient, say sugar. When asked to identify the sweeter sample, it is reasonable to expect that if the judge could distinguish between the samples (H_0 false), he would choose the sample with added sugar. Because we can predict the direction of the outcome, should H_0 be false, the test is one-tailed. On the other hand, we may not know enough about the processing to be able to make such a prediction. We may know that a change in processing may make the flavor of one food stronger but we may not know which one. In this case, we would not be able to predict the sample chosen as "stronger flavored" should H_0 be wrong. We would use a two-tailed test. So, a paired-comparison test can be one-tailed or two-tailed, depending on the experimental circumstances. The duo-trio test, however, is one-tailed. Should the judge be able to distinguish between the two samples, the one he will pick will be the one that matches the standard. We can thus predict the sample chosen should H_0 be false and so can use a one-tailed test.

The paired-comparison test can also be used to measure preference. Here the judge merely has to indicate which of two samples he prefers. In this case, we cannot predict which sample would be chosen should there be a significant preference (H_0 rejected). Because we cannot tell in advance the direction of the preference, we choose a two-tailed test.

The paired-comparison difference test is a directional test. The question is asked: "Which of the two samples has a greater amount of a given attribute?" Instead of doing this, it would also be theoretically possible to ask the judge whether the two samples were the "same" or "different." Here we would be seeing whether the response "same" or the response "different" occurred in a significant majority of the cases, as opposed to equally on H_0. Because we cannot predict which response would be most common should H_0 be wrong, we would use a two-tailed test in our binomial statistics. However, for psychological reasons, this approach should not be used. The data obtained will be of little use. The reasons for this are subtle but briefly it can be said that to ask whether two samples are the "same" or "different," is really to ask two questions. Firstly, it is to ask whether a difference is detected. Secondly, it is to ask a more subtle implied question. This is whether any difference detected is sufficiently large to be worthy of report. This second question may not seem very important but it is, in fact, sufficiently important to completely alter the responses of the judge. The answer to the second question depends on the judge's confidence level and willingness to take a risk; this is ever changing and introduces an unwelcome and uncontrolled variable into the testing procedure. So, the question is never phrased in this way.

An alternative to the duo-trio test, that also avoids the difficulty of having to define the attribute that is varying between the samples, is the *triangle test*. Rather than picking one of a pair that is the same as a standard, the judge has to pick out the odd sample from a group of three (two the same, one different). Again, a one-tailed binomial test is the appropriate statistical analysis, because should the judge be able to distinguish the sample, we can tell beforehand which sample he will pick. He will pick the odd sample. There is something new, however. In paired-comparison and duo-trio testing, the probability of picking the right or the wrong sample, by chance, is 1/2. This is, on the null hypotheses, $p = q = 1/2$. For the triangle test, the probability of picking the correct sample by chance is not 1/2; it is 1/3. Thus by H_0, $p = 1/3$, $q = 2/3$. Tables G.4.a and G.4.b assume $p = q = 1/2$. A new table, Table G.4.c, has been constructed for $p = 1/3$, $q = 2/3$; this gives one-tailed probabilities for triangle difference testing. A one-tailed test is used because, should judges be able to distinguish between the samples, we can predict which one will be chosen as the odd sample. Table 5.1 summarizes these sensory tests and their appropriate analyses.

We have been dealing solely with the statistical aspects of difference and preference testing. A brief summary of some of the more psychological aspects of these tests are given in Table G.6. Table G.4 is set up to give probabilities, should the null hypothesis be true, of getting the observed (or better) performance at distinguishing between the samples. We get a probability value and using this, we decide whether to reject H_0. We decide whether the judge can distinguish between the samples. Similarly, we can determine whether the judge shows a significant preference for one of the samples. Another approach would be to say, for a given level of significance ($p < 0.05$ or $p < 0.01$), how many of a given number of replicate tests a judge must choose correctly, to demonstrate that he can distinguish the samples (or show a definite preference). How many trials must he get right for us to reject the null hypothesis at $p < 0.05$ or $p < 0.01$. How this is done can be seen from the following examples.

Paired-Comparison Sensory Difference Test, One-Tailed

In 20 trials, how many correct responses, at the 5% level, are necessary to show that a judge can distinguish two samples in a paired-comparison test? In our example we pick the one-tailed case. We choose Table G.4.a (one-tailed,

Table 5.1 Sensory Tests and Their Analyses

	Difference tests	
Paired-comparison A vs. B		
Which is greater in a given attribute?	A or B?	One-tailed binomial test, $p = q = 1/2$; use Table G.4.a or two tailed, $p = q = 1/2$, use Table G.4.b
Is there a difference between the two?	Same or different?	Two-tailed binomial test, $p = q = 1/2$; use Table G.4.b; this test *not* used for reasons of psychological bias
Duo-trio A: A vs. B		
Which is the same as the standard?	A or B?	One-tailed binomial test, $p = q = 1/2$; use Table G.4.a
Triangle A B B		
Which is the odd sample?	A or B?	One-tailed binomial test, $p = 1/3$, $q = 2/3$; use Table G.4.c
	Preference tests	
Paired-comparison A vs. B		
Which do you prefer?	A or B?	Two-tailed binomial test, $p = q = 1/2$; use Table G.4.b

$p = q = 1/2$) and for $N = 20$ we find the minimum X number required to give a probability less than 0.05.

$X = 15$ gives a probability of 0.021. For $X = 14$ this rises to 0.058, above 5%. So at the 5% level of significance you need to get 15/20 trials correct to demonstrate that you can distinguish the samples (at 1% you need 16 correct). Rather than having to do this from Table G.4.a, a table can be constructed to do this directly. For a given number of judgments, Table G.5.a gives the number of correct judgments required to show significance at the 5% level. Merely pick your level of significance and read off the number of correct trials required for a given N. In the 0.05 column it is 15, in the 0.01 column it is 16.

Should we be using a two-tailed paired-comparison difference test, we would perform the same exercise with Table G.4.b and Table G.5.b, as illustrated next for the paired-comparison preference test.

Table 5.2 Guide to Tables for Sensory Testing

	To obtain probability on H_0, use:	To determine the requisite number of correct responses, use:
Paired-comparison: difference test	Table G.4.a or G.4.b	Table G.5.a or G.5.b
Paired-comparison: preference test	Table G.4.b	Table G.5.b
Duo-trio: difference test	Table G.4.a	Table G.5.a
Triangle: difference test	Table G.4.c	Table G.5.c

Paired-Comparison Preference Test, Two-Tailed

For a paired-comparison preference test the two-tailed Table G.4.b is used. We ask how many responses, out of 20, a judge must agree on to demonstrate a significant preference at the 5% level. We perform the same exercise. For $N = 20$ we find the largest probability below 0.05 (0.041) and read off the corresponding X value (15). Thus, at the 5% level of significance we would need 15/20 consistent preferences to reject H_0 and say that the judge had a preference for one of the products. Again, this information can be read directly; this time we use Table G.5.b (two-tailed table).

Duo-Trio Test, One-Tailed

Just as Table G.4.a can be used for the duo-trio test, as well as for the paired-comparison test, the same applies to Table G.5.a.

Triangle Test, One-Tailed

Just as Table G.4.c is adjusted for the fact that $p = 1/3$, $q = 2/3$, Table G.5.c is adjusted in the same way, so that the number of correct responses required to demonstrate that the judge can distinguish the samples can be read directly.

The use of the appropriate binomial tables for sensory testing is summarized in Table 5.2.

5.12 ASSUMPTIONS: USE AND ABUSE OF BINOMIAL STATISTICS IN SENSORY TESTING

Table G.4 is preferable to Table G.5 because it gives more information. It gives the exact probability of obtaining the given result on the null hypothesis rather than just a level of significance above or below which the answer falls. Although Table G.5 is simpler to use, it can be abused by inexperienced sensory analysts.

Consider the following example, where triangle tests are being conducted. Twelve have been conducted and seven have resulted in the odd sample being chosen; from Table G.4.c it can be seen that 7 out of 12 identifications of the odd sample can occur on H_0 with a probability of 6.6%. This is sufficient information to make a decision, but imagine that an inexperienced sensory analyst is using Table G.5.c. She is told to look under the column headed "0.05" and that if enough positives come out, there is a difference. She has only seven and needs eight, so keeps testing until she gets enough positives. Her results and suitably inexperienced thoughts are summarized in Table 5.3.

This is not the way to do sensory testing, desperately carrying on until there are enough "correct" triangles to prove a difference at $p < 0.05$, using Table G.5.c. Table G.4.c would tell us that the probability of getting the result on H_0 was wavering near the 5% level (around 5-10%) and this is enough information to make a decision; after all, there is nothing magic about 5%. Only rarely did the probability drop below 5%; the chances are that after more tests, the probability would again be above 5%. Unfortunately, all too many sensory analysts conduct their tests in this way.

Some discrepancies have been found in different versions of Table G.5. This is because if Table G.4 shows a probability only slightly greater than 5%, some authors (not this one) take it as near enough 5% to be significant. Such discrepancies can lead to quite different decisions being made. This would not occur if Table G.4 was used. Thus, in general, Table G.4 is recommended for use rather than Table G.5.

One further point regarding abuse in sensory testing. Should 10 judges each perform a difference test eight times, $N = 10$, not 80. You can use $N = 8$ to determine whether each judge can distinguish between samples (one-tailed). You then have 10 judges; use $N = 10$ to determine whether there are more distinguishers than nondistinguishers (two-tailed, because we cannot predict whether there would be more distinguishers or nondistinguishers). To combine replications and judges into one big N value is to assume that replications for a single judge are as independent of each other as judges; this is just not true.

The sensory analyst is advised to look at Section 5.14 for some thoughts on the philosophy of difference testing.

5.13 WORKED EXAMPLES: SOME BINOMIAL TESTS

Example 1

Tomato juice samples with added salt were tasted in paired comparison with the same juice sample without added salt by an experienced judge 12 times. The judge was asked to identify the saltier of the two samples and did so for 10 comparisons. Does this mean that the judge could distinguish the salted from the unsalted juice?

Table 5.3 Unfortunate Thoughts of an Inexperienced Sensory Analyst While Conducting a Triangle Test

Total number of triangle tests performed	Did the judge select the odd sample on the last triangle test?	Total number of triangle tests on which the judge selected the odd sample	Number of triangle tests on which the odd sample must be selected to be significant at 5% level (from Table G.5.c)	What the inexperienced sensory analyst is thinking	Probability of picking the odd sample in this number of triangle tests, on H_0 (from Table G.4.c)
12	Yes	7	8	OK! 7 out of 12. I need to get 8, so I'll do another one and keep my fingers crossed.	6.6%
13	No	7	8	No! If only this had been OK. I would have done it. Try again.	10.4%
14	Yes	8	9	Yes! But now I need 9. Try again.	5.8%
15	No	8	9	No! If only this had been OK. I would have got a difference. Try again.	8.8%
16	Yes	9	9	Good. 9 out of 16 is enough; I can stop now. I've shown there is a	5.0%

				difference. Stop! Hey! What's this? Who told the judge to do another triangle test?	
17	No	9	10	Oh no! 9 out of 17 isn't enough. Now I'll have to go on until I get enough again. Why didn't I stop while I could.	7.5%
18	No	9	10	This is terrible! I've got to get two right in a row now, to be able to stop.	10.8%
19	Yes	10	11	Yes! Keep going.	6.5%
20	Yes	11	11	Yes! I've done it. That proves it. A definite significant statistical difference. The judge can definitely tell the difference. Thank goodness, it was a close thing. I thought I'd never prove it!	3.8%

The paired-comparison difference test is a one-tailed test because if H_0 were not true, it can be predicted which sample would be chosen; it would be the one with the added salt. Table G.4.a will be used. From the table, the probability on H_0 of getting 10/12 paired comparisons correct = 0.019 (1.9%). This probability is small; thus H_0 may be rejected (with only a 1.9% chance of being wrong; i.e., making a Type I error). Thus it may be concluded that the judge could tell the difference. Note here that the judge was experienced and knew the taste effect of adding salt to tomato juice; small amounts of salt can alter the flavor of tomato juice while not necessarily making it taste salty.

Example 2

A judge was able to tell a reformulated high-fiber exploding candy product from the original product on six out of seven triangle tests. Does this indicate that the judge could tell the difference between the two products?

The triangle test is a one-tailed test because if H_0 were not true, it can be predicted that the sample chosen as different would actually be the odd sample of the three. Table G.4.c will be used. From the table, the probability of choosing the different sample by chance (on H_0) 6/7 times = 0.007 (0.7%). This probability is small; thus H_0 may be rejected (with only a 0.7% chance of making a Type I error). Thus it may be concluded that the judge could tell the difference between the two products.

Example 3

In a paired-comparison preference test, a judge, who could distinguish between two instantly effervescing, chewable multivitamin capsule products A and B, indicated that he preferred product A to product B six out of 10 times. Does this indicate a significant preference?

The paired-comparison preference test is a two-tailed test because if H_0 were not true it cannot be predicted beforehand whether product A or B would be preferred. From Table G.4.b, the probability of choosing product A, on H_0, 6/10 times = 0.754 (75.4%). This probability is high; thus H_0 cannot be rejected. Thus the results do not indicate a consistent preference for product A.

Example 4

Thirty-seven out of 50 experimental subjects improved their wine-testing ability while 13 got worse when a new psychophysical technique was adopted. Does this indicate that wine-tasting ability tended to improve by more than mere chance when the new psychophysical technique was adopted?

A two-tailed test is used because it cannot be predicted definitely whether the wine-tasting ability will improve or deteriorate with the new psychophysical technique. Thus we choose Table G.4.b. From the table, the probability on H_0 of having 37/50 subjects improve = 0.001 (0.1%). This probability is small; thus H_0 may be rejected (with only a 0.1% chance of making a Type I error).

Thus it may be concluded that subjects tended to improve their wine-tasting ability using the new psychophysical technique.

This test was really a sign test. Subjects tasted wine in two conditions: with and without the new psychophysical technique. Thus this is a related samples two-sample difference test. Thirty-seven subjects scored "+" for improving with the new psychophysical technique and 13 scored "–" for getting worse.

5.14 FINAL EXAMPLE: WHAT IS A FLAVOR DIFFERENCE? SOME THOUGHTS ON THE PHILOSOPHY OF DIFFERENCE TESTING

Twenty judges perform triangle tests to determine whether reformulation of a sauce has an effect on its flavor. Each judge performs 10 triangle tests. Nineteen judges perform at chance levels getting three triangle tests correct. (From Table G.4.c, the probability of getting 3/10 correct if H_0 is true = 70.1%. H_0 is not rejected.) Thus these 19 judges indicate no difference. The twentieth judge gets all 10 triangles correct. (From Table G.4.c, the probability of getting all 10 correct, if H_0 is true, is less than 0.3%. So reject H_0.) Thus this judge can consistently find a flavor difference.

Just to check, the testing is repeated with exactly the same results. Nineteen judges cannot tell any difference and one can, consistently without error. Thus a significant majority of judges cannot tell the difference. Table G.4.b indicates that the probability of getting a majority of 19/20 by chance is less than 0.3%. (H_0 states that there is no difference between the numbers of judges who can and who cannot tell a difference. H_0 is rejected.) There are thus significantly more people who cannot find a difference. Table G.4.b is used (two-tailed) because we cannot predict in advance whether there would be more who could distinguish a flavor difference or more who could not, should H_0 be wrong.

So one judge out of 19 can tell a difference. The question is: Is there a flavor difference?

The trouble with the question is that it is ambiguous. Often two things are meant by the question. It can mean:

1. Is there an actual flavor difference between the sauces? It may be very small and very few people will detect it, but we are interested in whether reformulation has affected the content, in the sauce, of those chemical or physical characteristics which affect the human senses.

or

2. Is there a flavor difference which is detectable in the general population?

These two questions are often confused in sensory testing. Which question is appropriate depends simply on the aim of the experiment. If the aim of the experiment was to see whether a flavor difference existed, regardless of whether people in the general population could detect it (question 1), highly trained judges would be used. We are not sampling the population. We use judges like laboratory instruments. One good judge, like one good gas chromatograph, is sufficient. If one good judge consistently detects a difference in flavor, a difference exists; the other judges are not sensitive or trained enough to detect the difference. The same is true for gas chromatographs. If one gas chromatograph consistently detects something that a second make of instrument cannot, we deduce that the second instrument is not as sensitive. In our example, a flavor difference existed and it was picked up consistently by one of our judges; the other 19 were just not as sensitive or as well trained.

Such an approach as this may be taken when wanting to research the parameters involved in reformulation, a change in processing, storage, or packaging procedures, or for investigating the possible options in product development. It may be apt when sensory evaluation is used as a screening test prior to chemical or physical analysis; changes in sensory characteristics can give valuable clues to the physical and chemical changes that have taken place and so save time.

On the other hand, should a flavor difference be known to exist, the question then becomes one of whether the general population of consumers will detect the difference (question 2). This is an entirely different question and is pertinent to quality control as well as issues of reformulation, processing, and packaging. Here, we do not necessarily want a highly trained panel because we want them to be equivalent to consumers. We want them to be as familiar with the product as the consumers so that they are equivalent to them and can be taken as a representative sample of them. But we want them to be familiar with sensory testing so that they can express their responses skillfully in test situations, a skill often lacking in the consumer population.

Now, if such a panel, a sample of the population, cannot distinguish the flavor differences, we deduce that the population cannot. Here, if 19 of the 20 panelists cannot tell the difference, a significant majority of the population will not. A cautious experimenter may like to increase a sample size to get a better estimate of what proportion cannot tell the difference in the population, so as to allow a more reliable sales forecast to be made. The tests conducted on the 20 judges indicated that only one in 20 people (5%) could tell the difference.

The moral of the story is to define precisely the aims of the testing for each product and use the statistics to answer this particular question. For further discussion, refer to Section 1.2.

6
Chi-Square

6.1 WHAT IS CHI-SQUARE?

We now examine a test called chi-square or chi-squared (also written as χ^2, where χ is the Greek lowercase letter chi); it is used to test hypotheses about frequency of occurrence. As the binomial test is used to test whether there may be more men or women in the university (a test of frequency of occurrence in the "men" and "women" categories), chi-square may be used for the same purpose. However, chi-square has more uses because it can test hypotheses about frequency of occurrence in more than two categories (e.g., dogs vs. cats vs. cows vs. horses). This is often used for categorizing responses to foods ("like" vs. "indifferent" vs. "dislike" or "too sweet" vs. "correct sweetness" vs. "not sweet enough").

Just as there is a normal and a binomial distribution, there is also a chi-square distribution, which can be used to calculate the probability of getting our particular results if the null hypothesis were true (see Section 6.6). In practice, a chi-square value is calculated and compared with the largest value that could occur on the null hypothesis (given in tables for various levels of significance); if the calculated value is larger than this value in the tables, H_0 is rejected. This procedure will become clearer with examples.

In general, chi-square is given by the formula

$$\text{Chi-square} = \Sigma\left[\frac{(O - E)^2}{E}\right]$$

where

O = observed frequency
E = expected frequency

We will now examine the application of this formula to various problems. First we look at the single-sample case, where we examine a sample to find out something about the population; this is the case in which a binomial test can also be used.

6.2 CHI-SQUARE: SINGLE-SAMPLE TEST—ONE-WAY CLASSIFICATION

In the example we used for the binomial test (Section 5.2) we were interested in whether there were different numbers of men and women on a university campus. Assume that we took a sample of 22 persons, of whom 16 were male and 6 were female. We use the same logic as with a binomial test. We calculate the probability of getting our result on H_0, and if it is small, we reject H_0. From Table G.4.b, the two-tailed binomial probability associated with this is 0.052, so we would not reject H_0 at $p < 0.05$. However, we can also set up a chi-square test. If H_0 is true, there is no difference in the numbers of men and women; the expected number of males and females from a sample of 22 is 11 each. Thus we have our observed frequencies (O = 16 and 6) and our expected frequencies (E = 11 and 11), which are what we need to calculate our value of chi-square. We can set up the following table:

	Men	Women
Expected frequency, E	11	11
Observed frequency, O	16	6
$O - E$	5	-5
$(O - E)^2$	25	25

Now the formula for chi-square is

$$\text{Chi-square, } \chi^2 = \Sigma\left[\frac{(O - E)^2}{E}\right]$$

which is the sum of the $\frac{(O - E)^2}{E}$ values for each category.

For men,

$$(O - E)^2 = 25 \text{ and } E = 11, \text{ so } \frac{(O - E)^2}{E} = \frac{25}{11}$$

For women,

$$(O - E)^2 = 25 \quad \text{and} \quad E = 11, \text{ so } \frac{(O - E)^2}{E} = \frac{25}{11}$$

Therefore,

$$\text{Chi-square}, \chi^2, = \frac{25}{11} + \frac{25}{11} = 2.3 + 2.3 = 4.6$$

The next question is whether we could get a chi-square value as big as this by chance. Obviously, if the observed frequencies (O) and the frequencies expected if H_0 were true (E) are similar, it is likely that H_0 is true. If this is so, and O nearly equals E, then $\frac{(O - E)^2}{E}$ and hence chi-square will be small. The greater the discrepancy between E and O, the greater the values of chi-square. The question is then: How big can chi-square be due to chance if H_0 is true? This is given for various levels of significance in Table G.7.

Just as before, our alternative hypothesis determines whether we use a one- or a two-tailed test. Here we will use a two-tailed test because we have no good reason for expecting more men or more women, should H_0 be untrue. We also look up the appropriate value *not* for the size of the sample as with the binomial test, but for the number of *degrees of freedom* (*df*) associated with the test. Here

$$\begin{aligned} \text{Degrees of freedom} &= \text{number of categories} - 1 \\ &= (\text{men vs. women}) - 1 = 2 - 1 \\ &= 1 \end{aligned}$$

Basically, the number of degrees of freedom refers to the number of categories to which data can be assigned freely, without being predetermined. This is seen most easily by example. Suppose that we have 100 animals, which must be either cats, dogs, horses, or cows. We can choose how many we have of each. So we begin choosing. We decide to have 20 cats (we freely choose that, so that was 1 degree of freedom). We decide to have 30 dogs (another free choice). We decide to have 40 horses. That is all we can decide freely. We cannot decide how many cows we have; it has been decided for us. Of the 100 animals, 20 are cats, 30 are dogs, and 40 are horses, so the remaining 10 animals must be cows. For the four categories, we could assign numbers freely to only three (cats, dogs, horses); for the fourth (cows) we had no choice. Thus there were 3 degrees of freedom (*df* = 3). With two categories, men and women, we can freely choose the number of men, but for a given sized population, the number of women have been decided for us. So for two categories there is 1 degree of freedom (*df* = 1). The number of degrees of freedom = number of categories − 1.

Thus in our problem about numbers of men and women on the campus, $df = 1$. So looking at Table G.7, the degrees of freedom are given on the left. Choose $df = 1$ (top row). Let us choose a two-tailed test (we cannot predict whether we will get more men or women if H_0 is false) and let us work at the 0.05 level. We select the appropriate column (third column of chi-square values) and find the value of 3.84. Our value, 4.6, is larger than 3.84 (the largest value obtainable on H_0 at the 5% level), so we reject H_0 (our result is more extreme than you would get should H_0 be true). Thus we reject the idea that there are equal numbers of men and women in the campus population; there were more men.

Note that we would not reject H_0 at the 1% level, where chi-square in Table G.7 = 6.64. Also note that should our value be only equal to that in the table, we still reject H_0.

Relative Power of Binomial and Chi-Square Tests

You may notice that when working at the 5% level, we rejected H_0 with the chi-square test ($p < 0.05$) but would not have rejected it using the binomial test ($p = 0.052$). So chi-square is more likely to reject H_0 than the binomial test; it is more powerful. In Section 5.3 we defined the power of the test as follows:

$$\text{Power} = 1 - \beta$$

where β is the probability of committing a Type II error (accepting H_0 when it is false). Thus

$$\text{Power} = \text{probability of rejecting } H_0 \text{ when it is false}$$

It can be seen that chi-square was better at rejecting H_0. For given data it has a higher probability of rejecting H_0. It is thus more powerful (more likely to reject H_0 when it is false) and is more likely to commit a Type I error (more likely to reject H_0 when it is true). So the binomial test needs a larger difference in the numbers of men and women before it will reject H_0. See Appendix F for further discussion of Type I and Type II errors and the power of tests.

However, the difference between the two tests, in this example, was small. If the probability of getting this result, should H_0 be true, on one test be slightly over 5% and on the other be slightly less, it is not a large difference. In each test, the probability was "near 5%" and this is sufficient to reject H_0. It is only when the 5% level is drawn as an absolute line dividing "different" from "not different" that difficulties arise. Note that if we were working at the 1% level, neither test would reject H_0.

Which test is better to use? This is a matter of opinion and experience. If one test does not reject H_0 and the other does, it is better to do more testing until the answer becomes clearer. If you are rejecting a null hypothesis, it is better to be able to do so on both tests. If you wish to be completely confident about rejecting H_0, it is safer to reject it on the test that is least likely to do so—the binomial test. If you are accepting H_0, is is safer to accept it on the test that is least likely to accept it—chi-square.

Extension to More Than Two Categories

Chi-square can be extended to as many categories as desired. We may, for instance, be interested in how a given ecological niche (area of land or forest) supports different species of birds which may be in competition for food. We could see whether there are more of one species than another. For four bird species, A, B, C, and D, we could take a sample of 44 birds. Assume that we found 16 each of A and B and 6 each of C and D. The expected frequencies by H_0 would be 11 of each (44/4 = 11). Once again the results are set up as before.

	Species of bird			
	A	B	C	D
Observed frequency, O	16	16	6	6
Expected frequency, E	11	11	11	11
$O - E$	5	5	-5	-5
$(O - E)^2$	25	25	25	25

$$\text{Chi-square} = \Sigma \frac{(O-E)^2}{E} = \frac{25}{11} + \frac{25}{11} + \frac{25}{11} + \frac{25}{11} = 9.2$$

Here $df = 4 - 1 = 3$.

We use a two-tailed test because we have no good reasons for predicting which of the four species there will be more or fewer of should H_0 be rejected. In Table G.7, for $df = 3$, two-tailed at the 0.05 level, the chi-square value in the tables = 7.82 (at 0.01 it is 11.34). Our value (9.2) is greater than the 0.05 value in the table (7.82), so we reject H_0. However, at the 1% level we would not reject H_0. Thus the probability of getting our result on H_0 is somewhere between 0.01 and 0.05. It is reasonable to say that our sample indicates that there are not equal numbers of the species A, B, C, and D in the ecological niche; there are more of species A and B.

Thus the advantage of the chi-square one-sample test over the binomial test is that you can use more than two categories. Responses are not confined to only two (like vs. dislike), you can use more (like vs. dislike vs. indifferent). It is interesting to note that the expected frequencies (E) were calculated using the null hypothesis. However, they can be estimated using any hypothesis, and chi-square will give you a measure of deviation from that hypothesis. As it is, the hypothesis generally used is H_0 and deviations from this are what are considered. Look up the one-sample chi-square in the Key to Statistical Tables to see its relationship to other tests.

6.3 CHI-SQUARE: TWO-WAY CLASSIFICATION

In the foregoing examples, the data were classified along one dimension: men vs. women or bird species A, B, C, D. However, the data for chi-square can be classified along two dimensions simultaneously. For instance, we could be interested in how men and women vary in their appreciation of a new synthetic leather textile product. We categorize people along one dimension as men vs. women. We also categorize them, on the second dimension, as to their approval: "approve," "disapprove," or "feel neutral." Let us assume that of 79 men sampled, 58 approved, 11 were neutral, and 10 disapproved, while of the 83 women sampled, 35 approved, 25 were neutral, and 23 disapproved. The results are classified in a table as follows:

	Men	Women	Total
Approve	58	35	93
Neutral	11	25	36
Disapprove	10	23	33
			Grand total
Total	79	83	162

We are interested in whether men and women differ in their attitude to the product. Looking at the data it would seem that they do. Men tend to approve (58) more than be neutral (11) or disapprove (10). Women, on the other hand, tend to fill the categories more equally (35, 25, 23) than men. However, we must ask whether this sample of data indicates real differences between men and women in the population or whether the differences were merely due to chance.

Thus we calculate chi-square. We calculate for each of the six categories the values

$$\frac{(O - E)^2}{E}$$

We have all our observed frequencies (O). We have yet to fill in the expected frequencies (E); we do this using the null hypothesis.

The null hypothesis assumes that there is no difference in the distribution of approval and disapproval of men and women. If this were true, expected frequencies must be calculated on this basis. Certainly we do not expect an equal number in each box if H_0 were true (162/6 = 27) because there are more women than men in the sample. So the expected frequencies must be weighted accordingly with higher values for women. Furthermore, there may be an overall tendency in subjects to approve more than disapprove; again the expected frequencies must be weighted accordingly. To incorporate these weightings, the expected frequencies in each of the six categories are computed using the geometric mean, as follows:

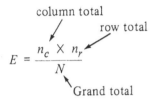

$$E = \frac{n_c \times n_r}{N}$$

(where n_c = column total, n_r = row total, N = Grand total)

The expected frequency, E for men who approve ($O = 58$) is

$$\frac{79 \times 93}{162} = 45.35$$

(79 = column total, 93 = row total, 162 = Grand total)

For men who are neutral ($O = 11$)

$$E = \frac{79 \times 36}{162} = 17.56$$

For men who disapprove ($O = 10$)

$$E = \frac{79 \times 33}{162} = 16.09$$

Or the latter value could be calculated knowing that the expected frequencies for men must add up to the total number of men as do the observed frequencies. Hence we can calculate E for men who disapprove, by subtraction:

$$79 - 45.35 - 17.56 = 16.09$$

The expected frequencies for women can be calculated in the same way. For women who approve ($O = 35$)

$$E = \frac{93 \times 83}{162} = 47.65$$

Or we could calculate it knowing that the expected frequencies for "approve" must add up to the observed frequencies for "approve." Thus we can calculate it by subtraction:

$$E = 93 - 45.35 = 47.65$$

In the same way we calculate the rest of the expected frequencies:

Women who are neutral:

$$E = \frac{36 \times 83}{162} = 18.44 \quad \text{or} \quad 36 - 17.56 = 18.44$$

Women who disapprove:

$$E = \frac{33 \times 83}{162} = 16.91 \quad \text{or} \quad 33 - 16.09 = 16.91$$

Thus the matrix, with the expected frequencies in parentheses, can now be written:

	Men	Women	Total
Approve	58 (45.35)	35 (47.65)	93
Neutral	11 (17.56)	25 (18.44)	36
Disapprove	10 (16.09)	23 (16.91)	33
			Grand total
Total	79	83	162

$\frac{(O - E)^2}{E}$ can then be calculated for each box and then summed to calculate chi-square.

$$\text{Chi-square} = \frac{(58 - 45.35)^2}{45.35} + \frac{(35 - 47.65)^2}{47.65} + \frac{(11 - 17.56)^2}{17.56}$$

$$+ \frac{(25 - 18.44)^2}{18.44} + \frac{(10 - 16.09)^2}{16.09} + \frac{(23 - 16.91)^2}{16.91}$$

$$= 16.17$$

We now want to compare our calculated chi-square value with the appropriate value in Table G.7, the highest value we can get should H_0 be true. To use Table G.7, we need to know the appropriate number of degrees of freedom. In this case it is the product of (the number of categories − 1) for each dimension. Thus

$$df = (\text{number of rows} - 1) \times (\text{number of columns} - 1)$$
$$= (3 - 1)(2 - 1) = 2$$

Using a two-tailed test (because we cannot predict the pattern of approval in advance), at the 5% level, two-tailed, $df = 2$, chi-square = 5.99. At the 1% level, chi-square = 9.21. Our calculated value is greater, so we reject H_0 at $p < 0.05$ and at $p < 0.01$. Thus we conclude that there is a difference in the pattern of approval for the new textile between men and women. We have shown that there is a difference in the pattern of approval. The data are inspected to see exactly what this pattern is. It would appear that men would be more disposed to approval.

Note that here we looked at the difference in pattern of approval between two independent groups of judges: men and women. In this way the test is an independent-samples, two-sample difference test. However, there is no need to confine this test to two samples; it can be applied to as many groups of people as desired. Thus the test can be represented in the chart on the inside cover as an independent-samples difference test for two samples or for more than two samples.

6.4 ASSUMPTIONS AND WHEN TO USE CHI-SQUARE

The data used in this test are nominal data or categories; people are put into categories depending on whether they are "men who approve," "women who approve," or whatever. Persons are not scored on how much they approve, using some type of numerical scale (interval of ratio scale); nor are they ranked in order of approval (ordinal scale). Chi-square uses nominal data. It is thus nonparametric. Certain conditions must be met:

1. Observations in each cell must be independent of each other. One cannot sometimes include multiple observations from one person; the frequencies in each cell are the numbers of independent cases falling into that cell.
2. In the application of the chi-square distribution to the calculation of probabilities for the statistical analysis, it is assumed that in each cell in the matrix of results, the distribution of observed frequencies about the expected frequency is normal. What this means is that should samples be drawn from the population, the observed frequencies (O) will be scattered around the expected frequency value (E). They will be distributed around (E), the mean, in the form of a normal distribution. This assumption has two consequences:
   a. When E is small (around 5), the distribution of observed frequencies in the cell cannot be normal. It can be seen from Figure 6.1 that it must be skewed, because for observed frequencies between the zero value and 5 there is insufficient room for the left-hand side of the distribution to fit. For the assumption of normality of observed frequencies about the

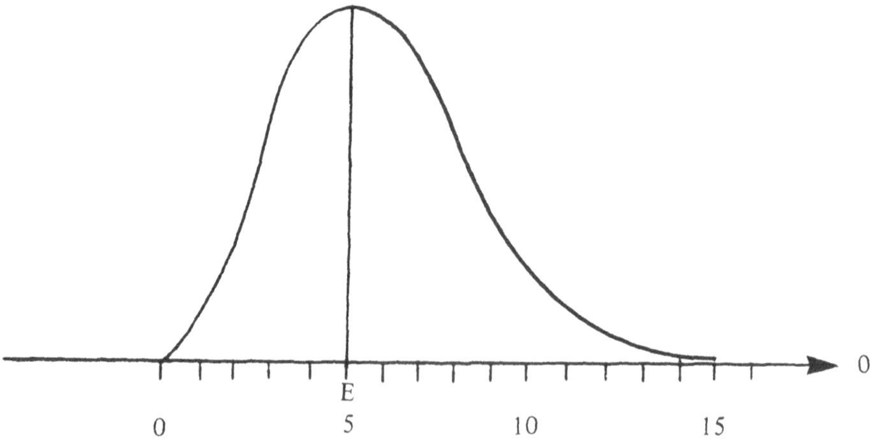

Figure 6.1 Distribution of observed frequencies.

expected frequency not to be seriously broken, *the expected frequency must be at least* 5. Some authorities say that it should be even greater (i.e., 10). If your E values are not large enough, you may combine categories so as to get large enough values of E.

b. For the observed frequencies (O) to be distributed normally about the expected frequency (E), they must be continuously distributed not discrete. A continuously distributed variable is one that can, in theory, have any value (e.g., 1, 2, 7.9345, 6.317645, etc.). Mostly, our variables are continuous (length, height, etc.; you can have any value of length or height). However, we sometimes encounter discrete variables, ones that can only have certain discrete values (e.g., 1, 2, or 3, but not 2.5 or 1.7). An example of a discrete variable is the number of children in a family; you cannot have 2.5 children. Another example of a discrete variable is the number of people falling into a given category as in Chi-Square (see Section 2.6).

Thus our assumption that the observed frequencies are normally distributed around the expected frequency implies that the observed frequencies must be continuously variable. This assumption of continuity is used in the derivation of the probability theory for the chi-square test. For simplicity, a mathematical operation (integration) is carried out which should only be carried out on continuous data; for discrete data an alternative operation (summation) should have been used. The result is that the chi-square value which is calculated from the experimental data will be too large. This can be corrected by subtracting a given amount from each ($O - E$) value, to reduce it.

This correction is generally not important and is not needed unless you have very few categories. It needs to be used when you have a 2 × 2 matrix. The correction most commonly encountered is called *Yates' correction*; here the given amount to be subtracted from each $(O - E)$ value is 1/2. The use of Yates' correction is, in fact, controversial and will be discussed later.

6.5 SPECIAL CASE OF CHI-SQUARE: THE 2 × 2 TABLE

With a 2 × 2 table, the calculation can be simplified by rearranging the formula to calculate chi-square, as follows:

A	*B*
C	*D*

A, *B*, *C*, and *D* are observed frequencies.
N is the total number $= (A + B + C + D)$

$$\text{Chi-square} = \frac{N\left(|AD - BC| - \frac{N}{2}\right)^2}{(A + B)(C + D)(A + C)(B + D)}$$

where $|AD - BC|$ is the value of $AD - BC$ taken only as a positive value. The number of degrees of freedom is calculated as before:

$$df = (2 - 1) \times (2 - 1) = 1$$

This formula is computationally simpler. It also includes Yates' correction for continuity, the $N/2$ part in the numerator.

Now let us consider an example, such as the prior one. Here we look at the pattern of approval for males and females for a new textile. But this time we consider only approval and disapproval and do not have the neutral category. We have a 2 × 2 (rather than a 2 × 3) table.

Let the observed frequencies be as follows:

	Men	Women	
Approve	$A = 28$	$B = 8$	$A + B = 36$
Disapprove	$C = 5$	$D = 24$	$C + D = 29$
	$A + C = 33$	$B + D = 32$	$N = 65$

Although we can calculate chi-square in the usual way, it is simpler to use our new formula. Hence

$$\text{Chi-square} = \frac{N\left(|AD - BC| - \frac{N}{2}\right)^2}{(A + B)(C + D)(A + C)(B + D)}$$

$$= \frac{65\left(|28 \times 24 - 8 \times 5| - \frac{65}{2}\right)^2}{36 \times 29 \times 33 \times 32}$$

$$= \frac{65(|672 - 40| - 32.5)^2}{36 \times 29 \times 33 \times 32} = \frac{65 \times 599.5^2}{36 \times 29 \times 33 \times 32} = 21.19$$

We now use Table G.7 to ascertain whether this value of chi-square is greater than could be obtained by chance (by H_0). Because we cannot predict the direction of the trends beforehand, we use a two-tailed test. We also have 1 degree of freedom ($df = 1$). From Table G.7, chi-square, two-tailed, for $p = 0.001$, $df = 1$, is 10.83. Our value is greater than the value in the tables (the largest value that could be obtained on H_0); thus we reject H_0.

Thus men and women differ in their approval patterns for the new textile; the observed frequencies show that the men tend to approve while the women tend to disapprove.

Caution and Controversy Surrounding the 2 X 2 Chi-Square Test: The Correction for Continuity—Free and Fixed Margins

We have given the traditional line regarding corrections for the 2 X 2 table, the use of Yates' correction in all cases. This is the recommendation given in most textbooks. However, the blind use of Yates' correction has been criticized and other corrections recommended. First, however, a point of design must be discussed. The totals in a 2 X 2 table may be FIXED or FREE depending on the experiment.

A	B	$(A + B)$
C	D	$(C + D)$
$(A + C)$	$(B + D)$	

It could be that values along the vertical margin, $(A + B)$ and $(C + D)$, are fixed by the experimenter or left free to vary. The same is true of the totals along the second margin, $(A + C)$ and $(B + D)$. In our previous example we examined the pattern of approval and disapproval for males and females for a new textile. Had we chosen a group of judges regardless of sex, without knowledge of their likes and dislikes, the numbers of men and of women as well as the numbers of approvers and disapprovers would not be controlled by us; they would merely depend on whom we happened to sample. Such totals are described as free; the dimension, or margin: men vs. women is said to be free, as is the margin: approve vs. disapprove. On the other hand, had we fixed the numbers of men and women we were going to pick beforehand (say, 30 of each), the totals along the men vs. women margin would now be fixed; it would be a fixed margin. The approve vs. disapprove margin would still be free, however. To fix both margins would be difficult. We have not only to fix the number of men and women we sample but also to fix the number of approvers and disapprovers, before we have even measured whether they approve or not. Believe it or not, it is possible to fix this margin. We could have instructed the judges to indicate their approval or disapproval on some type of scale. We could then choose a point on the scale such that if a given number of subjects had indicated scores above this point, they would be called approvers. The rest who score below the point would be called disapprovers. In this way we can control the numbers of subjects rated as approvers and the number rated as disapprovers; we can fix the margin.

An improved set of corrections can be used depending on whether two, one, or no margins are fixed. The corrections involve substituting the $N/2$ term in the previous formula by a range of terms depending on whether one, two, or no margins are fixed. The formula becomes

$$\chi^2 = \frac{N(|AD - BC| - K)^2}{(A+B)(C+D)(A+C)(B+D)}$$

where K is a correction term as follows:

$K = 0$: no margins fixed
$K = N/4$: one margin fixed
$K = N/2$: two margins fixed

So it can be seen that the usual Yates' correction applies only in the comparatively rare case where both margins are fixed.

It is worth considering a further example. We might be interested in the trend for wine drinkers to begin to favor drier wines as they get older. We may require subjects to indicate which of two wines they preferred: a dry or a sweet wine. We may pick two groups of wine drinkers: young adults and older drinkers. The people would be assigned to a position in the following matrix:

		Preference	
		Dry wine	Sweet wine
Wine drinkers	Young adults		
	Older drinkers		

Should the hypothesis be true, young adults will tend to be assigned more to the sweet wine category and the older adults more to the dry wine category. Chi-square would show whether this trend in our sample indicated a significant trend in the population (reject H_0) or whether it was merely due to chance (could occur if H_0 were true).

To set up this experiment we may merely have sampled people at random and seen how many people happened to fall into each category. In this case both margins would be free (the correction for continuity, $K = 0$). Should we have decided to have sampled a given number each of young adults and older people, we would have fixed this margin. With one margin fixed our correction for continuity would be $K = N/4$. Alternatively, we could have picked out a given number of people who each preferred dry or sweet wines (fixing this margin) and seen how many young and old people we happened to get in each category (leaving this margin free). This would still give us one margin fixed, albeit a different margin. We might in this latter case, however, alter our definitions of "young" and "old" so that we had equal numbers of people in each age category. We would now have fixed this second margin and so have two margins fixed (correction for continuity, $K = N/2$).

6.6 FURTHER NOTE ON THE CHI-SQUARE DISTRIBUTION

Just as there is a normal and a binomial distribution from which probability values can be calculated, there is also a chi-square distribution with its own probability values, which are used to calculate the critical values of chi-square in Table G.7. The distribution of chi-square depends on the number of degrees of freedom. For each number of degrees of freedom, there is a separate chi-square distribution. As the number of degrees of freedom increases, the distribution gets more symmetrical; as the number of degrees of freedom are reduced, the distribution gets more skewed. Because it would take up too much space to give each distribution for each number of degrees of freedom, selected values of chi-square are given for various degrees of freedom (1, 2, 3, etc.) at given probability levels (0.05, 0.01, etc.). Those are the values given in Table G.7.

6.7 THE McNEMAR TEST

Chi-square is generally an independent-samples test but there are related samples versions. The *McNemar test* is a two-sample related samples difference test which is an adaptation of chi-square. Judges are categorized into two categories in a "before" condition and then the same judges are recategorized in the same way in a second "after" condition. For example, judges may be categorized as "liking" or "disliking" a product. After a period of exposure to that product the same judges may be recategorized as "liking" or "disliking" the product. Judges who do not change—they stay "liking" or "disliking"—are ignored and dropped from the calculation. Let A judges change from "dislike" to "like" and B change from "like" to "dislike." If the period of exposure has no effect, then as many judges will change from "dislike" to "like" as will change from "like" to "dislike" ($A = B$). Should the period of exposure tend to make judges like the product, A will be greater than B ($A > B$); if the reverse were true, $A < B$. The question is whether the difference between A and B indicates a significant difference in the population or whether it is merely chance. Could such differences have occurred, should H_0 be true? (Here H_0 says that there was no difference in the liking/disliking for the product after the period of exposure and thus that $A = B$.) Chi-square is used to compute whether the change is greater than would occur by chance (on H_0) at a given significance level. The appropriate formula for chi-square, here, is

$$\text{Chi-square} = \frac{(|A - B| - 1)^2}{A + B} \qquad \text{for the McNemar test, } df = 1$$

Should A be larger than B (or vice versa), chi-square will be large. Should they be the same, chi-square will be small. Table G.7 gives largest values of chi-square which can occur by chance. If the calculated value exceeds or equals the value in the table, we reject H_0; the exposure period has had a significant effect. It is worth noting that the formula for chi-square contains a correction for continuity; such corrections have already been discussed.

Like the Sign test, the McNemar test ignores judges who show no change; this is only justified should this number be small (less than 25%). Now look at the Key to Statistical Tables to see the relationship of the McNemar test, a two-sample related-samples test, to the other tests.

6.8 THE COCHRAN Q TEST

The *Cochran Q test* extends the range of the McNemar test from a two-sample to a multisample, related-samples difference test. The situation can be the same as with the McNemar test, with judges being categorized as liking (+) or disliking (−) a product but with more periods of exposure to that product (periods *A, B, C, D, E,* etc.). Let us assume that we had four periods of exposure: *A, B, C,* and *D*. We will thus have like vs. dislike data for each subject under not two, but four conditions (or *treatments* as they are generally called in such multiple-sample tests). In general, we denote the number of treatments as k.

In our example there are four periods of time (four treatments, $k = 4$); the data can be displayed as follows:

	Period of exposure (treatments)							
Judge	A	B	C	D	...	k	Row total	(Row total)^2
S_1	–	–	–	+			R	R^2
S_2	–	+	+	+			R	R^2
S_3	–	–	+	+			R	R^2
	.	.	.	.			.	.
	.	.	.	.			.	.
	.	.	.	.			.	.
Column totals	C	C	C	C			ΣR	ΣR^2

The number of likes (+) in each row and column are added to give row totals (R) and column totals (C). [Alternatively, the – ("dislike") values could be added.]

Chi-square, or as Cochran called it, Q, is calculated using the formula

$$\text{Cochran's } Q \text{ or chi-square} = \frac{(k-1)\,[k\,\Sigma\,C^2 - (\Sigma\,C)^2]}{k\,\Sigma\,R - \Sigma\,R^2}$$

For the Cochran Q test, $df = k - 1$; here $df = 4 - 1 = 3$.

Should the period of exposure to the product have no effect, the + signs will be fairly equally distributed across the treatments and the column totals C will all be approximately equal. Should the period of exposure have effect, the + values will be concentrated toward the end of the columns, producing large and small C values. When C values are squared, the large C values become very large, increasing the value of chi-square. Thus if the period of exposure has effect, some of the C^2 values will be very large producing a large chi-square; if there are no differences in the periods of exposure (no differences between the treatments), the C^2 values will be smaller, producing a smaller chi-square. Thus the greater the difference between the treatments, the larger chi-square; the question becomes how large chi-square can be by chance. Could such differences occur, should H_0 be true? Table G.7 indicates how large chi-square may be by chance (by H_0). Should the calculated value be greater than or equal to the appropriate value in Table G.7, we reject H_0; we say that the like/dislike scores vary significantly between the treatments.

Treatments may be more than mere periods of time. They may be qualitatively different, such as difference processing treatments for various foods or different experimental situations. They may produce a whole range of "+" vs. "–" responses: "like" vs. "dislike," "has off-flavor" vs. "has no off-flavor," "can detect a taste" vs. "cannot detect a taste."

It is worth finding the Cochran Q test on the Key to Statistical Tables to see its relationship to other tests.

6.9 THE CONTINGENCY COEFFICIENT

So far, we have been using chi-square as a statistical test to determine whether differences occur between groups of people; for instance, whether males and females differ in their degree of approval for a given fabric. The data from a sample are examined to see whether the differences occurring in the sample indicate differences in the respective populations from which the samples were drawn (reject H_0).

However, there is an alternative use for chi-square. It can be used to compute the *coefficient of contingency, C*, which is a measure of the degree of association or relation between two groups of people. Instead of using chi-square as a measure, say of the degree of difference in the extent of approval for a fabric between groups of men and women, the contingency coefficient looks at the data from a different point of view; it is a measure of the degree of association in the approval patterns of men and women. Is there any relationship between patterns of approval (they could be the same or opposite) or are the patterns unrelated? Of course, the contingency coefficient is applicable to any two sets of attributes (dimensions on the matrix) not just sex (male vs. female) and pattern of approval (like vs. dislike). It could be used to examine the degree of association between the amount of training a judge has had (trained vs. untrained) and the consistency of his or her panel scores (consistent vs. inconsistent). It could be used to examine the association between the amount of salt consumption (high vs. medium vs. low) and blood pressure (high vs. medium vs. low).

The degree of association between sets of data is generally called *correlation*. Look at the Key to Statistical Tables to see the various correlation tests. The coefficient of contingency investigates associations for category data, Spearman's rank correlation coefficient is used in correlating ranked data, while Pearson's product-moment correlation coefficient is used for ratio or interval data. Correlation is discussed in detail later (Sections 15.1 to 15.9 and 16.6).

The contingency coefficient is given by

$$C = \sqrt{\frac{\chi^2}{N + \chi^2}}$$

where χ^2 (chi-square) is calculated as usual from the data, and N is the total number of people tested.

The contingency coefficient is significant if the chi-square it is calculated from is also significant. If the chi-square value for the sample values is significant, we may conclude that in the population the association between the two sets of categories (the two dimensions on our matrix: men vs. women and like vs. dislike) is not zero.

Generally, coefficients of correlation have a value of 1 for perfect association and zero when there is no correlation. C has a value of zero when there is no association but it never reaches a value of 1. For a matrix which has equal numbers of categories along each axis, the highest possible value of C can be calculated using the formula

$$C_{max} = \sqrt{\frac{k-1}{k}}$$

where k is the number of categories on each dimension. Thus for a 2 × 2 table, $k = 2$ and $C_{max} = \sqrt{1/2} = 0.707$ and for a 3 × 3 table, $k = 3$ and $C_{max} = \sqrt{2/3} = 0.816$. Because of the variation in the maximum value of C, the values are not comparable to each other (unless they come from identical-sized matrices) nor to the other correlation coefficients.

Thus chi-square can be seen to be the basis for several other nonparametric tests: the McNemar test, the Cochran Q test, and the contingency coefficient. It is also related to other tests which will be discussed in Chapter 16: the Kruskal-Wallis test (Section 16.8) and the Friedman test (Section 16.10). Look at the Key to Statistical Tables to see the relationship of these tests to each other.

6.10 WORKED EXAMPLES: CHI-SQUARE TESTS

Example 1

Sixty randomly selected customers in a store were asked to select which product they preferred from three shampoos with added beer, eggs, protein, herbs, and yoghurt. If 30 preferred product A, 18 product B, and 12 product C, would this represent significant differentiation in preference?

The 60 randomly selected customers preferred the products as follows:

		Product			
		A	B	C	
Number of Customers	O	30	18	12	Total = 60
Who Preferred Products	E	20	20	20	

Using $\frac{(O-E)^2}{E}$ for each of the categories to calculate χ^2, we get:

$$\chi^2 = \frac{(30-20)^2}{20} + \frac{(18-20)^2}{20} + \frac{(12-20)^2}{20}$$

$$= \frac{10^2}{20} + \frac{(-2)^2}{20} + \frac{(-8)^2}{20} = \frac{100+4+64}{20} = 8.4$$

df = number of categories − 1 = 3 − 1 = 2

There is no way of predicting beforehand any trend in preferences should H_0 be rejected, so a two-tailed test is used. In Table G.7, for $p = 0.05$, two-tailed, $df = 2$, chi square = 5.99. For $p = 0.02$, chi square = 7.82. For $p = 0.01$, chi square = 9.21.

Our calculated chi-square value (8.4) is greater than the $p = 0.02$ value, less than the $p = 0.01$ value. We thus reject H_0 (the hypothesis that states that the observed frequencies equal those expected should A, B, and C be equally liked). We reject H_0 at the 2% level but not at the 1% level. Thus, the trend is significant ($p < 0.02$); the results seen in our sample represent a trend in the population.

Example 2

In a consumer study investigating preferences for "organic" versus regular prune juice, 50 males and 50 females were chosen at a supermarket and asked which of the two they preferred. Ten males and 20 females preferred the organic prune juice. Was there a significant difference in the preferences between males and females?

Preferences for organic and regular prune juice for men and women are given in the following matrix:

	Organic	Regular	
Males	A 10	B 40	$(A + B)$ 50
Females	C 20	D 30	$(C + D)$ 50
	$(A + C)$ 30	$(B + D)$ 70	Total 100

$df = (2 - 1)(2 - 1) = 1$

H_0 states that there are not significant differences in the pattern of preference between males and females.

The numbers of males and females were chosen beforehand; thus one margin is fixed. Hence the correction factor $K = N/4$.

$$\text{Chi-square} = \frac{N\left(|AD - BC| - \frac{N}{4}\right)^2}{(A+B)(C+D)(A+C)(B+D)}$$

$$= \frac{100\left(|10 \times 30 - 40 \times 20| - \frac{100}{4}\right)^2}{50 \times 50 \times 30 \times 70}$$

$$= \frac{100(500 - 25)^2}{50 \times 50 \times 30 \times 70} = \frac{100 \times 475^2}{50 \times 50 \times 30 \times 70}$$

$$= \frac{100 \times 225{,}625}{50 \times 50 \times 30 \times 70} = 4.30$$

Because we cannot predict any pattern of preferences, we will use a two-tailed test. The chi-square values in Table G.7 (two-tailed, $df = 1$) are 3.84 at $p = 0.05$ and 5.41 at $p = 0.02$. Our value (4.3) exceeds the value in the table at $p = 0.05$, so we reject H_0. By inspection, we see that males in the sample favor the regular prune juice more than females. Because this trend is significant ($p < 0.05$), it represents a trend in the population and not just a chance effect of sampling.

7

Student's *t* Test

7.1 WHAT IS A *t* TEST?

We are now about to examine our first parametric test, Student's *t* test. It is designed to determine whether or not the means of two samples of data are significantly different. The data must be on an interval or ratio scale and the samples must be drawn from populations where the data are normally distributed. Basically, the *t* test investigates whether the differences between means observed in the samples of data indicate that these samples come from different populations (reject H_0) or whether the differences are only chance differences (not reject H_0) and the samples come from the same population.

There are several variations of the *t* test:

Two-sample difference tests:

related samples (also called paired or dependent): Tests for a significant difference between the means of two related samples (i.e., the scores from the same judges under different conditions).

independent samples (also called unpaired or unrelated): Tests for a significant difference between the means of two unrelated samples (i.e., the scores from the two samples under different conditions come from separate judges).

One-sample test: Tests whether a sample with a given mean came from a population with a known mean; is a sample mean significantly different from the population mean? The meaning of "one sample" here is not the same as the meaning of "one sample" when applied to the binomial or chi-square tests, where a one-sample test means to take one sample and use it to make inferences about the population. However, it is convenient to lump one-sample binomial and chi-square tests together with the one-sample *t* tests on the Key to Statistical Tables. It must just be borne in mind that "one sample" can mean more than one thing.

Look up these three *t* tests on the Key to Statistical Tables to see their relationship to the other tests.

Who Is Student?

Student is the pen name of William Sealy Gosset (1876–1937), who worked for the Guinness Brewery in Dublin. He published details of his test in *Biometrika* in 1908 under a pen name because the brewery did not want their rivals to realize that they were using statistics.

t Distribution

The *t* distribution resembles the normal distribution except that it is flatter (more platykurtic) and its shape alters with the size of the sample (actually the number of degrees of freedom).

7.2 THE ONE-SAMPLE *t* TEST: COMPUTATION

The one-sample *t* test is rarely used in sensory analysis and behavioral sciences, but it is included here for the sake of completion. In the one-sample *t* test we are testing whether a sample with a given mean, $\bar{X}$, came from a population with a given mean μ. In other words, is the mean of a sample ($\bar{X}$) significantly different from the mean of the population (μ)?; *t* is calculated using the formula

$$t = \frac{\bar{X} - \mu}{S_{\bar{X}}}$$

where

$\bar{X}$ = mean of the sample
μ = mean of the population
$S_{\bar{X}}$ = estimate from the sample of the standard error of the mean.

As discussed in the section on *z* tests, the standard error is the standard deviation of the sampling distribution, a theoretical distribution of means of samples of size N, drawn from the population of mean μ. This estimate is obtained from the sample S and N using the formula $S_{\bar{X}} = S/\sqrt{N}$.

Essentially, the calculation is one of seeing whether the difference between the means $(X - \mu)$ is large compared to some measure of how the scores might vary by chance [i.e., how they are spread $(S_{\bar{X}})$]. Should two means be 90 and 100, what would be a good basis for deciding whether the difference between them is really substantial? Should the scores making up the mean of 90 range from 89 to 91 and those around 100 range from 99 to 100, it would be reasonable to say that the means 90 and 100 came from two completely different groups of scores. The two sets of data do not even overlap. On the other hand, should the scores making up the mean of 90 range from 0 to 200 and the scores around 100 also range from 0 to 200, it would be reasonable to say that the difference between the means is a mere insignificant chance difference; a difference of only 10 is small when the scores in each group of numbers themselves vary by 200.

This is the general logic of difference tests. Is the difference between the means large compared to the general variation of the scores you would get anyway? The spread of scores is usually measured by a standard deviation or variance value. You can see that the formula for t is a ratio of the difference between the means to a measure of how the scores tend to spread out $(S_{\bar{X}})$ when they are collected (the general degree of chance variation that occurs as the data are collected). The same logic is used, later, for analysis of variance (Chapter 8). Note that this is very like the calculation of a z value:

$$z = \frac{x - \mu}{\sigma}$$ difference between a score and the population mean, counted in standard deviations

$$t = \frac{\bar{X} - \mu}{S/\sqrt{N}}$$ difference between a sample mean and the population mean, counted in estimated standard errors

Table G.1 (area under normal curve) would be used to find the probability of getting a value of z or greater. With a t test, we are doing essentially the same thing, except that we are now using the area under a t distribution curve. The relationship between z tests and t tests will be discussed in more detail in Section 7.10. First, however, we will concentrate on the computations required for t tests.

Looking at the formula for t, it can be seen that if the mean of the sample is very different from the population mean being tested, $\bar{X} - \mu$ and hence t will be large. The question becomes one of whether the difference between the sample mean and the population mean, and hence t, is so great as to indicate that the sample *was not* drawn from the population (reject H_0). Or was the difference between the two means (and hence t) sufficiently small to be only a chance difference (compatible with H_0), showing that the sample *was* drawn from the population? The greater the difference between μ and $\bar{X}$, the greater the calculated t value. Table G.8 indicates how large a value of t can occur by chance (H_0 being true) for various significance levels. If our calculated value of t is greater than or equal to the value in Table G.8, H_0 is rejected.

Associated with this t value are a number of degrees of freedom; this is important when comparing the calculated value with a value in the tables, the theoretically greatest value on H_0. The number of degrees of freedom is given by

$$df = \text{number of subjects in the sample} - 1 = N - 1$$

The t value calculated can be positive or negative depending on whether $\bar{X}$ is greater or smaller than μ, but the sign is ignored when looking up Table G.8. Thus we are ready to calculate t, using the formula

$$t = \frac{\bar{X} - \mu}{S/\sqrt{N}}$$

7.3 WORKED EXAMPLE: ONE-SAMPLE t TEST

The average rate of packing of pumpkins by regular workers was 16 per minute. However, a group of 10 packers who had newly joined the packing plant claimed to be significantly faster than average. They had a mean rate of 17 per minute with a standard deviation, $S = 3.13$. Were they significantly faster than average?

Our sample of new packers had $N = 10$, $\bar{X} = 17$, $S = 3.13$. The population had a mean value $\mu = 16$. Thus

$$t = \frac{\bar{X} - \mu}{S/\sqrt{N}} = \frac{17 - 16}{3.13/\sqrt{10}} = 1.01 \qquad \text{where } df = N - 1 = 9$$

We now need to test the significance of our calculated t value. We do this by comparing our value with the values given in Table G.8. These values are given for the appropriate number of degrees of freedom (df in the left-hand column) and appropriate one- or two-tailed levels of significance (select appropriate column of numbers). Because we did not know in advance whether the mean of the sample would be larger or smaller than that of the population, should we reject H_0 we will use a two-tailed test.

From Table G.8, our t value of 1.01 can be seen to be neither greater than nor equal to any of the tabular values for $df = 9$. It is not even greater than 1.383, the value for $p = 0.2$, two-tailed. Thus the data do not indicate any differences between the mean of the sample and the population. The 10 pumpkin packers in the sample were no different from the general population of pumpkin packers; they were drawn from that population.

Note that we also assumed that the data came from a normally distributed population. Also, note that sometimes t will be a negative value (when $\bar{X}$ is smaller than μ); the sign of t only indicates which mean is larger. When using Table G.8, ignore a negative sign; 1.01 and −1.01 are both treated as 1.01.

7.4 THE TWO-SAMPLE t TEST, RELATED SAMPLES: COMPUTATION

Besides using the t test to determine whether a given sample was drawn from a given population (one-sample test), it can also be used to determine whether two samples were drawn from the same population (means not significantly different) or from different populations (means significantly different). The latter applications are two-sample difference tests. The two-sample difference test can either have a related-samples design (same judges in two conditions) or an independent-samples design (different judges in each condition). We will deal with the related-samples design first.

Once again the assumption is made that the two samples are drawn from populations with normal distributions. The related-samples t test is similar in approach to the single-sample t test. The procedure adopted is to take the difference in scores between the two conditions for each judge. This gives us a sample of difference scores. Should there be no difference between the two conditions, the mean of these difference scores will be near zero. The more different the two conditions, the bigger the difference between the scores (the less near to zero will be the mean). The t test is used to determine whether these difference scores are near zero or not—whether they could have come from a population of scores with a mean of zero. This is essentially the one-sample test, where we are testing to see whether a sample of scores (difference scores) came from a population of scores whose mean is zero. The null hypothesis states that there is no difference between the two means (of the difference scores and the population mean of zero), thus that the mean of the difference scores is zero, thus that there is no difference between the means of the two samples of scores we are considering.

We adapt the formula for the one-sample t test accordingly:

$$t = \frac{\bar{X} - \mu}{S/\sqrt{N}}$$

Here the X scores under consideration are difference scores (d), so $\bar{X}$ is replaced by $\bar{d}$. The mean of the population (μ) we are testing is zero. So

$$t = \frac{\bar{d}}{S/\sqrt{N}}$$

Here S for the difference scores is given by the usual formula, using d values.

$$S = \sqrt{\frac{\Sigma d^2 - \frac{(\Sigma d)^2}{N}}{N - 1}}$$

Here the degrees of freedom are given by the expression

$$df = N - 1$$

where N is the number of people who were tested in the two conditions (number of difference scores).

We now have all the formulas we need, so we will demonstrate the related-samples t test with an example.

7.5 WORKED EXAMPLE: TWO-SAMPLE t TEST, RELATED SAMPLES

It was thought that viewing certain meats under red light might enhance judges' preference for meat. The same cuts of meat were viewed by judges under red and white light. They were rated on a complex preference scale which gave scores that we will assume came from a population of scores that were normally distributed. Thus we can use the t test. The results are as follows:

	Preference score			
Subject	Under white light	Under red light	Difference, d	d^2
1	20	22	2	4
2	18	19	1	1
3	19	17	-2	4
4	22	18	-4	16
5	17	21	4	16
6	20	23	3	9
7	19	19	0	0
8	16	20	4	16
9	21	22	1	1
10	19	20	1	1
$N = 10$			$\Sigma d = 10$ $\bar{d} = 1$	$\Sigma d^2 = 68$

We now substitute these values in the formulas for S and t, as follows:

$$S = \sqrt{\frac{\Sigma d^2 - \dfrac{(\Sigma d)^2}{N}}{N - 1}} = \sqrt{\frac{68 - \dfrac{100}{10}}{10 - 1}} = \sqrt{\frac{58}{9}}$$

Thus

$$t = \frac{\bar{d}}{S/\sqrt{N}} = \frac{1}{\sqrt{\frac{58}{9}} \Big/ \sqrt{10}} = \sqrt{\frac{90}{58}} = 1.25 \qquad df = 9$$

The null hypothesis states that there is no difference between the preference scores under red and white light—that the mean of the difference (d) scores is zero. The alternative could be that scores under a white light are greater than or smaller than scores under a red light. Should H_0 be rejected, there is no way of knowing whether the scores under red light or white light will be greater, so a two-tailed test is used.

The t value in Table G.8 for $df = 9$, two-tailed, $p = 0.05$, is 2.262. Our calculated value of 1.25 is smaller than this value in the tables and even the value for $p = 0.2$. So we cannot reject H_0, even at $p = 0.2$. Thus our data do not indicate that the d values are significantly different from zero. The preference scores under red light are not significantly different from those under white light. We indicate our level of significance by writing ($p > 0.2$), meaning: H_0 not rejected even at the 0.2 level.

7.6 THE TWO-SAMPLE *t* TEST, INDEPENDENT SAMPLES: COMPUTATION

The related-samples difference test examines whether the means of two sets of data differ significantly, the data coming from the same subjects under two conditions. The independent-samples test examines whether the means of two sets of data differ significantly, the data coming from two different sets of subjects under different conditions.

Again the test compares a difference between two means with some measure (standard error) of how the scores vary as they are collected, a chance level. It can be seen that the formulas have the same general pattern:

$$t = \frac{\text{difference between means}}{\text{standard error}}$$

For one sample,

$$t = \frac{\bar{X} - \mu}{S/\sqrt{N}}$$

For two related samples

$$t = \frac{\bar{d}}{S/\sqrt{N}}$$

where $S/\sqrt{N}$ is the standard error of the mean.

For two independent samples

$$t = \frac{\text{difference between means}}{\text{standard error of difference}}$$

$$= \frac{\bar{X} - \bar{Y}}{\text{standard error of difference}}$$

where $\bar{X}$ and $\bar{Y}$ are the mean scores for the two independent samples.

Again, if $\bar{X} - \bar{Y}$ is large compared to the standard error of the difference, t will be large. The standard error term can be seen as a measure of the chance random variation or experimental error involved in the experiment. So if differences between the two samples $(\bar{X} - \bar{Y})$ are large compared to experimental error (standard error term), the two samples are significantly different (t is large).

In this case,

$$df = N_1 + N_2 - 2$$

where N_1 and N_2 are the sizes of the two independent samples. N_1 for X scores, N_2 for Y scores. The standard error of difference is given by the formula

$$\sqrt{\frac{\Sigma (X - \bar{X})^2 + \Sigma (Y - \bar{Y})^2}{N_1 + N_2 - 2} \left(\frac{1}{N_1} + \frac{1}{N_2}\right)}$$

Thus

$$t = \frac{\bar{X} - \bar{Y}}{\sqrt{\frac{\Sigma (X - \bar{X})^2 + \Sigma (Y - \bar{Y})^2}{N_1 + N_2 - 2} \left(\frac{1}{N_1} + \frac{1}{N_2}\right)}}$$

This formula is inconvenient, so we rearrange it into a computationally simpler form:

$$t = \frac{(\bar{X} - \bar{Y}) \sqrt{\frac{N_1 N_2 (N_1 + N_2 - 2)}{N_1 + N_2}}}{\sqrt{\Sigma X^2 - \frac{(\Sigma X)^2}{N_1} + \Sigma Y^2 - \frac{(\Sigma Y)^2}{N_2}}}$$

If our calculated t value exceeds or is equal to the t value given in Table G.8 (for the appropriate df and chosen significance level), we reject H_0; the calculated t has exceeded the largest t value that can be obtained by H_0 (H_0 says that there is no difference between the two means; the samples were taken from the same population). Therefore, we reject the idea that the means came from the same population. The difference in means is a significant difference; it indicates that the samples came from different populations; there are real differences between X and Y scores.

The formula above is a considerable simplication. Values of X and Y and X^2 and Y^2 must be calculated. The expression

$$\sqrt{\frac{N_1 N_2 (N_1 + N_2 - 2)}{N_1 + N_2}}$$

is given in Table G.9 for when N_1 and N_2 are large (greater than 10). For example: When $N_1 = 34$ and $N_2 = 10$, the value for the expression is 18.02. The table can be used either way around (e.g., $N_2 = 34, N_1 = 10$).

At this point it is well to compare the related and independent samples designs. Both t formulas take the form

$$t = \frac{\text{difference between means}}{\text{standard error value}}$$

For a related-samples design, the standard error is computed in the normal way using difference (d) scores. For an independent-samples design it is calculated using a different formula (the standard error of the difference) which gives a larger value. This larger denominator will result in a smaller value of t; it will therefore be less likely to exceed the appropriate value in Table G.8. So for a related-samples design it is easier to reject H_0. This makes sense. In the related-samples design, the differences between the two samples of scores will be due only to the variables in the experiment; they will *not* be due to the fact that the people in the two samples are different anyway. We can accept a smaller difference between the means as an indication of a significant difference. For an independent-samples design, the means must be more different to allow for the fact that the difference could be due to individual differences between the sets of subjects in each of the two samples, as well as to the different experimental conditions.

7.7 WORKED EXAMPLE: TWO-SAMPLE *t* TEST, INDEPENDENT SAMPLES

To illustrate the use of an independent or unrelated samples t test, we will now consider an example.

Are trained judges on a panel better than untrained judges?

Trained and untrained judges were given a series of tests and the average scores over the tests are given below for each subject. Note that for an unrelated-samples test, the samples need not be exactly the same size.

Untrained subjects, X	Trained subjects, Y
3	5
5	8
6	9
4	6
3	12
3	9
4	7
	6
$\Sigma X = 28$	$\Sigma Y = 62$
$\Sigma X^2 = 120$	$\Sigma Y^2 = 516$
$N_1 = 7$	$N_2 = 8$
$\bar{X} = 4$	$\bar{Y} = 7.75$

We have the requisite values from the data, so we now fill in the formula:

$$t = \frac{(\bar{X} - \bar{Y})\sqrt{\dfrac{N_1 N_2 (N_1 + N_2 - 2)}{N_1 + N_2}}}{\sqrt{\Sigma X^2 - \dfrac{(\Sigma X)^2}{N_1} + \Sigma Y^2 - \dfrac{(\Sigma Y)^2}{N_2}}}$$

Here $N_1 = 7$ and $N_2 = 8$; these values are sufficiently small for Table G.9 not to be used. We now fill in the values for various parts of the formula:

$$(\bar{X} - \bar{Y}) = 4 - 7.75 = -3.75$$

$$\sqrt{\frac{N_1 N_2 (N_1 + N_2 - 2)}{N_1 + N_2}} = \sqrt{\frac{7 \times 8(7 + 8 - 2)}{7 + 8}} = \sqrt{\frac{7 \times 8 \times 13}{15}} = 6.97$$

$$\sqrt{\Sigma X^2 - \frac{(\Sigma X)^2}{N_1} + \Sigma Y^2 - \frac{(\Sigma Y)^2}{N_2}} = \sqrt{120 - \frac{28^2}{7} + 516 - \frac{62^2}{8}}$$

$$= \sqrt{43.5}$$

Thus

$$t = \frac{-3.75 \times 6.97}{\sqrt{43.5}} = -3.96$$

and

$$df = N_1 + N_2 - 2 = 7 + 8 - 2 = 13$$

We now compare our calculated t value with the values in Table G.8 for df = 13. Shall we use a one- or a two-tailed test? We could argue that we can predict before the experiment that if H_0 were rejected, it would be because the trained judges were better (scored higher) than the untrained judges. Thus we could use a one-tailed test. This seems quite a reasonable expectation. On the other hand, experience with behavioral measurement tells us that this is not always the case. In some cases untrained people can do better; perhaps they try harder or perhaps training was actually a hindrance. So perhaps we should use a two-tailed test. We leave the decision to you, dear reader. Personally, I will choose a two-tailed test because I am cautious and want to make sure that the difference I detect is really there (I don't want to make a Type I error); therefore, I will use the tougher two-tailed test. The t values in the table are higher for two-tailed tests for a given significance level, so the difference between the means has to be bigger for it to be exceeded. I choose a two-tailed test, but you can see that such a choice is open to discussion.

In Table G.8 the two-tailed values for df = 13 are 2.160 ($p = 0.05$), 3.012 ($p = 0.01$), and 4.221 ($p = 0.001$). Our value, 3.96 (remember that we ignore the negative sign when using the table), exceeds the 1% value but not the 0.1% value. Thus we reject H_0 with our probability being less than 1%, indicated by $p < 0.01$.

So the trained judges had a significantly higher mean score ($p < 0.01$) than the untrained judges. The two samples of data were not drawn from the same population.

7.8 ASSUMPTIONS AND WHEN TO USE THE t TEST

It is important to note the assumptions made when deriving these tests and the resulting restrictions on their use; we have already mentioned some of these, such as data coming from normally distributed populations. However, it is useful to have them stated formally, even though common sense would dictate that one never did some of the things that are forbidden.

t tests can be used for any size sample, although when $N > 30$, the t distribution and the normal distribution become so similar that a z test can be used to determine whether means are different. This will be discussed in more depth later (Section 7.10). In fact, the t test can be completely replaced by the more general analysis-of-variance procedure. We only need study the t test because it is still used by many researchers and because it is the basis of one of the multiple-comparison tests, Fisher's LSD test (Section 9.5), which we will encounter later.

The assumptions for *t* tests are as follows:

1. Related-samples test
   a. The *d* scores come from a population of *d* scores which are normally distributed, and this implies that data must be on an interval or ratio scale.
   b. The *d* scores are randomly sampled and are independent of each other. Each *d* score must come from a different randomly sampled person; no person can contribute more than one score.
   c. *d* scores are the differences between matched pairs of subjects. You would hardly take one person's score from that of a completely different person.

For a one-sample test, the first two of these assumptions must be fulfilled, except that the scores are now actual scores rather than differences between scores (*d* scores).

2. Independent-samples test
   a. Both samples are random samples taken from their respective populations.
   b. The two samples must be independent of each other, and all scores within a sample must also be independent. (All scores must come from separate judges; no judge can contribute more than one score.)
   c. Both samples of scores must come from normally distributed populations of scores and this implies that data must be on an interval or ratio scale.
   d. Both samples must come from populations with the same variance (homoscedasticity).

Naturally, data that come from a normally distributed population must be continuous (see Section 2.6).

These assumptions are important when deciding whether results can be analyzed using a *t* test. For instance, how do we know that the data were sampled from a population that was normally distributed? There are tests which can be applied to the data but they will not be dealt with here. We can, however, use a useful rule of thumb for behavioral data. Should the numbers we get be the result of a lot of interacting factors, the distribution in the population will generally be normal. Degree of liking for a given food is determined by many factors (age, parents' eating habits, place of residence, time of day, recent history, contact with other cultures, etc.) and so is likely to be normally distributed in the population. Age in the general population, however, is determined by only one factor, day of birth. It is certainly not normally distributed. The older the age, the fewer people there are; the most common (mode) age is somewhere around "just born" (see Section 2.7). Ranked data do not fall on a normal distribution either; there is one person for each score (one 1st, one 2nd, one 3rd, etc.), so the distribution is a horizontal line. These two examples are illustrated in Figure 7.1.

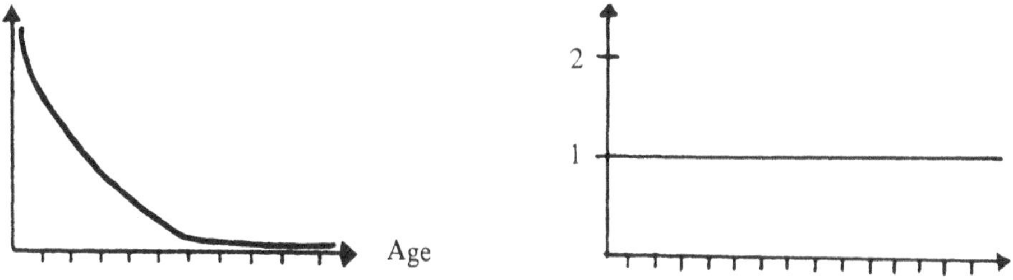

Figure 7.1 Age distribution (left), and ranked data (right).

Homoscedasticity (equal variance) can also be a problem and should be considered carefully. How do we know whether both our samples of data come from populations with equal variance? Generally, we assume that they do, unless we have good reason to assume they do not. Assume that we test one sample of subjects and then test the second sample of subjects under the same conditions, except for a change in the variable we are investigating. It is reasonable to assume that the scatter (variance) of their scores will be the same in both cases (unless we think our variable under study itself could produce a change in variance). On the other hand, should a mild distraction interfere with the second group of subjects (one judge screaming to the others that the scores will vary much more than usual, or an earthquake which may frighten judges so that they score semirandomly all the way along their scale), we might surmise that the variance of that second group may be different from that of the first group. We may then decide to retest. There are sometimes good reasons for suspecting that the assumption of homoscedasticity is broken when subjects use rating scales; these will be considered later (see Section 7.11).

For sensory testing, a related-samples *t* test is preferable to an independent-samples *t* test, because it minimizes the effects of differences due to the fact that different subjects are tested in each condition. These differences may be due to differences in subjects per se or due to their idiosyncratic use of the measuring procedures employed, such as intensity scaling. Thus unrelated-samples designs will be less common in this field. However, should all the data come entirely from only one judge, an unrelated-samples design is appropriate, because the scores are just as independent between the samples as they are within the samples. This point is discussed further in Section. 16.20.

7.9 THE INDEPENDENT-SAMPLES *t* TEST WITH UNEQUAL VARIANCES

For the independent-samples *t* test, it is assumed that the variances of the populations, from which the scores in the samples came, are the same; this property is called *homoscedasticity*. If the variances are not the same, the computation of *t* is more complex. In fact, in this case, we have only an approximate test. Once again the value of t is given by

$$t = \frac{\bar{X} - \bar{Y}}{\text{standard error of difference}}$$

$$\text{Standard error of difference} = \sqrt{\frac{S_X^2}{N_1} + \frac{S_Y^2}{N_2}}$$

where S_X and S_Y are standard deviation values (estimate of population value from the sample) for the X and Y samples, whose sizes are N_1 and N_2, respectively.

If the two distributions are normal, the number of degrees of freedom is given by a complex expression.

$$df = \frac{\left(\frac{S_X^2}{N_1} + \frac{S_Y^2}{N_2}\right)^2}{\frac{\left(\frac{S_X^2}{N_1}\right)^2}{N_1 - 1} + \frac{\left(\frac{S_Y^2}{N_2}\right)^2}{N_2 - 1}} - 2$$

This value will not always be a whole number; it may be a fraction and Table G.8 has to be interpolated.

Most computer programs for t tests will give, for two-sample tests, a computation for

Related samples
Independent samples, equal variance
Independent samples, unequal variance

There are procedures for testing whether the variances are equal or not, but they will not be dealt with here. On the whole, however, unless there is good reason for expecting different variances, they may be assumed to be the same.

7.10 FURTHER STATISTICAL THEORY: THE RELATIONSHIP BETWEEN *t* DISTRIBUTIONS AND NORMAL DISTRIBUTIONS, *t* TESTS AND *z* TESTS

At this point we will examine in more detail some of the theory behind the t test. We will look at the relation between the t distribution and the normal distribution; we will consider sampling distributions, the relationship between t and z, and how the assumption of normality is involved in the t test. The aim here is to give a background to the t test. Should you be struggling somewhat with the text, return to this section when you are more familiar with t tests.

In Section 4.2 concerning z tests, we noted that for a normally distributed population, a given score (X) can be represented in terms of how many standard deviations it is above or below the mean (μ). This is called a z score. If X has a z score of 1.5, it is 1.5 standard deviations above the mean.

Table G.1, a table of probabilities for the normal distribution, can be used to determine from the z score, the chance probability (if H_0 is true) of getting a score so high or higher in the population (or a score so low or even lower). This exercise is called a z test. If this probability is very low, it is unlikely that the large difference between X and μ is merely due to chance. It is more likely that X does not belong to the population of which μ is the mean. We reject a null hypothesis saying that X and μ are from the same population.

In Section 4.5 we repeated this line of reasoning except that instead of examining a score (X) and a normally distributed population with a mean of μ, we examined a mean of a sample ($\bar{X}$) and a normally distributed *sampling distribution* with a mean μ. We used Table G.1 in the same way to see the probability of getting a sample mean as high as $\bar{X}$ or higher on the null hypothesis. In effect, we used Table G.1 to determine whether $\bar{X}$ was significantly different from μ, whether the sample mean actually belonged to the sampling distribution, or in other words, whether the sample was drawn from that population.

In the single-score case (X), the population (mean = μ, standard deviation = σ) must be normal for Table G.1 to be used. In the sample mean case ($\bar{X}$), the sampling distribution will always be normal (by the central limit theorem) regardless of the population, as long as the sample size (N) is not too small. The sampling distribution has the same mean (μ) as the population, but being a distribution of means, its standard deviation (called the standard error $\sigma/\sqrt{N}$) is smaller than that of the population (σ). However, to calculate a z score and hence use Table G.1, the standard deviation of the population (σ) must be known. It must be known to be able to determine the number of standard deviations (σ) a score (X) is above μ or the number of standard errors ($\sigma/\sqrt{N}$) a sample mean ($\bar{X}$) is above μ. And here is the difficulty.

We rarely know σ and thus estimate it from the sample using S (see Section 2.4). Thus to determine whether $\bar{X}$ and μ are significantly different, we do not have a value for the standard error ($\sigma/\sqrt{N}$) and cannot calculate a z score. Instead, we have only ($S/\sqrt{N}$). What we calculate is now called a t score. A t score is the number of *estimated* standard errors a sample mean ($\bar{X}$) is above μ $\left(t = \dfrac{(\bar{X} - \mu)}{S/\sqrt{N}}\right)$. Furthermore, the sampling distribution will no longer be a normal distribution because S is not an exact value for the population standard deviation; it will be a t distribution. Thus we cannot calculate probabilities using a z score with Table G.1 and the normal distribution, we have to use a t score with Table G.8 and the t distribution. And this is just what we did for the one-sample t test. Table G.8, however, is set up a little differently from Table G.1 because although there is only one normal distribution, there are many t distributions, one for each number of degrees of freedom. Thus:

For a given population: z is the difference between a given score (X) and the population mean (μ), counted in standard deviations (σ) for that population.

$$z = \frac{X - \mu}{\sigma}$$

For a given sampling distribution: z is the difference between a given sample mean ($\bar{X}$) and the population mean (μ), counted in standard errors, that is, counted in standard deviations ($\sigma/\sqrt{N}$) of the sampling distribution.

$$z = \frac{\bar{X} - \mu}{\sigma/\sqrt{N}}$$

For a given sampling distribution: t is the difference between a given sample mean ($\bar{X}$) and the population mean (μ), counted in estimated standard errors, that is, counted in estimated standard deviations ($S/\sqrt{N}$) of the sampling distribution.

$$t = \frac{\bar{X} - \mu}{S/\sqrt{N}}$$

At this point, we can see where the assumption of normality fits in. A z value can be used to compute probabilities for means on the normally distributed sampling distribution. For the sampling distribution to be normal, the population can have any distribution. A t value can be used to compute probabilities for means on a sampling distribution, which is distributed as a t distribution. For the sampling distribution to have a t distribution, the population *must* have a normal distribution; it *cannot* be just anything. This is where the assumption that samples must be drawn from populations that are distributed normally is introduced. The same type of argument also applies to analysis of variance (Chapter 8).

The t distribution is symmetrical and bell-shaped like the normal distribution but has a greater spread. Like the chi-square distribution, the t distribution depends on the number of degrees of freedom. Instead of giving distributions for each number of degrees of freedom, selected t values are quoted for various degrees of freedom at given probability values (0.05, 0.01, etc.). These are the values given in Table G.8.

Once sample sizes get large ($N > 30$), the difference between σ and S diminishes. The t distribution then approximates to the z distribution (if $S = \sigma$ then $t = z$). The normal distribution (Table G.1) can then be used to compute probabilities for t values.

It is interesting to note that some statisticians claim that if N is fairly large ($N > 10$) and the samples are the same size, the t test is robust against small deviations from normality and the assumptions of equal variance.

Having now dealt with the sampling distribution, it is worth considering *confidence intervals* in Appendix D; they provide another way of looking at difference testing.

7.11 CAN YOU USE *t* TESTS TO ANALYZE SCALING DATA?

The application of parametric t tests to data obtained from rating scales should be considered carefully. There are two main assumptions required before t tests can be used. First, the sample of data must have been drawn from a population where such data are normally distributed. Second, the variances of the two populations from which the samples were taken must be equal (homoscedasticity). Before proceeding, the readers should remind themselves of the definition of ratio, interval, and ordinal scales (Section 2.6).

There is a lot of evidence to show that judges are not very skillful at using rating scales. It is unwise to expect them to use them as equal-interval scales. For example, judges are reluctant to use the ends of a scale. It is as though the judge is more reluctant to pass from the penultimate category to the end category than to move from the middle, say, to an adjoining category. It is as though the psychological distances between categories are not the same at the end of the scale as in the center. Thus the judges' responses are not on an interval scale. Perhaps the most that can be said is that the high scale value is, at least, higher than a low scale value; the data can be ranked. Thus if scaling data are merely ordinal, it cannot be normally distributed; scaled data break the assumption of normality. This argument applies to category scales (e.g., select one category on a 9-point category scale ranging from, say, very sweet = 9 to very unsweet = 1) and graphic scales (e.g., a 10-cm line labeled at each end with say "very sweet" and "very unsweet"; the judge indicates his response by marking the line at an appropriate place). Magnitude estimation is slightly different. The judge is given a standard sample and told to score it as 10. He then scores other samples in ratio to the standard (if twice as strong it is scored 20, half as strong 5, three times as strong 30). The idea is that stimuli can be given any score and not be restrained by categories. Unfortunately, there is a tendency for the subject to respond mainly in simple multiples of the standard's value (standard = 10, tendency to respond 5, 10, 20, 30, etc.; standard = 100, tendency to respond 50, 100, 200, 300, etc.). This would mean that the population would not be normally distributed but would be a set of minidistributions around those favored numbers. Thus for all these rating procedures there are doubts about whether the data are drawn from normally distributed populations.

Homoscedasticity can also be a problem. Imagine that a point in the center of the scale is the mean of a sample of data. The scores in this sample are scattered around the mean value with a given variance (let us hope that the distribution is normal). Now imagine a mean value right at the end of the scale. The scatter at this point on the scale will not be the same. The scores can spread toward the center of the scale as before but there is not sufficient room on the scale for them to spread outward, toward the end of the scale. Thus the spread or variance of the sample will not be so great at the end of the scale. To compare the mean values from the center and from the end of the scale will be to compare means of samples drawn from populations with different variances, breaking the assumption of homoscedasticity. Furthermore, the skewing of the distribution for a mean at the end of the scale means that it is no longer normal. It also biases the mean. Means are a suitable measure of central tendency for symmetrical distributions but not for skewed distributions. A test which compares means would not seem to be the best statistical analysis when the means are biased.

Should *t* tests be used, they will be used with the assumptions broken. Thus they cannot be used with the precision intended; they can only be used in a rough-and-ready manner. In practice, it means that the values in Table G.8 can only be treated as rough values to give an indication of whether differences are significant. This is a further argument for not attributing too much importance to an exact 5% cutoff point for rejecting H_0. These lines of argument are somewhat controversial because, in general, parametric tests are nearly always applied to data from rating scales. It is often said that *t* tests are sufficiently robust that breaking the assumptions is not important. However, exactly how robust *t* tests have to be seems rather uncertain.

So what is one to do? There are two alternatives. The first is to use the parametric *t* test and regard the results as approximate. The second is to convert the scale values into ranks and use a nonparametric test for ranked data (see the Key to Statistical Tables). The appropriate test for a related samples design is the Wilcoxon test, and for the independent samples design it is the Mann-Whitney U test. If you take both approaches, the data should generally agree. The *t* tests may be more powerful than the nonparametric tests, but both should give a similar result—unless you use a strict 5% cutoff. Then one test may have an H_0 probability of 4% and the other 6% and you may be led to believe that the first test rejects H_0 and the second does not.

The arguments that we have discussed for *t* tests also apply to analysis of variance (Chapter 8). For further discussion of this topic, read "Some assumptions and difficulties with common statistics," *Food Technology*, 1982, Vol. 36, pp. 75-82.

Remember that a *t* test (or analysis of variance) is only one way of looking at the data and that this way can be distorted because the statistical assumptions may have been broken. If differences are found using a *t* test, check to see whether a majority of judges found these differences (binomial test). Do these differences occur if we use a nonparametric test such as the Wilcoxon or Mann-Whitney U test? A set of data from an experiment is like a statue. To look at it from only one direction can give a distorted view; you should walk around it and examine it from several angles.

7.12 ADDITIONAL WORKED EXAMPLES: *t* TESTS

Example 1

A panel of highly trained French judges scored all Californian Chenin Blanc wines produced in 1986 on a rather complex 9-point scale representing varietal nature (mean for all wines = 8.2). Permission was sought for a vintage 1975 wine derived from varietal grapes and sugar cane grown in Jamaica to be called Chenin Blanc. It was decided that if the "varietal nature" scores of the Jamaican wine were not significantly different from the California Chenin Blanc scores, permission would be given. Samples of the Jamaican wine were scored for varietal nature as follows: 8.0, 6.2, 8.1, 6.1, 6.5, 8.2, 8.0, 6.9, 8.9, and 8.8. Winemakers and analytical chemists were not able to predict in advance whether the mean of the Jamaican wine scores, if it varied from 8.2, would be higher or lower. Were the varietal scores of the Jamaican wine different from California Chenin Blanc scores? You may assume that all populations are normal.

This is a one-sample *t* test; a sample of scores is being tested to see if it comes from a given population. Because it was impossible to predict in advance whether the Jamaican wine would have higher or lower scores, if H_0 were not true, a two-tailed test will be used.

For a one-sample test

$$t = \frac{\bar{X} - \mu}{S/\sqrt{N}} \qquad \text{where } \mu = 8.2$$

and

$$S = \sqrt{\frac{\sum X^2 - \frac{(\sum X)^2}{N}}{N - 1}}$$

The varietal scores for the Jamaican wines were as follows:

X	$X2$
8.0	64.0
6.2	38.44
8.1	65.61
6.1	37.21
6.5	42.25
8.2	67.24
8.0	64.0
6.9	47.61
8.9	79.21
8.8	77.44
$\Sigma X = 75.7$	$\Sigma X^2 = 583.01$

$$N = 10$$
$$\bar{X} = 7.57$$
$$(\Sigma X)^2 = 5730.49$$

Now the values can be filled in to the formula for t.

$$t = \frac{\bar{X} - \mu}{S/\sqrt{N}}$$

$$S = \sqrt{\frac{\Sigma X^2 - \frac{(\Sigma X)^2}{N}}{N - 1}} = \sqrt{\frac{583.01 - \frac{5730.49}{10}}{10 - 1}}$$

$$= \sqrt{\frac{583.01 - 573.049}{10 - 1}} = \sqrt{1.107} = 1.05$$

$$\bar{X} - \mu = 7.57 - 8.2 = -0.63$$

$$t = -\frac{0.63}{1.05/\sqrt{10}} = -1.9$$

$$df = N - 1 = 10 - 1 = 9$$

From Table G.8, our two-tailed value of t, $df = 9$, is

1.833 ($p = 0.1$)
2.262 ($p = 0.05$)
3.250 ($p = 0.01$)

The calculated value of t is less than the 5% value in the tables but not the 10% value. But as 10% is generally regarded as too high, we will stick to the 5% level and not reject H_0. Thus we conclude that the scores for the Jamaican wine were not significantly different from those of the California Chenin Blanc wines, as far as the French judges were concerned.

It is worth noting that no significant differences were found between the two wines, so we rejected H_0; we did *not* accept H_0. We could not say the wines were the same; we could only say that they were not different. The t test is designed to seek differences, not similarities. The lack of difference could be because the wines were, in fact, the same. However, it could be that H_0 was not rejected, because the sample of data was too small to reach significance. This is more likely when the t value is only just smaller than the appropriate value in Table G.8.

Example 2

Seven sky-diving experts scored the durability for each of two new parachute fabrics as shown below. On the basis of these data, is there a significant difference between the mean scores for the two fabrics? You may assume that d scores were normally distributed in the population.

Expert	Fabric X	Fabric Y
A	15	14
B	12	14
C	14	15
D	17	14
E	11	11
F	16	14
G	15	13

Because the same testers were used for each fabric, a related-samples test is used. A two-tailed test is used. This is because it could not be predicted in advance which fabric, if any, would have the higher score should H_0 be rejected. The data are arranged as follows:

Tester	Fabric X	Fabric Y	d	d^2
A	15	14	1	1
B	12	14	-2	4
C	14	15	-1	1
D	17	14	3	9
E	11	11	0	0
F	16	14	2	4
G	15	13	2	4
$N = 7$			$\Sigma d = 5$	$\Sigma d^2 = 23$
			Mean, $\bar{d} = \frac{5}{7} = 0.71$	

We now have the data to compute t.

$$t = \frac{\bar{d}}{S/\sqrt{N}}$$

$$S = \sqrt{\frac{\Sigma d^2 - \frac{(\Sigma d)^2}{N}}{N-1}} = \sqrt{\frac{23 - \frac{5^2}{7}}{7-1}} = \sqrt{\frac{23 - 3.57}{6}} = \sqrt{3.24}$$

Therefore,

$$t = \frac{0.71}{\sqrt{3.24}/\sqrt{7}} = \frac{0.71 \times \sqrt{7}}{\sqrt{3.24}} = 1.04$$

$$df = 7 - 1 = 6$$

From Table G.8, for a two-tailed test, $df = 6$; our t value exceeds no value in the table, not even for the 20% level of significance. So we do not reject H_0. Thus, on the basis of the data given, we cannot conclude that there is a significant difference between the mean durability scores for the two fabrics.

Example 3

In an experiment to determine the effect of cigarette smoking on differential taste sensitivity to sucrose in citrus drinks, seven smokers and five nonsmokers were presented with 10 pairs of samples containing different levels of sucrose, compared to a standard, and asked to identify the sweeter sample within each pair. The number of correct identifications were as follows:

Nonsmokers: 6, 8, 7, 9, 8
Smokers: 4, 7, 5, 4, 5, 6, 4

Did smokers and nonsmokers correctly identify significantly different numbers of the sweeter sample? Assume that samples of scores came from normal distributions of equal variance.

As different judges were used in each sample (smokers and nonsmokers), an independent samples test is used. Whether we can predict, should we reject H_0, if nonsmokers will have a greater or lesser differential sensitivity to sucrose in citric drinks is a difficult question. It is often thought that smoking impairs taste sensitivity, but does it impair differential sensitivity? If in doubt, it is better to be cautious and choose a two-tailed test. The data are set up as follows:

Number of Correct Identifications

Nonsmokers		Smokers	
X	X^2	Y	Y^2
6	36	4	16
8	64	7	49
7	49	5	25
9	81	4	16
8	64	5	25
		6	36
		4	16
$\Sigma X = 38$	$\Sigma X^2 = 294$	$\Sigma Y = 35$	$\Sigma Y^2 = 183$
$N_1 = 5$	$\bar{X} = 7.6$	$N_2 = 7$	$\bar{Y} = 7$
$(\Sigma X)^2 = 1444$		$(\Sigma Y)^2 = 1225$	

We now have the data to compute a value of t.

$$t = \frac{(\bar{X} - \bar{Y})\sqrt{\dfrac{N_1 N_2 (N_1 + N_2 - 2)}{N_1 + N_2}}}{\sqrt{\Sigma X^2 - \dfrac{(\Sigma X)^2}{N_1} + \Sigma Y^2 - \dfrac{(\Sigma Y)^2}{N_2}}}$$

Substituting the numbers, we get:

$$(\bar{X} - \bar{Y}) = (7.6 - 5) = 2.6$$

$$\sqrt{\frac{N_1 N_2 (N_1 + N_2 - 2)}{N_1 + N_2}} = \sqrt{\frac{5 \times 7(5 + 7 - 2)}{5 + 7}} = \sqrt{\frac{35 \times 10}{12}} = 5.4$$

$$\sqrt{\Sigma X^2 - \frac{(\Sigma X)^2}{N_1} + \Sigma Y^2 - \frac{(\Sigma Y)^2}{N_2}} = \sqrt{294 - \frac{1444}{5} + 183 - \frac{1225}{7}}$$

$$= \sqrt{294 - 288.8 + 183 - 175}$$

$$= \sqrt{13.2} = 3.63$$

Thus,

$$t = \frac{2.6 \times 5.4}{3.63} = 3.87$$

$$df = N_1 + N_2 - 2 = 5 + 7 - 2 = 10$$

From Table G.8, $df = 10$, for two-tailed tests, our value exceeds the value in the table for $p = 0.01$ ($t = 3.169$) but not for $p = 0.001$ ($t = 4.587$). Hence we reject H_0 ($p < 0.01$). So nonsmokers correctly identified more of the sweeter samples than did smokers.

8

Introduction to Analysis of Variance and the One-Factor Completely Randomized Design

8.1 WHY YOU DON'T USE MULTIPLE t TESTS

Using t tests it was seen that differences between the means of two samples of data can be tested. The two samples could come from the same people (related-samples design) or from different people (independent-samples design). For three samples of data A, B, and C, three t tests are required to test all the differences in means between the samples: A vs. B, A vs. C, B vs. C. The number of t tests that can be involved rises rapidly with the number of samples.

If a t test is performed at the 5% level, the probability of rejection H_0 when it is true (making a Type I error) is 5%, 1 in 20. Over 20 tests it is expected, therefore, that one of the tests will make a Type I error. This is a situation to be avoided.

Because a whole batch of t tests, when taken together as one analysis, give an unacceptably high level of significance (probability of making a Type I error), an alternative procedure is used. This is called analysis of variance (often abbreviated ANOVA or A.O.V.). This technique compares the means from several samples and tests whether they are all (within experimental error) the same, or whether one or more of them are significantly different.

Just as with t tests, analyses of variance can be applied to various experimental designs. There is an independent-samples type of design where the data in each sample come from different people (completely randomized design) and there are related-samples designs where the data from several different samples all come from the same subjects (repeated-measures or randomized block design). Unfortunately, different statisticians and textbooks have different names for all these designs and the subject can become confusing. Table G.10 lists the names used in the major behavioral textbooks for analysis of variance. Look at the Key to Statistical Tables to see the relationship of ANOVA to other statistical tests.

8.2 LOGIC BEHIND ANALYSIS OF VARIANCE AND THE *F* RATIO

Consider three groups of subjects *A*, *B*, and *C*, each in a different experimental condition.

	A	*B*	*C*
	X	X	X
	X	X	X
	X	X	X
	X	X	X
	X	X	X
	X	X	X
	X	X	X
Mean, X =	90	100	110

The three different means, 90, 100, and 110, may indicate that the three treatments *A*, *B*, and *C* have a significant effect on scores. On the other hand, there may be no significant difference between the treatments and the differences in the mean values may merely be due to chance. How can we tell? We use the same logic as that used for the t test (Section 7.6).

We look at the variation of the scores within any group or treatment. The variation of scores within a treatment is due to chance variations in the population, because of differences between people and uncontrolled experimental variables. They can be thought of as "experimental error." We compare the differences between treatments with the differences within treatments (experimental error).

If the differences between treatments are large compared with differences within treatments (if the range of treatment means is large compared with the range of scores within a treatment), it would be reasonable to suppose that the treatments differ. That is, if there is little or no overlap between the three sets of scores, it is reasonable to assume that they differ significantly and their means are significantly different. Consider the following data:

	A	*B*	*C*
Range	89-91	99-101	109-111
Mean *X*	90	100	110

Here the range of treatment means is larger than the range of scores within a treatment (experimental error). The means differ significantly.

If, on the other hand, the differences between treatments were comparable to differences within treatments (the range of treatment means was comparable to experimental error) it is reasonable to suppose that the treatments do not differ. That is, if the three sets of data overlap sufficiently, it is reasonable to assume that their means are not significantly different.

	A	*B*	*C*
Range	80-120	80-120	80-120
Mean *X*	90	100	110

Here the range of treatment means is not larger than the range of scores within a treatment (experimental error). The means do not differ significantly. So the argument can be summarized as follows:

If

between-treatments differences (range of treatment means)	$>$	within-treatments differences (range of scores within a treatment)

the means *are* significantly different; the treatments have effect.

If

between-treatments differences (range of treatment means)	$\triangleq$	within-treatments differences (range of scores within a treatment)

the means *are not* significantly different; the treatments have *no* effect.

This is the basic logic of analysis of variance. The argument is modified a little in practice but the basic approach of comparing between-treatments and within-treatments variation remains the same. We can compare the range of scores between and within treatments, but the range is not really the best measure because the odd "wildcat" extreme value may give an exaggerated appearance of overlap of the scores. So a middle-range value, the variance, is used instead.

From earlier (Section 2.4) it should be remembered that the variance, S^2, is given by

$$S^2 = \frac{\Sigma (X - \bar{X})^2}{N - 1} = \frac{\Sigma X^2 - \frac{(\Sigma X)^2}{N}}{N - 1}$$

The argument can be repeated as follows: If the between-treatments variance is large compared with the within-treatments variance, the treatment means are significantly different. So all that is required is to find the variance of the treatment means (S^2_{between}) and compare it with the variance of scores within the treatments (S^2_{within}) and see if the value is sufficiently large for it not to have occurred by chance. We examine the ratio

$$\frac{\text{between-treatments variance}}{\text{within-treatments or error variance}}$$

Actually, to calculate the within-treatments variance and the between-treatments variance is not too straightforward. Which treatment do we use to calculate the within-treatments or error variance? We ought to incorporate them all, in some way. Also, do we just put the treatment means into the variance formula to calculate between-treatments variance? It would then appear that we had only three people (*A*, *B*, and *C*). So what we do is calculate the ratio of the total variance, if it were made up of only between-treatments effects, to the total variance, if it were made up of only within-treatments or error effects. We call this the *F ratio*. Thus

$$F = \frac{\text{total variance (if only between-treatments effects)}}{\text{total variance (if only within-treatments or error effects)}}$$

We will explain further. Imagine the data in three treatments (*A*, *B*, and *C*, whose means are 90, 100, and 110) as follows:

	A	*B*	*C*
	86	98	107
	90	100	110
	94	102	113
	90	100	110
$\bar{X}$ =	90	100	110

Now there are actually 12 scores present (N = 12), each from one of 12 subjects. If we used only the treatment means to calculate the between-treatment effect, it would seem that we only had three subjects. So we must incorporate all 12 subjects somehow. We do this by assuming that the scores are as follows:

	A	*B*	*C*
	90	100	110
	90	100	110
	90	100	110
	90	100	110
$\bar{X}$ =	90	100	110

Now we have allowed for the fact that there are four people in each treatment. We put all 12 scores into the variance formula. In effect, what we are doing is calculating the total variance (variance of all the numbers) if it were due only to between-treatments effects. The same argument holds for within-treatments (error) effects. Which treatment do we use to calculate the within-treatments effects: *A*, *B*, or *C*? Surely, we should incorporate them all. The only trouble is that the scores also vary because the treatments all have different means, so we need to allow for this by arranging for them all to have the same mean. Let us arrange that they all vary around 100 (actually, the mean of all the numbers). Hence in treatment **A**, 86 becomes 96, 90 becomes 100, etc. So we adjust the scores to be as follows:

	A	*B*	*C*
	96	98	97
	100	100	100
	104	102	103
	100	100	100
$\bar{X}$ =	100	100	100

Now we have adjusted all the scores so that they vary around a common mean, 100. So we now put all 12 adjusted scores (one from each subject) into the variance formula. In effect, what we are doing is calculating the total variance if it were due only to within-treatments (error) effects. So we calculate the following ratio of variances:

$$F = \frac{\text{total variance (if it were due only to between-treatments effects)}}{\text{total variance (if it were due only to within-treatments or error effects)}}$$

We could write this as

$$F = \frac{S^2_{\text{total}}(\text{only between})}{S^2_{\text{total}}(\text{only within or error})}$$

$$F = \frac{MS_B}{MS_E}$$

where MS_B and MS_E are the generally used symbols for these estimates of total variance.

If the treatments have no effect, the variance between treatments will be mere error and the same as the within-treatments variance. Thus any estimates of total variance based on these values will be the same and $F = 1$. If, however, treatments have an effect, the between-treatments variance will be greater than mere error and the estimate of the total variance based on it will be greater and $F > 1$. Table G.11 indicates the largest values F can have by chance at various significance levels. If your calculated F is equal to or exceeds the value of F given in the tables, the between-treatments estimate of the total variance is larger than the within-treatments (mere experimental error) estimate. The between-treatments differences are more than mere error, so the means for the treatments are significantly different.

Of course, what we are really doing is seeing whether the differences between the means of the samples are large enough for us to say that the samples were drawn from different populations with different means; the treatments had such an effect that people under different treatments could no longer be regarded as coming from the same population. The null hypothesis states that the sample means do not differ significantly enough for us to say that they were drawn from populations with different means; they were all drawn from the same population and the treatments had no effect.

The critical values of F in Table G.11 are calculated using a probability distribution called the *F distribution*. Each F distribution depends on not one but two degrees of freedom. It would take up too much room to give each separate F distribution for each set of pairs of degrees of freedom, so specific values of F are given for specific probability levels (0.05, 0.01, 0.001, etc.) for given pairs of degrees of freedom (Table G.11). However, calculators and computers are now available which, for a given pair of degrees of freedom, will give an exact probability for getting a given F value on the null hypothesis.

8.3 HOW TO COMPUTE *F*

To perform an ANOVA test, we merely compute two variances and compare their ratio (F) to see whether it exceeds a theoretical maximum value (on H_0) in the tables. If it does, we reject H_0, the hypothesis which says that the means of the samples under each treatment are the same; we deduce that the means for the treatments are different.

However, the calculation of the variances is streamlined so that the same routine can be used for all applications of analysis of variance, even the most complex. For this first simple design it may seem overcomplicated, but the routine pays off later. The streamlining uses two major shortcuts which, although they simplify the mathematics, make the calculation appear to have little to do with variances. So bear in mind throughout that we are merely calculating two variance values for our F ratio.

Shortcut 1: Rearrange Formula for Variance

It would be too tedious to calculate each of the estimated total variances separately and compare them, so a shortcut procedure is used. The formula for the variance is pulled apart, different parts calculated separately, and only at the end is the formula reconstructed. Consider the formula

$$S^2 = \frac{\Sigma X^2 - \dfrac{(\Sigma X)^2}{N}}{N - 1}$$

Each numerator is calculated separately and is called a *sum of squares* (*SS* or *SOS*).

$$SS = \Sigma X^2 - \frac{(\Sigma X)^2}{N}$$

This is altered even more. $(\Sigma X)^2/N$ crops up a lot and is calculated separately and called the *correction term* or *correction factor, C.* So

$$SS = \Sigma X^2 - C$$

For different variances appropriate ΣX^2 values are calculated and C subtracted to give the numerator. The denominators for all the variances are divided in at the end of the calculation by choosing an appropriate $N - 1$ value; the denominator is called *the number of degrees of freedom, df.* The final estimated variance value (estimate of total variance) is called the *mean-square estimate* or *mean square* (*MS*). Thus our original formula

$$S^2 = \frac{\Sigma X^2 - \dfrac{(\Sigma X)^2}{N}}{N - 1}$$

becomes

$$\text{Mean square} = \frac{\text{sum of squares}}{\text{degrees of freedom}} \qquad \text{or} \qquad MS = \frac{SS}{df}$$

where *SS* involves a correction term, C.

Although the formulas appear quite different, they are, in fact, equivalent. Remember this; do not lose sight of what you are really doing in ANOVA.

Shortcut 2: Use the Fact that Total Variance is Made Up of Between- and Within-Treatments Variance

In a given experiment, the total variance will be made up of between-treatments variance and within-treatments or error variance. More exactly, the total variance about the grand mean of all the scores is caused by different treatment conditions (between-treatments variance) and also by the within-treatments or error variance of the scores around each treatment mean. Thus it seems intuitively correct to state:

$$\text{Total variance} = \begin{array}{c}\text{between-treatments}\\ \text{variance}\end{array} + \begin{array}{c}\text{within-treatments (error)}\\ \text{variance}\end{array}$$

If this is so, it is obvious that we cannot write

$$\text{Total } SS = \text{between-treatments } SS + \text{within-treatments (error) } SS$$

This is because the denominators for the various variance terms would be different. However, if you remember, we do not calculate actual variances for the F ratio; we calculate an estimate of the total variance should it be due solely to between-treatments effects or due solely to within-treatments (error) effects. Because of this, we can write such an expression for the sums of squares. Thus

$$\text{Total } SS = \begin{array}{l}\text{between-treatments } SS\\ \text{(the total } SS\text{, should}\\ \text{it be due solely to}\\ \text{between-treatments}\\ \text{effects)}\end{array} + \begin{array}{l}\text{within-treatments (error) } SS\\ \text{(the total } SS\text{, should it be}\\ \text{due solely due to within-}\\ \text{treatments, error, effects)}\end{array}$$

We denote those values by SS_T, SS_B, and SS_E, respectively, the subscripts standing for "total," "between," and "error." Thus we write

$$SS_T = SS_B + SS_E$$

If you can accept this as intuitively correct, all well and good. If not, a mathematical proof is given in Appendix C.

The SS values that are important for us to find are:

SS_B for the between-samples estimate of the total variance,

$$MS_B = \frac{SS_B}{df_B}$$

SS_E for the within-samples (or error) estimate of the total variance,

$$MS_E = \frac{SS_E}{df_E}$$

However, SS_E often turns out to be inconvenient to calculate directly, especially when the ANOVA designs get more complex. On the other hand, SS_T is always easy to calculate, so we compute SS_E by subtracting SS_B from SS_T. So this is our second shortcut.

SS_E	=	SS_T	-	SS_B
needed to compute F, less easy to calculate		not needed but easy to calculate		needed to compute F, easy to calculate

To summarize the shortcuts:

1. Break down the formula for S^2, calling the numerator SS and the denominator df. Calculate all SS values first. Then divide by the appropriate df at the end to obtain MS values for the F ratio.
2. SS_E is not calculated directly; it is obtained by the subtraction $SS_T - SS_B$. So we calculate SS_T not for its own sake but as a means of finding SS_E.

8.4 COMPUTATIONAL FORMULAS FOR SUMS OF SQUARES AND DEGREES OF FREEDOM

Armed with these shortcuts, we now set about finding the computational formulas for the ANOVA. First we look at the formulas for the sums of squares.

Consider three conditions or treatments: A, B, and C. In general, we say there are k treatments. Imagine separate groups of subjects assigned to each treatment; with different subjects in each condition, this unrelated samples design is called a *completely randomized design*. Imagine that there are n subjects in each treatment while the total number of subjects in the whole experiment (all three treatments) is N. Then $N = n \times 3$; in general, $N = nk$.

The scores for each subject are represented by X values. The grand total of all these values is T; the totals under each treatment are denoted by T_A, T_B, and T_C. The means for each treatment are $\bar{X}_A$, $\bar{X}_B$, and $\bar{X}_C$, while the grand mean of all the X values in the whole experiment is m. These are represented as follows:

	A	B	C	$\cdots$ k	
	X	X	X		There are k treatments
	X	X	X		n = number of scores per treatment
	.	.	.		Overall number of scores, $N = nk$
	.	.	.		
	.	.	.		
	X	X	X		
Total	T_A	T_B	T_C	T =	Grand total = $T_A + T_B + T_C$
Mean	$\bar{X}_A$	$\bar{X}_B$	$\bar{X}_C$	m =	Grand mean

The formula for SS is

$$\Sigma (X - \bar{X})^2 = \Sigma X^2 - \frac{(\Sigma X)^2}{N}$$

First we calculate SS_T, the total sum of squares:

$$SS_T = \Sigma X^2 - \frac{(\Sigma X)^2}{N}$$

We have denoted the sum of all the X values by T, so

$$SS_T = \Sigma X^2 - \frac{T^2}{N}$$

and we have decided to denote T^2/N as the correction term, C. So

$$SS_T = \Sigma X^2 - C$$

Thus we add up the X values; we square the total and divide by N to get C. We subtract this from the sum of the squares of the X values. This is merely the ordinary numerator for S^2.

Next we calculate the between-treatments sum of squares. If can be shown that

$$SS_B = \frac{T_A^2 + T_B^2 + T_C^2}{n} - C$$

We will not bother to prove it here, but should the reader be interested, the proof is given in Appendix C. Note that the correction term (C) also appears in this expression. We generally calculate C first because it appears in all the expressions we want. Although we do not generally calculate it directly, it can also be shown (see Appendix C) that

$$SS_E = \Sigma X^2 - \frac{T_A^2 + T_B^2 + T_C^2}{n}$$

Thus it can be seen that our second shortcut is, in fact, correct:

$$\overbrace{X^2 - C}^{SS_T} = \overbrace{\frac{T_A^2 + T_B^2 + T_C^2}{n} - C}^{SS_B} + \overbrace{\Sigma X^2 - \frac{T_A^2 + T_B^2 + T_C^2}{n}}^{SS_E}$$

compute directly — compute directly — compute by subtraction

The mean-square estimates (MS) are now obtained by dividing by the appropriate df. As before, degrees of freedom (df) are always given by an $N - 1$ value.

Total $df = N - 1$ because the total number of subjects is N
Between-treatments $df = k - 1$ because the total number of treatments is k

The error or within-treatments df is obtained by subtraction in the same manner as SS_E,

$$\text{Error } df\ (df_E) = \underset{\text{total}}{(N-1)} - \underset{\text{between treatments}}{(k-1)} = N - k$$

This seems intuitively correct. Some degrees of freedom in the total are due to between-treatments effects, so the rest will be due to error. Another way of approaching this is to say that within each treatment there are n scores, thus $n - 1$ degrees of freedom. There are k such treatments, so the total within-treatments (error) $df = k(n-1) = kn - k = N - k$.

Appropriate MS values are found using $MS = SS/df$ and then compared to give F.

$$MS_B = \frac{SS_B}{k-1}$$

$$MS_E = \frac{SS_E}{df_E}$$

$$F = \frac{MS_B}{MS_E}$$

to be compared with an appropriate value in Table G.11.

8.5 COMPUTATIONAL SUMMARY FOR THE ONE-FACTOR, COMPLETELY RANDOMIZED DESIGN

The steps in the computation of F for the completely randomized design are represented schematically as follows:

	A	B	$C \ldots k$	
	X	X	X	k treatments, n scores in each treatment
	X	X	X	
	X	X	X	Total number of scores $= N = nk$
	X	X	X	
	X	X	X	
	X	X	X	
Total	T_A	T_B	T_C	T = Grand total $= T_A + T_B + T_C$
Mean	$\bar{X}_A$	$\bar{X}_B$	$\bar{X}_C$	m = Grand mean

Correction term, $C = \dfrac{T^2}{N}$

Total sum of squares, $SS_T = \Sigma X^2 - C$ with $df = N - 1$

Between-treatments sum of squares, $SS_B = \dfrac{T_A^2 + T_B^2 + T_C^2}{n} - C$; $df = k - 1$

Error sum of squares,

$SS_E = SS_T - SS_B$ with $df_E = df$(total) $- df$(between treatments)
$(N - 1) - (k - 1)$

Table 8.1 shows the traditional method for presenting the results of analysis of variance.

It is worth noting that the error mean square is sometimes called the *residual mean square.* "Residual" merely means the variance left over once the variance due to the treatments has been accounted for. It is another way of thinking of the experimental error, the variation in scores not due to any manipulation of the conditions of the experiment.

Having computed a value for $F = MS_B/MS_E$, it is compared with a value in Table G.11, which indicates the maximum value F can have for a given significance level. If our calculated F value is greater than or equal to the tabulated value, the treatment means are significantly different. The appropriate value is found for the 5%, 1%, and 0.1% levels by finding the F value in the tables corresponding to the number of degrees of freedom for the numerator (for $MS_B = k - 1$) and the denominator [for $MS_E = (N - 1) - (k - 1)$].

There are several types of ANOVA designs but all work essentially the same way. The various treatment effects are usually compared to an error term and the calculated F value compared with a value in Table G.11. It is worth noting that the confidence-interval approach (Appendix D) can also be used for F, just as it can for t (Section 7.10).

Table 8.1 One-Factor ANOVA Table

Source of variation	Analysis of variance			Calculated
	SS	df	MS	F
Total	SS_T	$(N - 1)$		
Between treatments	SS_B	$(k - 1)$	$MS_B = \dfrac{SS_B}{k - 1}$	$\dfrac{MS_B}{MS_E}$
Error	SS_E	$(N - 1) - (k - 1)$	$MS_E = \dfrac{SS_E}{(N - 1) - (k - 1)}$	

8.6 WORKED EXAMPLE: ONE-FACTOR, COMPLETELY RANDOMIZED DESIGN

Imagine three treatments *A*, *B*, and *C*. There are four separate subjects in each treatment ($n = 4; N = 12$). We will compute an F value to see whether the means of the three samples are significantly different (whether the means vary enough for them to have been drawn from different populations). H_0 states that the means vary only by chance (that the means do not vary significantly; the samples all came from the same population).

When would we use such an experimental design? The treatments could be different lighting conditions under which the judges were required to perform some visual task or even taste food samples. We may be interested here in the effect on perception of changed lighting conditions. However, here we are using a completely randomized design, with different subjects under each treatment. Surely, it would be better to make each subject perform under each treatment condition so as to obtain a better comparison of conditions. The same is true should the three treatments be the tasting of a given food, processed in three different ways. A related samples design would be more desirable. So when would we use the completely randomized design? The answer is, generally, not if we can arrange to test the same people under all treatments. It is only when we cannot do this that we use the completely randomized design. Perhaps we are only able to test each judge once. This may occur in consumer testing (too little time for extensive testing) or in medical studies (each treatment may be a different disease or a different cure). Perhaps experience of one treatment may carry over and affect the next treatment. There is no general rule; each case is decided on its merits.

In the example we will merely refer to judges in three treatment conditions and test to see whether the treatments have any effect on the judges' scores (the treatment means differ). The data are displayed as follows:

	A	*B*	*C*	
	10	8	5	Number of treatments, $k = 3$
	8	7	7	Number of scores in treatment, $n = 4$
	9	9	6	
	9	8	5	Total number of scores, $N = 12$
Total	36	32	23	
Mean	9	8	5.75	Grand total, $T = 91$

The null hypothesis H_0 is that the treatment means are all equal (no difference between them).

$$\text{Correction term, } C = \frac{T^2}{N} = \frac{91^2}{12} = 690.08$$

$$SS_T = \Sigma X^2 - C = 10^2 + 8^2 + 5^2 + 8^2 + \cdots + 5^2 - C$$

$$= 719 - 690.08 = 28.92 \qquad \text{with } df = N - 1 = 12 - 1 = 11$$

$$SS_B = \frac{T_A^2 + T_B^2 + T_C^2}{n} - C = \frac{36^2 + 32^2 + 23^2}{4} - C$$

$$= 712.25 - 690.08 = 22.17 \qquad \text{with } df = k - 1 = 3 - 1 = 2$$

$$SS_E = SS_T - SS_B = 28.92 - 22.17 = 6.75$$

$$\text{with } df = df_{\text{total}} - df_{\text{between treatments}}$$

$$= 11 - 2 = 9$$

The analysis of variance table is as follows:

Source	*SS*	*df*	*MS*	*F*
Total	28.92	11		
Between treatments	22.17	2	$\frac{22.17}{2} = 11.09$	$\frac{11.09}{0.75} = 14.78^{**}$
Error	6.75	9	$\frac{6.75}{9} = 0.75$	

We now use Table G.11 to determine whether our F value is large enough for us to reject H_0. We need two sets of degrees of freedom to look up an F value: df for the numerator (= 2), displayed across the top of the table, and df for the denominator (= 9) displayed down the side of the table. Select the appropriate column (2) and row (9) to find the appropriate tabular value. You will find a pair of values; the upper value is that for the 5% level and the lower value is for the 1% level. Over the page in a separate table you will find values for the 0.1% level. From Table G.11 for: numerator $df = 2$, denominator $df = 9$

$F = 4.26$ at 5%
$F = 8.02$ at 1%
$F = 16.39$ at 0.1%

Our value exceeds the 1% value but not the 0.1% value. Thus we reject the null hypothesis with a less than 1% chance of being wrong (making a Type I error). We say that the treatment means differ significantly ($p < 0.01$).

There is a shorthand notation which is used to denote whether F is significant or not. One asterisk (*) is placed by the F value if it is significant at the 5% level. Two asterisks (**) are used if it is significant at the 1% level (as in the example above) and three asterisks (***) for the 0.1% level.

Now that we have rejected H_0 and established that the means are different, the question arises as to which are the means causing the difference. Is it that 5.75 is significantly smaller than the other two, or are all three means significantly different? ANOVA does not tell us this; it only tells us that some difference exists. We need further tests (multiple comparisons: Chapter 9) to establish where the differences are.

8.7 COMPUTATION WHEN SAMPLE SIZES ARE UNEQUAL

For simplicity, we have always assumed that the number of people under each treatment is the same. This, of course, is not necessarily true for a completely randomized design. When the number of people under each treatment condition varies, the computation of F is the same except that the calculation of the between-treatments sum of squares varies slightly; each treatment total (e.g., T_A) is divided by its own sample size value (n_A) rather than by a general sample size value (n). Instead of

$$SS_B = \frac{T_A^2 + T_B^2 + T_C^2}{n} - C$$

we have

$$SS_B = \frac{T_A^2}{n_A} + \frac{T_B^2}{n_B} + \frac{T_C^2}{n_C} - C$$

This is represented schematically below. We have treatments A, B, and C with n_A scores in the first treatment, n_B in the second, and n_C in the third. Other terms are as before:

A	B	C	$\cdots k$
X	X	X	
X	X	X	
X	X	X	
X	X	n_C X	
n_A X	X		
	X		
	X		$N = n_A + n_B + n_C$
	n_B X		
T_A	T_B	T_C	

The calculation proceeds as expected:

$$\text{Correction term, } C = \frac{T^2}{N} \qquad \text{as before}$$

$$SS_T = \Sigma X^2 - C \qquad \text{as before and } df = N - 1$$

$$SS_B = \frac{T_A^2}{n_A} + \frac{T_B^2}{n_B} + \frac{T_C^2}{n_C} - C \qquad \text{with } df = k - 1$$

$$SS_E = SS_T - SS_B \qquad \text{as before and } df_E = df(\text{total}) - df(\text{between treatments})$$

In the case where the sample sizes were the same, we mentioned that we could calculate df_E by noting that there were $(n - 1)$ degrees of freedom for each sample so that the total within-treatments (error) df was given by $df_E = k(n - 1) = N - k$. When all the sample sizes differ, the expression becomes

$$df_E = (n_A - 1) + (n_B - 1) + (n_C - 1)$$

8.8 IS ANALYSIS OF VARIANCE A ONE- OR A TWO-TAILED TEST?

Table G.11 shows cutoff points along the F distribution (Figure 8.1) in the same way as for normal and t distributions. At the 5% level, the largest 5% of the F values that could occur by chance are to the right of (greater than) this point. Note that only one tail of the distribution is used. Thus the ratio $F = MS_B/MS_E$ gives a one-tailed test for comparing variances. If the null hypothesis is not true and the means differ, the between-samples variance MS_B, and thus F, can only get larger; F cannot get smaller. This is a one-tailed situation.

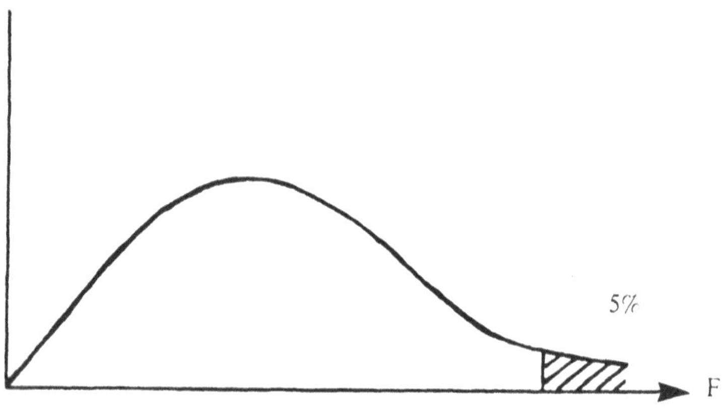

Figure 8.1 F distribution.

However, although MS_B only increases as the treatment means ($\bar{X}_A$, $\bar{X}_B$, $\bar{X}_C$, etc.) become more different (one-tailed), the means themselves could increase or decrease. This is a two-tailed situation. So ANOVA can be seen not only as a one-tailed test of variance change but also as a two-tailed test as far as differences between means are concerned. It is a two-tailed test of differences between means which works via a one-tailed test of variance change. Because we use ANOVA to test differences between means, we treat it, in practice, as a two-tailed test.

8.9 HOW *F* IS RELATED TO *t*

The F and t tests are essentially the same. For the completely randomized ANOVA design where there are only two treatments and the two-tailed independent-samples t test, the following is true:

$$F = t^2$$

This is discussed further in Section 10.4, where the same is seen to be true for the related-samples t and two-factor repeated-measures ANOVA designs.

8.10 ASSUMPTIONS AND WHEN TO USE ANALYSIS OF VARIANCE

It is worth considering formally the assumptions for use of the one-factor completely randomized ANOVA. It is called "one-factor" because we are only varying one set of treatments. The assumptions are essentially the same as for the independent-samples t test (Section 7.8). They are:

1. Samples taken under each treatment must be randomly picked from their respective populations.
2. The treatments must be independent of each other, and all scores within a treatment must also be independent. (Each treatment must have different judges and within each treatment each score must come from a separate judge; no judge can contribute more than one score.)
3. Samples of scores under each treatment must come from normally distributed populations of scores and this implies that data must be on an interval or ratio scale.
4. Samples of scores under each treatment must come from populations with the same variance (homoscedasticity).

As with a t test, the same difficulties arise as far as ascertaining whether the data were samples from normally distributed populations (Section 7.8). Age or ranked data break these assumptions, while it is generally assumed that if a given score is the result of many interacting variables, the population it came from will be normally distributed.

It is also of interest to state formally the mathematical model used in the derivation of the ANOVA test. The model assumes that any score is equal to the overall population mean, with an amount added (or subtracted) to account for the effect on the score due to the treatment it is under, and an amount added to account for the effects of experimental error. The model is an additive model, expressed as follows:

X	$=$	μ	$+$	β	$+$	e
↑		↑		↑		↑
a given score		population mean value		effect due to treatment		effect due to error

The X variable above is normally distributed and, as such, is continuous (see Section 2.6).

There are procedures available for testing whether the data came from a normally distributed population, but these are beyond the scope of this text.

8.11 CAN YOU USE ANALYSIS OF VARIANCE FOR SCALING DATA?

As with the t test, ANOVA makes the same assumptions of normality and homoscedasticity; these cause the same problems for rating scales. (Refer to Section 7.11 for a full discussion of this.) The fact that scaling data may be on an interval scale (violating the assumption of normality) and that end effects alter the variance (violating the assumption of homoscedasticity) means that ANOVA can be used only as an approximate analysis.

As with t tests, some statisticans state that ANOVA is robust as far as breaking these assumptions is concerned. Exactly how robust seems unclear, but the assumptions become more important as the ANOVA design becomes more complex. Some statisticians state that the assumption of homoscedasticity is the important one, while derivations from normality are of secondary importance.

An alternative approach is to regard data from a rating scale as being ordinal. The scores are thus ranked and a nonparametric ranked analysis of variance used. The Key to Statistical Tables indicates that the appropriate nonparametric test is the Kruskal-Wallis one-factor ranked ANOVA (Section 16.8).

Again with the decision as to whether to use a parametric or a nonparametric test, or both, rests with the experimenter. The reader is referred to Section 7.11 for a discussion on this point. See also Section 16.20 and Appendix F for further discussion.

9
Multiple Comparisons

9.1 LOGIC BEHIND MULTIPLE COMPARISONS

ANOVA tells you only whether differences between means exist or not; it does not tell you where these differences are. Should the ANOVA tell you that differences exist (reject H_0), you need to do some multiple-comparison tests to establish which are the means that are actually differing from each other. Should the ANOVA not reject H_0, you need not bother with multiple comparisons. You always do an ANOVA first to see whether or not you need multiple comparisons. Merely performing multiple-comparison tests without a prior ANOVA can cause the same trouble as using many t tests; you may pick up differences that are not there (reject H_0 when it is true: make a Type I error). The ANOVA can be seen as a screening test, protecting against Type I error.

Of course, should you be testing only two treatments and your ANOVA says that they are significantly different, you need not perform any multiple comparisons; you have only two means, so these must be the ones that differ. Quite often beginners waste time performing multiple-comparison tests on only two means.

The testing sequence can be represented thus:

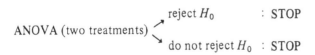

$$\text{ANOVA (two treatments)} \begin{array}{l} \nearrow \text{reject } H_0 \quad : \text{STOP} \\ \searrow \text{do not reject } H_0 \quad : \text{STOP} \end{array}$$

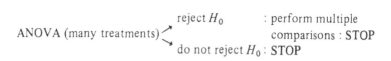

ANOVA (many treatments) → reject H_0 : perform multiple comparisons : STOP
ANOVA (many treatments) → do not reject H_0 : STOP

There is a plethora of multiple-comparison tests, generally named after the statisticians who developed them. The following are some of the most common multiple-comparison procedures:

Sheffé test
Tukey HSD (honestly significant difference) test
Newman-Keuls test
Duncan's multiple-range test
Fisher's LSD (least significant difference) test
Dunn test
Dunnett test

(↓ more powerful; ↑ more conservative)

Once the ANOVA has shown a significant F value (rejected H_0), the Sheffé, Tukey, Newman-Keuls, Duncan, or LSD tests can be used to find out which pairs of means are significantly different. The Dunn and Dunnett tests have a more specialized use. The use of these tests will be discussed later; first we will examine the logic behind them.

All the procedures involve calculating a range that is matched to the difference between the means. If the range value is larger than the difference between the means, the means are not significantly different. If the range value is smaller, the means are significantly different. This can be seen diagrammatically. Consider a set of means: X_1, X_2, X_3, X_4, and X_5:

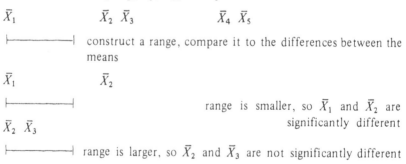

$\bar{X}_1$ $\bar{X}_2$ $\bar{X}_3$ $\bar{X}_4$ $\bar{X}_5$

|————| construct a range, compare it to the differences between the means

$\bar{X}_1$ $\bar{X}_2$

|————| range is smaller, so $\bar{X}_1$ and $\bar{X}_2$ are significantly different

$\bar{X}_2$ $\bar{X}_3$

|————| range is larger, so $\bar{X}_2$ and $\bar{X}_3$ are not significantly different

A more powerful test (Fisher's LSD) will have a smaller range, which makes it easier for two means to be found significantly different. A more conservative test (Sheffé) will produce a larger range, which will make it more difficult to find two means which are significantly different. Again, this can be seen diagrammatically:

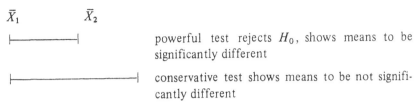

$\bar{X}_1$ $\bar{X}_2$

|——————| powerful test rejects H_0, shows means to be significantly different

|————————————| conservative test shows means to be not significantly different

Thus the more conservative the test, the larger the range value and the harder it is to reject H_0.

Some tests also have a set of ranges (multiple-range tests) depending on how many means are between those being tested. They increase their conservatism by increasing some, but not all, ranges. The means are arranged in order of increasing magnitude and tested with a set of ranges:

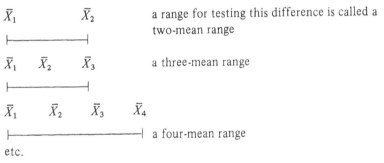

$\bar{X}_1$ $\bar{X}_2$ a range for testing this difference is called a two-mean range

|——————|

$\bar{X}_1$ $\bar{X}_2$ $\bar{X}_3$ a three-mean range

|——————|

$\bar{X}_1$ $\bar{X}_2$ $\bar{X}_3$ $\bar{X}_4$

|————————————| a four-mean range

etc.

Multiple-comparison tests are generally two-tailed so as to correspond to the two-tailed nature of the ANOVA (ANOVA is a two-tailed test of differences between means; Section 8.8). There are exceptions to this two-tailed rule, which will be discussed later.

Traditionally, the multiple-comparison tests are performed at the same level of significance as the ANOVA. For an ANOVA significant at the 5% level (not 1%), we perform multiple comparisons at 5%. For an ANOVA significant at 0.1%, we chose the same level for multiple comparisons. This is the procedure that will be adopted in this text. This is only a tradition, however. Some authors may perform multiple comparisons at 5% when the ANOVA is significant at the 1% level; this might be done for the sake of clarity in the data. What should not be done, however, is to perform multiple comparisons at 1% when the ANOVA only shows differences at 5%.

Interestingly enough, the range values for the multiple comparisons are all given by similar formulas. Generally, these are given by two types of formulas: a type based on the t statistic (or F, which is related to t; $F = t^2$ for a two-tailed test) and a further type with a similar formula. We will now prepare to consider these formulas.

Assume that we have performed ANOVA which indicates significant differences between means and that we decide to perform multiple comparisons. Let us assume the following:

k	=	number of treatments in the ANOVA
n	=	number of scores in each treatment sample
MS_E	=	error mean square for the ANOVA
df_E	=	number of degrees of freedom associated with MS_E (df_E is the denominator for MS_E)

The multiple-comparison formulas involve these terms and they are summarized in the following section.

9.2 VARIOUS MULTIPLE-COMPARISON TESTS: FORMULAS FOR CALCULATING RANGES

The tests for types based on t are as follows:

Fisher's LSD test (LSD = least significant difference):

$$\text{single range} = t\sqrt{\frac{2\,MS_E}{n}}$$

t is taken from Table G.8 for the level of significance of the ANOVA, where $df = df_E$, two tailed.

Sheffé test:

$$\text{single range} = S\sqrt{\frac{2\,MS_E}{n}}$$

Here $S = \sqrt{(k-1)F}$. F is taken from Table G.11 for the level of significance of the ANOVA, for the numerator $df = k - 1$ and denominator $df = df_E$.

Dunnett test:

$$\text{single range} = D\sqrt{\frac{2\,MS_E}{n}}$$

D is an adjusted t value taken from Table G.14 for the level of significance of the ANOVA. Table G.14 gives a D value for $df = df_E$ and the number of means in the ANOVA. (These means include a control mean, and because of the special nature of this test, it can be one- or two-tailed.)

For types based on other forms:

Newman-Keuls test:

$$\text{multiple ranges} = Q\sqrt{\frac{MS_E}{n}}$$

Q is taken from Table G.12 for the level of significance of the ANOVA. Table G.12 gives a Q value for $df = df_E$ and the number of means in the range being tested (a two- mean or three-mean range, etc.).

Duncan's test:

$$\text{multiple ranges} = Q_D \sqrt{\frac{MS_E}{n}}$$

Q_D is taken from Table G.13 in exactly the same manner as Q is taken from Table G.12.

Tukey HSD test (HSD = honestly significant difference):

$$\text{single range} = Q_{\max} \sqrt{\frac{MS_E}{n}}$$

where $Q_{\max}$ is the maximum value of Q taken from the Newman-Keuls test, using Table G.12 at the level of significance of the ANOVA (for $df = df_E$ and the maximum number of means in the analysis).

Dunn test:

$$\text{single range} = D \sqrt{\frac{MS_E}{n}}$$

D is taken from Table G.15 for the level of significance of the ANOVA. Table G.15 gives D values for $df = df_E$ and the number of comparisons being made. (This test is used only for a few planned comparisons.) Table 9.1 summarizes the formulas.

Table 9.1 Formulas for Calculating Ranges

LSD range	$t \sqrt{\frac{2\,MS_E}{n}}$	t from Table G.8
Sheffé range	$S \sqrt{\frac{2\,MS_E}{n}}$	$S = \sqrt{(k-1)F}$ (F from Table G.11)
Dunnett range	$D \sqrt{\frac{2\,MS_E}{n}}$	D from Table G.14
Newman-Keuls range	$Q \sqrt{\frac{MS_E}{n}}$	Q from Table G.12
Duncan range	$Q_D \sqrt{\frac{MS_E}{n}}$	Q_D from Table G.13
Tukey HSD range	$Q_{\max} \sqrt{\frac{MS_E}{n}}$	$Q_{\max}$ from Table G.12
Dunn range	$D \sqrt{\frac{MS_E}{n}}$	D from Table G.15

9.3 WHICH MULTIPLE-COMPARISON TEST SHOULD YOU USE?

There are many multiple-comparison tests and this text does not claim to cover them all; it merely mentions some of the more common procedures used in behavioral measurement. The rules governing the use of these tests can be complex, with statisticians disagreeing with one another. However, the following principles form a guide for choosing between them.

Basically, the tests are improvements on the t test. We saw that if one t test was performed, the level of significance was, say, 0.05. If several are performed, the level of significance for all the tests taken as a batch will be way above 0.05. So tests are needed so that after a whole set of comparisons are made, the level of significance for the batch, taken as a whole, will be 0.05 or not far above it. These tests are designed to solve such problems and they do so in slightly different ways, which in turn determines when they are to be used.

Roughly, though, the tests get an overall level of significance of 0.05 (or 0.01) for all the tests taken as a batch, by making all the individual comparisons stricter than 0.05. Thus, when they are all combined, the overall level of significance comes to 0.05. Making them more strict or more conservative means, in effect, increasing the calculated range value which is used to compare whether the difference between means is significant. The larger the range, the more difficult it is to reject H_0, thus the more conservative the test for those two means. Multiple-range tests increase the range for some comparisons but not others. The aim is not to increase the range value for adjacent means because these are the closest and their differences are most likely to be missed (Type I error). So three-mean ranges and greater are increased instead.

The *Dunnett test* is used only when all the means are to be compared to one mean of, say, a control group. It could be that several new varieties of rice are to be compared to the regular rice (and only to that regular rice, not among themselves); the Dunnett test would then be the appropriate multiple-comparison procedure.

The *Dunn test* is used only for comparisons planned before the ANOVA has actually been performed. If you had some theoretical reason for comparing only a few specific mean values with each other, you could use the Dunn test. You might have subjects tested under several treatments but not be interested in comparing all the means to see whether they differed; you may only be interested in comparing the means from one or two important treatments. This feature, whereby specific comparisons are planned before performing the ANOVA, is an essential difference between the Dunn test and the other multiple comparisons. The other tests are performed *a posteriori*, after a significant F value is found and the experimenter wants to know which differences between means were significant. These a posteriori comparisons (post hoc comparison, data snooping, data sifting) can be made only if a significant F value is found.

On the other hand, the planned comparisons (*a priori* comparisons) for the Dunn test do not require an ANOVA beforehand. It is rather like planning to perform only one or two t tests, in which case the level of significance of the batch of tests taken as a whole would not be raised much. We do not need to perform an initial screening ANOVA.

The Dunn test is designed for simple comparisons such as

$$\bar{X}_1 \text{ vs. } \bar{X}_2$$

$$\bar{X}_3 \text{ vs. } \bar{X}_4$$

and complex comparisons, comparisons between combinations of means such as

$$\frac{\bar{X}_1 + \bar{X}_2}{2} \text{ vs. } \frac{\bar{X}_3 + \bar{X}_4}{2}$$

$$\bar{X}_1 \text{ vs. } \frac{\bar{X}_1 + \bar{X}_2 + \bar{X}_3}{3}$$

The remaining tests are *post hoc* tests, made after a significant F has been found.

The *LSD test* is really a t test and should be used only when very few comparisons are made. If the level of significance is, say, 0.05 when one comparison is made, the level of significance will be higher for that batch of tests should several comparisons be made. This leads to the possibility of error (Type I error, rejecting a true H_0). Because of this, it is safer to use this test at lower levels of significance ($p = 0.01$), so that the overall significance level does not become too high. The LSD is the least conservative test (most powerful, smallest range value) and is most likely to find significant differences (and commit Type I errors).

The *Newman-Keuls* and *Duncan tests* use multiple ranges (larger ranges when more means fall within the comparison being made) and adjust the levels of significance so that for each comparison being made, the level of significance is 0.05 (or 0.01) and stays there. The overall level of significance for all the tests taken as a batch will be above 0.05 but not as far as it was for the LSD test. Thus the Newman-Keuls and Duncan multiple-range tests are more conservative than the LSD test. When it was introduced the Duncan test caused a controversy among statisticians because of some of the assumptions made in its derivation.

The *Sheffé* and *Tukey HSD* (honestly significant difference) *tests* are even more conservative. They are adjusted so that after all comparisons (both simple pairwise and complex) the overall level of significance will be 0.05 (or 0.01). Thus they are the most conservative tests (have the largest range values, so that it is more difficult to reject H_0). The Sheffé test is more conservative than the Tukey test.

If you are very worried about Type I errors (rejecting H_0 when it is true), you may use these tests but if you are only making simple comparisons (not complex), these tests are too conservative and may not reject H_0 when they should; they may miss differences that another test would detect.

The multiple-comparison tests are summarized in Table 9.2. Again, it is worth noting that the multiple comparisons are all two-tailed tests, except for Dunnett's test. In the latter case, we are not attempting to match the two-tailed nature of the ANOVA; we are making a limited number of planned comparisons and it is reasonable that we may expect some of the differences to be one-tailed in nature.

It can be seen that the procedures for the multiple-comparison tests are very similar. We will only pick two, a single-range and a multiple-range test, to examine in detail; use of the others can easily be understood from a consideration of these two. We will choose tests designed for simple pairwise comparisons, these being the most common in sensory testing. We will deal with Fisher's LSD test as an example of a single-range test and Duncan's test as an example of a multiple-range test.

It is worth noting that there are other procedures beyond the scope of this book that can be carried out after ANOVA. Instead of searching to find which means are significantly different, the means could be examined to see how they are related to each other: linearly or curvilinearly, and so on. This procedure is called *trend analysis*.

Table 9.2 Summary of Multiple-Comparison Tests

Test		Comments	
Dunnett		When comparing all conditions with a given control	
Dunn		A priori comparisons only: simple and complex	
Sheffé Tukey HSD	↑ more conservative	Post hoc, simple and complex comparisons	Too conservative for simple comparisons only
Newman-Keuls Duncan		Post hoc, simple comparisons only	
LSD		Post hoc, simple comparisons	Few comparisons

9.4 HOW TO COMPUTE FISHER'S LSD TEST

This test is used for pairwise comparisons only when a significant F value has been obtained. It is best to use it at low levels of significance ($p < 0.01$) because the overall level of significance for a batch of several tests will be higher.

In general, the t test is given by the formula

$$t = \frac{\bar{X} - \bar{Y}}{\text{standard error of the difference}} \qquad \text{(See Section 7.6)}$$

The standard error of the difference is given by the formula

$$\text{SE} = \sqrt{\frac{\sigma_1^2}{n_1} + \frac{\sigma_2^2}{n_2}}$$

where n_1 and n_2 are sample sizes and σ_1^2 and σ_2^2 are variances of the samples whose means are being compared. We used a simpler, computational formula in the section on t tests using S^2 values rather than σ^2.

One of the assumptions for the F test is that the samples should have equal variance:

$$\sigma_1^2 = \sigma_2^2 = \sigma^2$$

Thus

$$\text{SE} = \sqrt{\sigma^2\left(\frac{1}{n_1} + \frac{1}{n_2}\right)}$$

This is simplified further if the sample sizes are the same. Consider:

$$\text{SE} = \sqrt{\sigma^2\left(\frac{1}{n_1} + \frac{1}{n_2}\right)} \qquad \text{for unequal sample sizes}$$

If $n_1 = n_2 = n$, then

$$\left(\frac{1}{n_1} + \frac{1}{n_2}\right) = \frac{2}{n};$$

and

$$\text{SE} = \sqrt{\frac{2\sigma^2}{n}} \qquad \text{for equal sample sizes}$$

The population variance σ^2 is not usually known; it is generally estimated from the sample by using S^2 (see Section 2.4). From the ANOVA, MS_E, the estimate of the total variance, should it be solely due to within-samples (error) effects, is a good estimate of σ^2. Thus we can replace σ^2 by MS_E in the formula for SE. Thus

$$\text{SE} = \sqrt{MS_E\left(\frac{1}{n_1} + \frac{1}{n_2}\right)} \qquad \text{for unequal sample sizes}$$

$$\text{SE} = \sqrt{\frac{2MS_E}{n}} \qquad \text{for equal sample sizes}$$

Returning to the formula for t, we have

$$t = \frac{\bar{X} - \bar{Y}}{\text{SE}} \quad \begin{array}{l} \longleftarrow \text{difference between means} \\ \longleftarrow \text{standard error of the difference} \end{array}$$

Remembering that

$$\text{SE} = \sqrt{\frac{2\,MS_E}{n}}$$

we have

$$t = \frac{\bar{X} - \bar{Y}}{\sqrt{\dfrac{2MS_E}{n}}}$$

Thus

$$\bar{X} - \bar{Y} = t\sqrt{\frac{2MS_E}{n}}$$

If a given level of significance is chosen (e.g., $p < 0.01$), a value of t can be found in Table G.8 corresponding to df_E, the number of degrees of freedom associated with MS_E. This is the critical value of t (for $p < 0.01$), the smallest value of t that is associated with a significant difference between the means. Should this critical value of t be used in the expression

$$t\sqrt{\frac{2MS_E}{n}}$$

then $\bar{X} - \bar{Y}$ will correspond to the smallest difference between the means that is still significant (at $p < 0.01$). This is precisely the yardstick (range value) required for the LSD test. Thus

$$\text{LSD} = t\sqrt{\frac{2MS_E}{n}}$$

You merely look up t for $df = df_E$, at the level of significance found for the ANOVA and you have your yardstick or range value for Fisher's LSD test.

9.5 WORKED EXAMPLE: FISHER'S LSD TEST

Let us go back to the worked example used for the ANOVA (Section 8.6) and use the data for the calculation of the range value for the LSD test.

(See page 147.)

	A	B	C
	10	8	5
	8	7	7
	9	9	6
$n = 4$	9	8	5
Mean	9	8	5.75

The ANOVA table was as follows:

Source	SS	df	MS	F
Total	28.92	11		
Between	22.17	2	$MS_B = 11.08$	14.78**
Error	6.75	9	$MS_E = 0.75$	

Having found a significant F value, we now proceed to investigate which are the means that are different from each other. We calculate the LSD range value, thus:

$$\text{LSD} = t\sqrt{\frac{2MS_E}{n}} = t\sqrt{\frac{2 \times 0.75}{4}}$$

From Table G.8, for the 1% level of significance (corresponding to the ANOVA), the t value for $df = 9$ (df_E) is 3.250. Thus

$$\text{LSD} = 3.25\sqrt{\frac{2 \times 0.75}{4}} = 1.99$$

Thus we see that means 9 and 8 are not significantly different, as they are less than 1.99 apart ($9 - 8 = 1 < 1.99$), while both are significantly different from 5.75. This can be represented diagrammatically as follows:

Means <u>9 8</u> 5.75

A horizontal line underlines the means which are not significantly different. Thus this reads: The LSD test shows that means 9 and 8 are not significantly different from each other, while both are significantly different from 5.75. (Note that the overall level of significance of this statement is somewhat higher than 0.01.)

An alternative way of representing this is to use the same superscripts for means that are not significantly different:

9^a 8^a 5.75^b

Should there be several means, with a complicated pattern of differences, these can be represented as follows:

$$\bar{X}_A \qquad \bar{X}_B \qquad \bar{X}_C \qquad \bar{X}_D \qquad \bar{X}_E$$

$$\bar{X}_A^a \qquad \bar{X}_B^{ab} \qquad \bar{X}_C^{bc} \qquad \bar{X}_C^c \qquad \bar{X}_E^c$$

These are interpreted as: $\bar{X}_A$ and $\bar{X}_B$ are not significantly different, nor are $\bar{X}_B$ and $\bar{X}_C$, nor are $\bar{X}_C$, $\bar{X}_D$, and $\bar{X}_E$. $\bar{X}_A$ is significantly different from $\bar{X}_C$, $\bar{X}_D$, and $\bar{X}_E$. $\bar{X}_B$ is significantly different from $\bar{X}_D$ and $\bar{X}_E$. You may have philosophical difficulties stating that $\bar{X}_A$ and $\bar{X}_C$ differ significantly, when neither differs significantly from $\bar{X}_B$. There is no logical mistake. It merely means that sufficient testing has been done to identify differences between $\bar{X}_A$ and $\bar{X}_C$ but not between $\bar{X}_A$ and $\bar{X}_B$ or $\bar{X}_B$ and $\bar{X}_C$. Perhaps further testing might enlarge the sample enough to pick up such smaller differences at this level of significance.

If the samples being compared have unequal sizes, the other formula for SE is used.

$$\text{SE equal sample size} = \sqrt{\frac{2MS_E}{n}} \qquad \therefore \qquad \text{LSD} = t\sqrt{\frac{2MS_E}{n}}$$

$$\text{SE unequal sample size} = \sqrt{MS_E\left(\frac{1}{n_1} + \frac{1}{n_2}\right)} \qquad \therefore \qquad \text{LSD} = t\sqrt{MS_E\left(\frac{1}{n_1} + \frac{1}{n_2}\right)}$$

The calculation is performed in essentially the same way except that the unequal sample sizes will yield a different LSD value for each pair of means being compared.

9.6 HOW TO COMPUTE DUNCAN'S MULTIPLE-RANGE TEST

This is a multiple-range test and the formula for these ranges is given by

$$\text{Duncan's ranges} = Q_D\sqrt{\frac{MS_E}{n}}$$

or if the samples are of unequal size,

$$\text{Duncan's ranges} = Q_D\sqrt{\frac{MS_E}{2}\left(\frac{1}{n_1} + \frac{1}{n_2}\right)}$$

This formula is equivalent to taking the harmonic mean of the n values, that is,

$$\frac{1}{n}\ (\text{harmonic mean}) = \frac{1}{2}\left(\frac{1}{n_1} + \frac{1}{n_2}\right)$$

The calculation is straightforward and similar to that for an LSD test.

A value of Q_D is obtained from Table G.13 for $df = df_E$ at the level of significance corresponding to the ANOVA. Q_D values are then noted for the appropriate numbers of means in the comparison range. Should there be three means $\bar{X}_1, \bar{X}_2$, and $\bar{X}_3$, arranged in increasing order, then

$$\bar{X}_1 \qquad\qquad \bar{X}_2$$
$$\uparrow\!\!\_\_\_\ \text{versus}\ \_\_\_\!\!\uparrow$$

represents a comparison with two means in the range and

$$\bar{X}_1 \qquad \bar{X}_2 \qquad \bar{X}_3$$
$$\uparrow\!\!\_\_\_\ \text{versus}\ \_\_\_\!\!\uparrow$$

represents a comparison with three means in the range. Q_D values are found in Table G.13 for two- and three-mean ranges, which are then computed.

9.7 WORKED EXAMPLE: DUNCAN'S MULTIPLE-RANGE TEST

Again, we can apply the Duncan test to the same ANOVA data that we used for the LSD example (see Section 9.5). The means were as follows:

$$\bar{X}_A = 9 \qquad \bar{X}_B = 8 \qquad \bar{X}_C = 5.75$$

while in the ANOVA

$$MS_E = 0.75 \text{ with } df = 9 \qquad\qquad n = 4 \text{ for all samples}$$

The means are arranged in order of size:

Sample:	*C*	*B*	*A*
Means:	5.75	8	9

and the range values then computed:

$$\text{Range} = Q_D\sqrt{\frac{MS_E}{n}} = Q_D\sqrt{\frac{0.75}{4}}$$

Comparisons $\bar{X}_C$ vs. $\bar{X}_B$ and $\bar{X}_B$ vs. $\bar{X}_A$ have a range of two means

Comparisons $\bar{X}_C$ vs. $\bar{X}_A$ has a range of three means

Table G.13 gives for $p = 0.01$ (ANOVA significance level), $df_E = 9$, Q_D values as below:

Two means: 4.596
Three means: 4.787

Thus the range values are

Two means in range:

$$Q_D \sqrt{\frac{0.75}{4}} = 4.596 \sqrt{\frac{0.75}{4}} = 1.99$$

Three means in range:

$$Q_D \sqrt{\frac{0.75}{4}} = 4.787 \sqrt{\frac{0.75}{4}} = 2.07$$

Now examining the means:

C	*B*	*A*
5.75	8	9

The three-mean difference is

├── 2.07 ──┤ three means

and the two-mean difference is

├── 1.99 ──┤ two means

Applying the three-mean range (2.07) to *A* and *C*, it can be seen that these means are significantly different (9.00 - 5.75 > 2.07). Applying the two-mean range (1.99), it can be seen that *C* is significantly lower than *B*, which itself does not differ significantly from *A*.

So the overall conclusion is, once again, that 8 and 9 do not differ significantly but both differ significantly from 5.75. Note that the two-mean comparison range for Duncan's test is the same as that for the LSD range—it is only when three-mean comparisons and greater are used that the values differ.

Whe using multiple-comparison tests, it is best to start first with comparisons for the most separate means. If these are not significantly different, closer means will not be either and we need not bother testing them. So in the example above, three-mean differences were tested first and two-mean differences next.

9.8 ADDITIONAL WORKED EXAMPLE: A COMPLETELY RANDOMIZED DESIGN WITH MULTIPLE COMPARISONS

Barrels of California Cabernet Sauvignon wine from four separate wineries, *A*, *B*, *C*, and *D*, were tested for tannin content. The tannin content of each barrel was represented on a scale ranging from 0 to 30 (30 represents a high tannin content). The tannin scores for the wines were as follows:

Wine A: 8, 6, 5, 7, 6, 7, 7, 8, 5, 6, 7, 7
Wine B: 9, 7, 6, 8, 8, 7, 8, 8, 7, 9
Wine C: 8, 5, 6, 6, 7, 7, 7, 5, 8
Wine D: 1, 2, 1, 0, 0, 2, 0

Were the mean tannin scores for wines *A*, *B*, *C*, and *D* significantly different? If so, which ones differed? Use LSD tests if required. You may assume that the samples of tannin scores came from normal, homoscedastic populations.

As each score came from a completely different barrel of wine, the design is completely randomized. It is equivalent to having four separate sets of judges under four separate treatments. First the data can be put in a more conventional form:

Wine *A*	Wine *B*	Wine *C*	Wine *D*
8	9	8	1
6	7	5	2
5	6	6	1
7	8	6	0
6	8	7	0
7	7	7	2
7	8	7	0
8	8	5	6
5	7	8	$n = 7$
6	9	59	$\bar{X} = 0.86$
7	77	$n = 9$	
7	$n = 10$	$\bar{X} = 6.56$	$T = 221$
79	$\bar{X} = 7.7$		$N = 38$
$n = 12$			
$\bar{X} = 6.58$			

$$\text{Correction term, } C = \frac{(221)^2}{38} = \frac{48{,}841}{38} = 1285.29$$

$$\text{Total } SS = 8^2 + 6^2 + 5^2 + 7^2 + \cdots + 2^2 + 0^2 - C$$
$$= 1539 - 1285.29 = 253.71$$

$$\text{Total } df = N - 1 = 38 - 1 = 37$$

$$\text{Between-wines } SS = \frac{79^2}{12} + \frac{77^2}{10} + \frac{59^2}{9} + \frac{6^2}{7} - C$$

$$= 520.08 + 592.90 + 386.78 + 5.14 - C$$

$$= 1504.90 - 1285.29 = 219.61$$

$$\text{Between-wines } df = k - 1 = 4 - 1 = 3$$

$$\text{Error } SS = \text{total } SS - \text{between-wines } SS$$
$$= 253.71 - 219.61 = 34.1$$

$$\text{Error } df = \text{total } df - \text{between-wines } df$$
$$= 37 - 3 = 34$$

These values can now be presented in an ANOVA table:

Source	*SS*	*df*	*MS*	*F*
Total	253.71	37		
Between wines	219.61	3	73.2	73.2***
Error	34.10	34	1.0	

The F values quoted in Table G.11 are

$$(df = 3, 34) = 2.88 \qquad p = 0.05$$
$$= 4.42 \qquad p = 0.01$$
$$= 7.05 \qquad p = 0.001$$

The calculated F value exceeds all these values in the table, so we reject H_0; the mean tannin scores for different wines are not the same ($p < 0.001$). The means are:

	Wine			
	D	*C*	*A*	*B*
Mean	0.86	6.56	6.58	7.7
n	7	9	12	10

LSD tests can be used with a significance level of $p = 0.001$, the level of significance of the ANOVA test. At this low level of significance, LSD tests are unlikely to result in Type I error.

$$\text{LSD} = t\sqrt{MS_E\left(\frac{1}{n_1} + \frac{1}{n_2}\right)} \qquad MS_E = 1.0$$

From Table G.8, $t(df = 34, p = 0.1\%) = 3.646$; actually, $df = 30$ is nearest.

Compare wine C and wine D:

Means $6.56 - 0.86 = 5.7$

$$\text{LSD} = 3.646\sqrt{1 \times \left(\frac{1}{7} + \frac{1}{9}\right)} = 1.84$$

Thus wines D and C (with A or B) differ significantly ($p < 0.001$).

Compare wine B and wine C:

Means $7.7 - 6.56 = 1.14$

$$\text{LSD} = 3.646\sqrt{1 \times \left(\frac{1}{9} + \frac{1}{10}\right)} = 1.67$$

Thus wines B and C (with A) do not differ significantly ($p < 0.001$). Thus

D	C	A	B
0.86	6.56	6.58	7.7

(C, A, and B are underlined together.)

Wines A, B, and C did not differ significantly from each other as far as tannin content is concerned, while wine D's mean tannin score was significantly lower than the others ($p < 0.001$).

10

Analysis of Variance: Two-Factor Design Without Interaction, Repeated Measures

10.1 LOGIC BEHIND THE TWO-FACTOR DESIGN WITHOUT INTERACTION, REPEATED MEASURES

In the completely randomized ANOVA design (Chapter 9), separate samples of subjects are used for each treatment or condition; this is an independent-samples design. On the other hand, the treatments could be investigated, making the same sample of subjects undergo all treatments; this would be a related-samples design.

If the treatments *A*, *B*, and *C* are to be investigated, a group of subjects could be treated in condition *A*, then *B*, and then *C*. This has the advantage that the same subjects are in each group and so the variation between the groups is more likely to be due to the treatments received, rather than the fact that the different groups of subjects under each condition may differ substantially anyway.

This repeated-measures type of design controls for individual differences between people. However, precautions are necessary. The order of treatments must be counterbalanced to control for practice or fatigue effects. All orders of treatments should occur equally (*ABC, ACB, BAC, BCA, CAB, CBA*) to balance out these effects. Thus, if three different foods were to be tested by a panel of judges, the order of presentation of the foods should be varied for the different panelists, or over replicate tastings for a given judge, so as to balance out any order effects that might occur. Aspects of related- and independent-samples design are discussed more fully in Sections 5.8 and 16.20.

However, if a related-samples or repeated-measures design is chosen, it does have the advantage that, because the same subjects are used in each treatment, the effect of individual differences between subjects can be extracted out from the results. In the completely randomized ANOVA, what happens is that a between-treatments variance estimate is calculated. Normally this would be divided by the error estimate to get a value for F. However, this error term includes all experimental error—not only uncontrolled variables in the experiment but also the effects of differences between subjects. If the effect of differences between subjects was pulled out, as in the between-treatments variance, the error term would be far smaller. When this smaller error estimate is divided into the between-subjects estimate, the calculated F value will be larger. This larger F value is more likely to exceed the maximum value possible on the null hypothesis at the chosen level of significance; it is more likely to indicate that significant differences exist between the treatment means. The test is, in effect, more powerful. This can be represented diagrammatically:

Completely randomized:

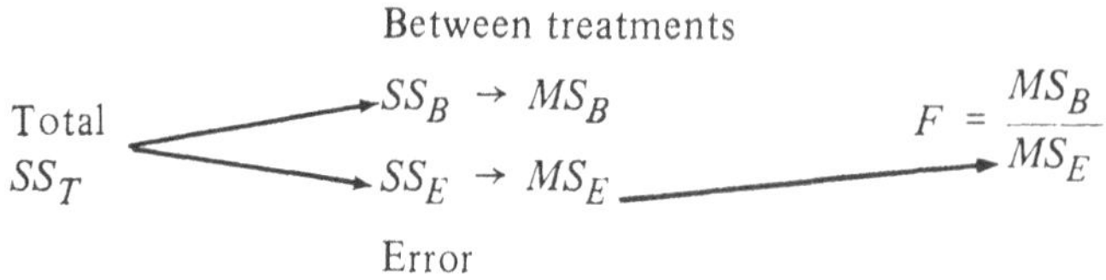

Two-factor repeated measures:

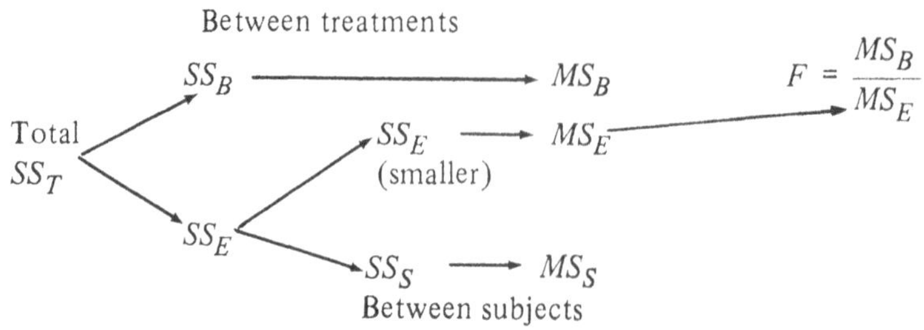

Not only can the effect of different treatments be studied,

$$F = \frac{MS_B}{MS_E}$$

but also the effect of different subjects, using an estimate of the total variance from the variance due to different subjects MS_S, to calculate F. M_S is computed in the same way as M_B except that instead of using treatment totals (T_A, T_B, and T_C), totals for each subject are used.

$$F = \frac{MS_S}{MS_E}$$

Thus we can see whether the mean score over all treatments, for each given subject, varied significantly. For instance, should we be analyzing the results of scaling data, differences in the means for the subjects may indicate whether some subjects habitually tend to use the high end of the scale (giving high mean scores) while others tend to use the low end (giving low mean scores). Often we do not take this analysis further (multiple comparisons to indicate which subjects are "scoring high" or "scoring low") because such individual differences are hardly surprising and are not generally the focus of the experiment. Differences between treatments are usually of greater interest.

So in this design a value of SS_S (subjects) and MS_S is pulled out just like values for SS_B and MS_B, and in exactly the same way. Because two factors (treatments and subjects) are pulled out like this, the design is called a *two-factor design*. The design is also said to be *without interaction*, for reasons which will become clear in Chapter 11. Because the measures are repeated on the same subject in each treatment, it is also called a *repeated-measures design*. So the full name is *Two-Factor ANOVA Without Interaction, Repeated Measures.*

Sometimes the repeated-measures design is given different names (see Table G.10). For instance, it might be called a *Randomized Complete Block Design* because it is completely randomized as far as one factor (subjects) is concerned but is a repeated-measures design for the other factor (experimental treatments); each randomly selected block (subject) is said to complete all treatments. A "block" is merely the statisticians' name for a set of matched scores, in our case, a judge or subject. We will not refer to people as "blocks"; we will call them "judges" or "subjects."

Now, compare the two-factor design (of which the repeated-measures design is a special case) with the other tests, using the Key to Statistical Tables. The two-factor design, in general, falls in the two-or-more-factor column. The two-factor repeated-measures design could be placed here or it could be considered to be a one-factor related-samples design; the one factor would be the two or more experimental treatments while the second "subjects" factor would merely indicate that the design is a related-samples design. We have chosen the latter. Naturally, the three-or-more-factor designs can have "subjects" as one of their factors also; we might indicate this by adding to the name of the design the phrase "one-factor subjects."

10.2 COMPUTATIONAL SUMMARY FOR THE TWO-FACTOR DESIGN WITHOUT INTERACTION, REPEATED MEASURES

We will now examine the computation for a Two-Factor ANOVA Without Interaction, Repeated Measures. The computation is represented as for the completely randomized design (Section 8.5). There are k treatments (A, B, and C; $k = 3$)

with n scores (X values) in each treatment. The n scores, this time, come from n subjects who are tested under each treatment ($S_1, S_2, S_3, \ldots, S_n$). The totals for the scores under treatments A, B, and C are T_A, T_B, and T_C (column totals). The totals for each subject are T_1 for S_1, T_2 for S_2, T_3 for $S_3, \ldots, T_n$ for S_n (row totals). There are N scores ($N = nk$) in the whole experiment; their grand total is T. The means under each treatment are $\bar{X}_A$, $\bar{X}_B$, and $\bar{X}_C$; H_0 states that these do not differ significantly. The ANOVA computation scheme is outlined as follows:

		Treatment				
		A	B	$C \cdots$	k	
Subject	S_1	X	X	X	T_1	
	S_2	X	X	X	T_2	
	S_3	X	X	X	T_3	
	.	.	.	.	.	
	.	.	.	.	.	
	.	.	.	.	.	
	S_n	X	X	X	T_n	$N = nk$
	n	T_A	T_B	T_C	T	
		$\bar{X}_A$	$\bar{X}_B$	$\bar{X}_C$		

Just as before, we calculate C, SS_T, and SS_B.

$$C = \frac{T^2}{N}$$

Total, $SS_T = \Sigma X^2 - C$ with $df = N - 1$

Between treatments, $SS_B = \dfrac{T_A^2 + T_B^2 + T_C^2}{n} - C$ with $df = k - 1$

Before calculating SS_E, we now calculate a between-subjects sum of squares SS_S in the way that we calculated SS_B, except summing across rows rather than down columns.

Between subjects, $SS_S = \dfrac{T_1^2 + T_2^2 + T_3^2 + T_4^2 + \cdots + T_n^2}{k} - C$ with $df = n - 1$

Here the denominator is k (the number of scores making up the total T_1 or T_2, etc.) rather than n (the number of scores making up the total T_A or T_B, etc.). df is the "number of subjects − 1" rather than the "number of treatments − 1."

Finally, the error term is obtained in the usual way by subtraction; this time, SS_S is also subtracted from SS_T.

Error, $SS_E = SS_T - SS_B - SS_S$

df is calculated in the same way, by subtraction:

$$df_E = df_T - df_B - df_S$$

where $df_E = (N - 1) - (k - 1) - (n - 1)$; then

$$MS_E = \frac{SS_E}{df_E}$$

We then present the ANOVA table of results as before, except that this time we have an extra source of variance due to subjects.

Source of variance	*SS*	*df*	*MS*	*F*
Total	SS_T	$(N - 1)$		
Between treatments	SS_B	$(k - 1)$	$MS_B = \frac{SS_B}{k - 1}$	$F = \frac{MS_B}{MS_E}$
Between subjects	SS_S	$(n - 1)$	$MS_S = \frac{SS_S}{n - 1}$	$F = \frac{MS_S}{MS_E}$
Error	SS_E	df_E	$MS_E = \frac{SS_E}{df_E}$	

where $df_E = (N - 1) - (k - 1) - (n - 1)$

df_E is sometimes calculated using the formula

$$df_E = df_B \times df_S = (k - 1) \times (n - 1)$$

If no significant difference is found between the treatment means (*F* not significant), that is the end of the calculation. If a difference is found, multiple-comparison techniques can be used as before. For instance,

$$\text{LSD} = t\sqrt{\frac{2MS_E}{n}}$$

$$\text{Duncan's range} = Q_D\sqrt{\frac{MS_E}{n}}$$

We could also do the same for the subject means, should we wish

$$\text{LSD} = t\sqrt{\frac{2MS_E}{k}}$$

Here we divide by k, the number of figures in a row (that made up the subjects mean score) rather than n, the number of figures in a column (that made up the treatments mean score). However, for routine sensory evaluation, we rarely wish to study differences between subjects in this way.

10.3 WORKED EXAMPLE: TWO-FACTOR DESIGN WITHOUT INTERACTION, REPEATED MEASURES

For clarity, it is worth considering the following worked example.

Ten wine tasters scored for quality three low-alcohol, vitamin-fortified, holistic, low-calorie "natural" wines derived from vegetable by-products. The three "wines" were called X, Y, and Z. The quality scores are shown below; we assume homoscedasticity and normality (Section 8.10) for the data. Were there significant differences between the quality scores for the three wines and if so, what were they? Use LSD test for multiple comparisons if necessary.

	Wine			
Taster	X	Y	Z	Total
A	13	14	13	40
B	14	15	12	41
C	14	15	13	42
D	12	14	12	38
E	14	14	12	40
F	15	13	13	41
G	13	15	12	40
H	13	12	12	37
I	14	14	12	40
J	15	14	13	42
Total	137	140	124	401
Mean	13.7	14.0	12.4	

H_0 states that the means are not significantly different.

$$\text{Correction term, } C = \frac{T^2}{N} = \frac{401^2}{30} = 5360.03$$

$$\text{Total, } SS_T = \Sigma X^2 - C = 13^2 + 14^2 + 13^2 + 14^2 + 15^2 + \cdots + 13^2 - C$$

$$= 5393.00 - 5360.03$$

$$= 32.97 \qquad \text{with } df = N - 1 = 30 - 1 = 29$$

$$\text{Between} \begin{cases} \text{wines} \\ \text{treatments,} \end{cases} SS_B = \frac{137^2 + 140^2 + 124^2}{10} - C$$

$$= 5374.50 - 5360.03 = 14.47$$

$$\text{with } df = k - 1 = 3 - 1 = 2$$

$$\text{Between} \begin{cases} \text{tasters} \\ \text{subjects,} \end{cases} SS_S = \frac{40^2 + 41^2 + 42^2 + \cdots + 42^2}{3} - C$$

$$= 5367.67 - 5360.03 = 7.64$$

$$\text{with } df = n - 1 = 10 - 1 = 9$$

$$\text{Error, } SS_E = SS_T - SS_B - SS_S = 32.97 - 14.47 - 7.64 = 10.86$$

$$\text{with } df = (N-1) - (k-1) - (n-1) = 29 - 2 - 9 = 18$$

Having worked out the relevant *SS* and *df* values, we now present the ANOVA table:

Source	*SS*	*df*	Mean square	*F*
Total	$SS_T = 32.97$	29		
Between wines	$SS_B = 14.47$	2	$MS_B = 7.24$	12.0***
Between tasters	$SS_S = 7.64$	9	$MS_S = 0.849$	1.41
Error	$SS_E = 10.86$	18	$MS_E = 0.603$	

For differences between wines, the calculated F value (12.0) exceeds the values in Table G.11 (*df*: 2,18) for the 5% (3.55), 1% (6.01), and 0.1% (10.39) levels. Thus we say that we reject H_0; the mean wine quality scores for these rather unusual wines show some significant differences ($p < 0.001$). Interestingly, the wine tasters showed no significant differences in their mean scores (calculated F: $1.41 < 2.46$, 5% level, *df*: 9,18). This would suggest that the wine tasters were using the wine quality scale in the same way.

The LSD test is relatively safe to use at the 0.1% level, to calculate which of the means for the wines, 13.7, 14.0, or 12.4, were significantly different. Hence

$$\text{LSD} = t\sqrt{\frac{2MS_E}{n}}$$

From the ANOVA table, $MS_E = 0.603$, while there are 10 scores under each treatment, so $n = 10$. From Table G.8, at the level of significance of the ANOVA (0.1%), for $df = 18$ (the *df* for MS_E), two-tailed, $t = 3.922$. Thus

$$\text{LSD} = 3.922\sqrt{\frac{2 \times 0.603}{10}} = 1.362$$

10.4 ASSUMPTIONS AND WHEN TO USE THE TWO-FACTOR ANALYSIS OF VARIANCE WITHOUT INTERACTION

It is worth considering formally the assumptions for use of the two-factor repeated-measures ANOVA, without interaction. These are similar to those for the completely randomized design (Section 8.10). They are:

1. Samples must be randomly picked from their respective populations.
2. The scores within a treatment must be independent, while scores should be matched across treatments. (The same judges must be tested under each treatment condition, while within each treatment, each score must come from a separate judge; no judge can contribute more than one score per treatment condition.)
3. Samples of scores under each treatment must come from normally distributed populations of scores and this implies that data must be on an interval or ratio scale.
4. Samples of scores under each treatment must come from populations with the same variance (homoscedasticity).

The assumptions discussed here can be generalized to all other ANOVA designs.

It is also of interest to state formally the mathematical model used in the derivation of the ANOVA test. The model assumes that any score is equal to the overall population mean, with an amount added (or subtracted) to account for the effect on the score due to the treatment it is under, an amount added to account for the effect on the score due to the subject being tested, and an amount added to account for the effects of experimental error. The model is called an *additive model* and is expressed as follows:

X	=	μ	+	β	+	S	+	ϵ
a given score		population mean value		effect due to treatment		effect due to subject		effect due to error

X is a normally distributed and thus continuous variable (see Section 2.6). This additive model is generalized to other ANOVA designs. A three- or four-factor design will simply have more terms, to account for the further factors and interactions (see Chapter 11) which occur.

The two-factor ANOVA makes the same assumptions of normality and homoscedasticity as the one-factor completely randomized design; these can cause problems for testing the data from rating scales (see Sections 7.10 and 8.11, and Appendix F). Should a nonparametric analysis be chosen instead, the Friedman two-way ANOVA and the Kramer test are alternatives for ranked data (see the Key to Statistical Tables). There is also discussion among statisticians regarding the robustness of ANOVA to the breaking of the assumptions of normality and homoscedasticity (see Section 8.11).

Relationship Between F and t

In Section 8.9 we discussed the relationship between t and F tests. For the completely randomized one-factor ANOVA with two treatments and the independent-samples t test, we stated that

$$F = t^2$$

The same is true for the related-samples designs. For a two-factor ANOVA, repeated-measures design with only two treatments and the related-samples t test, the same relationship ($F = t^2$) is true; the tests are essentially the same. In this case, the ANOVA is a preferable test to use because it gives you information regarding whether the judges differ significantly or not; the t test tells you only about differences between treatments.

10.5 A FURTHER LOOK AT THE TWO-FACTOR ANOVA: THE COMPLETELY RANDOMIZED DESIGN

In this chapter our examination of the two-factor ANOVA has centered on the repeated-measures design, where all subjects are tested under each treatment. Imagine three treatments A, B, and C; all subjects (S_1, S_2, S_3) are tested under each treatment. This is represented as follows:

	A	B	C	
S_1	X	X	X	T_1
S_2	X	X	X	T_2
S_3	X	X	X	T_3
	T_A	T_B	T_C	

and $$SS_S = \frac{T_1^2 + T_2^2 + T_3^2}{n} - C \qquad n = 3$$

A mathematically similar design is the *completely randomized two-factor design*, where quite different subjects are tested under conditions A, B, C and simultaneously under further conditions α, β, γ. Here the X values are scores from different subjects in each combination of treatment conditions: $A\alpha$, $B\alpha$, $C\alpha$, $A\beta$, etc.

	A	B	C	
α	X	X	X	T_α
β	X	X	X	T_β
γ	X	X	X	T_γ
	T	T	T	

and $$SS_\alpha = \frac{T_\alpha + T_\alpha + T_\alpha}{n} - C \qquad n = 3$$

Just as we extracted sums of squares SS_T, SS_B, SS_S, and SS_E, we now have a value SS_α to replace SS_S. We go on to calculate MS_α instead of MS_S. The mathematics are the same, but the experimental design is quite different.

A, *B*, and *C* could be different wines and they could all three be scored for quality by the same judges (repeated measures) or the wines *A*, *B*, and *C* could be tested under different tasting conditions α, β, γ (e.g., under white, red, and green light) and different judges assigned to each wine-tasting combination of treatments (completely randomized). Thus each combination of conditions—$A\alpha$, $A\beta$, $A\gamma$, $B\alpha$, $B\beta$, $B\gamma$, $C\alpha$, $C\beta$, $C\gamma$—is tasted by a separate judge. The assumptions for ANOVA given in Section 10.4 need to be modified accordingly for this test.

A repeated-measures design is preferable to a completely randomized design for sensory testing because then the treatment means (for wines *A*, *B*, and *C*) are more likely to vary as the result of differences between treatments rather than differences between subjects. Thus the repeated-measures design is the one generally used in this situation. This is especially important for scaling data. People are so variable in their use of rating scales that it is very likely that differences obtained between two groups of judges could be caused merely by the differences in how they used the scales. Accordingly, completely randomized designs are comparatively rare in sensory evaluation.

10.6 ADDITIONAL WORKED EXAMPLE: TWO-FACTOR DESIGN WITHOUT INTERACTION, REPEATED MEASURES

Eight judges assessed the softness of a new perfumed toilet tissue on a scale where a standard paper was given a softness value of 30. Each judge retested the toilet tissue on six successive days (six replications). Was the overall performance of the judges significantly different? Was the panel consistent over replications? You may assume that the data were sampled from populations that were normally distributed with equal variance. The softness scores were as follows:

	Replications						
Judge	1	2	3	4	5	6	Total
A	7	4	6	9	8	6	40
B	6	3	5	7	8	7	36
C	5	7	8	8	7	7	42
D	6	8	8	6	7	6	41
E	7	6	8	5	6	5	37
F	8	5	7	7	9	6	42
G	7	7	7	8	7	5	41
H	6	8	6	7	7	5	39
	52	48	55	57	59	47	318

We now perform the analysis in the routine manner.

$$\text{Correction term} = \frac{(318)^2}{48} = \frac{101{,}124}{48} = 2106.75 = C$$

$$\text{Total } SS = 2128 - C = 75.25 \qquad df = 48 - 1 = 47$$

$$\text{Between-judges } SS = \frac{(40)^2 + (36)^2 + \cdots + (39)^2}{6} - C$$

$$= \frac{12{,}676}{6} - C = 2112.67 - 2106.75 = 5.92$$

$$df = 8 - 1 = 7$$

$$\text{Between-replications } SS = \frac{(52)^2 + (48)^2 + \cdots + (47)^2}{8} - C$$

$$= \frac{16{,}972}{8} - C = 2121.5 - 2106.75 = 14.7$$

$$df = 6 - 1 = 5$$

$$\text{Error } SS = 75.25 - 5.92 - 14.75 = 54.58$$

$$df = 47 - 7 - 5 = 35$$

We now have all the information for the ANOVA table:

Source	*SS*	*df*	*MS*	*F* [a]
Total	75.25	47		
Replications	14.76	5	2.95	1.89 (NS)
Judges	5.92	7	0.84	0.54 (NS)
Error	54.58	35	1.56	

[a] NS, not significant at $p = 0.05$.

The calculated F values are not significant at the 5% level when compared with theoretical values in Table G.11 for $df = 5{,}35$ and $df = 7{,}35$. However, the table does not give exact values for $df = 35$ in the denominator; there are only values for $df = 34$ and 36. However, these tabular values all exceed the calculated value. Thus $1.89 < 3.58$ ($df = 5{,}36$), $1.89 < 3.61$ ($df = 5{,}34$) and $0.54 < 3.18$ ($df = 7{,}36$), $0.54 < 3.21$ ($df = 7{,}34$): all values at $p = 0.05$.

Thus we do not reject H_0 for replications or judges ($p > 0.05$). The overall performance of the judges was not significantly different, while the panel was consistent over replications. No multiple-comparison tests are required because the F values are not significant.

Had the judges been different over replications, LSD tests would have been used to find which replications were different. The t value for the LSD test would be the two-tailed value at the level of significance of the ANOVA for the degrees of freedom associated with MS_E ($df_E = 35$). Thus

$$\text{LSD} = t\sqrt{\frac{2MS_E}{n}}$$

where

$$\begin{aligned} n &= \text{number of judges} \\ &= \text{number of values per replication} \\ &= 8 \end{aligned}$$

Had the judges been different, LSD tests would again distinguish which ones.

$$\text{LSD} = t\sqrt{\frac{2MS_E}{k}}$$

where

$$\begin{aligned} k &= \text{number of replications} \\ &= \text{number of values per judge} \\ &= 6 \end{aligned}$$

11

Analysis of Variance: Two-Factor Design with Interaction

11.1 WHAT IS INTERACTION?

We will now consider the two-factor ANOVA with interaction. Before considering this design in detail, we will examine exactly what we mean by "interaction." Interaction takes place between factors. Two factors interact if the trend between treatments in one factor varies depending on which treatment is present in the other factor.

It is important to note that the name "two-factor ANOVA with interaction" does not imply that interaction actually occurs; it merely means that interaction is considered in the calculation. The name "two-factor ANOVA without interaction" considered in Chapter 10 does not imply that there is no interaction; it merely means that interaction is not examined in the calculation.

We could be comparing three wines, *A, B,* and *C*. Should all judges rate *A* the worst and *C* the best, the trend between wines is the same for all judges. The trend does not depend on which judge is doing the testing; in statistical jargon, "wines" and "judges" are said not to interact. On the other hand, if some judges rate *A* the worst wine and *C* the best, while others rate *C* the worst and *A* the best, the trend between the wines is not the same for all judges; there is no consistency over the judges. Should you ask how the wines are rated, the answer would depend on which judge was doing the rating. The trend over wines depends on which particular judge does the tasting; "wines" and "judges" are said to interact.

Interaction can be understood by considering a graph of the mean values under various combinations of treatments. Consider our wine example. Let us imagine that three judges (or subjects S_1, S_2, S_3) taste wines *A*, *B*, and *C* and rate them according to their quality (using a scale that fulfills assumptions of normality and homoscedasticity, so that we can use ANOVA). Let us suppose that the judges use the quality rating scale differently: S_1 tends to give high scores, S_3 low scores, and S_2 middle-range scores. A graph of scores for S_1 will be above that of S_2, which in turn will be above that of S_3. Let us suppose that all judges rate wine *C* as the best, *A* the worst, and *B* in the middle, so that *C* scores are the highest and *A* scores the lowest. A graph of the results would be as follows:

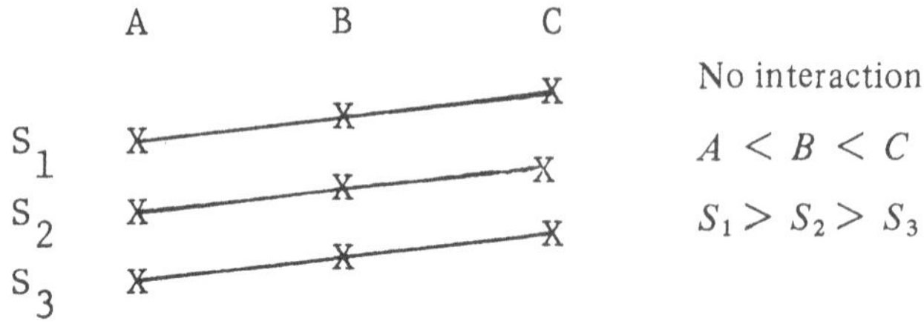

It can be seen that all three judges rate *C* the highest and *A* the lowest despite the fact that S_1 always gives higher scores and S_3 lower scores. So an ANOVA would find significant differences between the wines when mean values were taken for each wine over all the judges; LSD tests would indicate the overall mean score for *C* was highest and *A* lowest ($A < B < C$). In the same way, if the mean score over all three wines was taken for each judge, S_1's scores would be the highest, S_3's the lowest (significant between judges effect: $S_1 > S_2 > S_3$). Furthermore, the trend over the wines ($A < B < C$) is the same for each judge; the lines joining the points are parallel. The trend over the wines is consistent for each judge; there is no interaction between "wines" and "judges."

Now consider the case when all judges rated the wines as the same. The ratings for *A*, *B*, and *C* would be the same for a given judge. Thus, if mean values over all three judges were taken for each wine, they would be the same (no significant between-treatments effect for wines, $A = B = C$). Again, if S_1 characteristically tended to give higher scores and S_3 lower scores, S_1's graph would be above S_2's and S_3's would be lower (significant between-subjects effect: $S_1 > S_2 > S_3$). The graph would be as follows:

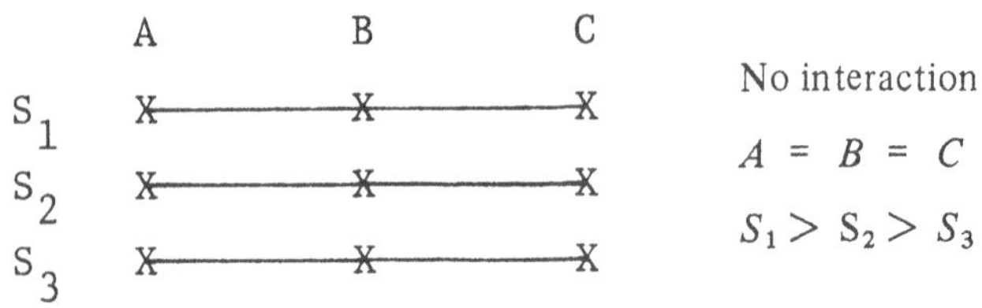

Again, the lines on the graph are parallel. The trend over wines is the same for each judge. This consistency is expressed by saying there is no "judges X wines interaction."

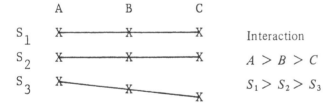

A further case can be considered should all judges consistently rate A as the best wine, C the worst, and B intermediate (significant between-wines effect: $A > B > C$). Again, S_1 tends to score high, S_3 low (significant between-judges effect: $S_1 > S_2 > S_3$). Again, all judges are consistent in their rating trend for wines A, B, and C. The graphs are parallel; there is no "wines X judges interaction" as follows:

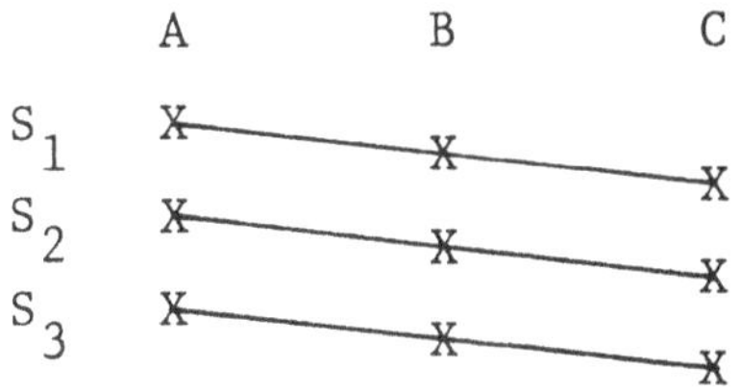

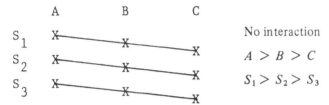

No interaction

$A > B > C$

$S_1 > S_2 > S_3$

In these three examples, the trend over the wines has been the same for each judge. If one judge says that A is better than B, all judges will. The trend over wines is not specific to any given judge; it is a general trend for all judges. All judges are consistent. In statistical jargon, there is no wines X judges interaction.

But imagine the first two judges (S_1 and S_2) rating the wines as equal but the third judge (S_3) rating the wines differently: A best, B intermediate, and C worst. Again, if S_1 tended to give higher scores, S_2 intermediate, and S_3 lower scores, the graph would be as follows:

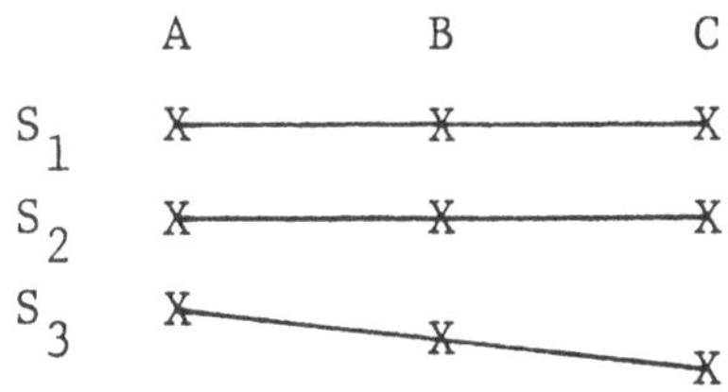

Interaction

$A > B > C$

$S_1 > S_2 > S_3$

The graphs are no longer parallel. The graphs for S_1 and S_2 are parallel (both rated the wines equally) but the graph for S_3 is not parallel (S_3 rated the wines as different). Thus, if you asked how the wines were rated, the answer would vary depending on which judge's ratings you were using. There is no consistency over judges—there is *interaction between wines and judges*.

Again, the mean of the scores over all the wines for S_1 will be higher than for S_2, whose mean will in turn be higher than for S_3 (significant between-judges effect $S_1 > S_2 > S_3$). What about between-treatments effects for wines?

S_1 and S_2 rate them as the same, but S_3 rates them $A > B > C$. Should the mean ratings for S_1, S_2, and S_3 be taken, the mean for A will be the highest and for C the lowest, with B intermediate. Should they be significantly different (the effect due to S_3 be sufficiently large), there will be a significant between-treatments effect for wines in the ANOVA ($A > B > C$).

Naturally, there are many ways that S_1, S_2, and S_3 can be inconsistent in their ratings of wines and give a judges X wines interaction. Consider the example below, where S_1 rates them: $A < B < C$, S_2 says: $A = B = C$, and S_3 states: $A > B > C$; again S_1's tends to give higher values, S_2 intermediate, and S_3 lower values. Thus

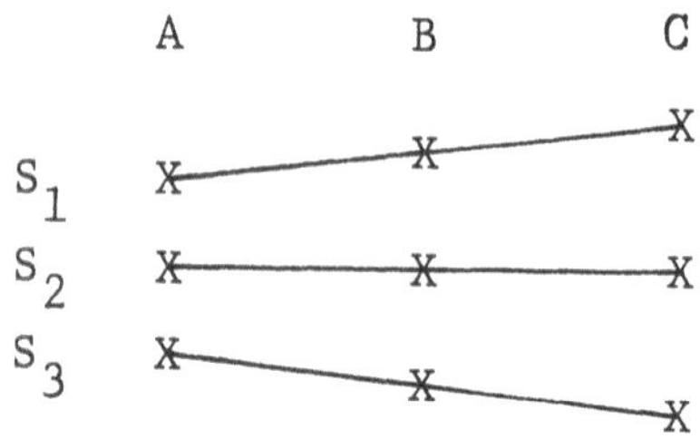

Interaction

$A = B = C$

$S_1 > S_2 > S_3$

Again, there will be a significant between-judges effect ($S_1 > S_2 > S_3$) and a significant judges X wines interaction. Should mean scores over the three judges be taken for each wine, it is possible that the opposing trends shown by judges S_1 and S_3 will cancel so that the mean scores will not differ; there will be no significant between-treatments effect for wines ($A = B = C$).

Interaction can be conceptualized as *lack of consistency*. If there is no consistency in how wines are rated over the judges, there is interaction. If all judges rate the wines with the same trend (e.g., $A < B < C$: graphs parallel), they are consistent; there is no interaction.

Note that in our example the judges were not consistent in how they rated the wines (S_1 says $A < B < C$; S_2 says $A = B = C$; S_3 says $A > B > C$; etc.). But for all wines, there was a consistency in that S_1 always gave higher ratings than S_2 than S_3. Thus interaction occurs because trends for wine are not consistent over judges. Hence

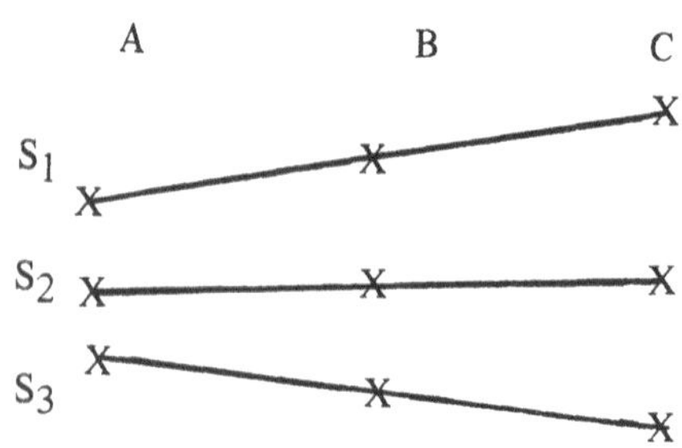

But interaction is also said to occur if both the trend for wines is inconsistent over judges and the trend for judges is inconsistent over wines. Hence

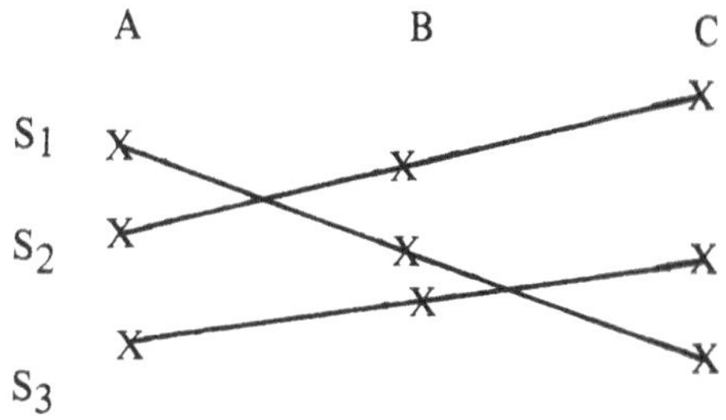

Here S_1 says: $A > B > C$ while S_2 says: $A < B < C$. But also for wine A, $S_1 > S_2 > S_3$ while for wine C, $S_2 > S_3 > S_1$. So, inconsistency can occur for one or for both factors. Both are registered by the ANOVA merely as interaction.

Thus, should the graphs for each subject not be parallel, there is interaction. An ANOVA will also detect interaction, as well as differences between treatments, as we shall see later in this chapter. For interaction, H_0 states that there is no interaction, that the graphs are parallel. The ANOVA tests whether the interaction (or lack of consistency over judges or the deviation of the graphs from parallel) is significant or just a chance effect.

We have concentrated on a single example of judges rating wines; one factor was wines (A, B, C) and the other was judges (S_1, S_2, S_3). But the argument is, of course, general. A, B, and C can represent experimental treatments but the second factor need not be subjects; it could be another set of treatments for a completely randomized design (different subjects in each combination of treatments). For instance, fabrics could be rated for comfort in different climates. In a cold climate, the thick fabric may be rated as being more comfortable than the thin fabric; in a hot climate the thin fabric may be rated as being more comfortable. Thus we have fabrics as one factor and climate as the second factor. We will have a fabrics × climate interaction; the trend for comfort over the fabrics will vary depending on the climate "treatment" we are in. On average, there may be no overall differences between fabrics, the advantage of the thick fabric in the cold weather being offset by the disadvantage in the hot weather. But although there may be no overall significant between-fabrics effect, the significant interaction alerts us to the fact that the fabrics did in fact have a profound effect; it just so happened that the effect was different in different climates, and the different trends balanced each other out in our ANOVA.

We can thus draw the same graphs that we drew for the wines (A, B, C) and the subjects (S_1, S_2, S_3) in a more general form. One factor is A, B, C (say fabrics) and the second factor is 1, 2, 3 (say climates). A, B, C and 1, 2, 3 can be any combination of factors. Should both factors be experimental treatments the design will be completely randomized; should one factor be subjects, the design will be repeated measures. (See Section 16.20 for further discussion of this point.)

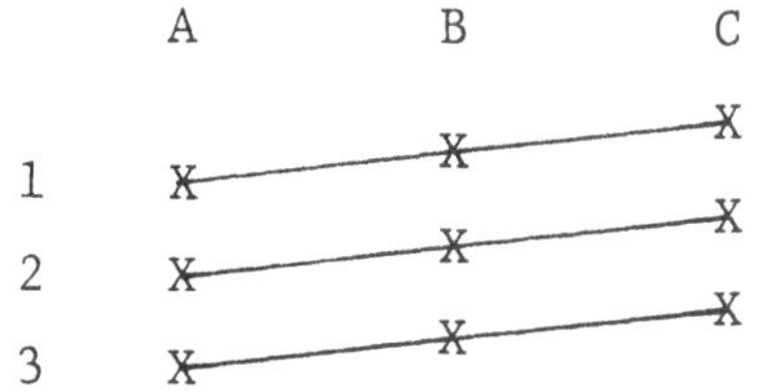

No interaction

Whether subjects are in condition 1, 2, or 3, they still give higher scores in condition *C* than in *B* than in *A*, so there is no interaction. Again

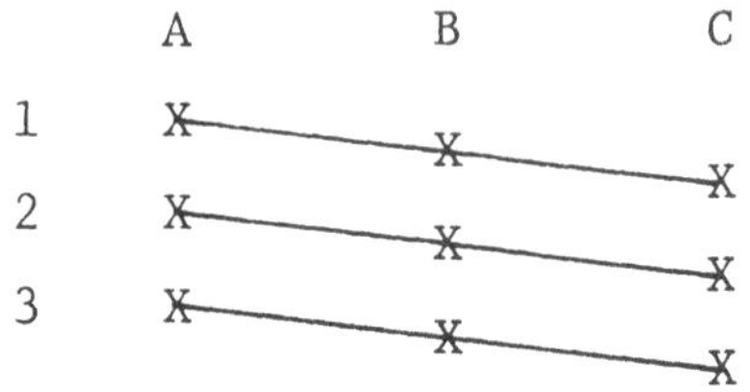

No interaction

The trend is the same over *A*, *B*, and *C* regardless of conditions 1, 2, and 3, so there is no interaction. But consider

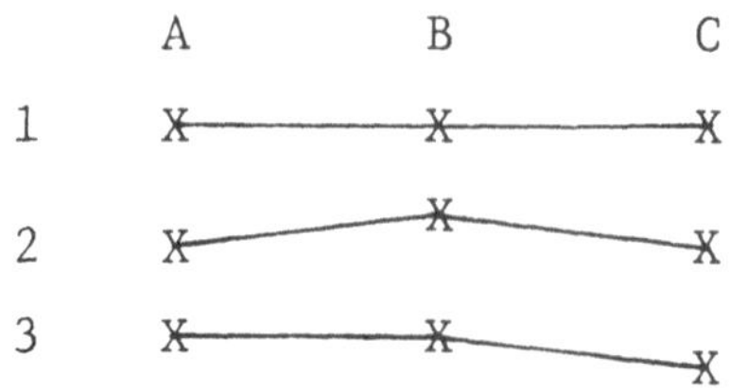

Interaction

Here the trend over *A*, *B*, *C* depends on the condition 1, 2, or 3. In condition 1 there is no change; in condition 2 there is an increase from *A* to *B* followed by a decrease down to *C*; in condition 3 the only change is from *B* to *C*.

11.2 INTERPRETATION OF INTERACTION AND BETWEEN-TREATMENTS *F* VALUES

In ANOVA, of course, we get F values which tell us whether or not we have differences between treatments. As we shall see later, the ANOVA also provides F values telling us whether we have any significant interactions or not. What we have to do in interpreting the ANOVA is to be able to reconstruct the graph in our mind or from our data and interpret the trends from a set of significant or nonsignificant F values. It is worth considering a few cases to assist us in making these interpretations.

Consider the results for subjects in three treatments A, B, and C for one factor and simultaneously in three treatments 1, 2, and 3 for a second factor. We will talk here in general terms: A, B, C and 1, 2, 3. Let the scores in treatment 1 be higher than in treatment 2 than in treatment 3. Let the scores for A, B, and C be the same in treatment 1, increase for treatment 2 ($C > B > A$), and decrease for treatment 3 ($C < B < A$). Thus there is interaction; the trend over A, B, C varies with treatment 1, 2, 3. This is represented below, together with scores over treatments 1, 2, and 3 for each treatment A, B, and C. The mean for A is computed by averaging the A values in treatment 1, 2, and 3; the same is done for B and for C. We can see that on average the mean for treatment A is greater than for B than for C, giving significant between-treatments effects for this factor ($A > B > C$). But the interaction tells us that $A > B > C$ is not true for each treatment 1, 2, or 3. This is represented in Figure 11.1.

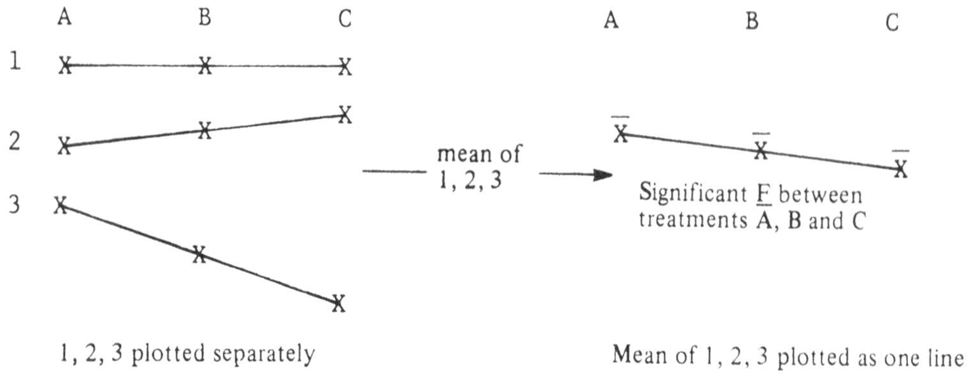

Figure 11.1 Mean values for A, B, and C.

Similarly, we can take mean scores for each treatment 1, 2, and 3 (mean over treatments *A*, *B*, and *C*). As the scores in treatment 1 were always higher than in treatment 2, which in turn were higher than in treatment 3, the mean for treatment 1 will be the highest and for 3 the lowest, with 2 intermediate. This will give a significant between-treatments effect for treatments 1, 2, and 3 ($1 > 2 > 3$). This is represented in Figure 11.2, where there is a significant between-treatments effect for *A*, *B*, *C* ($A > B > C$) and for 1, 2, 3 ($1 > 2 > 3$), and a significant interaction between the two factors.

ANOVA indicates only whether or not the between-treatments effects and the interaction are significant. This information allows us to interpret the graph of our data. To help acquire this skill, it is worth considering the examples in Figure 11.3 before proceeding further. Note that in the three last examples there were no overall between-treatments effects for *A*, *B*, *C*. However, in the first two there was interaction, while in the last one there was not. Only in the last example were all the values equal; only in this case with no interaction did the lack of between-treatments effects for *A*, *B*, and *C* indicate that all these treatment means were always equal in treatments 1, 2, and 3. In the other cases, the equality was due to opposite trends canceling; a trend in treatment 1 might cancel a trend in treatment 3, so that overall the means showed no differences. Also, note that in two of these examples $1 = 2 = 3$ and $A = B = C$. But the two cases are very different; one had interaction, the other did not.

Thus, just because there is no between-treatments effect for a given factor, it does not mean that the treatment means are always the same for each

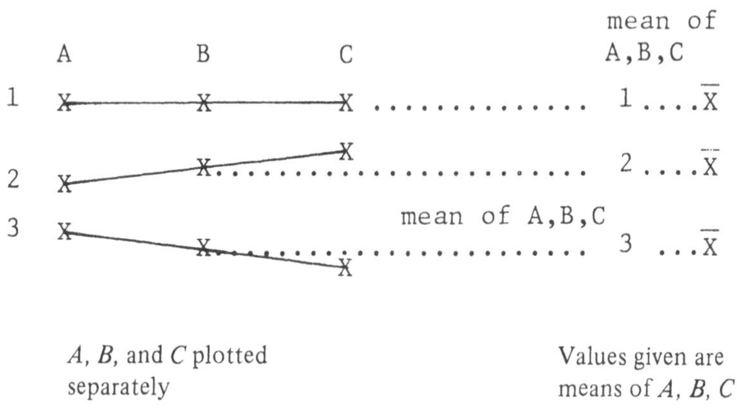

Figure 11.2 Mean values for 1, 2, and 3.

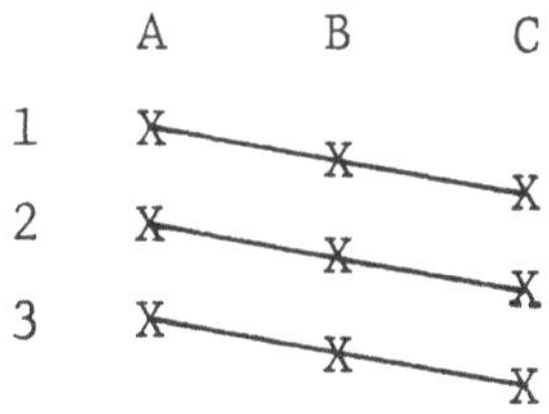

1 > 2 > 3	between (1, 2, 3) F significant
$A > B > C$	between (A, B, C) F significant
No interaction	interaction F not significant

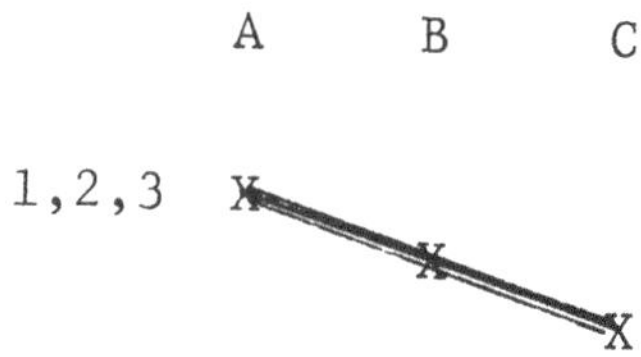

1 = 2 = 3	between (1, 2, 3) F not significant
$A > B > C$	between (A, B, C) F significant
No interaction	interaction F not significant

All three values fall on top of each other

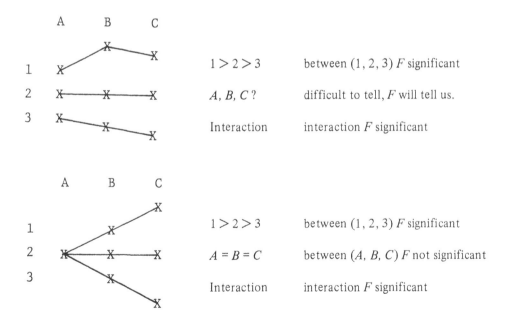

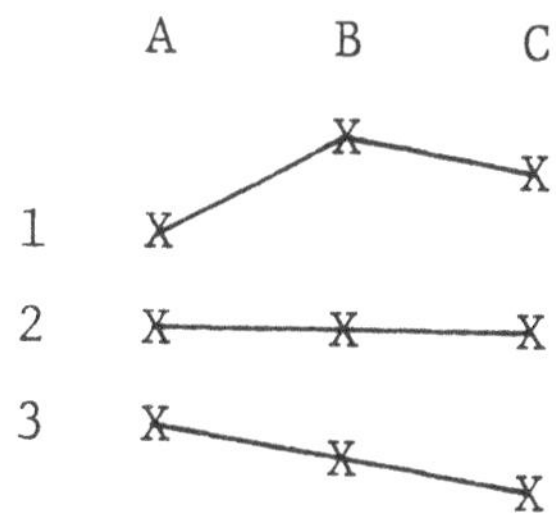

1 > 2 > 3	between (1, 2, 3) F significant
A, B, C ?	difficult to tell, F will tell us.
Interaction	interaction F significant

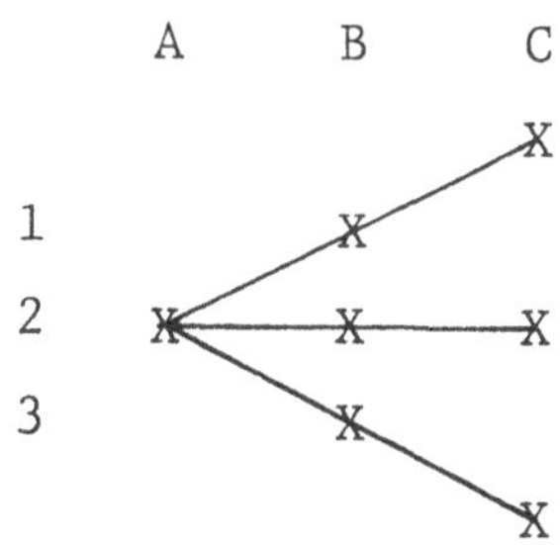

1 > 2 > 3	between (1, 2, 3) F significant
$A = B = C$	between (A, B, C) F not significant
Interaction	interaction F significant

Figure 11.3 Examples of interaction and between-treatments F values.

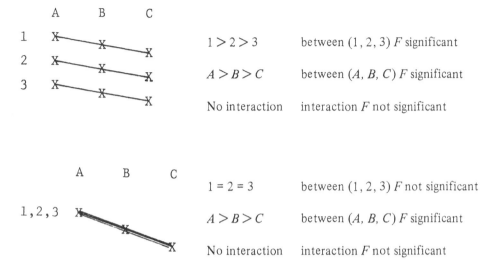

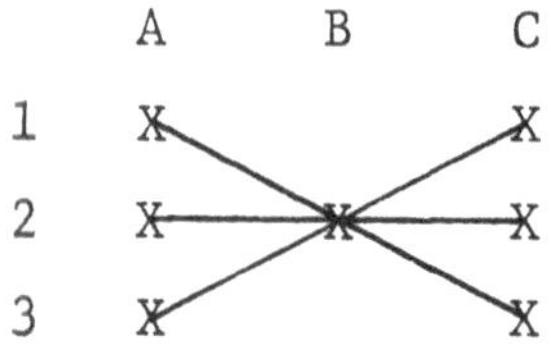

1 = 2 = 3 between (1, 2, 3) F not significan

$A = B = C$ between (A, B, C) F not significa

Interaction interaction F significant

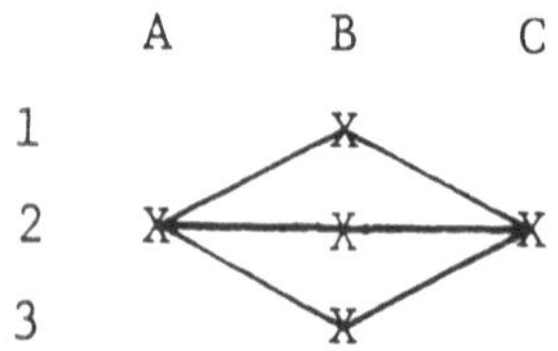

1 > 2 > 3 between (1, 2, 3) F significant

$A = B = C$ between (A, B, C) F not significa

Interaction interaction F significant

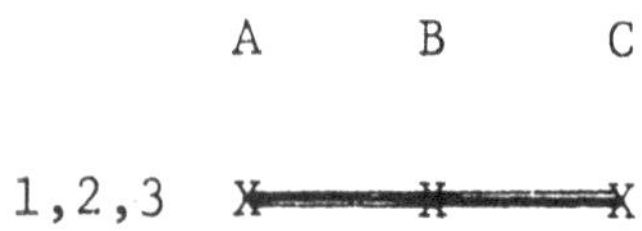

1 = 2 = 3 between (1, 2, 3) F not significan

$A = B = C$ between (A, B, C) F not significa

No interaction interaction F not significant

Figure 11.3 *(continued)*

condition of the other factor; interaction indicates whether they are not. It is important to remember this when interpreting the results of ANOVA. An insignificant between-treatments effect does not necessarily mean that the treatments had no effect on our data; they may have had profound effects which just happened to cancel each other out when overall mean values were taken.

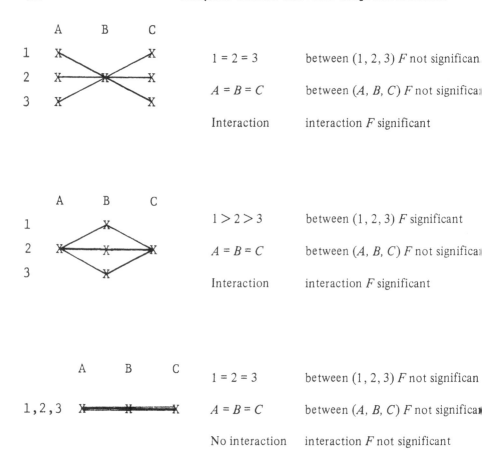

11.3 HOW TO COMPUTE A TWO-FACTOR ANOVA WITH INTERACTION

We have discussed the concept of interaction at great length with many examples. This is because the beginner often finds this concept difficult. Having discussed the concept, it is now time to discuss how the ANOVA indicates whether the interaction is significant (lack of consistency in our samples of data indicates a lack of consistency in the population) or whether the interaction is not significant (any lack of consistency in our samples of data is merely a chance variation and does not indicate any significant interaction in the population).

In fact, the ANOVA merely produces and extra F value for interaction, which is treated just like between-treatments F values. Should it be equal to or exceed the appropriate value in Table G.11 (for its respective df values), the interaction is significant (reject H_0). All that remains is to see how we get the appropriate F value from the data. To do this it is worth considering how these ANOVA designs have developed.

The *one-factor completely randomized design* compares the variation between treatments *A, B, C* with the variation within treatments; if it is greater than the within-treatments variation, it is said to be significantly large. The variation within treatments occurs for several reasons; it occurs because scores come from different subjects and also because of the many other influences that might occur in the experiment, which are not due to treatments *A, B*, and *C* (e.g., fatigue, distractions, motivation, subject-experimenter interaction, variations in the instructions, etc.). In fact, any variables which are not deliberately manipulated by the experimenter will give within-treatments or error variation. The term *error* is preferable because it indicates experimental error or variation in the scores not deliberately manipulated by the experimenter. Further designs merely pull out parts of the error for separate examination.

The *two-factor repeated-measures design* pulls out from the error the effect of controlled variation due to judges. We then have an effect due to treatments *A, B, C* and a second effect due to variation between judges, which we then compare to the remaining (smaller) error.

Again, the error remaining is due to all those remaining variables (after extracting *A, B, C* and judges) which could alter the score. These will be distractions, variation in the presentation of instructions, etc. One other source of variation included in error is the interaction. Scores may vary due to *A, B, C* and due to judges, but they will also vary due to how these two interact. For example, let us assume that we have four trained judges rating two toothpastes *A* and *B*. Three of the judges may rate *A* better than *B*, while the fourth thinks the reverse is true. Two judges give high scores and two give low scores. There would be a significant between-toothpastes effect (*A* better than *B*, despite the disagreement of the fourth judge). There would be a significant between-judges effect with two judges always giving high scores and two always giving low scores (two high means, two low means). There would also be a significant interaction because the fourth judge disagreed with the other two. Thus any given score would vary depending on whether it was for toothpaste A or toothpaste B (tends to be higher for toothpaste A), and also whether it comes from the two high-scoring or two low-scoring judges. It would also vary depending on the toothpaste $\times$ judges interaction; that is, it would tend to be higher for toothpaste A unless the scores came from the fourth judge, in which case the score would be lower for A. The score would not depend only on whether the toothpaste was A or B and whether the judges were the low-scoring or high-scoring judges; it would also depend on the particular set of treatments present in these two interacting factors. It would not be enough to specify whether the score came from the toothpastes or judges who generally give high (or low) scores; the particular judge and toothpaste must be specified for us to tell whether the score will be high or low. There is no consistency in the between-treatments effects over all the judges. This inconsistency or interaction is a further source of variation in the scores which can be studied independently once it has been extracted from the error term. In the two-factor design without interaction, the interaction effect on the scores remains part of the error together with other variables not manipulated by the experimenter. In the two-factor ANOVA with interaction, the interaction term is pulled out from the error together with the between-subjects effect, leaving an even smaller error term. Again, this smaller error term makes the calculated F values larger, giving an even more powerful test.

Thus the two-factor design without interaction leaves the effects of interaction in the error term; the two-factor design with interaction partitions out the interaction from the error term so that it can be examined as a separate entity. Thus, as we have developed new designs, we have gradually pulled more factors out of the error term. This can be represented diagrammatically (see Figure 11.4). Imagine that SS_A is the between-treatments sum of squares for the factor ABC, SS_S is for subjects, SS_E is the error sum of squares, and $SS_{A\times S}$ is the interaction sum of squares for the two factors "A, B, C" and "subjects."

Of course, the second factor need not be judges as we have said here; it may be a second set of experimental treatments. Furthermore—and this is worth repeating because it is a common source of error—the *two-factor ANOVA without interaction does not mean that there is no interaction; it merely means that the interaction is not calculated. The two-factor ANOVA with interaction does not mean that interaction occurs; it merely means that the interaction term is calculated.*

The question arises as to why the interaction is not calculated for the two-factor design without interaction. Is it sheer laziness? No! It is generally because there is insufficient data for the calculation to be made. We have generally

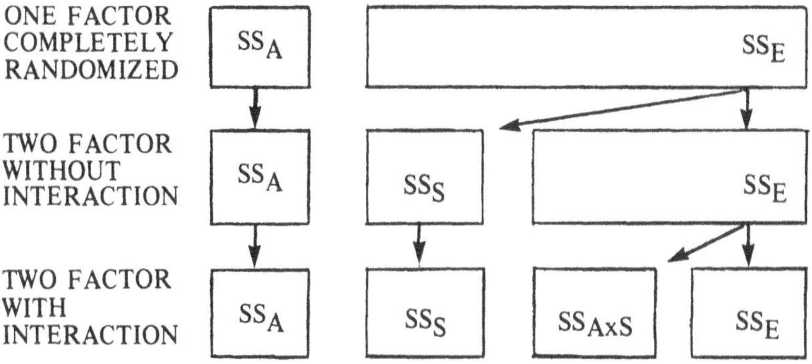

Figure 11.4 Extraction of terms from the error.

dealt only with the case where there is only one score for each combination of conditions:

	A	*B*	*C*
1	X	X	X
2	X	X	X
3	X	X	X

For interaction to be calculated we must have more than one score for each combination of conditions.

	A	*B*	*C*
1	XXX	XXX	XXX
2	XXX	XXX	XXX
3	XXX	XXX	XXX

We will consider why this is so later (Section 11.7). But first let us see how to calculate this interaction F value.

11.4 WORKED EXAMPLE: TWO-FACTOR ANOVA WITH INTERACTION

We will demonstrate the calculation with an example. Let us suppose that three wines A, B, and C were assessed under two tasting conditions α and β. (α and β may be two different temperatures or they may be two separate occasions, such as a formal tasting and a dinner party.) There were three tasters per wine per condition, generating three scores for each combination of conditions. (Alternatively, α and β could be two different wine tasters, each making three replicate tastings of each wine.) More than one score in each combination of conditions allows us to calculate an F value for interaction. The scores were as follows:

	A	B	C
α	8, 4, 0	10, 8, 6	8, 6, 4
	$\bar{X} = 4$	$\bar{X} = 8$	$\bar{X} = 6$
β	14, 10, 6	4, 2, 0	15, 12, 9
	$\bar{X} = 10$	$\bar{X} = 2$	$\bar{X} = 12$

The data can be represented simply by Figure 11.5, showing the means for each combination of treatments: αA, αB, αC, βA, βB, βC. It can be seen that there is interaction—the trend over ABC is different for α and β—but is it significant? Furthermore, β scores on the whole are higher than α scores, but is this between-treatments effect significant? Also, scores in condition C are highest and in condition B lowest, but is this between-treatment effect significant? The calculation follows in a predictable manner.

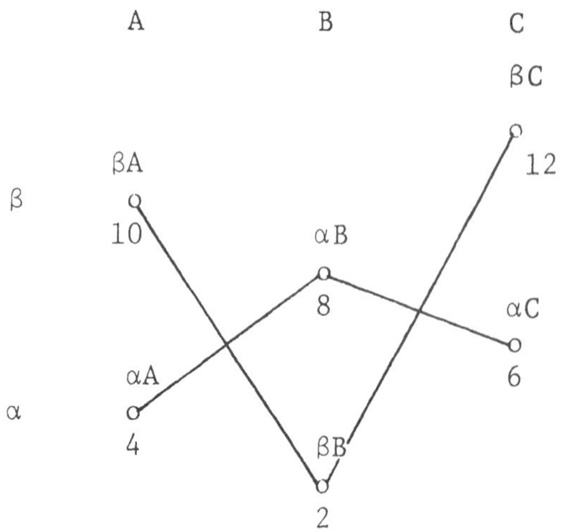

Figure 11.5 Combinations of treatments.

We display the data as we did before, in a matrix of all the original data (this is called the *whole matrix*):

		A	B	C	...	
$k_\alpha = 2$	α	8, 4, 0	10, 8, 6	8, 6, 4	$k_A = 3$ $T_\alpha = 54$	$n_\alpha = 9$
	β	14, 10, 6	4, 2, 0	15, 12, 9	$T_\beta = 72$	
		$T_A = 42$ $n_A = 6$	$T_B = 30$	$T_C = 54$	$T = 126$ $N = 18$	

However, we now introduce a new step. We add the scores for each combination of conditions ($8 + 4 + 0 = 12$; $10 + 8 + 6 = 24$; etc.) and present the totals in the cells of a *totals matrix* (this is sometimes called the *$A\alpha$ two-way matrix*):

	A	B	C
α	12	24	18
β	30	6	36

$n_{cell} = 3$

Again, the various totals T_A, T_B, T_C, T_α, T_β, etc., can be obtained from this matrix. Each total is the total of three values, hence $n_{cell} = 3$. There are six cells, hence $k_{cell} = 6$.

We then proceed with the calculation in the usual way. First, we go to the original data, the whole matrix:

$$\text{Correction term, } C = \frac{T^2}{N} = \frac{126^2}{18} = 882$$

$$\text{Total, } SS_T = 8^2 + 4^2 + 0^2 + \cdots + 9^2 - C = 1198 - 882 = 316$$

$$df_T = N - 1 = 18 - 1 = 17 \qquad df_T = \text{total number of scores} - 1$$

$$ABC,\ SS_A{}^* = \frac{T_A^2 + T_B^2 + T_C^2}{n_A} - C = \frac{42^2 + 30^2 + 54^2}{6} - C$$

$$= 930 - 882 = 48$$

$$df_A = k_A - 1 = 3 - 1 = 2$$

$$df_A = \text{number of treatments } (ABC) - 1$$

---

*We will now use SS_A to denote the between-treatments sum of squares rather than SS_B. The B in SS_B stood for "between treatments." The A in SS_A stands for treatments A, B, C. This notation is more suitable for advanced ANOVA designs where there may be several factors [e.g., α, β, γ (SS_α), a, b, c, (SS_a)].

So

$$SS_\alpha = \frac{T_\alpha^2 + T_\beta^2}{n_\alpha} - C = \frac{54^2 + 72^2}{9} - 882 = 900 - 882 = 18$$

$$df_\alpha = k_\alpha - 1 = 2 - 1 = 1 \qquad df_\alpha = \text{number of treatments } (\alpha\beta) - 1$$

It is worth noting that n_A is the number of scores from the original data (whole matrix) which are added together to give the totals T_A, T_B, or T_C (here $n_A = 6$), while n_α is the number of scores added to obtain T_α or T_β ($n_\alpha = 9$). This is obvious should you obtain your totals from the whole matrix. However, the totals (T_A, T_B, etc.) can also be obtained from the totals matrix; here it is not always obvious how many scores made up each total. So n values should be chosen with care.

Now we change the procedure somewhat. We go to the totals matrix. These totals are the result of three sources of variation. They are determined by which treatment they are in for the ABC factor and they are also determined by which treatment they are in for the $\alpha\beta$ factor. Furthermore, they are also determined by interaction effects between the ABC and the $\alpha\beta$ factor. We have seen before that the sums of squares are additive; we have used this to obtain SS_E by subtraction (see Section 8.3). Well, in a similar manner we can see that the sums of squares derived from the cell totals in the matrix (cell-total SS) are made up of SS_A (ABC sums of squares), SS_α ($\alpha\beta$ sums of squares), and $SS_{\alpha\times A}$ (interaction sums of squares). Thus

$$\text{Cell-total } SS = SS_A + SS_\alpha + SS_{\alpha\times A}$$

We can calculate the cell-total SS from the cell totals in the totals matrix; we thus have a way of obtaining the interaction sum of squares by subtraction (knowing SS_A and SS_α).

$$(\alpha A)\text{ Cell-total } SS = \frac{T_{\alpha A}^2 + T_{\alpha B}^2 + T_{\alpha C}^2 + T_{\beta A}^2 + T_{\beta B}^2 + T_{\beta C}^2}{n_{\text{cell}}} - C$$

$$= \frac{12^2 + 24^2 + 18^2 + 30^2 + 6^2 + 36^2}{3} - 882$$

$$= 1092 - 882 = 210$$

$$\text{Cell-total } df = \text{number of cells in the totals matrix} - 1$$

$$= 6 - 1 = 5$$

The interaction sum of squares is then obtained by subtraction:

Interaction ($\alpha \times A$), $SS_{\alpha\times A} = \text{cell-total } SS - SS_A - SS_\alpha = 210 - 48 - 18 = 144$

The interaction df, $df_{\alpha\times A}$, is obtained by subtraction in the same way:

$$df_{\alpha\times A} = df_{\text{cell}} - df_A - df_\alpha = 5 - 2 - 1 = 2$$

Alternatively, $df_{\alpha \times A}$ can be calculated by multiplying df for the factors involved in the interaction. Thus

$$df_{\alpha \times A} = df_\alpha \times df_A = 1 \times 2 = 2$$

This is a general rule for all interactions.

The error is obtained in the usual way, subtracting all terms from the total.

$$\text{Error, } SS_E = SS_T - SS_A - SS_\alpha - SS_{\alpha \times A} = 316 - 48 - 18 - 144 = 106$$

$$df_E = df_T - df_A - df_\alpha - df_{\alpha \times A} = 17 - 2 - 1 - 2 = 12$$

The ANOVA table can be expressed theoretically as shown in Table 11.1.

Table 11.1 Two-Factor ANOVA Table

Source	SS	df	MS	F
Total	SS_T	df_T		
(ABC)	SS_A	df_A	$MS_A = \dfrac{SS_A}{df_A}$	$\dfrac{MS_A}{MS_E}$
$(\alpha\beta)$	SS_α	df_α	$MS_\alpha = \dfrac{SS_\alpha}{df_\alpha}$	$\dfrac{MS_\alpha}{MS_E}$
$(ABC) \times (\alpha\beta)$	$SS_{\alpha \times A}$	$df_{\alpha \times A}$	$MS_{\alpha \times A} = \dfrac{SS_{\alpha \times A}}{df_{\alpha \times A}}$	$\dfrac{MS_{\alpha \times A}}{MS_E}$
Error	SS_E	df_E	$MS_E = \dfrac{SS_E}{df_E}$	

For this example, the table is as follows:

Source	SS	df	MS	F
Total	316	17		
(ABC)	48	2	24	24/8.83 = 2.72
$(\alpha\beta)$	18	1	18	18/8.83 = 2.04
$(ABC) \times (\alpha\beta)$	144	2	72	72/8.83 = 8.15**
Error	106	12	8.83	

The F values in Table G.11 that are relevant here are those for $df = 2$ and 12 and $df = 1$ and 12. These are as follows:

df	5%	1%	0.1%
2, 12	3.88	6.93	12.97
1, 12	4.75	9.33	18.64

Only the value for the interaction is significant at $p < 0.01$.

Thus there is a significant interaction but the between-treatments effects are not significant. Thus, when averaged over both tasting conditions (α and β), there is no difference between the wines, while when averaged over the three wines (A, B, and C), the tasting conditions have no significant effect. The interaction, though, is significant. This indicates that the trend over the wines A, B, and C is not the same for the conditions α and β, or alternatively the trend for α and β is not the same for wines A, B, and C. This can be seen from the diagram of the mean scores. Even though the means for the three wines do not differ on average it does not mean that the wines are all the same. The trend for the wines A, B, and C, in conditions α and β, are very different but cancel each other out. We have an example where the wines and tasting conditions do have a significant effect, but when averaged over all conditions of this experiment, the effects cancel.

11.5 ASSUMPTIONS AND WHEN TO USE THE TWO-FACTOR DESIGN WITH INTERACTION

It should be noted that the assumptions for the two-factor ANOVA with interaction are the same as for the two-factor ANOVA without interaction (see Section 10.4). The assumptions are general for these and all higher-order designs. The position of this design on the Key to Statistical Tables should be noted. As with the design without interaction (see Section 10.1), it falls into the general class of two or more factor designs. If α and β were judges, the test would be classified with the Repeated Measures ANOVA.

11.6 WHAT IF THE INTERACTION IS NOT SIGNIFICANT?

If the interaction (or interactions in higher-order designs) is *not significant*, it is customary to pool it back into the error. The sum of squares is added to SS_E and the degrees of freedom added to df_E. This gives a new MS_E which is now used as a new denominator for all the F tests; a new F is recalculated for all effects. Should this process result in further interactions becoming insignificant (for higher-order designs), the process is repeated, giving an even newer MS_E which is used to recalculate additional, even newer F values.

The logic behind this is that if an interaction F value is not significant, there was no interaction in the population. This being so, no interaction term should have been separated out from the error. Thus the relevant sums of squares and degrees of freedom are returned to the error term and the new MS_E used in the F tests. This procedure is not used for insignificant between-treatments effects which are of a different nature, because they are deliberately designed into the experiment.

Pooling the insignificant interaction terms is generally only a slight adjustment. The new MS_E term may be a little larger or smaller, while the probabilities of getting F on the null hypothesis will generally change only a little. This should not cause problems unless there is a slavish adherence to 5% or 1% cutoff points.

It is worth stressing that the pooling of insignificant interactions with the error is only a tradition and is not customary with all researchers. Some even oppose it strongly, arguing that insignificant interactions are not especially different from insignificant between-treatments effects and should not be treated differently. To pool or not to pool is a decision we will leave to you, dear reader. We will, however, follow this tradition in the text, so that readers can become familiar with the procedure.

$$\text{Pooled } SS_E = SS_E + SS_{\alpha \times A}$$

$$\text{Pooled } df = df_E + df_{\alpha \times A}$$

$$\text{New } MS_E = \frac{\text{pooled } SS_E}{\text{pooled } df}$$

11.7 SOMETIMES YOU CANNOT CALCULATE INTERACTION MEAN SQUARES

We stated earlier that you could not calculate the interaction term unless you had more than one score in each combination of conditions (Section 11.3). Having completed an ANOVA with interaction, we are now in a position to see why.

When we do *not* calculate the interaction terms, we do not use a totals matrix; we work only with the original data in the whole matrix. During the calculation, we subtract between-treatments sums of squares from the total sum of squares to obtain the error sum of squares:

	A	B	C
α	X	X	X
β	X	X	X

$$SS_E = SS_T - SS_A - SS_\alpha$$

This is true whether or not we have more than one score in each combination of treatment conditions, as long as we do not calculate interactions.

	A	B	C
α	XXX	XXX	XXX
β	XXX	XXX	XXX

$$SS_E = SS_T - SS_A - SS_\alpha$$

If we wish to calculate interactions, we must have more than one score in each combination of conditions. We must add the scores for each combination of conditions in the original data to give a set of totals; these we place in a totals matrix.

	A	B	C
α	XXX	XXX	XXX
β	XXX	XXX	XXX

	A	B	C
α	$T_{\alpha A}$	$T_{\alpha B}$	$T_{\alpha C}$
β	$T_{\beta A}$	$T_{\beta B}$	$T_{\beta C}$

The cell totals yield a cell total SS which is *different from* the total sum of squares, SS_T. To find the interaction SS, we take SS_A and SS_α from the cell total sum of squares:

$$\text{Interaction } SS_{\alpha \times A} = \text{cell-total } SS - SS_A - SS_\alpha$$

Should we take these values from the total sum of squares; we merely obtain the error value for the without-interaction design:

$$SS_E = SS_T - SS_A - SS_\alpha$$

If we now tried to obtain an interaction term from a design where there was only one score in each combination of conditions, the totals matrix would have exactly the same numbers as the whole matrix (the total of one number is the same number). The totals matrix would be the same as the whole matrix (original data). If SS_A and SS_α were subtracted, we would be taking these values from the original data (whole matrix) whether we liked it or not; the result would be the error term SS_E, for a without-interaction design.

Thus, unless we have a totals matrix that has numbers that are different from those in the original data, subtracting SS_A and SS_α will give the same answer: an error term. The two matrices must be different to allow the extra step of calculating $SS_{\alpha \times A}$, to partition the interaction out from the error.

How is more than one score obtained for each combination of conditions? In the repeated-measures design, where one factor (*A, B, C*) is experimental treatments and the second (S_1, S_2, S_3) is subjects, multiple scores are obtained by requiring the subject to repeat his performance in each condition; the subject will perform several trials or replicates (replications):

	A	*B*	*C*
S_1	XXX	XXX	XXX
S_2	XXX	XXX	XXX
S_3	XXX	XXX	XXX

replications under a given treatment

Should the design be completely randomized, with two sets of treatments (*A, B, C*) and ($\alpha\beta\gamma$), replicates are obtained by using more than one subject in each treatment:

	A	*B*	*C*
α	S_1 S_2 S_3	S_4 S_5 S_6	S_7 S_8 S_9
β	S_{10} S_{11} S_{12}	S_{13} S_{14} S_{15}	S_{16} S_{17} S_{18}
γ	S_{19} S_{20} S_{21}	S_{22} S_{23} S_{24}	S_{25} S_{26} S_{27}

more than one subject in each combination of treatments

Such a completely randomized design is not common in sensory evaluation, however.

We will see in Chapter 12 how these replications (or subjects) can be extracted as a third factor to give a three-factor ANOVA.

11.8 A CAUTIONARY TALE

In general, it could be said that a parametric test is generally more powerful than a nonparametric test; it will be better at detecting differences. However, imagine a two-factor design without interaction, used in a situation where there was in fact a high degree of interaction. Assume that there were five foods and consumers were being asked to rate them for like vs. dislike on some hedonic scale. Assume also that each consumer tasted each food only once; there were no replications. The statistical analysis would be a two-factor (consumers and foods) ANOVA without interaction, because there would be only one score in each combination of treatments. Now, if there was in fact a high degree of interaction, which is likely for hedonic testing (consumers are unlikely to agree on which foods they like the best), and the interaction sum of squares cannot be calculated, it will remain in the error term. This will have the effect of making SS_E, and sometimes MS_E, a large value so that the between-foods F value MS_{foods}/MS_E may be small and a between-foods effect not picked up as significant. Had the interaction been separated out from the error, the mean scores for the foods might have been significantly different, but as it had not been separated out, the mean scores for foods would not be significantly different. In this case, a significant majority of consumers may have rated the foods in a specific order of preference. The fact that this was a significant majority could be found from a binomial test. Furthermore, if the data were ranked in order of preference for each consumer, a Kramer test (see the Key to Statistical Tables) may also pick up significant differences between the foods. So in this case, ANOVA may not be as good at detecting differences as a nonparametric test.

One should watch out for this effect in hedonic testing, where there is likely to be a high degree of interaction. What can one do about it? Each consumer could be required to taste and rate each food sample more than once, so that the interaction term could be calculated. However, to spend this amount of time in consumer testing may be impractical. Furthermore, in the replications, the consumers may not really reconsider their liking for the food; they may merely remember their previous ratings and repeat them, suspicious that this was some sort of consistency check, which it would be.

Added to this are the difficulties of analyzing scaling data with parametric statistics (see Section 7.11 and Appendix F), compounded by the lack of skill that untrained consumers might have when performing such a task. It is better not to ask untrained consumers to use complex hedonic scales. An alternative is merely to ask consumers to rank the foods in order of preference. Ranking is fast and simple; we are generally familiar with and well practiced in ranking. Suitable nonparametric statistics could then be used for the data analysis.

11.9 ADDITIONAL WORKED EXAMPLE: TWO-FACTOR ANALYSIS OF VARIANCE WITH INTERACTION

The effect of irradiation on mold growth on genetically engineered 10-lb eating snails was studied. Snails from the same genetic precursor were treated with four different types of radiation, A, X, Y, and Z. They were examined later by a panel of nine untrained judges who assessed the mold growth visually by making simple paired comparisons with standard pictures. Their data were analyzed, producing scale values which ranged from −10 for mold growth much worse than a nonirradiated reference, through 0 for mold growth for the same as the reference, to +10 for mold growth far better than the reference. The judges rated the mold growth on two snails for each type of radiation. The mold growth scores are given below. It is assumed that these come from normally distributed populations with equal variance. Could the untrained judges detect changes in mold growth with irradiation treatment? Did all the judges agree on the effects of irradiation?

Original Data: Whole Matrix

Judge	Irradiation Treatments A		X		Y		Z		Total	
1	5	3	6	7	8	9	5	3	46	$n = 8$ per judge
2	5	4	7	5	7	8	3	4	43	
3	5	3	6	8	8	7	5	6	48	
4	6	6	7	8	8	8	4	5	52	
5	7	4	5	7	8	9	4	3	47	
6	5	5	6	8	7	9	3	5	48	
7	4	6	7	7	8	8	6	4	50	
8	5	5	8	8	7	9	4	3	49	
9	3	3	7	8	7	8	2	3	41	
	84		125		143		72			Grand total = 424
	$n = 18$ per radiation treatment									$N = 72$

A totals matrix can be constructed for each treatment, for example, for radiation treatment A, the total for judge 1 is $5 + 3 = 8$; judge 2, $5 + 4 = 9$; judge 3, $5 + 3 = 8$; etc.

Cell Totals Matrix

Judge	A	X	Y	Z	Total	
1	8	13	17	8	46	
2	9	12	15	7	43	
3	8	14	15	11	48	
4	12	15	16	9	52	number per cell = 2
5	11	12	17	7	47	
6	10	14	16	8	48	number of cells = 36
7	10	14	16	10	50	
8	10	16	16	7	49	
9	6	15	15	5	41	
Total	84	125	143	72		
Mean	4.67	6.94	7.94	4.00		

We now proceed with the ANOVA in the usual manner.

$$\text{Correction term, } C = \frac{(424)^2}{72} = \frac{179{,}776}{72} = 2496.89$$

$$\text{Total } SS = 5^2 + 3^2 + 6^2 + 7^2 + \cdots + 2^2 + 3^2 - C$$

$$= 2756 - 2496.89 = 259.11$$

$$\text{Total } df = N - 1 = 72 - 1 = 71$$

$$\text{Judges } SS = \frac{46^2 + 43^2 + 48^2 + \cdots + 41^2}{8} - C = \frac{20{,}068}{8} - C$$

$$= 2508.5 - 2496.89 = 11.61$$

$$\text{Judges } df = 9 - 1 = 8$$

$$\text{Treatments } SS = \frac{84^2 + 125^2 + 143^2 + 72^2}{18} - C = \frac{48{,}314}{18} - C$$

$$= 2684.11 - 2496.89 = 187.22$$

$$\text{Treatments } df = 4 - 1 = 3$$

$$\text{Cell-total } SS = \frac{8^2 + 13^2 + 17^2 + 8^2 + \cdots + 5^2}{2} - C$$

$$= \frac{5440}{2} - C = 2720 - 2496.89 = 223.11$$

$$\text{Cell-total } df = \text{number of cells} - 1 = 36 - 1 = 35$$

$$\begin{aligned}\text{Interaction } SS &= \text{cell-total } SS - \text{treatments } SS - \text{judges } SS\\ \text{(judges} \times \text{treatments)} &= 223.11 - 187.22 - 11.61 = 24.28\end{aligned}$$

$$\begin{aligned}\text{Interaction } df &= \text{cell-total } df - \text{treatment } df - \text{judges } df\\ \text{(judges} \times \text{treatments)} &= 35 - 3 - 8 = 24\end{aligned}$$

or, alternatively, the "judges × treatments interaction" df can be calculated by multiplying together the df for the respective factors.

$$df_{\text{treatments}} \times df_{\text{judges}} = 3 \times 8 = 24$$

$$\text{Error } SS = \text{total } SS - \text{treatments } SS - \text{judges } SS - \text{interaction } SS$$
$$= 259.11 - 187.22 - 11.61 - 24.28 = 36.00$$

$$\text{Error } df = \text{total } df - \text{treatments } df - \text{judges } df - \text{interaction } df$$
$$= 71 - 3 - 8 - 24 = 36$$

The ANOVA table is as follows:

Source	*SS*	*df*	*MS*	Calculated *F*
Total	259.11	71		
Judges	11.61	8	1.45	1.45 (NS)
Treatments	187.22	3	62.41	62.41***
Judges × treatments interaction	24.28	24	1.01	1.01 (NS)
Error	36.00	36	1.00	

The *F* values from Table G.11 are as follows:

df	5%	1%	0.1%
8, 36	2.21		
3, 36	2.86	4.38	6.78 approx.
24, 36	1.82		

Only the between-treatments *F* is significant. The between-judges and interaction *F* values are insignificant. Because the interaction is not significant, the interaction sum of squares and degrees of freedom can be pooled with the error term. Note that pooling is only a tradition; many researchers do not pool.

$$\text{New error } SS_E = SS_E + \text{interaction } SS = 36.00 + 24.28 = 60.28$$

$$\text{New error } df_E = df_E + \text{interaction } df = 36 + 24 = 60$$

This gives a new MS_E term (60.28/60 = 1.00), slightly larger yet hardly any different from the prior MS_E value. Thus we have a new ANOVA table:

Source	*SS*	*df*	*MS*	Calculated *F*
Total	259.11	71		
Judges	11.61	8	1.45	1.45 (NS)
Treatments	187.22	3	62.41	62.41***
Error	60.28	60	1.00	

The F values from Table G.11 are as follows:

df	5%	1%	0.1%
8, 60	2.10		
3, 60	2.76	4.13	6.17

Again, the between-treatments F value is significant ($p < 0.001$) while the between-judges value is not.

The question now becomes: Which are the treatment means that are different?

Irradiation treatment:	Z	A	X	Y
Mold-growth mean:	4.00	4.67	6.94	7.94

An LSD test will be used to determine which means are different.

$$\text{LSD} = t\sqrt{\frac{2MS_E}{n}} \qquad \text{where } MS_E = 1.01$$
$$n = 18, \text{ number of scores under each treatment}$$

$$t\ (\text{at } p = 0.001 \quad \text{and} \quad df_E = 60, \text{two-tailed}) = 3.460$$

Thus

$$\text{LSD} = 3.460\sqrt{\frac{2 \times 1.01}{18}} = 1.159$$

This is the yardstick used to determine which means are significantly different from each other. Applying this yardstick, we get

Irradiation treatment:	Z	A	X	Y
Mold-growth mean:	4.00	4.67	6.94	7.94

Thus the mold growth for irradiation treatments X and Y is significantly better than for treatments A and Z. However, there were no differences in means scores for judges and there was no "judges × irradiation treatment" interaction.

Now, to answer the specific questions. Did irradiation affect the mold growth on the snails? Because a zero score is given to a reference snail receiving no irradiation and all our irradiated snails received positive scores, all irradiation treatments had some effect; they all produced mold growth which was better than the reference. However, the treatments X and Y had a significantly greater effect ($p < 0.001$) than treatments Z and A.

There was no significant between-judges effect, so their mean scores over all the treatments did not differ significantly; this means that they all made mold-growth judgments in a similar way. This is to be expected because the judgments were simple comparisons. Did all the judges agree on the effects of irradiation? There was no "irradiation treatments × judges interaction," which means that all judges agreed on the trend for mold-growth scores, over the four irradiation treatments. They all rated treatments X and Y as yielding better mold-growth scores than Z or A. Thus the judges agreed on the effects of irradiation. Again, if these effects were sufficiently clear-cut, we would expect judges to agree.

12

Analysis of Variance: Three- and Four-Factor Designs

12.1 HOW TO COMPUTE A THREE-FACTOR ANOVA

With two factors (*ABC* and $\alpha\beta\gamma$) there are mean-square terms MS_A (between-treatments *ABC*), MS_α (between-treatments $\alpha\beta\gamma$), and an interaction $MS_{\alpha \times A}$ (interaction between *ABC* and $\alpha\beta\gamma$). It is possible to have three factors (*ABC*, $\alpha\beta\gamma$, and *abc*). For instance, in a completely randomized design judges could be scoring wines *A*, *B*, and *C* for quality under different lighting conditions α, β, γ (white, red, and green light) on consecutive days *a*, *b*, *c* (replications effect). Alternatively, one of the factors could be "subjects." Besides the three between-treatments effects, we can now have three two-way interactions $\alpha \times A$, $\alpha \times a$, $a \times A$, and even a three-way interaction $\alpha \times A \times a$. The two-way interactions are exactly what you would expect from the two-factor analysis of variance, while three-way interactions are a little more complex and will be dealt with in more detail later (Section 12.2). First, we will run through the calculation and then interpret the results in detail. The calculation is an extension of the calculation for the two-factor ANOVA. There we had a table of the original data (whole matrix) and also a table of the totals of the scores from each combination of treatments (totals matrix or two-way matrix). The totals matrix is used as a means of obtaining the two-way interaction. For a three-factor ANOVA, we have a table of the original data (whole matrix), but we find that we can break down the totals into more than one two-way matrix; we even find that we have a three-way matrix of totals. We will now see how this is done.

Let us assume that we are obtaining scores (*X* values) under three combinations of factors: one with treatments *A*, *B*, *C*, a second with treatments α, β, γ, and a third with treatments *a, b, c*. One of these factors may or may not be subjects (e.g., three subjects named *a, b,* and *c*). There must be more than one score in each combination of treatments for us to be able to get the highest-order (three-way) interaction. In fact, we can generalize and say that we need more than one score in each combination of factors to get the highest possible order interaction (n-way interaction in an n-factor design).

Whole Matrix

A table of the original data (X values) is first displayed; this is the whole matrix:

	A			*B*			*C*		
	α	β	γ	α	β	γ	α	β	γ
	abc	*abc*	*abc*	*abc*	*abc*	*abc*	*abc*	*abc*	*abc*
	XXX	XXX	XXX	XXX	XXX	XXX	XXX	XXX	XXX
	XXX	XXX	XXX	XXX	XXX	XXX	XXX	XXX	XXX
	XXX	XXX	XXX	XXX	XXX	XXX	XXX	XXX	XXX
Total	T.T.T.	T.T.T.	T.T.T.	T.T.T.	T.T.T.	T.T.T.	T.T.T.	T.T.T.	T.T.T.

Grand total = T

Number of readings = N here $N = 81$

Number in each treatment = $n_A = n_\alpha = n_a = 27$

$A\alpha a$ Three-Way Matrix

From the whole matrix it can be seen that there are three scores (X values) for each combination of conditions. These three values can be added to give total values ($T.$). We can now present these totals in a table; this is a three-way matrix because all three factors are represented in the table. One way of arranging the totals ($T.$ values) is as follows:

	A			*B*			*C*			
	α	β	γ	α	β	γ	α	β	γ	Total
a	T.	T.	T.	T.	T.	T.	T.	T.	T.	T_a
b	T.	T.	T.	T.	T.	T.	T.	T.	T.	T_b
c	T.	T.	T.	T.	T.	T.	T.	T.	T.	T_c
Total	$T_{\alpha A}$	$T_{\beta A}$	$T_{\gamma A}$	$T_{\alpha B}$	$T_{\beta B}$	$T_{\gamma B}$	$T_{\alpha C}$	$T_{\beta C}$	$T_{\gamma C}$	

n per cell = 3

Two-Way Matrices

From this three-way matrix of totals ($T.$ values), we can also construct two-way matrices like we had in the two-factor ANOVA. We could add the three totals for the α, β, and γ conditions to give totals for each combination of factors A, B, C and a, b, c (totals for Aa, Ab, Ac, Ba, Bb, Bc, Ca, Cb, Cc). This would be the Aa two-way matrix. In the same manner we could combine the three totals for the a, b, and c conditions to give a new set of totals for each combination of factors A, B, C and α, β, γ; this would give the $A\alpha$ two-way matrix. Finally, we could add the three totals for the A, B, and C conditions to give yet another set of totals for each combination of factors a, b, c and α, β, γ; this would give the αa two-way matrix. The three two-way matrices are as follows:

Aa Matrix

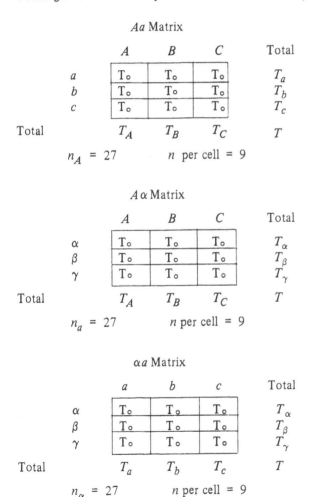

		A	*B*	*C*	Total
	a	T∘	T∘	T∘	T_a
	b	T∘	T∘	T∘	T_b
	c	T∘	T∘	T∘	T_c
Total		T_A	T_B	T_C	T

$n_A = 27$ $\qquad n$ per cell $= 9$

*A*α Matrix

		A	*B*	*C*	Total
	α	T∘	T∘	T∘	T_α
	β	T∘	T∘	T∘	T_β
	γ	T∘	T∘	T∘	T_γ
Total		T_A	T_B	T_C	T

$n_a = 27$ $\qquad n$ per cell $= 9$

α*a* Matrix

		a	*b*	*c*	Total
	α	T∘	T∘	T∘	T_α
	β	T∘	T∘	T∘	T_β
	γ	T∘	T∘	T∘	T_γ
Total		T_a	T_b	T_c	T

$n_\alpha = 27$ $\qquad n$ per cell $= 9$

Three-Factor ANOVA Computation Scheme

The calculation proceeds as expected. The correction term C and the total sum of squares SS_T are obtained from the whole matrix as usual. The between-treatments sums of squares, SS_A, SS_α, and SS_a are also obtained from the whole matrix in the usual way. Of course, SS_A, SS_α, and SS_a can also be calculated using the two- or three-way matrices; the between-treatments totals (T_A, T_B, T_C, etc.) are all readily obtained from them. One word of caution, however: Make sure that the denominator you use to calculate the between-treatments sum of squares in the number of original scores (from the whole matrix) that were added to make up that total. For example, there are 27 scores (X values) which make up the total T_A. Thus

$$SS_A = \frac{T_A^2 + T_B^2 + T_C^2}{n_a} - C$$

$$n_A = 27\text{, the number of scores that make up each total: } T_A,\ T_B,\text{ or } T_C$$

Having obtained the total sum of squares and the between-treatments sums of squares (and their associated degrees of freedom), the interactions are the next to be calculated. You start at the simplest (two-way) matrices to get the simplest (two-way) interactions. This is done exactly as before in the two-factor ANOVA. You then work up to higher-order matrices (three-way) so as to calculate higher-order (three-way) interactions. We will discuss three-way interactions later (Section 12.2). This strategy is general; for higher-order designs (four-, five-, six-factor) we start with the two-way matrices and then work up to three-way, four-way, five-way matrices, etc.

More of higher-order designs later. We will now proceed with our humble three-factor ANOVA. The calculation proceeds in the usual way:

From any matrix:

$$\text{Correction term, } C = \frac{T^2}{N} = \frac{T^2}{81}$$

From the whole matrix:

$$\text{Total, } SS_T = X^2 + X^2 + X^2 + X^2 + \cdots + X^2 - C$$

$$df_T = N - 1 = 81 - 1 = 80$$

where N = total number of scores so that df_T = total number of scores − 1

From any matrix (but safer to use the whole matrix):

Between-treatments ABC:

$$SS_A = \frac{T_A^2 + T_B^2 + T_C^2}{n_A} - C \qquad n_A = 27$$

$$df_A = \text{number of treatments} - 1 = 3 - 1 = 2$$

Between treatments $\alpha\beta\gamma$:

$$SS_\alpha = \frac{T_\alpha^2 + T_\beta^2 + T_\gamma^2}{n_\alpha} - C \qquad n_\alpha = 27$$

$$df_\alpha = \text{number of treatments} - 1 = 3 - 1 = 2$$

Between treatments abc:

$$SS_a = \frac{T_a^2 + T_b^2 + T_c^2}{n_a} - C \qquad n_a = 27$$

$$df_a = \text{number of treatments} - 1 = 3 - 1 = 2$$

So far, so good. Now we choose the simplest matrix to start calculating interactions. The simplest matrices are the two-way matrices. There are three of these; let us choose the $A\alpha$ matrix first.

From the Two-Way $A\alpha$ Matrix

The calculation proceeds just as it does for the two-factor ANOVA. We could have calculated the two between-treatments sums of squares from the totals in this matrix had we wished, rather than doing so from the whole matrix.

We calculate the sum of squares for the interaction between ABC and $\alpha\beta\gamma$, $SS_{\alpha \times A}$. This is done in the usual way, from the cell-total SS:

$$\text{Cell-total } SS = \frac{T_\circ^2 + T_\circ^2 + T_\circ^2 + \cdots + T_\circ^2}{n \text{ per cell}} - C \qquad n \text{ per cell} = 9$$

$$\text{Cell-total } df = \text{number of cells} - 1 = 9 - 1 = 8$$

$$(ABC) \times (\alpha\beta\gamma) \text{ interaction, } SS_{\alpha \times A} = \text{cell-total } SS - SS_A - SS_\alpha$$

$$\text{Interaction } df = \text{cell-total } df - df_A - df_\alpha = 8 - 2 - 2 = 4$$

or

$$= df_A \times df_\alpha \qquad = 2 \times 2 = 4$$

Thus, after dealing with the two-way $A\alpha$ matrix, we have

$$SS_A, SS_\alpha, SS_{\alpha \times A} \qquad \text{and} \qquad df_A, df_\alpha, df_{\alpha \times A}$$

In the same way after considering the other two-way matrices, we could have

Aa Matrix:	SS_A	SS_a	$SS_{a \times A}$	df_A	df_a	$df_{a \times A}$
aα Matrix:	SS_a	SS_α	$SS_{\alpha \times a}$	df_a	df_α	$df_{\alpha \times a}$

In this example

$$df_A = df_a = df_\alpha = 2 \qquad \text{and} \qquad df_{\alpha \times A} = df_{\alpha \times a} = df_{a \times A} = 4$$

Naturally, once we have the between-treatments effects SS_A, SS_α, and SS_a, the two-way matrices are useful only for giving us the two-way interactions $SS_{\alpha \times A}$, $SS_{a \times A}$, and $SS_{\alpha \times a}$.

Having finished with the two-way matrices, we now proceed to the next matrix, the three-way αAa matrix. It is from this matrix that we will obtain a term for the three-way interaction. We will consider exactly what a three-way interaction is later; at the moment we will just concentrate on computing it.

From the Three-Way αAa Matrix

We noted with a two-way matrix that the cell totals were a result of the relevant between-treatments effects and the two-way interaction. In the same way, the cell totals in a three-way matrix are the result of the relevant between-treatments effects, the relevant two-way interactions, and a three-way interaction. Thus, just as a two-way matrix can be used to calculate a two-way interaction, the three-way matrix can be used to calculate a three-way interaction. The three-way cell-total sum of squares is made up this time of between-treatments sums of squares, two-way-interaction sums of squares, and a three-way-interaction sum of squares; the latter is thus obtained by subtraction. The same is true for degrees of freedom. Thus

$$(\alpha Aa)\text{ Cell-total } SS = \frac{T_.^2 + T_.^2 + T_.^2 + T_.^2 + \cdots + T_.^2}{n \text{ per cell}} - C$$

n per cell $= 3$

$$\alpha Aa \text{ cell-total } df = \text{number of cells} - 1 = 27 - 1 = 26$$

From the cell-total SS and df, the three-way interaction can be calculated:

$$(ABC) \times (\alpha\beta\gamma) \times (abc) \text{ interaction, } SS_{\alpha \times A \times a}$$

$$= \text{cell-total } SS - \underbrace{SS_A - SS_\alpha - SS_a}_{\text{between treatments}} - \underbrace{SS_{\alpha \times A} - SS_{\alpha \times a} - SS_{a \times A}}_{\text{two-way interaction}}$$

and

$$df_{\alpha \times A \times a} = \text{cell-total } df - df_A - df_\alpha - df_a - df_{\alpha \times A} - df_{\alpha \times a} - df_{a \times A}$$

$$= 26 - 2 - 2 - 2 - 4 - 4 - 4 = 8$$

or alternatively, the interaction df can be obtained by multiplying together df for each factor involved in the interaction. Thus

$$df_{\alpha \times A \times a} = df_\alpha \times df_A \times df_a = 2 \times 2 \times 2 = 8$$

Finally, the error SS and df can be calculated, as before, by subtraction of all SS and df terms from the total values, SS_T and df_T.

$$SS_E = SS_T - SS_\alpha - SS_A - SS_a - SS_{\alpha \times A} - SS_{\alpha \times a} - SS_{a \times A} - SS_{\alpha \times A \times a}$$

and

$$df_E = df_T - df_\alpha - df_A - df_a - df_{\alpha \times A} - df_{\alpha \times a} - df_{a \times A} - df_{\alpha \times A \times a}$$

Actually, as all the between-treatments SS, two-way interaction SS, and the three-way-interaction SS sum to give the three-way-matrix cell-total SS, we can write, more conveniently:

$$SS_E = SS_T - \text{cell-total } SS \text{ (for three-way matrix)}$$

and similarly,

$$df_E = df_T - \text{cell-total } df \text{ (for three-way matrix)}$$

We now have all the terms for the three-factor ANOVA table (Table 12.1).

Table 12.1 Three-Factor ANOVA Table

Source	df	SS	MS	F
Total	df_T	SS_T		
ABC	df_A	SS_A	$MS_A = \frac{SS_A}{df_A}$	$\frac{MS_A}{MS_E}$
$\alpha\beta\gamma$	df_α	SS_α	$MS_\alpha = \frac{SS_\alpha}{df_\alpha}$	$\frac{MS_\alpha}{MS_E}$
abc	df_a	SS_a	$MS_a = \frac{SS_a}{df_a}$	$\frac{MS_a}{MS_E}$
$(\alpha\beta\gamma) \times (ABC)$	$df_{\alpha \times A}$	$SS_{\alpha \times A}$	$MS_{\alpha \times A} = \frac{SS_{\alpha \times A}}{df_{\alpha \times A}}$	$\frac{MS_{\alpha \times A}}{MS_E}$
$(abc) \times (ABC)$	$df_{a \times A}$	$SS_{a \times A}$	$MS_{a \times A} = \frac{SS_{a \times A}}{df_{a \times A}}$	$\frac{MS_{a \times A}}{MS_E}$
$(\alpha\beta\gamma) \times (abc)$	$df_{\alpha \times a}$	$SS_{\alpha \times a}$	$MS_{\alpha \times a} = \frac{SS_{\alpha \times a}}{df_{\alpha \times a}}$	$\frac{MS_{\alpha \times a}}{MS_E}$
$(\alpha\beta\gamma) \times (ABC) \times (abc)$	$df_{\alpha \times A \times a}$	$SS_{\alpha \times A \times a}$	$MS_{\alpha \times A \times a} = \frac{SS_{\alpha \times A \times a}}{df_{\alpha \times A \times a}}$	$\frac{MS_{\alpha \times A \times a}}{MS_E}$
Error	df_E	SS_E	$MS_E = \frac{SS_E}{df_E}$	

Just as in the two-factor design (Section 11.6), it is traditional to pool all interactions that are not significant back into the error.

The assumptions for the three-factor and higher-order (four-factor, five-factor, etc.) ANOVA designs are the same as for the two-factor ANOVA with interaction (Section 10.4). Now look at the Key to Statistical Tables to see where this design and higher-order designs fall, in relation to other statistical tests.

12.2 HOW TO INTERPRET TWO- AND THREE-WAY INTERACTIONS

We have mentioned and seen how to calculate an F value for a three-way interaction; it is time to examine what a three-way or higher-order interaction actually means. The three-factor (ABC, $\alpha\beta\gamma$, abc) example we considered provided several interactions. These can be represented once again, graphically.

Two-Way Interactions

Two-way interactions are calculated by collapsing all the data on to just two dimensions [e.g., for an $(ABC) \times (\alpha\beta\gamma)$ table all the data are combined over conditions a, b, and c]. The graphs in Figure 12.1 represent two-way interactions.

Three-Way Interactions

To consider the three-way interaction, let us first pick one of the two-way interactions, say, the αA matrix (Figure 12.2). Here all data are pooled over conditions a, b, and c. Now consider what would happen if the data for

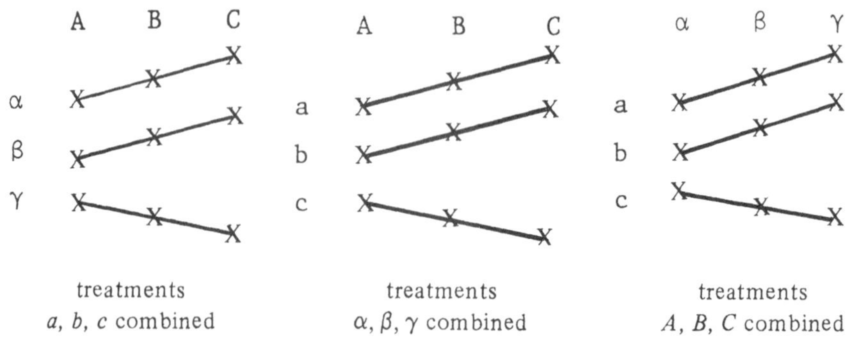

Figure 12.1 Two-way interactions.

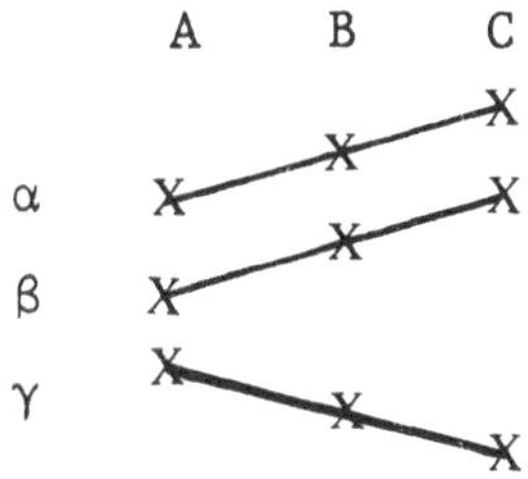

Figure 12.2 αA-matrix interactions.

conditions *a*, *b*, and *c* were treated separately (Figure 12.3). Here all the trends over *A*, *B*, and *C* for treatments α, β, and γ are the same for conditions *a*, *b*, and *c*. There is said to be no three-way interaction. Now consider Figure 12.4. Here the trends over *A*, *B*, and *C* for treatments α, β, and γ are different over *a*, *b*, and *c*. They are the same for treatment *a* and *c* but not for *b*. Because of this lack of consistency, there is said to be a three-way interaction. Three-way interactions can occur whether one of the conditions or all of the conditions are different, as shown in Figure 12.5.

What does a three-way interaction mean? As an example, let us imagine quality scores taken for wines *A*, *B*, *C* under different lighting conditions α, β, γ (white, red, green light) on consecutive days *a*, *b*, *c* (three replications).

Let us assume that there is a significant between-treatments effect for the wines with *A* getting the highest mean scores, *C* the lowest, and *B* an intermediate value ($A > B > C$). Let this be true for red and white light, but under green light the wines appear strange and they all get the same quality scores. Thus although on average there is a trend ($A > B > C$) for the wines, this trend is

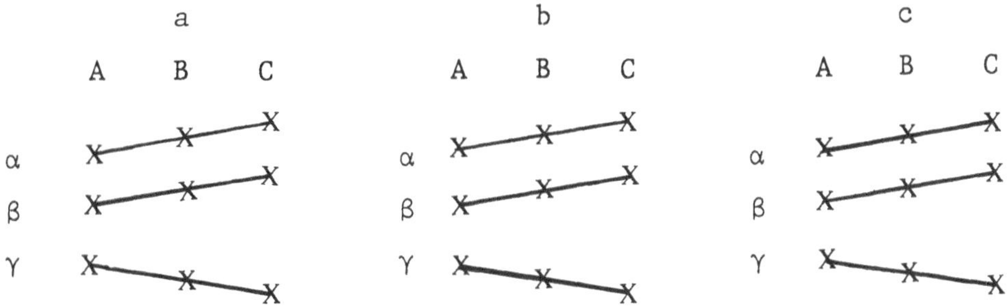

Figure 12.3 αA-matrices with conditions *a*, *b*, and *c* treated separately; consistent trend. No three-way ($\alpha A a$) interaction.

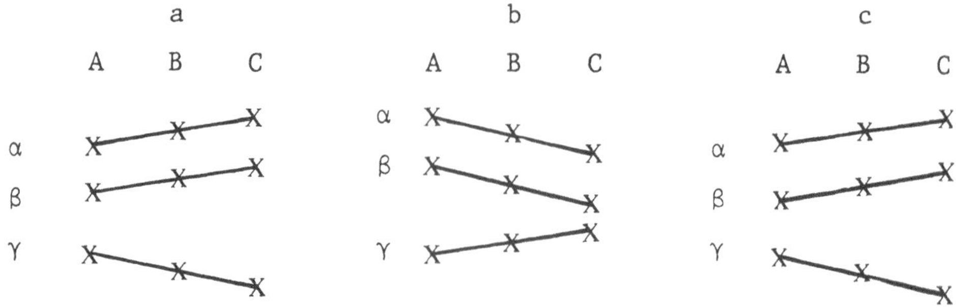

Figure 12.4 Inconsistent trends. Three-way ($\alpha A a$) interaction.

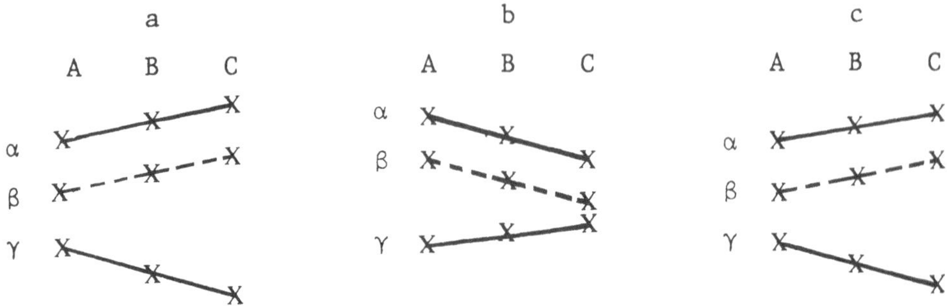

Three-way interaction. Condition b is different.

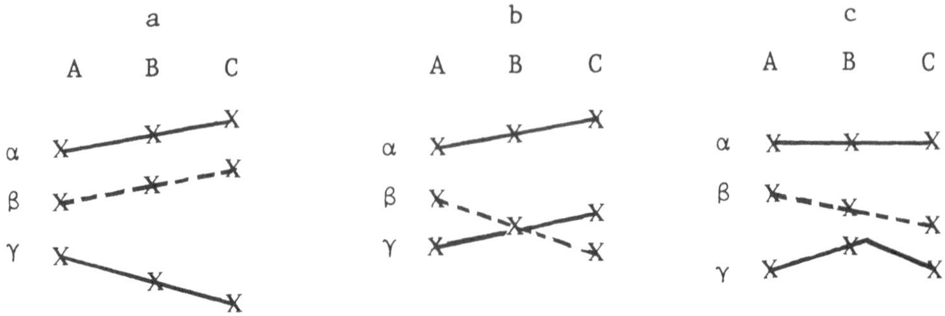

Three-way interaction. All conditions different.

Figure 12.5 Examples of three-way interactions.

not consistent over all lighting conditions (only for red and white light, not green); thus there is a "wines × lighting conditions two-way interaction."

Now consider the case under green light. Here the scores for the wines were the same. However, the experiment was repeated on three consecutive days. On the first two days the green light confused the tasters, so all wines were rated the same. However, by the third day the judges had got used to the green light and indicated some significant differences ($A > B > C$). This trend was insufficient, however, to cause significant differences in the means taken over the whole three days. (In this case, tasting should have been continued and due account been taken of practice effects, before drawing any conclusions.) The trend for the wines under green light was not consistent over the three days and this is expressed in statistical jargon as having a "wines × lighting conditions × days three-way interaction." Thus the trend for the wines scores which is inconsistent over lighting conditions (two-way interaction) is inconsistent over days for a given lighting condition (three-way interaction).

Considering the green light case again, it could be that the light confused the judges so much that they never got used to it. They showed a different trend on each day ($A = B = C; A > B > C; A < B < C$), such that over the three days the trends canceled each other, giving a resulting equality ($A = B = C$). Once again this is described as a "wines × lighting conditions × days three-way interaction."

Should the inconsistency over days occur only for the green light condition or for any one, two, or three lighting conditions, we say that there is a three-way interaction. There will only be no three-way interaction should the trend for the wines under every lighting condition be perfectly consistent over days.

What if we now brought judges into the picture? We stated that the pattern for the wines under green light in the third day showed significant differences ($A > B > C$). If this occurred for most of the subjects (to give the overall trend) but not all, there would now be a "wines × lighting conditions × days × subjects four-way interaction." We could go even further. Should the scores for each subject be the result of several replicate tastings, but the subjects be inconsistent in their rating trend for the wines over these replicates, we would have a "wines × lighting conditions × days × subjects × replications five-way interaction." And so it could continue. You need absolute consistency over all these factors to avoid such interactions. Look at the worked problem (Section 12.5) for more examples of interactions.

12.3 SOMETIMES YOU CANNOT COMPUTE HIGHER-ORDER INTERACTIONS

We have discussed for the two-factor ANOVA how there must be more than one score in each combination of treatments from the two factors to allow calculation of a two-way interaction (Section 11.7). Only in this way will the cell total

SS for the two-way matrix to be different from SS_T. In the same way, there must be more than one score in each combination of treatments from the three factors, for calculation of a three-way interaction. Only then will the cell total *SS* for the three-way matrix be different from SS_T. Without more than one score in each combination of factors, the three-way interaction cannot be calculated. The two-way interactions can be, however.

Thus, in general, you will not get your highest-order interaction unless there is more than one score in each combination of treatments from all the factors involved in the design. In a four-factor ANOVA there must be more than one score in each combination of treatments from the four factors or else your cannot calculate a four-way interaction; you will still get two- and three-way interactions, however. In general, for the *n*-factor ANOVA, you need more than one score in each combination of treatments from the *n* factors to be able to calculate an *n*-way interaction.

12.4 HOW TO COMPUTE A FOUR-FACTOR ANOVA

We now consider the strategy for a four-factor ANOVA. Let the four factors be *ABC*, $\alpha\beta\gamma$, *abc*, and 123. For instance, quality scores could be obtained for three soup formulations (*A*, *B*, *C*) under different lighting conditions (α, β, γ) on consecutive days (*a*, *b*, *c*) at three different temperatures (1, 2, 3). Alternatively, one of the factors could be subjects.

The calculation is merely an extension of the three-factor ANOVA. The original data are placed in a whole matrix, while there are six two-way matrices, four three-way matrices, and a four-way matrix. The strategy is the same as before. The total sum of squares (SS_T) and the four between-treatments sums of squares (SS_A, SS_α, SS_a, SS_1) are obtained from the whole matrix (the same is done for the degrees of freedom). Then the two-way interaction sums of squares (between-treatments sums of squares, if you wish) are obtained from the two-way matrices (the same is done for *df* values). The three-way interactions are then obtained from the three-way matrices, the four-way interaction from the four-way matrix. The computation passes from the lower (two-way) up to the highest (four-way) matrix; this is a general strategy for all higher-order ANOVA designs.

First the original data are displayed in a table (Figure 12.6) with separate columns for each combination of the treatments from the four factors, *ABC*, $\alpha\beta\gamma$, *abc*, and 123. There is more than one score in each combination of treatments, so a four-way interaction can be calculated. These scores (*X* values) can be added to give a total (*T*.) for each combination of treatments. The totals (*T*.) can be presented in a four-way matrix (Figure 12.7).

From the four-way matrix, totals can be obtained for every combination of three factors by adding the totals in all treatments of the fourth factor. For

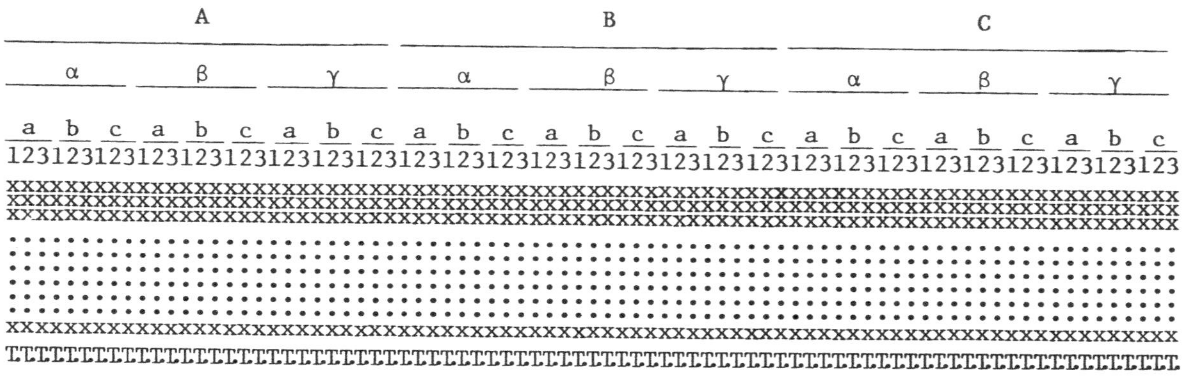

Figure 12.6 Whole matrix.

	A									B									C									
	α			β			γ			α			β			γ			α			β			γ			
	a	b	c	a	b	c	a	b	c	a	b	c	a	b	c	a	b	c	a	b	c	a	b	c	a	b	c	
1	T.	T.	T.	T.	T.	T.	T.	T.	T.	T.	T.	T.	T.	T.	T.	T.	T.	T.	T.	T.	T.	T.	T.	T.	T.	T.	T.	T_1
2	T.	T.	T.	T.	T.	T.	T.	T.	T.	T.	T.	T.	T.	T.	T.	T.	T.	T.	T.	T.	T.	T.	T.	T.	T.	T.	T.	T_2
3	T.	T.	T.	T.	T.	T.	T.	T.	T.	T.	T.	T.	T.	T.	T.	T.	T.	T.	T.	T.	T.	T.	T.	T.	T.	T.	T.	T_3

Figure 12.7 *Aαa* 1 matrix.

example, the *T.* values for treatments 1, 2, and 3 could be added, giving a total (*T*) for each combination of *ABC*, $\alpha\beta\gamma$, and *abc* treatments (e.g., $A\alpha a$, $A\alpha b$, $A\alpha c$, $A\beta a$, $A\beta b$, etc.). This produces the αAa three-way matrix. This can be done, adding together the scores for treatments 1, 2, and 3, or for *A*, *B*, and *C*, or for α, β, and γ, or for *a*, *b*, and *c*. This produces four possible three-way matrices (Figure 12.8).

Two-way matrices can then be obtained from the three-way matrices in the same way. For example, consider the αAa three-way matrix. We can add the totals for treatments *a*, *b*, and *c* to obtain the $A\alpha$ two-way matrix. We can add the totals for treatments α, β, and γ to obtain the *Aa* two-way matrix, and we can add the totals for treatments *A*, *B*, and *C* to obtain the αa two-way matrix. Similar exercises on the other three-way matrices will produce the remaining two-way matrices (*A* 1, α1, and *a* 1 matrices). The two-way matrices are shown in Figure 12.9.

The computation scheme for *SS* and *df*, using the relevant matrices, is outlined in Table 12.2.

αAa

	A	B	C
	αβγ	αβγ	αβγ
a	TTT	TTT	TTT
b	TTT	TTT	TTT
c	TTT	TTT	TTT

αA1

	A	B	C
	αβγ	αβγ	αβγ
1	TTT	TTT	TTT
2	TTT	TTT	TTT
3	TTT	TTT	TTT

αa1

	1	2	3
	αβγ	αβγ	αβγ
a	TTT	TTT	TTT
b	TTT	TTT	TTT
c	TTT	TTT	TTT

aA1

	A	B	C
	123	123	123
a	TTT	TTT	TTT
b	TTT	TTT	TTT
c	TTT	TTT	TTT

Figure 12.8 Three-way matrices.

$A\alpha$

	A	B	C	
α	T	T	T	T_α
β	T	T	T	T_β
γ	T	T	T	T_γ
	T_A	T_B	T_C	

Aa

	A	B	C	
a	T	T	T	T_a
b	T	T	T	T_b
c	T	T	T	T_c
	T_A	T_B	T_C	

A1

	A	B	C	
1	T	T	T	T_1
2	T	T	T	T_2
3	T	T	T	T_3
	T_A	T_B	T_C	

αa

	α	β	γ	
a	T	T	T	T_a
b	T	T	T	T_b
c	T	T	T	T_c
	T_α	T_β	T_γ	

α1

	α	β	γ	
1	T	T	T	T_1
2	T	T	T	T_2
3	T	T	T	T_3
	T_α	T_β	T_γ	

a1

	a	b	c	
1	T	T	T	T_1
2	T	T	T	T_2
3	T	T	T	T_3
	T_a	T_b	T_c	

Figure 12.9 Two-way matrices.

Table 12.2 Schemes for Computing *SS* and *df*

Whole matrix	Two-way matrices (two-way interactions)	Three-way matrices (three-way interactions)	Four-way matrices (four-way interactions)	By subtraction
C	$A\alpha$	$A\alpha a$	$A\alpha a1$	Error
Total	Aa	$A\alpha 1$		
A	$A1$	$\alpha a1$		
α	αa	$Aa1$		
a	$\alpha 1$			
1	$a1$ (A, α, a, 1) could be calculated here			

How many two-, three-, and four-way matrices do you get for a given ANOVA? For a two-factor ANOVA, you get one two-way matrix. For a three-factor ANOVA, you get three two-way matrices and one three-way matrix. For a four-factor ANOVA, there are six two-way, four three-way, and one four-way matrix.

The number of matrices obtained from a given factor ANOVA is calculated using a combination (see Section 3.7). For an N-factor ANOVA, there are ${}_NC_n$ n-way matrices. For example: How many two-way matrices are there in a four-factor ANOVA? This is given by ${}_4C_2$:

$$ {}_4C_2 = \frac{4!}{(4-2)!\,2!} = \frac{4 \times 3 \times 2 \times 1}{2 \times 1 \times 2 \times 1} = 6 $$

As we have seen, there were six two-way matrices for the four-factor ANOVA.

How many four-way matrices are there in a seven-factor ANOVA? This is given by ${}_7C_4$.

$$ {}_7C_4 = \frac{7!}{(7-4)!\,4!} = \frac{7 \times 6 \times 5 \times 4 \times 3 \times 2 \times 1}{3 \times 2 \times 1 \times 4 \times 3 \times 2 \times 1} = 35 $$

Thus there are 35 four-way matrices in a seven-factor ANOVA. It is no wonder that ANOVA F values are generally calculated using computers!

12.5 WORKED EXAMPLES: COMPUTATION AND INTERPRETATION OF ANOVA WITH TWO OR MORE FACTORS

It is now worth considering a few examples of higher ANOVAs and their interpretation.

Example 1

Six judges scored the crunchiness of three different cereals A, B, and C derived from chitin. They scored the crunchiness on a 20-point scale by comparison with standard crunchiness stimuli. Higher scores indicate greater crunchiness. We can assume that their scores came from populations of scores which were normally distributed with equal variances. The cereals were eaten with and without added milk; there were no replications per judge. From the data given below, answer the following questions:

a. Which cereals were crunchier?
b. Did the presence of milk affect crunchiness scores?
c. Was the trend over the cereals A, B, C for crunchiness affected by the presence of milk?
d. Were there any differences between overall judges' scores?
e. Was the trend for judges' scores affected by the presence of milk?
f. Were there any three-way interactions?

There are three factors: judges, cereals (A, B, C), milk vs. no milk. The data for the three-factor ANOVA are as follows:

Crunchiness Scores

	Cereal					
	Without milk			With milk		
Judge	*A*	*B*	*C*	*A*	*B*	*C*
1	7	3	1	4	2	1
2	6	4	2	5	6	1
3	7	7	1	6	1	0
4	6	2	3	6	3	2
5	5	4	7	4	1	5
6	5	2	2	3	3	1

n per judge = 6
n per cereal = 12
n per milk treatment = 18
N total = 36

Grand total T = 128

These are the original data and will be called the whole matrix. It can be seen that there is only one score per combination of treatments, so that there is no three-way interaction and no three-way matrix. The three two-way matrices can be constructed by adding the milk condition scores (for the cereal × judges matrix), adding the scores for the cereals (for the milk × judges matrix), and adding the scores for the judges (for milk × cereals matrix). These are given below:

Cereals × Judges Matrix

		Cereal			
		A	*B*	*C*	Total
Judge	1	11	5	2	18
	2	11	10	3	24
	3	13	8	1	22
	4	12	5	5	22
	5	9	5	12	26
	6	8	5	3	16
Total		64	38	26	128

Milk × Judges Matrix

		Milk treatment		
		Without	With	Total
Judge	1	11	7	18
	2	12	12	24
	3	15	7	22
	4	11	11	22
	5	16	10	26
	6	9	7	16
Total		74	54	128

Milk × Cereals Matrix

		Cereal			
		A	*B*	*C*	Total
Milk Treatment	Without	36	22	16	74
	With	28	16	10	54
	Total	64	38	26	128

Having established the whole matrix and the three two-way matrices, we can now start the calculation. This proceeds in the usual way. We first calculate some values from the whole matrix:

$$\text{Correction term}, C = \frac{T^2}{N} = \frac{128^2}{36} = 455.11$$

$$\text{Total } SS = 7^2 + 3^2 + 1^2 + 4^2 + 2^2 + 1^2 + 6^2 + \cdots + 3^2 + 3^2 + 1^2 - C$$

$$= 616 - 455.11 = 160.89$$

$$df = N - 1 = 36 - 1 = 35$$

$$\text{Between-cereals } SS = \frac{64^2 + 38^2 + 26^2}{12} - C$$

$$= 518 - 455.11 = 62.89$$

$$df = k_{\text{cereals}} - 1 = 3 - 1 = 2$$

$$\text{Between-milk-treatments } SS = \frac{74^2 + 54^2}{18} - C$$

$$= 466.22 - 455.11 = 11.11$$

$$df = k_{\text{milk treatments}} - 1 = 2 - 1 = 1$$

$$\text{Between-judges } SS = \frac{18^2 + 24^2 + 22^2 + 22^2 + 26^2 + 16^2}{6} - C$$

$$= 466.67 - 455.11 = 11.56$$

$$df = n_{\text{judges}} - 1 = 6 - 1 = 5$$

We now turn to the cereals × judges two-way matrix. We use the cell totals to calculate the cereals × judges interaction:

$$\text{Cereals} \times \text{judges cell-total } SS = \frac{11^2 + 5^2 + 2^2 + 11^2 + 10^2 + \cdots + 3^2}{2} - C$$

$$= 578 - 455.11 = 122.89$$

$$df = \text{number of cells} - 1 = 18 - 1 = 17$$

Cereals × judges interaction SS = cereals × judges cell-total SS − between-cereals SS − between-judges SS

$$= 122.89 - 62.89 - 11.56 = 48.44$$

Cereals × judges df = cereals × judges cell-total df − between-cereals df − between-judges df

$$= 17 - 2 - 5 = 10$$

Next, we turn to the milk X judges two-way matrix and repeat the same operation.

$$\text{Milk} \times \text{judges cell-total } SS = \frac{11^2 + 7^2 + 12^2 + 12^2 + 15^2 + \cdots + 7^2}{3} - C$$

$$= 486.67 - 455.11 = 31.56$$

$$df = \text{number of cells} - 1 = 12 - 1 = 11$$

Milk X judges interaction SS = milk X judges cell-total SS - between-milk treatments SS - between-judges SS

$$= 31.56 - 11.11 - 11.56 = 8.89$$

Milk X judges df = milk X judges cell-total df - between-milk treatments df - between-judges df

$$= 11 - 1 - 5 = 5$$

And now we turn to the milk X cereals two-way matrix.

$$\text{Milk} \times \text{cereals cell-total } SS = \frac{36^2 + 22^2 + 16^2 + 28^2 + 16^2 + 10^2}{6} - C$$

$$= 529.33 - 455.11 = 74.22$$

$$df = \text{number of cells} - 1 = 6 - 1 = 5$$

Milk X cereals interaction SS = milk X cereals cell-total SS - between-milk-treatments SS - between-cereals SS

$$= 74.22 - 11.11 - 62.89 = 0.22$$

Milk X cereals df = milk X cereals cell-total df - between-milk-treatments df - between-cereals df

$$= 5 - 1 - 2 = 2$$

There being no three-way matrix, we do not look for a three-way interaction.

Finally, the error term is obtained by subtracting the between-treatments and two-way interaction terms from the total value (SS and df).

Error SS = total SS - between-cereals SS - between-milk treatments SS - between-judges SS - cereals X judges SS - milk X judges SS - milk X cereals SS

$$= 160.89 - 62.89 - 11.11 - 11.56 - 48.44 - 8.89 - 0.22 = 17.78$$

Error df = total df - between-cereals df - between-milk treatments df - between-judges df - cereals X judges df - milk X judges df - milk X cereals df

$$= 35 - 2 - 1 - 5 - 10 - 5 - 2 = 10$$

We now have all the information to construct an ANOVA table.

Source	*SS*	*df*	*MS*	*F*
Total	160.89	35		
Between cereals	62.89	2	62.89/2 = 31.45	31.45/1.78 = 17.67***
Between-milk treatments	11.11	1	11.11/1 = 11.11	11.11/1.78 = 6.24*
Between judges	11.56	5	11.56/5 = 2.31	2.31/1.78 = 1.30 (NS)
Cereals X milk treatment interaction	0.22	2	0.22/2 = 0.11	0.11/1.78 = 0.06 (NS)
Judges X cereals interaction	48.44	10	48.44/10 = 4.84	4.84/1.78 = 2.72 (NS) (just)
Judges X milk treatment interaction	8.89	5	8.89/5 = 1.78	1.78/1.78 = 1.00 (NS)
Error	17.78	10	17.78/10 = 1.78	

We now use Table G.11 to look up *F* values for the relevant number of degrees of freedom. These are as follows:

df	5%	1%	0.1%
1, 10	4.96	10.04	21.04
2, 10	4.10	7.56	14.91
5, 10	3.33	5.64	10.48
10, 10	2.97	4.85	< 9.20

It can be seen that only the between-treatments effects for cereals ($p < 0.001$) and milk treatments ($p < 0.05$) are significant. None of the three interaction terms are significant. We will decide to pool nonsignificant interactions with the error. The sums of squares for the insignificant interactions are added to the error sum of squares (0.22 + 48.44 + 8.89 + 17.78 = 75.33), while the same is done for the degrees of freedom (2 + 10 + 5 + 10 = 27). The new ANOVA table becomes

Source	SS	df	MS	F
Total	160.89	35		
Between cereals	62.89	2	31.45	31.45/2.79 = 11.27***
Between-milk treatments	11.11	1	11.11	11.11/2.79 = 3.98 (NS)
Between judges	11.56	5	2.31	2.31/2.79 = 0.83 (NS)
Error	75.33	27	75.33/27 = 2.79	

We now use Table G.11 once again to look up F values for the relevant numbers of degrees of freedom. These are as follows:

df	5%	1%	0.1%
1, 27	4.21	7.68	13.61
2, 27	3.35	5.49	9.02
5, 27	2.57	3.79	5.73

Again the between-cereals F value is significant ($p < 0.001$) and the between-judges effect is not significant. The only change is that the between-milk-treatments effect is no longer significant at $p < 0.05$.

The cereals A, B, C differed significantly ($p < 0.001$). Cereal means are

A	B	C
5.33	3.17	2.17

We use the LSD test to determine which cereals are significantly different. At $p < 0.001$, the level of significance for the ANOVA, the LSD is comparatively safe against Type I error. The formula for the LSD is given by

$$\text{LSD} = t\sqrt{\frac{2MS_E}{n}}$$

From the ANOVA, $MS_E = 2.79$. The number of scores for each cereal mean, $n = 12$. From Table G.8, t (at $p = 0.001$, $df_E = 27$, two-tailed) = 3.690.

$$\text{LSD} = 3.690\sqrt{\frac{2 \times 2.79}{12}} = 2.52$$

Thus the means of the crunchiness scores for the three cereals A, B, and C can be represented as follows:

A	*B*	*C*
5.33	3.17	2.17

The answers to the specific questions are as follows:

a. Which cereals were crunchier?
For these judges, cereal *A* was significantly more crunchy than *C*. Neither *A* nor *C* were significantly different from *B* ($p < 0.001$).

b. Did the presence of milk affect crunchiness scores?
The presence of milk did not affect crunchiness scores ($p > 0.05$). However, if variances were not pooled, cereals without milk were significantly crunchier ($p < 0.05$). Thus there was a trend here for milk to reduce crunchiness scores; whether it was significant at $p = 0.05$ depends on the exact ANOVA technique used. Thus its significance level was near 5%. It would also be a good idea to examine this further with nonparametric analyses or collect more data.

c. Was the trend over the cereals *A, B, C* for crunchiness affected by the presence of milk?
There was no significant cereals × milk treatments interaction; thus the trend over cereals *A, B, C* for crunchiness was not itself affected by the presence of milk. *A* was always crunchier than *C*, with *B* as an intermediate and different from neither.

d. Were there any differences between overall judges' scores?
There were no differences between the overall judges scores; the between-judges effect was not significant.

e. Was the trend for judges' scores affected by the presence of milk?
There was no judges × milk treatments interaction; thus the trend for judges' scores was not affected by the absence or presence of milk.

f. Were there any three-way interactions?
There may or may not have been a three-way interaction but the ANOVA could not examine this; there was only one score in each combination of the three conditions, precluding calculation of a three-way interaction *MS*. The only way of getting any clue would be to examine the original data.
Also, it would be wise to be careful about concluding that there was definitely not a judges × cereals interaction. Remember that for much sensory work, ANOVA is only a rough analysis because the assumptions are broken. This interaction was only just "insignificant" ($p > 0.05$) in this calculation.

In general, if ANOVA indicates trends, it is a good idea to confirm them by seeing whether a majority of judges show the trend (binomial test). It is also a good idea to see whether these trends emerge with a nonparametric ranking analysis (see Chapter 16).

Example 2

In Example 1, how would you interpret the results if all between-treatments effects, all two-way interactions, and the one three-way interaction (this time there were replications in each combination of three factors) were significant.

Source		*F*
a.	Between cereals *A, B, C*	Significant
b.	Between-milk treatment (milk vs. no milk)	Significant
c.	Between judges	Significant
d.	Cereals × milk treatments interaction	Significant
e.	Cereals × judges interaction	Significant
f.	Judges × milk treatments interaction	Significant
g.	Judges × cereals × milk treatments interaction	Significant

Significant between-treatment effects mean that:

a. Summed over all judges and milk treatments, cereals *A*, *B*, and *C* vary in crunchiness.
b. Summed over all judges and cereals, addition of milk alters the crunchiness. Milk makes cereals less crunchy.
c. Summed over all cereals and milk treatments, judges differed significantly in their scores. They may be differentially responsive to crunchiness or merely use the scale differently.
d. A significant cereals × milk treatments interaction means that the order of crunchiness of *A*, *B*, and *C* (summed over all judges) varies when milk is or is not added. Alternatively, it could mean that the trend for crunchiness-change (summed over all judges) when milk is added is not consistent for cereals *A*, *B*, and *C*. Finally, both of the above could be true.
e. A significant cereals × judges interaction means that the order of crunchiness of *A*, *B*, and *C* (summed over both milk conditions) varies with judges. Alternatively, it could mean that the trend for the differences in mean scores for judges (summed over milk conditions) varies with cereals. Finally, both of the above could be true.
f. A significant judges × milk treatments interaction means that when crunchiness scores are summed over all three cereals, the trend for the milk treatments is not consistent over judges, or the trend for scores over judges is not the same when milk is added, or both of the above.

g. A significant three-way interaction means that how two factors vary with each other will itself depend on the third factor. For example, it could mean that the way the order of crunchiness of *A, B, C* varied with the judges itself varied with the addition of milk. Or we could say that the way the effect of the addition of milk varied the judges' scores was not consistent over cereals. Or the way the crunchiness of *A, B, C* varied with milk was not consistent over judges. Or any two or all three of these possibilities.

Example 3

The following are a set of scores for a two-factor ANOVA (*ABC* and $\alpha\beta\gamma$). They are shown in the following matrices:

(I)

	α	β	γ	Total
A	10	8	6	24
B	8	6	4	18
C	6	4	2	12
Total	24	18	12	

(II)

	α	β	γ	Total
A	10	11	6	27
B	8	6	4	18
C	6	4	2	12
Total	24	21	12	

(III)

	α	β	γ	Total
A	10	8	6	24
B	11	6	4	21
C	6	4	2	12
Total	27	18	12	

(IV)

	α	β	γ	Total
A	10	11	6	27
B	11	6	4	21
C	6	4	2	12
Total	27	21	12	

(V)

	α	β	γ	Total
A	10	10	10	30
B	8	8	8	24
C	6	6	6	18
Total	24	24	24	

(VI)

	α	β	γ	Total
A	11	10	10	31
B	8	8	8	24
C	5	6	6	17
Total	24	24	24	

(VII)

	α	β	γ	Total
A	10	10	8	28
B	8	8	10	26
C	6	6	6	18
Total	24	24	24	

a. What are the trends between treatments for each of the matrices?
For each of the two factors (ABC and $\alpha\beta\gamma$) the trends for the mean scores over the various treatments for the two factors can be obtained from the totals. The trends are as follows:

For matrices (I), (II), (III), and IV):

$$A > B > C \quad \text{and} \quad \alpha > \beta > \gamma$$

For matrices (V), (VI), and (VII):

$$A > B > C \quad \text{and} \quad \alpha = \beta = \gamma$$

b. In matrices (I), (II), (III), and (IV) the trends over the three treatments for the two factors are the same. How do the matrices differ?
The matrices differ in their two-way interactions.

Matrix (I):
The trend $A > B > C$ is consistent for conditions α, β, and γ.
The trend $\alpha > \beta > \gamma$ is consistent for conditions A, B, and C.
There is thus no two-way interaction.

Matrix (II):
The trend $A > B > C$ is consistent for conditions α, β, and γ.
But the trend $\alpha > \beta > \gamma$ is only true for conditions B and C, not for condition A. Thus there is a two-way interaction.

Matrix (III):
The trend $\alpha > \beta > \gamma$ is true for all conditions A, B, and C.
The trend $A > B > C$ is only true for conditions β and γ but not for condition α.
Thus there is a two-way interaction.

Matrix (IV):
The trend $A > B > C$ is true only for conditions β and γ, not for condition α.
The trend $\alpha > \beta > \gamma$ is true only for conditions B and C, not for condition A.
Thus there is a two-way interaction.

From matrices (II), (III), and (IV) it can be seen that a two-way interaction can be caused by inconsistency over the $\alpha\beta\gamma$ treatments, the ABC treatments, or both. There are thus three manners in which a two-way interaction can occur.

c. In matrices (V), (VI), and (VII), the trends over the three treatments for the two factors are the same. How do the matrices differ?

The matrices differ in their two-way interactions.

Matrix (V):

The trend $A > B > C$ is true for conditions α, β, and γ.
The trend $\alpha = \beta = \gamma$ is true for conditions A, B, and C.
There is thus no two-way interaction.

Matrix (VI):

The trend $A > B > C$ is true for conditions α, β, and γ.
The trend $\alpha = \beta = \gamma$ is not true for all conditions A, B, and C.
Thus there is a two-way interaction.

Matrix (VII):

The trend $A > B > C$ is not true for all conditions α, β, and γ.
Thus there is a two-way interaction.
If $A > B > C$ is not true for all conditions α, β, and γ, there is no way that $\alpha = \beta = \gamma$ for all conditions A, B, and C.

Thus when treatments in one factor are equal, a two-way interaction can be caused by inconsistency in trend for the equal treatments, or both this and inconsistency in the unequal treatments. There are only two manners in which a two-way interaction can occur if the treatment means for one factor are equal.

Should all treatments be equal in both factors, a two-way interaction can occur in only one manner. Inconsistency over treatments must occur for both factors.

No Interaction

	α	β	γ	Total
A	10	10	10	30
B	10	10	10	30
C	10	10	10	30
Total	30	30	30	

Two-Way Interaction

	α	β	γ	Total
A	11	10	10	31
B	10	11	10	31
C	10	10	11	31
Total	31	31	31	

It is as well to remember the different manner in which two-way interactions can occur when interpreting ANOVA data:

$A > B > C$ and $\alpha > \beta > \gamma$; there are three manners in which a two-way interaction can occur.
$A > B > C$ and $\alpha = \beta = \gamma$; there are two manners in which a two-way interaction can occur.
$A = B = C$ and $\alpha = \beta = \gamma$; there is only one manner in which a two-way interaction can occur.

Example 4

A group of subjects rated seven concentrations of salt water for taste intensity. The concentrations were 10, 50, 100, 200, 500, 1000, and 1500 millimolar. The higher the concentration, the higher was the mean rating obtained; ANOVA and LSD tests indicated that all these mean ratings were significantly different from each other ($p < 0.001$). Each subject tasted several replicates of each concentration. That is all the information you are given, except the following incomplete ANOVA table:

Source	N	df	F
Subjects		9	1.95
Concentrations	7		12.3
Replications			1.0
Subjects X concentrations			1.40
Subjects X replications			1.35
Concentration X replications		54	1.0
Residual			

a. Indicate the levels of significance (using asterisks) for the F values (NS, $p < 0.05$; *, $p < 0.05$; **, $p < 0.01$; ***, $p < 0.001$).

To do this, we need to know the number of degrees of freedom for each source of variance. To do this we have to piece the evidence together. The subjects $df = 9$; thus there were 10 subjects. There were seven concentrations, so $df = 6$. How many replications were there? The concentrations X replications $df = 54$; this is the product of degrees of freedom for concentrations and replications, so the replications $df = 9$ ($6 \times 9 = 54$). Thus there were 10 replicate salt solutions at each concentration given to each subject. While we are about it, we can note that the total number of scores in the experiment is $10 \times 7 \times 10 = 700$, with $df = 699$.

Considering the two-way interactions, the numbers of degrees of freedom can be obtained from products of the degrees of freedom of each factor in the interaction. Thus for subjects X concentrations, $df = 9 \times 6 = 54$; subjects X replications, $df = 9 \times 9 = 81$.

We note that the error term is called by its alternative name, the *residual* (see Section 8.5). The degrees of freedom associated with the error is obtained by subtraction, thus: $df_E = 699 - 9 - 6 - 9 - 54 - 81 - 54 = 486$.

We now have all the df values required for testing the significance of the F values in Table G.11. We thus give the ANOVA table in full:

Source	N	df	F
Total	700	699	
Subjects	10	9	1.95*
Concentrations	7	6	12.3***
Replications	10	9	1.0 (NS)
Subjects X concentrations		54	1.40*
Subjects X replications		81	1.35*
Concentrations X replications		54	1.0 (NS)
Residual or error		486	

We will now examine in detail the level of significance for all the F values using Table G.11. The table does not give values for the exact number of degrees of freedom that we require but we can still use the nearest values in the table to check significance. For instance, at $p = 0.05$ and $p = 0.01$ we do not have values for the denominator degrees of freedom of 486; we have to use values for $df = 400$ and 1000 and interpolate.

For $df = 9$ and 486, the between-subjects F value (1.95) exceeds the relevant values in Table G.11 at $p = 0.05$ only ($df = 9$, 400, $F = 1.90$; $df = 9$, 1000, $F = 1.89$); the between-replications F value (1.0) does not exceed these values. The between-concentrations F value (12.3, $df = 6$, 486) exceeds the relevant values in Table G.11 at $p = 0.001$ ($df = 6$, 120, $F = 4.04$; $df = 6, \infty, F = 3.74$).

We consider the interactions. The $df = 54$ and 486, the subjects X concentrations F value (1.40), exceeds the nearest values in Table G.11 only at $p = 0.05$ ($df = 50$, 400, $F = 1.38$; $df = 50$, 1000, $F = 1.36$; $df = 75$, 400, $F = 1.32$; $df = 75$, 1000, $F = 1.30$); the concentrations X replications F value (1.0) did not exceed these values. The subjects X replications F value (1.35, $df = 81$, 486) exceeds the relevant values in Table G.11, $p = 0.05$ only ($df = 75$, 400, $F = 1.32$; $df = 75$, 1000, $F = 1.30$; $df = 100$, 400, $F = 1.28$; $df = 100$, 1000, $F = 1.26$).

In testing for significance, we found that our F values clearly exceeded all the nearest values in Table G.11; we never had to interpolate F values. Thus our value of F for $df = 81$ and 486 (1.35) exceeded the nearest values in table (at $p = 0.05$) which bracketed our degrees of freedom:

F	df
1.32	75, 400
1.30	75, 1000
1.28	100, 400
1.26	100, 1000

75 and 100 bracket 81, while 400 and 1000 bracket 486. Actually, if our value of 1.35 exceeds the highest of these values in the tables (1.32) which has the lowest numbers degrees of freedom (75 and 400), it will exceed the other values (with higher numbers of degrees of freedom). Thus we could have saved ourselves time by merely comparing our value with the near value in the table associated with the fewest degrees of freedom; if it exceeded this, we would not have had to check the other values. If it did not exceed this but did exceed the other near values, we would have had to interpolate.

b. Where did the three-way interaction go?

A three-way interaction is not mentioned at all. However, because replications are extracted as a separate factor, it is hard to see how more than one score could be obtained for each combination of concentrations, subjects, and replications. Thus, with one score in each combination of conditions, we are not able to compute an F value for a three-way interaction. There may be one but we cannot calculate it. A three-way interaction would have had $df = 9 \times 6 \times 9 = 486$. This is the same as the number of degrees of freedom for the error term, showing that they are one and the same thing. So no terms have been left out of the table by mistake.

c. Interpret the between-treatments effects.

Concentrations: When the ratings were averaged over all subjects and replications, the mean values for each concentration showed significant differences. Actually, in the original question, we were told that all these means were significantly different from each other at $p < 0.001$ (there is a less than 0.1% chance of the means not being different, of our rejecting H_0 when it is correct, of committing a Type I error). Thus the higher the concentration of the solution, the higher was its mean intensity rating.

Subjects: When the ratings were averaged over all concentrations and replications, the mean values for each subject showed significant differences. (The significance level was $p < 0.05$. Thus we have a less than 5% chance of rejecting H_0 in error; we have a less than 5% chance of the means being the same.) Thus some subjects habitually gave higher intensity ratings, while others gave lower intensity ratings. Some subjects tended to use the higher end of the scale; some used the lower end. Such inconsistency in the use of rating scales is common. Alternatively, some subjects may have experienced generally stronger sensations than others. There is no way of knowing.

Replications: When ratings were averaged over all concentrations and subjects, the mean values for each replication did not show significant differences ($p > 0.05$). The chance of there being no differences, of H_0 being true, is greater than 5%. Thus there appeared to be no drift in the scores as the experiment proceeded. The judges did not experience taste fatigue or "warm up," and did not change in their use of the scale during the experiment (i.e., start by using one end of the scale and gradually drift to using the other end). We say that our data are not sufficient to reject H_0; the chance of H_0 being false is low because $F = 1.0$, the chance value.

d. What does the significant subjects × concentrations interaction mean?

This significant two-way interaction can be interpreted in three different ways:

Either The trend for the mean taste intensity scores (mean of all 10 replicates), whereby scores increased with higher concentrations, was not followed consistently by each subject. Thus the trend was an average trend only.

Or The differences that occurred between the mean scores (over all 10 replicates) for each subject were not consistent for each salt concentration; it was only an average trend that emerged when the data from all concentrations were pooled.

Or Both of the above.

There are three manners in which the interaction can occur when treatments in both factors are unequal (see Example 3). An examination of the data would be required to determine which was the case here.

e. What does the significant subjects × replications interaction mean?

This interaction can also be interpreted in more than one way. Again it indicates that the between-treatments trends were not consistent.

Either The lack of difference in the mean scores for each replication (mean for all concentrations) was not a consistent effect for each subject.

Or The differences that occurred between the mean scores (over all seven concentrations) for each subject were not consistent from replication to replication. For this to happen, the lack of difference in mean scores for each replication would also not be consistent for each subject.

There are two manners in which the two-way interaction can occur here (see Example 3). An examination of the original data would be required to determine which was the case.

f. Interpret the insignificant concentrations × replications interaction.

This lack of interaction has two implications:

The trend for the concentration means (over all 10 subjects), whereby the mean scores get significantly greater as the solution concentration increases, was consistent over each of the 10 replications.

And The trend whereby the mean score (over all 10 subjects) for each replication is not significantly different was consistently true for each solution concentration.

g. The trend whereby the higher solution concentrations gave higher mean intensity ratings was consistent for each replication, when the data were averaged over all 10 subjects. However, was this consistency of trend over all replications followed by each subject individually?

We are asking whether there was a three-way interaction. Was the concentration-replication pattern of scores the same for each subject? We do not know. We have not calculated whether a three-way interaction is significant or not. You could, however, inspect the original data to get a clue.

h. The trend for higher ratings to be given to higher concentrations was not consistent over subjects (significant subjects × concentrations interaction); some subjects gave higher (or equal) scores to lower concentrations. Was the pattern of scores given to concentrations by all the subjects consistent from replication to replication?

Again we are asking whether there was a three-way interaction. We are asking whether the pattern of scores for concentrations and judges on the first replication was the same for all replications. We have not calculated a three-way interaction F value and so do not know the answer to this question. Again, you could inspect the original data for a clue.

i. Have you any comments on the validity of use of the ANOVA in this experiment?

The data used in the ANOVA came from some sort of taste-intensity scaling procedure; we are not told which. All we can do is bear in mind the limitations on the use of ANOVA for scaling data (breaking assumptions of homoscedasticity and normality) outlined in Section 8.11 and suspect that the ANOVA may be only an approximate test in this case.

j. (Do not answer this question until you have read Chapter 13.) We have used a fixed-effects model here. However, we may wish to extend our findings for these subjects to other subjects in the population; we may wish to treat subjects as a random effect. How would this affect the calculation?

The calculation would change only as far as the denominator used for the calculated F value is concerned. Instead of always dividing the appropriate mean-square value by the error mean square, an interaction mean square is substituted as seen in the following scheme:

Source	Mean square	Denominator for F value
Concentrations (fixed effect)	MS_{conc}	$MS_{\text{conc} \times S}$
Replications (fixed effect)	MS_{rep}	$MS_{\text{rep} \times S}$
Subjects (random effect)	MS_S	MS_E

Source	Mean square	Denominator for F value
Concentrations × subjects	$MS_{\text{conc} \times S}$	MS_E
Replications × subjects	$MS_{\text{rep} \times S}$	MS_E
Concentrations × replications	$MS_{\text{conc} \times \text{rep}}$	Cannot calculate
Error	MS_E	

We cannot calculate the F value for the concentrations × replications two-way interaction because the appropriate denominator is the mean square for the three-way interaction. In this design we have not calculated such a value.

k. (Do not answer this question until you have read Chapter 13.) What if you now treated subjects and replications as random effects? What if you wished to extend your conclusions to the population of subjects our sample was drawn from and the population of replications our samples were drawn from? We wish to say that our results hold for all further subjects and all further replicates that we might test. How would this affect the ANOVA?

Again the change would only affect the denominators for the F values. This can be seen in the following scheme:

Source	Mean square	Denominator for F value
Concentrations (fixed effect)	MS_{conc}	None
Replications (random effect)	MS_{rep}	$MS_{\text{rep} \times S}$
Subjects (random effect)	MS_S	$MS_{\text{rep} \times S}$
Concentrations × subjects	$MS_{\text{conc} \times S}$	Cannot calculate
Concentrations × replications	$MS_{\text{conc} \times \text{rep}}$	Cannot calculate
Replications × subjects	$MS_{\text{rep} \times S}$	MS_E

The concentrations × subjects and concentrations × replications two-way interaction F values cannot be calculated because the denominator required is the mean square for the three-way interaction; we have not calculated one here. The between-concentrations F value cannot be calculated because the ANOVA is incapable of doing so when the other two factors are random effects. As more factors become random effects, the calculation becomes more limited.

l. (Do not answer this question until you have read Chapter 15.) The experimenter noted that as the concentration of each solution increased, the intensity rating also increased significantly. He concluded that there was a significant correlation between intensity scores and concentration. Was he right?

No! He was not! For there to be a significant correlation, the relationship between the intensity scores and concentrations would have to be linear. You need to calculate a correlation coefficient to establish this. Just because the mean intensity scores increase significantly at higher concentrations, it does not mean that intensity and concentration are linearly related.

Example 5

Four reformulations of cookies were tasted by a panel of judges who rated them for a set of flavor attributes. Cookies were baked at different cooking temperatures and stored for different lengths of time. Each judge tasted several cookies under each condition. A table indicating the levels of significance of F values for the various treatments and some interactions is given below for various flavor attributes rated by the judges. (Asterisks indicate the levels of significance.)

	df	Sweetness	Malt flavor	Baked flavor	Chocolate flavor	Overall flavor persistence	Crunchiness
Treatment:							
Reformulations	3	NS	**	*	NS	***	NS
Cooking temperatures	6	NS	*	**	*	*	***
Storage times	3	NS	NS	NS	*	*	***
Replications	5	*	*	NS	NS	NS	NS
Judges		NS	NS	*	*	NS	NS
Two-way interactions:							
Reformulations × judges	57	NS	*	*	NS	NS	NS
Reformulations × cooking temperatures	18	NS	*	NS	NS	NS	NS
Reformulations × storage times	9	NS	*	*	NS	*	*

a. How many cooking temperatures, storage times, and replications were used in this experiment?

Cooking temperatures: $df = 6$
Storage times: $df = 3$
Replications: $df = 5$

Thus there were seven cooking temperatures, four storage times, and six replications.

b. How many judges were used on the panel?

The reformulations X judges $df = 57$ with the reformulations $df = 3$, the judges $df = 57/3 = 19$. (Two-way interaction df = product of df values for each interactng factor.) Thus there were 20 judges.

c. What attributes changed the most from reformulation to reformulation?

Overall flavor persistence varied the most ($p < 0.001$, ***). Malt flavor ($p < 0.01$) and baked flavor ($p < 0.05$) also changed significantly.

d. For what attributes did reformulations have the least effect?

Sweetness, chocolate flavor, and crunchiness showed least change from reformulation to reformulation; in all cases there were no significant differences.

e. What factors had a significant effect on crunchiness?

Cooking temperature ($p < 0.001$) and storage time ($p < 0.001$) had significant effects on crunchiness scores. There was also a significant two-way reformulations X storage times interaction ($p < 0.05$). This means that the lack of difference in crunchiness scores between reformulations was not consistent over cookies stored for different lengths of time; alternatively, it could mean that the trend for the effect of storage time on crunchiness varied with the reformulation considered, in which case the first alternative would also have to be true. There are only two manners in which this two-way interaction can occur (see Example 3).

f. Did sweetness alter with reformulations, cooking temperature, or storage time?

In all cases F was not significant. Our data do not indicate any effect.

g. What attributes did cooking temperature affect?

Cooking temperature altered malt flavor ($p < 0.05$), baked flavor ($p < 0.01$), chocolate flavor ($p < 0.05$), overall flavor persistence ($p < 0.05$), and crunchiness ($p < 0.001$).

h. The malt flavor altered significantly from reformulation to reformulation. Was the trend for malt flavor scores over the reformulations the same for all judges, for all cooking temperatures, and for storage times?

The significant ($p < 0.05$) two-way interactions for reformulations X judges, for reformulations X cooking temperatures, and for reformulations X storage times can suggest a variety of things. It could mean that the lack of difference in malt flavor scores for different storage times and for judges was not consistent from reformulation to reformulation. It could also mean that the trend for malt flavor scores over reformulations was not consistent over storage times or judges. (Should the latter happen, the former must also; there are only two manners in which a two-way interaction can occur here; see Example 3.)

It could mean that the trend for malt flavor scores over cooking temperatures may vary with reformulations or the trend over reformulations may vary with cooking temperature, or both. (Here the two-way interaction may occur in three different ways (see Example 3).

An examination of the original data would indicate which alternatives are likely.

i. There were no changes in sweetness or chocolate flavor from reformulation to reformulation. Was the lack of change with reformulations consistent over judges, various storage times and cooking temperatures.

The two-way interactions for reformulations X judges, reformulations X cooking temperatures, and reformulations X storage times were not significant for sweetness or chocolate flavor. Thus the reported lack of difference in sweetness scores and chocolate flavor scores from reformulation to reformulation occurred regardless of the judge making the judgment, the cooking temperature, or the storage time; there was consistentency over all these treatments.

j. Was there any tendency for judges as a whole to change in their use of the rating scales as the experiment proceeded?

Should there be a change in the way the panel of judges used the rating scales as the experiment proceeded, it would show up in the mean scores averaged over all treatments for each replication. A significant between-replications effect would indicate such a change. This occurred only for ratings of sweetness and malt flavor. If individual judges changed on the other attribute scales, the insignificant between-replications effect indicates that such changes were balanced out by other judges changing in an opposing manner.

k. For sweetness scores: there was not a between-treatments effect for reformulations; there were no reformulations X cooking temperatures, no reformulations X cooking temperatures X storage times, and no reformulations X cooking temperatures X storage times X judges interactions. What does this mean? Not all of these data occur in the table.

No between-treatments effects for reformulations means that the mean sweetness scores for each reformulation, averaged over the 20 judges, six replications, four storage times, and seven cooking temperatures, do not differ. These mean sweetness scores are the same for reformulations regardless of cooking temperatures (insignificant reformulations X cooking temperatures two-way interaction). The lack of difference between reformulations is consistent for all seven cooking temperatures and this is true regardless of storage times (insignificant reformulations X cooking temperatures X storage times three-way interaction). For sweetness scores, the lack of effect of storage time on the consistency over cooking temperatures of scores from different reformulations is itself consistent for each judge. (Insignificant reformulations X cooking temperatures X storage times X judges four-way interaction.)

l. How many degrees of freedom are there associated with the reformulations X cooking temperatures X storage times X judges four-way interaction?

Reformulations: $df = 3$
Cooking temperatures: $df = 6$
Storage times: $df = 3$
Judges: $df = 19$

For the four-way interaction between these factors, $df = 3 \times 6 \times 3 \times 19 = 1026$.

m. How many two-way, three-way, and four-way interactions are possible in this study?

There are five factors: judges, replications, reformulations, cooking temperatures, and storage times. Thus the number of two-way interactions is given by

$$_5C_2 = \frac{5!}{(5-2)!\,2!} = \frac{5 \times 4 \times 3 \times 2 \times 1}{3 \times 2 \times 1 \times 2 \times 1} = 10$$

The number of three-way interactions is given by

$$_5C_3 = \frac{5!}{(5-3)!\,3!} = \frac{5 \times 4 \times 3 \times 2 \times 1}{2 \times 1 \times 3 \times 2 \times 1} = 10$$

The number of four-way interactions is given by

$$_5C_4 = \frac{5!}{(5-4)!\,4!} = \frac{5 \times 4 \times 3 \times 2 \times 1}{1 \times 4 \times 3 \times 2 \times 1} = 5$$

n. What are the 10 possible three-way interactions, and how many degrees of freedom are associated with each?

The three-way interactions, with associated numbers of degrees of freedom, are as follows:

Interaction	*df*
Judges X replications X reformulations	19 X 5 X 3 = 285
Judges X replications X cooking temperatures	19 X 5 X 6 = 570
Judges X replications X storage times	19 X 5 X 3 = 285
Judges X reformulations X cooking temperatures	19 X 3 X 6 = 342
Judges X reformulations X storage times	19 X 3 X 3 = 171
Judges X cooking temperatures X storage times	19 X 6 X 3 = 342
Replications X reformulations X cooking temperatures	5 X 3 X 6 = 90
Replications X reformulations X storage times	5 X 3 X 3 = 45
Replications X cooking temperatures X storage times	5 X 6 X 3 = 90
Reformulations X cooking temperatures X storage times	3 X 6 X 3 = 54

If you have just worked through Examples 1 to 5, you deserve a rest. Why not indulge your hankering for a little cup of something?

13

Fixed- and Random-Effects Models

13.1 WHAT ARE FIXED- AND RANDOM-EFFECTS MODELS?

Use of Fixed-, Random-, and Mixed-Effects Models

The ANOVA calculations that we have performed so far have been according to a *fixed-effects model*; there is another model called the *random-effects model*. With the fixed-effects model, the conclusions apply only to the treatment levels tested. With the random-effects model, the treatment levels tested are only a sample of all the possible levels that we wish to consider; the conclusions then apply to all these possible levels. There is also a *mixed-effects model*, where some factors are fixed and others are random. For instance, we may be interested in the effect of an increase in baking temperature on frozen apple pies. We sample a few apple pies baked at the higher temperature for comparison. We then extend our results to a whole population of apple pies that would be baked at the higher temperature. Here, the replicate samples of apple pie are a random-effects factor.

The most common differentiation between fixed and random effects, however, concerns human subjects. If human subjects are being tested as a sample of the population, the conclusions drawn from these subjects will be extended to the whole population. Subjects, in this case, are a random-effects factor. However, sometimes the conclusions drawn from the subjects are not extended beyond the subjects actually tested; in this case the subjects are a fixed-effects factor. Before proceeding, the reader is now advised to look again at Section 1.2 and note the different uses of human beings in sensory evaluation, psychophysics, and consumer testing.

Imagine a case in sensory analysis where judges are used as highly trained instruments to measure the flavor of a food, say detect a change in flavor due to a change in processing. This could be part of a study of the variables involved in the processing procedure. The panel of judges might be previously screened and highly trained and not representative of the general population of people. Thus the conclusions made for the panel would not be extended to people in general. If the panel detected a difference, it would not mean that the general population would detect a difference. But this does not matter. Here the panel is used as a sophisticated instrument for testing the food. We would not expect all instruments to detect the flavor change, just as we do not expect all gas chromatographs to be as sensitive as the state-of-the-art models. Thus the judges, here, are a fixed effect.

If, on the other hand, we wished to know whether a flavor difference could be detected by the general population, we would take a sample of judges, representative of the population, and test them. Here, the judges are a random effect.

In the case of sensory psychophysics, where we wish to study how the senses work, we take a sample of human subjects and place them in an experimental situation. We then generalize our conclusions about our sample to people in general. Our people here are a random effect. For consumer testing, we test the likes and dislikes of a sample of consumers and then generalize the results to the population. Consumers are a random effect.

In sensory evaluation, replicate samples are generally a random effect. We usually wish to generalize our results beyond the replicates tested, to the whole population of samples of the food prepared in this way. The same would be generally true for consumer testing. If we determined the degree of liking for replicate samples of food tasted by the consumer, we would wish to generalize beyond these replicate samples. However, this occasion may not arise, because consumers are unsophisticated in the use of scaling. Perhaps ranking or paired-comparisons tests might be simpler and more reliable for consumers, in which case the data would not be analyzed by ANOVA. For psychophysics, the various experimental conditions that a person is subjected to are generally a fixed effect. The experimenter would not usually extend his or her conclusions beyond those conditions tested.

Computational Difference Is Merely a Change in the Denominator

The difference between fixed-effects, random-effects, and mixed-effects models merely involves a difference in the final step of the ANOVA; it involves merely

Table 13.1 Fixed-Effects Model

Source	Mean square	F
ABC	MS_A	$\dfrac{MS_A}{MS_E}$ Fixed
$\alpha\beta\gamma$	MS_α	$\dfrac{MS_\alpha}{MS_E}$ Fixed
Interaction $ABC \times \alpha\beta\gamma$	$MS_{\alpha \times A}$	$\dfrac{MS_{\alpha \times A}}{MS_E}$
Error	MS_E	

the use of an interaction mean square for some of the denominators for the F values, rather than error mean squares.

This is best seen by a consideration of the two-factor ANOVA with interaction. Consider a two-factor design with factors A, B, C and α, β, γ. We can represent the denominators by the scheme in Table 13.1. Consider the familiar fixed-effects model (the only one we have discussed so far). The factors A, B, C, and α, β, γ are fixed and the denominator is always the error mean square.

Now consider the random effects model (Table 13.2) where factor A, B, C and α, β, γ are random-effects factors. Here the denominator is no longer the error mean square; it is the interaction mean square. The reason for this will be discussed in Section 13.2. The interaction F value, however, still uses the error mean square.

Table 13.2 Random-Effects Model

Source	Mean square	F
ABC	MS_A	$\dfrac{MS_A}{MS_{\alpha \times A}}$ Random
$\alpha\beta\gamma$	MS_α	$\dfrac{MS_\alpha}{MS_{\alpha \times A}}$ Random
Interaction $ABC \times \alpha\beta\gamma$	$MS_{\alpha \times A}$	$\dfrac{MS_{\alpha \times A}}{MS_E}$
Error	MS_E	

Table 13.3 Mixed-Effects Model

Source	Mean square	F	
ABC	MS_A	$\dfrac{MS_A}{MS_{\alpha \times A}}$	Fixed
$\alpha\beta\gamma$	MS_α	$\dfrac{MS_\alpha}{MS_E}$	Random
Interaction $ABC \times \alpha\beta\gamma$	$MS_{\alpha \times A}$		
Error	MS_E		

Now consider a third model, the mixed-effects model (Table 13.3), where one of the factors is fixed and the other random. Here, one denominator is the error mean square and the other is the interaction mean square. The fixed-effects factor uses the interaction term ($MS_{A \times \alpha}$) unlike in the fixed-effects model where MS_E is used. The random-effects factor uses MS_E unlike in the random-effects model. Thus in the mixed-effects model, the denominators are counterintuitive; they are the opposite of what is used for these terms in the fixed-effects or random-effects models.

It is now worth examining why the various denominators are chosen.

13.2 THEORY BEHIND FIXED- AND RANDOM-EFFECTS MODELS

The best approach to understanding the reasons for the different denominators in the various ANOVA models is a mathematical one. In deriving the probability theory for ANOVA, theoretical mean square (*MS*) values are calculated; these are called *expected mean square* (or EMS) *values*. It is in these EMS values that the reason can be seen for the variety of denominators in the ANOVA models.

The approach to the probability calculations runs as follows: A given score (X) is said to be compounded of a population mean (μ), an effect due to the *ABC* factor (A), an effect due to the $\alpha\beta\gamma$ factor (α), an effect due to the interaction between the *ABC* and $\alpha\beta\gamma$ factors (αA), and any leftover effects due to none of these factors, the experimental error (e). The population mean (μ) is altered by all these effects working on it; an amount is added (or subtracted) because of factor *ABC*, another amount because of factor $\alpha\beta\gamma$, another amount because of the interaction, and a final amount because of the error. It is assumed that these effects all add to (or subtract from) the population mean to give a final value. This is called an additive model. Thus we can write

X	=	μ	+	A	+	α	+	αA	+	e
observed score		population mean		effect of ABC factor		effect of $\alpha\beta\gamma$ factor		effect of $ABC \times \alpha\beta\gamma$ interaction		effect of error

We now let the variance of the A values be σ_A^2, of α values be σ_α^2, of the interaction (αA) values be $\sigma_{\alpha A}^2$, and of the error (e) values be σ^2. Further, we let the ABC factor have k_A treatments and the $\alpha\beta\gamma$ factors have k_α treatments. Finally, we let there be n observations for each combination of treatments; the total number of scores is then $nk_A k_\alpha$. This can be represented as follows:

	A	B	C	$\cdots$	k_A
α	n	n	n		
β	n	n	n		
γ	n	n	n		
$\vdots$					
k_α					

Fixed-Effects Model

For a fixed-effects model, the EMS values can be derived as shown in Table 13.4.

Now consider the equation

X	=	μ	+	A	+	α	+	αA	+	e
observed score		population mean		effect of ABC		effect of $\alpha\beta\gamma$		interaction effect		error effect

Table 13.4 EMS Calculations for the Fixed-Effects Model

Treatments		EMS Values
A	k_A treatments	$\sigma^2 + nk_\alpha \sigma_A^2$
α	k_α treatments	$\sigma^2 + nk_A \sigma_\alpha^2$
AB interaction		$\sigma^2 + n\sigma_{\alpha A}^2$
Error		σ^2

If the treatments in the factor ABC have an effect on the score (alter X from the population mean value μ), a range of A values will be added to μ to give a range of X values. These A values will have a variance σ_A^2. The greater the effect of the ABC factor, the more the A values (and hence the X values) will vary and thus the greater will be σ_A^2. On the other hand, if the factor ABC has no effect on X, there will be no spread of A values and σ_A^2 will be zero. H_0 states that the factor ABC has no effect and thus that $\sigma_A^2 = 0$. The same is true of the factor $\alpha\beta\gamma$. Thus, if H_0 is true for both ABC and $\alpha\beta\gamma$:

$\sigma_A^2 = 0$: factor ABC has no effect.
$\sigma_\alpha^2 = 0$: factor $\alpha\beta\gamma$ has no effect.

We now go on to consider the F values. For the factor ABC, the theoretical F value in terms of EMS values is given by

$$F = \frac{MS_A}{MS_E} = \frac{\sigma^2 + nk_\alpha \sigma_A^2}{\sigma^2}$$

If the factor ABC has no effect, $\sigma_A^2 = 0$ and the F fraction $(\sigma^2 + nk_\alpha \sigma_A^2)/\sigma^2$ becomes $\sigma^2/\sigma^2 = 1$. If factor ABC has an effect, σ_A^2 will get larger, the greater the effect of this factor. The larger σ_A^2 gets, the larger F fraction $(\sigma^2 + nk_\alpha \sigma_A^2)/\sigma^2$ becomes.

The same is true for the $\alpha\beta\gamma$ factor:

$$F = \frac{MS_\alpha}{MS_E} = \frac{\sigma^2 + nk_A \sigma_\alpha^2}{\sigma^2}$$

The greater the effect of factor $\alpha\beta\gamma$, the larger will be σ_α^2 and the larger will be F. Should factor $\alpha\beta\gamma$ have no effect, $F = 1$ (σ^2/σ^2). It can be seen that the same argument also applies to the interaction term.

Thus when MS_E is the denominator, F is unity should the factors have no effect, while F gets larger the greater the effect of the factors. How much larger than unity the F value can get by chance is calculated and given in tables for comparison. If our calculated value is greater than the chance value, we reject H_0. Table G.11 gives a set of the smallest F values at which H_0 can just be rejected. Thus, in the fixed-effects model, MS_E is a convenient denominator.

Now consider the EMS values for the random-effects model. These can be derived as follows:

Random-Effects Model

Here the expected mean squares can be calculated as shown in Table 13.5.

Here the EMS value for factor A is

$$\sigma^2 + n\sigma_{\alpha A}^2 + nk_\alpha \sigma_A^2$$

Table 13.5 EMS Calculations for the Random-Effects Model

Treatments		EMS Values
A	k_A treatments	$\sigma^2 + n\sigma^2_{\alpha A} + nk_\alpha \sigma^2_A$
α	k_α treatments	$\sigma^2 + n\sigma^2_{\alpha A} + nk_A \sigma^2_\alpha$
$A\alpha$		$\sigma^2 + n\sigma^2_{\alpha A}$
Error		σ^2

Here the expression for the EMS value has in it a value for both $\sigma^2_{\alpha A}$ and σ^2_A. When testing for the effect of factor A, we are interested only in whether σ^2_A is zero (H_0 states that $\sigma^2_A = 0$). If $\sigma^2(MS_E)$ was the denominator, the F value could be greater than unity even if σ^2_A was zero, because of the interaction term $\sigma^2_{\alpha A}$, which may also be greater than zero.

Thus an inappropriate F fraction would be

$$\frac{MS_A}{MS_E} = \frac{\sigma^2 + \sigma^2_{\alpha A} + nk_\alpha \sigma^2_A}{\sigma^2}$$

When σ^2_A was zero (factor ABC had no effect) the fraction would become $(\sigma^2 + \sigma_{\alpha A})/\sigma^2$, not $\sigma^2/\sigma^2 = 1$.

We want the fraction to equal unity when the factor ABC has no effect; we need a different denominator. If we use the interaction term as the denominator our problems are solved. Then

$$\text{Factor } ABC,\ F = \frac{MS_A}{MS_{\alpha\times A}} = \frac{\sigma^2 + n\sigma^2_{\alpha A} + nk_\alpha \sigma^2_A}{\sigma^2 + n\sigma^2_{\alpha A}}$$

Should σ^2_A be zero (factor ABC have no effect), the fraction becomes unity:

$$\frac{\sigma^2 + n\sigma^2_{\alpha A}}{\sigma^2 + n\sigma^2_{\alpha A}} = 1$$

Should the factor ABC have an effect, σ^2_A will exceed unity. The greater the effect of this factor, the greater σ^2_A and the greater F.

The same is true for the $\alpha\beta\gamma$ factor. The interaction term is the appropriate denominator. Once again, $F = 1$ when $\sigma^2_\alpha = 0$.

$$\text{Factor } \alpha\beta\gamma, F = \frac{MS_\alpha}{MS_{\alpha\times A}} = \frac{\sigma^2 + n\sigma^2_{\alpha A} + nk_A \sigma^2_\alpha}{\sigma^2 + n\sigma^2_{\alpha A}}$$

Of course, for the interaction term, the appropriate denominator will be MS_E, for the same reasons.

$$\text{Interaction factor } ABC \times \alpha\beta\gamma, F = \frac{\sigma^2 + n\sigma^2_{\alpha A}}{\sigma^2}$$

We summarize the argument so far in Table 13.6.

Table 13.6 Expected Mean Squares for the Fixed- and Random-Effects Models

Source of variance	Expected mean square	Appropriate F
	Fixed effects	
ABC	$MS_A = \sigma^2 + nk_\alpha \sigma^2_A$	$\frac{MS_A}{MS_E}$
$\alpha\beta\gamma$	$MS_\alpha = \sigma^2 + nk_A \sigma^2_\alpha$	$\frac{MS_\alpha}{MS_E}$
$A \times \alpha$	$MS_{\alpha \times A} = \sigma^2 + n\sigma^2_{\alpha A}$	$\frac{MS_{\alpha \times A}}{MS_E}$
Error	$MS_E = \sigma^2$	
	Random effects	
ABC	$MS_A = \sigma^2 + n\sigma^2_{\alpha A} + nk_\alpha \sigma^2_A$	$\frac{MS_A}{MS_{\alpha \times A}}$
$\alpha\beta\gamma$	$MS_\alpha = \sigma^2 + n\sigma^2_{\alpha A} + nk_A \sigma^2_\alpha$	$\frac{MS_\alpha}{MS_{\alpha \times A}}$
$A \times \alpha$	$MS_{\alpha \times A} = \sigma^2 + n\sigma^2_{\alpha A}$	$\frac{MS_{\alpha \times A}}{MS_E}$
Error	$MS_E = \sigma^2$	

Mixed-Effects Model

The same reasoning can be applied to the mixed model. Should *ABC* be the fixed-effects factor and $\alpha\beta\gamma$ be the random-effects factors, the EMS value for *ABC* is given by $\sigma^2 + n\sigma^2_{\alpha A} + nk_\alpha \sigma^2_A$; this requires the interaction term $(\sigma^2 + n\sigma^2_{\alpha A})$ to be the denominator. The random-effects factor $(\alpha\beta\gamma)$ EMS is $\sigma^2 + nk_A \sigma^2_\alpha$; the error term is a suitable denominator here. The error term is also suitable for interaction. Thus we have Table 13.7.

This type of argument can be applied to all ANOVA designs to determine the appropriate denominators. To do so here would be tedious, so we will merely summarize the results. The task becomes merely one of selecting the appropriate denominator for *F*, depending on which factors are fixed effects and which are random; we will give some tables for this.

There are two points to note. First, for more complex designs (three factors and higher), the EMS values for the numerator can get so complicated that it is not always possible to find an appropriate denominator which will give an *F* value of unity when the appropriate factor has no effect. Second, in designs where no interaction terms are calculated, the error mean square is always the appropriate denominator; fixed-, random-, and mixed-effects models turn out to be the same.

Table 13.7 Expected Mean Squares for the Mixed-Effects Model

Mixed effects	Expected mean square	Appropriate *F*
ABC (fixed)	$MS_A = \sigma^2 + n\sigma^2_{\alpha A} + nk_\alpha \sigma^2_A$	$\dfrac{MS_A}{MS_{\alpha \times A}}$
$\alpha\beta\gamma$ (random)	$MS_\alpha = \sigma^2 + nk_A \sigma^2_\alpha$	$\dfrac{MS_\alpha}{MS_E}$
$A \times \alpha$	$MS_{\alpha \times A} = \sigma^2 + n\sigma^2_{\alpha A}$	$\dfrac{MS_{\alpha \times A}}{MS_E}$
Error	$MS_E = \sigma$	

13.3 SUMMARY OF THE DENOMINATORS FOR *F* RATIOS FOR FIXED AND RANDOM EFFECTS

The appropriate denominators for the F values for various ANOVA designs are summarized in Tables 13.8 to 13.11. In Table 13.8, the only factor is ABC (A). In Table 13.9 one factor is ABC (A) while the second is $S_1S_2S_3$ (S). This second factor ($S_1S_2S_3$) could be "subjects" for a repeated-measures design, or a set of experimental treatments for a completely randomized design. In Table 13.10, again one factor is ABC (A); the second is $S_1S_2S_3$ (S), "subjects" or a set of experimental treatments. Finally, in Table 13.11, the three factors are ABC (A), $\alpha\beta\gamma$ (α), and $S_1S_2S_3$ (S). The latter can be "subjects" or a set of experimental treatments. In this table, where "none" is indicated, there is no exact value for an expected mean square.

Table 13.8 One-Factor ANOVA, Completely Randomized

Source	Mean square	Denominator for F: A fixed or random
A	MS_A	MS_E
Error	MS_E	

Table 13.9 Two-Factor ANOVA Without Interaction

Source	Mean square	Denominator for F: A fixed or random, S fixed or random
A	MS_A	MS_E
S	MS_S	MS_E
Error	MS_E	

Table 13.10 Two-Factor ANOVA With Interaction

		Denominator for F			
Source	Mean square	A fixed, S fixed	A fixed, S random	A random, S fixed	A random, S random
A	MS_A	MS_E	$MS_{A\times S}$	MS_E	$MS_{A\times S}$
S	MS_S	MS_E	MS_E	$MS_{A\times S}$	$MS_{A\times S}$
$A \times S$	$MS_{A\times S}$	MS_E	MS_E	MS_E	MS_E
Error	MS_E				

Table 13.11 Three-Factor ANOVA

		Denominator for F			
Source	Mean square	All fixed	A, α fixed; S random	A fixed; α, S random	All random
A	MS_A	MS_E	$MS_{A\times S}$	None	None
α	MS_α	MS_E	$MS_{\alpha\times S}$	$MS_{\alpha\times S}$	None
S	MS_S	MS_E	MS_E	$MS_{\alpha\times S}$	None
$A \times \alpha$	$MS_{\alpha\times A}$	MS_E	$MS_{\alpha\times A\times S}$	$MS_{\alpha\times A\times S}$	$MS_{\alpha\times A\times S}$
$A \times S$	$MS_{A\times S}$	MS_E	MS_E	$MS_{\alpha\times A\times S}$	$MS_{\alpha\times A\times S}$
$\alpha \times S$	$MS_{\alpha\times S}$	MS_E	MS_E	MS_E	$MS_{\alpha\times A\times S}$
$A \times \alpha \times S$	$MS_{\alpha\times A\times S}$	MS_E	MS_E	MS_E	MS_E
Error	MS_E				

14
Split-Plot Design

14.1 HOW THE SPLIT-PLOT DESIGN IS RELATED TO OTHER DESIGNS

The conventional two-factor (without interaction) repeated-measures design is familiar, where one factor is *ABC* and a second factor is "subjects." Because there are no replications per subject in each condition *A, B, C*, there is no two-way interaction. The design can be represented as follows:

	A	B	C	
S_1	X	X	X	T_1
S_2	X	X	X	T_2
S_3	X	X	X	T_3
S_4	X	X	X	T_4
S_5	X	X	X	T_5
S_6	X	X	X	T_6
S_7	X	X	X	T_7
	T_A	T_B	T_C	T

SS_A, SS_S, and SS_E are calculated as usual. For fixed- or random-effects models, F values are calculated in the same way, using MS_E as the denominator.

The conventional three-factor design is also familiar; where one factor is *ABC*, another could be $\alpha\beta$ and a third may be "subjects." Let us assume that each subject is tested once in each combination of conditions *A*, *B*, *C* and α, β. Because there are no replications per subject there will be no three-way interaction. This design can be represented, as follows:

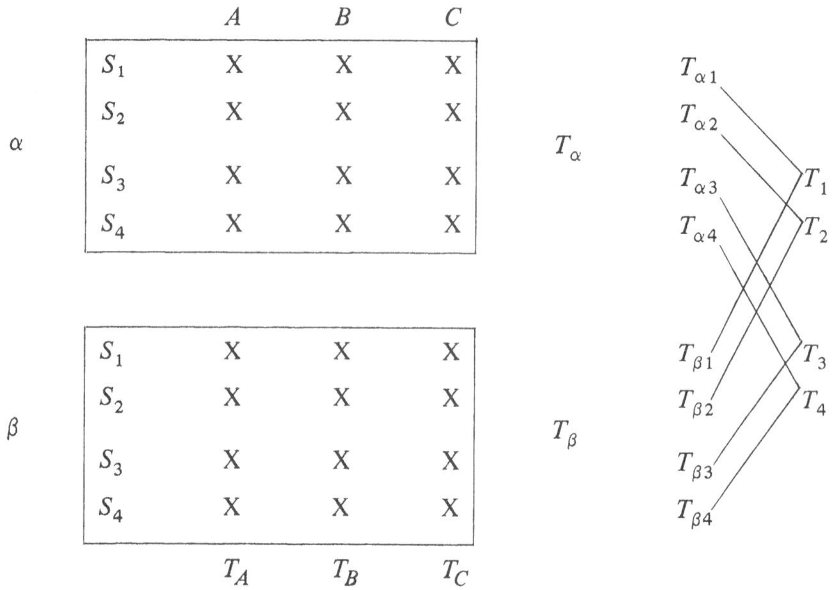

The between-treatments effects SS_A, SS_α, and SS_S, the two-way interactions $SS_{\alpha\times A}$, $SS_{\alpha\times S}$, and $SS_{A\times S}$, and the error term SS_E are calculated in the usual way. In this particular example, there is no three-way interaction because there are no replications for subjects in each combination of conditions *A*, *B*, *C* and α, β. For the fixed-effects model, the between-treatments and interaction *F* values are calculated by dividing by the error term. If the subjects are only a sample of the population being investigated, a mixed design may be used with factors *ABC* and $\alpha\beta$ being a fixed effect and subjects being a random effect. It is also possible that *A*, *B*, and *C* could be replicate food samples, from which it is necessary to generalize to a population of food samples; these, then, would also be a random effect (see Section 13.1).

However, consider the following case, where there are three factors: *ABC*, $\alpha\beta$, and "subjects" but not all subjects are tested in all conditions. All are tested in conditions *A*, *B*, and *C* but only some are tested in condition α, while the rest are tested in condition β. For example, α could be an experimental group and β the control group, as follows:

		A	B	C		
α: Experimental group	S_1	X	X	X	$T_{\alpha 1}$	
	S_2	X	X	X	$T_{\alpha 2}$	
	S_3	X	X	X	$T_{\alpha 3}$	T_α
	S_4	X	X	X	$T_{\alpha 4}$	
		$T_{\alpha A}$	$T_{\alpha B}$	$T_{\alpha C}$		
β: Control group	S_5	X	X	X	$T_{\beta 5}$	
	S_6	X	X	X	$T_{\beta 6}$	T_β
	S_7	X	X	X	$T_{\beta 7}$	
		$T_{\beta A}$	$T_{\beta B}$	$T_{\beta C}$		
		T_A	T_B	T_C		

A, B, C, fixed
α, β fixed

Subjects can be fixed or random

It can be seen that this design is similar to the three-factor design except that not all subjects are tested in each condition. It is also rather like a repeated-measures design except that the subjects are divided or split into two groups, creating an extra factor; hence the name *split plot*. Factor ABC is a repeated measures factor where A, B, and C have a related-samples design; on the other hand, α and β have an independent-samples design. This mixed up design is a little complicated, especially as far as choosing the error terms for the denominators. The factors that this design are set up to test are ABC and $\alpha\beta$ and any interaction between the two ($ABC \times \alpha\beta$). The factors ABC and $\alpha\beta$ are considered here as *fixed-effects factors*. The design is not set up to examine an F value for the "subjects" factor, so it makes no difference whether "subjects" are fixed or random. This design is commonly encountered in behavioral research.

14.2 LOGIC OF THE SPLIT-PLOT DESIGN

The logic of the split-plot design can best be understood by examining the computation sequence. For simplicity, let us assume that we have seven subjects who all undergo treatments ABC, while four undergo them in treatment α and three in β. This can be represented as follows:

		A	B	C		
	S_1	X	X	X	$T_{\alpha 1}$	
	S_2	X	X	X	$T_{\alpha 2}$	
α	S_3	X	X	X	$T_{\alpha 3}$	T_α
	S_4	X	X	X	$T_{\alpha 4}$	
		$T_{\alpha A}$	$T_{\alpha B}$	$T_{\alpha C}$		
	S_5	X	X	X	$T_{\beta 5}$	
β	S_6	X	X	X	$T_{\beta 6}$	T_β
	S_7	X	X	X	$T_{\beta 7}$	
		$T_{\beta A}$	$T_{\beta B}$	$T_{\beta C}$		
		T_A	T_B	T_C	Grand total, T	

n_A = number of scores in treatment A, B, or C; n_α = number of scores in treatment α; n_β = number of scores in treatment β; N = total number of scores, Here, $n_A = 7, n_\alpha = 12, n_\beta = 9, N = 21$.

The scheme above, depicting the original data, is sometimes called the *whole matrix* (see Sections 12.1 and 12.4). From this whole matrix we can determine the total and between-treatments effects in the usual way.

$$\text{Correction term, } C = \frac{T^2}{N} \qquad \text{here } N = 21$$

$$\text{Total sum of squares, } SS_T = \Sigma X^2 - C \qquad \text{with } df = N - 1$$

The between-treatments effects for the factors ABC and $\alpha\beta$ are then computed as usual from the respective totals for each treatment; once again, the denominator in each case (n_A for ABC; n_α or n_β for $\alpha\beta$) is the number of scores in the original data making up the total under consideration.

For factor ABC,

$$SS_A = \frac{T_A^2 + T_B^2 + T_C^2}{n_A} - C \qquad \text{here } n_A = 7$$

$$df = \text{number of treatments} - 1 = 3 - 1 = 2$$

For factor $\alpha\beta$,

$$SS_\alpha = \frac{T_\alpha^2}{n_\alpha} + \frac{T_\beta^2}{n_\beta} - C \qquad \text{here } n_\alpha = 12 \text{ and } n_\beta = 9$$

$$df_\alpha = \text{number of treatments} - 1 = 2 - 1 = 1$$

We do not examine the subjects factor in this design. This is a deviation from the regular three-factor format.

We also have two sets of subtotals ($T_{\alpha A}, T_{\alpha B}, T_{\alpha C}, T_{\beta A}, T_{\beta B}, T_{\beta C}$ and $T_{\alpha 1}$, $T_{\alpha 2}$, $T_{\alpha 3}$, $T_{\alpha 4}$, $T_{\beta 5}$, $T_{\beta 6}$, $T_{\beta 7}$) which can be used to construct appropriate matrices for calculating interactions (see Sections 11.3 and 12.1). The first matrix is the regular two-way $A\alpha$ matrix:

	A	B	C	
α	$T_{\alpha A}$	$T_{\alpha B}$	$T_{\alpha C}$	T_α
β	$T_{\beta A}$	$T_{\beta B}$	$T_{\beta C}$	T_β
	T_A	T_B	T_C	

As usual, the cell totals are due to the effects of the factors ABC and $\alpha\beta$ and the interaction between the two (Sections 11.3, 12.1, and 12.4); the interaction sum of squares can thus be obtained by subtraction:

$$ABC \times \alpha\beta\gamma \text{ interaction}, SS_{\alpha \times A} = \text{cell-total } SS - SS_A - SS_\alpha$$

$$df_{\alpha \times A} = \text{cell-total } df - df_A - df_\alpha$$

The calculation of the cell-total sum of squares is complicated by the fact that not all the totals are the sums of the same number of scores in the original data; there are more scores under condition α than under condition β. Hence

$$\text{Cell-total } SS = \frac{T_{\alpha A}^2}{n_{\alpha A}} + \frac{T_{\alpha B}^2}{n_{\alpha B}} + \frac{T_{\alpha C}^2}{n_{\alpha C}} + \frac{T_{\beta A}^2}{n_{\beta A}} + \frac{T_{\beta B}^2}{n_{\beta B}} + \frac{T_{\beta C}^2}{n_{\beta C}} - C$$

where

$$n_{\alpha A} = n_{\alpha B} = n_{\alpha C} = 4$$

$$n_{\beta A} = n_{\beta B} = n_{\beta C} = 3$$

$$\text{Cell-total } df = \text{number of cells} - 1 = 6 - 1$$

So far we have done nothing different from a regular two-factor ANOVA with interaction. We have gone to the appropriate totals for factors *ABC* and $\alpha\beta$ and the αA two-way matrix for the $ABC \times \alpha\beta$ interaction. The only complication is that the numbers of scores that make up the totals vary somewhat.

We now construct the second possible subtotals matrix, but it is now that the computation deviates from the regular format. Our second set of totals make up the matrix as follows:

α	$T_{\alpha 1}$	$T_{\alpha 2}$	$T_{\alpha 3}$	$T_{\alpha 4}$
β	$T_{\beta 5}$	$T_{\beta 6}$	$T_{\beta 7}$	

This matrix is quite unlike the regular matrices that we have constructed before (Sections 12.1 and 12.4). This is not a two-way matrix because it is not a set of totals that are simultaneously due to variations in subjects and treatments α and β. Such a matrix would be as follows:

	S_1	S_2	S_3	S_4	S_5	S_6	S_7	
α	$T_{\alpha 1}$	$T_{\alpha 2}$	$T_{\alpha 3}$	$T_{\alpha 4}$	$T_{\alpha 5}$	$T_{\alpha 6}$	$T_{\alpha 7}$	T_{α}
β	$T_{\beta 1}$	$T_{\beta 2}$	$T_{\beta 3}$	$T_{\beta 4}$	$T_{\beta 5}$	$T_{\beta 6}$	$T_{\beta 7}$	T_{β}
	T_1	T_2	T_3	T_4	T_5	T_6	T_7	

In this familiar two-way matrix all the totals are the result of the simultaneous action of the "subjects" and the $\alpha\beta$ factor as well as the two-way interaction between the two; the interaction can be calculated using this matrix. Each subject exercises influence under both treatments α and β; both treatments α and β influence each subject. This is not the case with the split-plot matrix. Here, each subject only exercises influence under one of the treatments: α or β; each treatment influences separate subjects. Clearly, this is not a two-way $\alpha\beta \times$ subjects matrix; the totals are not due to the simultaneous effect of "subjects," $\alpha\beta$, and their two-way interaction. The matrix must be treated differently.

One thing is true about the matrix, however; all the totals are either from condition α or from condition β. But the subjects in each condition are quite different. It is rather like an independent samples or completely randomized version of a regular two-way matrix. If you recall, the effects of treatments from only one factor are separated out (MS_A) in the completely randomized design (see Chapter 8); the rest of the effects (uncontrolled experimental variables, effects due to different subjects under each treatment, etc.) are included under an error term (MS_E).

As it was with a completely randomized ANOVA, so it is with our matrix. The effects of only one factor can be separated out (MS_α) while the rest of the effects on the totals are included under an error term. Thus we write

$$\text{Cell-total } SS = SS_\alpha + \text{an error } SS$$

Contrast this with the regular two-way matrix:

$$\text{Cell-total } SS = SS_\alpha + SS_S + SS_{\alpha \times S}$$

We need to ask ourselves exactly what type of error we have here. It is certainly not the regular error (MS_E) that is due to all factors not deliberately manipulated in the experiment and is obtained by subtracting all the effects from the total sum of squares. It depends on which subjects are in conditions α and β. It can be seen as a type of error term for the $\alpha\beta$ factor due to the fact that the effects of factor $\alpha\beta$ are determined using a completely randomized approach. In fact, it is not surprising that it is this error term that is used as the denominator, when determining the F value for the effects of the $\alpha\beta$ factor. The regular error term MS_E would be inappropriate because not all subjects were tested in both conditions α and β. The choice of this error term for the denominator can actually be justified in terms of appropriate EMS values (see Section 13.2). Because of its dependence on differences due to subjects, the error term is called the *between-subjects error term*. Thus we can now write

$$\text{Cell-total } SS = SS_\alpha + \text{between-subjects error } SS$$

We can use $SS(BSE)$ to denote the between-subjects error SS (we can use the same notation for df and MS). Thus we can write

$$\text{Cell-total } SS = SS_\alpha + SS(BSE)$$

In the same way,

$$\text{Cell total } df = df_\alpha + \text{between-subjects error } df$$
$$= df_\alpha + df(BSE)$$

We use these terms to compute the between-subjects error mean square, $MS(BSE)$, which will be used as the denominator for F when determining the effects of factor $\alpha\beta$.

$$\text{Factor } \alpha\beta,\ F = \frac{MS_\alpha}{\text{between-subjects error } MS} = \frac{MS_\alpha}{MS(BSE)}$$

On the other hand, the denominator for the F ratios for the treatments ABC and the $ABC \times \alpha\beta$ interaction is the regular error term MS_E. This is the term representing all uncontrolled error not deliberately manipulated by the experimenter. Again the argument for MS_E being the appropriate denominator depends on the appropriateness of EMS values (see Section 13.2). MS_E is computed from SS_E and df_E. SS_E is calculated, in the usual way, by subtracting all sums of squares values from SS_T; df_E is calculated in the same manner. Thus

$$SS_E = SS_T - SS_A - SS_\alpha - SS_{\alpha \times A} - SS(BSE)$$

$$df_E = df_T - df_A - df_\alpha - df_{\alpha \times A} - df(BSE)$$

We now have all the SS and df terms required to calculate the mean-square values for the ANOVA. We have the effect due to the factor $\alpha\beta$ (MS_α) and its denominator in the F ratio, the between-subjects error effect $[MS(BSE)]$. We have the effects due to factor ABC (MS_A), the interaction between factors ABC and $\alpha\beta$ ($MS_{\alpha \times A}$), and their denominator in the F ratios, the error term (MS_E). For simplicity, these terms will now be summarized in the computation scheme.

14.3 SPLIT-PLOT COMPUTATION SCHEME

Let there be two factors: ABC with k_A treatments and $\alpha\beta\gamma$ with k_α treatments. While all subjects undergo all treatments from factor ABC (repeated measures), different sets of subjects undergo treatments in factor $\alpha\beta\gamma$ (completely randomized). This can be represented on the opposite page (whole matrix).

The calculation proceeds in the usual way. From the whole matrix:

$$\text{Correction term, } C = \frac{T^2}{N}$$

$$\text{Total, } SS_T = \Sigma X^2 - C \qquad \text{with } df_T = N - 1$$

For the ABC factor:

$$SS_A = \frac{T_A^2}{n_A} + \frac{T_B^2}{n_B} + \frac{T_C^2}{n_C} - C$$

$$= \frac{T_A^2 + T_B^2 + T_C^2}{7} - C \qquad \text{with } df_A = k_A - 1 = 3 - 1 = 2$$

		A	B	C	$\cdots k_A$		
	S_1	X	X	X		Subtotal:	$T_{\alpha 1}$
	S_2	X	X	X			$T_{\alpha 2}$
α	S_3	X	X	X			$T_{\alpha 3}$
	S_4	X	X	X			$T_{\alpha 4}$
	$\vdots$						$\vdots$
	$S_{\alpha n}$						$T_{\alpha n}$
	Subtotals	$T_{\alpha A}$	$T_{\alpha B}$	$T_{\alpha C}$		Total	T_{α}

Total T_{α}

$n_S = 3$ = number of scores making up totals $T_{\alpha 1}$, $T_{\alpha 2}$, $T_{\alpha 3}$, etc.

$n_{\alpha} = 12$ = total number of scores in condition α

$n_{\alpha A} = n_{\alpha B} = n_{\alpha C} = 4$ = number of scores making up totals $T_{\alpha A}$, $T_{\alpha B}$, $T_{\alpha C}$
= number of subjects in condition α

		A	B	C	$\cdots k_A$		
	S_5	X	X	X		Subtotal:	$T_{\beta 5}$
β	S_6	X	X	X			$T_{\beta 6}$
$\vdots$	S_7	X	X	X			$T_{\beta 7}$
$\vdots$	$\vdots$						$\vdots$
$\vdots$	$S_{\beta n}$						$T_{\beta n}$
$\vdots$	Subtotals	$T_{\beta A}$	$T_{\beta B}$	$T_{\beta C}$		Total	T_{β}
k_{α}							

Total T_{β}

$n_S = 3$ = number of scores making up totals $T_{\beta 1}$, $T_{\beta 2}$, $T_{\beta 3}$, etc.

$n_{\beta} = 9$ = total number of scores in condition β

$n_{\beta A} = n_{\beta B} = n_{\beta C} = 3$ = number of scores making up totals $T_{\beta A}$, $T_{\beta B}$, $T_{\beta C}$
= number of subjects in condition β

Totals T_A T_B T_C

$n_A = n_B = n_C = 7$ = number of scores making up totals T_A, T_B, T_C
= number of subjects in the experiment.

Grand total, $T = T_A + T_B + T_C = T_{\alpha} + T_{\beta}$

$N = 21$ = total number of scores

For the $\alpha\beta\gamma$ factor:

$$SS_\alpha = \frac{T_\alpha^2}{n_\alpha} + \frac{T_\beta^2}{n_\beta} - C$$

$$= \frac{T_\alpha^2}{12} + \frac{T_\beta^2}{9} - C \qquad \text{with } df_\alpha = k_\alpha - 1 = 2 - 1 = 1$$

We can now use the subtotals to construct appropriate matrices. First the $A\alpha$ two-way matrix:

	A	B	C	
α	$T_{\alpha A}$	$T_{\alpha B}$	$T_{\alpha C}$	T_α
β	$T_{\beta A}$	$T_{\beta B}$	$T_{\beta C}$	T_β
	T_A	T_B	T_C	

$$A\alpha \text{ Cell-total } SS = \frac{T_{\alpha A}^2}{n_{\alpha A}} + \frac{T_{\alpha B}^2}{n_{\alpha B}} + \frac{T_{\alpha C}^2}{n_{\alpha C}} + \frac{T_{\beta A}^2}{n_{\beta A}} + \frac{T_{\beta B}^2}{n_{\beta B}} + \frac{T_{\beta C}^2}{n_{\beta C}} - C$$

$$= \frac{T_{\alpha A}^2}{4} + \frac{T_{\alpha B}^2}{4} + \frac{T_{\alpha C}^2}{4} + \frac{T_{\beta A}^2}{3} + \frac{T_{\beta B}^2}{3} + \frac{T_{\beta C}^2}{3} - C$$

$A\alpha$ cell-total df = number of cells − 1 = 6 − 1 = 5

$$ABC \times \alpha\beta \text{ interaction, } SS_{\alpha \times A} = A\alpha \text{ cell-total } SS - SS_A - SS_\alpha$$

$$df_{\alpha \times A} = A\alpha \text{ cell-total } df - df_A - df_\alpha = 5 - 2 - 1 = 2$$

The second matrix is not a conventional two-way matrix, it is the special completely randomized matrix.

α	$T_{\alpha 1}$	$T_{\alpha 2}$	$T_{\alpha 3}$	$T_{\alpha 4}$	...	$T_{\alpha n}$
β	$T_{\beta 5}$	$T_{\beta 6}$	$T_{\beta 7}$	...	...	$T_{\beta n}$

$$\text{Cell-total } SS = \frac{T_{\alpha 1}^2 + T_{\alpha 2}^2 + T_{\alpha 3}^2 + T_{\alpha 4}^2 + T_{\beta 5}^2 + T_{\beta 6}^2 + T_{\beta 7}^2}{n_{\text{cell}}} - C$$

$$= \frac{T_{\alpha 1}^2 + T_{\alpha 2}^2 + T_{\alpha 3}^2 + T_{\alpha 4}^2 + T_{\beta 5}^2 + T_{\beta 6}^2 + T_{\beta 7}^2}{3} - C$$

cell-total df = number of cells − 1 = 7 − 1 = 6

Between-subjects error, $SS(BSE)$ = cell-total $SS - SS_\alpha$

with between-subjects error, $df(BSE)$ = cell-total $df - df_\alpha = 6 - 1 = 5$

Now the error terms are calculated by subtraction:

$$SS_E = SS_T - SS_A - SS_\alpha - SS_{\alpha \times A} - SS(BSE)$$

$$df_E = df_T - df_A - df_\alpha - df_{\alpha \times A} - df(BSE)$$

We now have all the SS and df terms required for the ANOVA table (Table 14.1). The F values are tested for significance in the usual way using Table G.11. Note that we do not test between-subjects effects for significance.

Any significant differences between means are calculated using multiple comparisons (Chapter 9). However, it should be noted that the error-mean-square term used for computing single- or multiple-range scores varies for factors ABC and $\alpha\beta\gamma$. For ABC the appropriate error terms is MS_E, while for $\alpha\beta\gamma$ the appropriate term is $MS(BSE)$. The appropriate error term is the one used to determine the significance of F. For example, with the LSD test (see Section 9.5), the ranges are given by

Factor ABC, LSD $= t\sqrt{\dfrac{2MS_E}{n_A}}$ where n_A = number of scores making up the mean values $\bar{X}_A$ or $\bar{X}_B$ or $\bar{X}_C$ $= n_B = n_C = 7$

Factor $\alpha\beta$, LSD $= t\sqrt{MS(BSE)\left(\dfrac{1}{n_\alpha} + \dfrac{1}{n_\beta}\right)}$ where n_α and n_β are the number of scores making up the respective means $\bar{X}_\alpha$ and $\bar{X}_\beta$; $n_\alpha = 12$, $n_\beta = 9$.

The same adjustment is necessary in the error mean square for all multiple comparison tests.

This computation scheme has used three treatments in factor ABC and only two in factor $\alpha\beta\gamma$. Naturally, the computation can be extended to as many treatments as desired by using exactly the same strategies as outlined above. For instance, should there be an extra two subjects in a further condition γ, we would adjust the computation accordingly.

For between-treatments effects:

$$SS_\alpha = \frac{T_\alpha^2}{n_\alpha} + \frac{T_\beta^2}{n_\beta} + \frac{T_\gamma^2}{n_\gamma} - C$$

where n_γ is the number of scores in treatment γ; for two subjects in treatment $\gamma, n = 6$.

$$= \frac{T_\alpha^2}{12} + \frac{T_\beta^2}{9} + \frac{T_\gamma^2}{6} - C$$

with $df_\alpha = k_\alpha - 1 = 3 - 1 = 2$

Table 14.1 Split-Plot ANOVA Table

Source	SS	df	MS	F
Total	Total SS	Total $df = N - 1$		
Treatments $\alpha\beta$	SS_α	df_α	$MS_\alpha = \frac{SS_\alpha}{df_\alpha}$	$\frac{MS_\alpha}{MS\,(BSE)}$
Between-subjects error	$SS\,(BSE)$	$df\,(BSE)$	$MS\,(BSE) = \frac{SS\,(BSE)}{df\,(BSE)}$	
Treatments ABC	SS_A	df_A	$MS_A = \frac{SS_A}{df_A}$	$\frac{MS_A}{MS_E}$
Interaction $ABC \times \alpha\beta$	$SS_{\alpha \times A}$	$df_{\alpha \times A}$	$MS_{\alpha \times A} = \frac{SS_{\alpha \times A}}{df_{\alpha \times A}}$	$\frac{MS_{\alpha \times A}}{MS_E}$
Error	SS_E	df_E	$MS_E = \frac{SS_E}{df_E}$	

The $A\alpha$ two-way matrix will be

	A	B	C
α	$T_{\alpha A}$	$T_{\alpha B}$	$T_{\alpha C}$
β	$T_{\beta A}$	$T_{\beta B}$	$T_{\beta C}$
γ	$T_{\gamma A}$	$T_{\gamma B}$	$T_{\gamma C}$

with the $A\alpha$ cell-total $SS = \dfrac{T^2_{\alpha A}}{n_{\alpha A}} + \dfrac{T^2_{\alpha B}}{n_{\alpha B}} + \dfrac{T^2_{\alpha C}}{n_{\alpha C}} + \dfrac{T^2_{\beta A}}{n_{\beta A}} + \dfrac{T^2_{\beta B}}{n_{\beta B}} + \dfrac{T^2_{\beta C}}{n_{\beta C}}$

$$+ \frac{T^2_{\gamma A}}{n_{\gamma A}} + \frac{T^2_{\gamma B}}{n_{\gamma B}} + \frac{T^2_{\gamma C}}{n_{\gamma C}} - C$$

where $n_{\gamma A} = n_{\gamma B} = n_{\gamma C}$ = number of scores making up totals $T_{\gamma A}$, $T_{\gamma B}$, $T_{\gamma C}$, respectively

= number of subjects in condition γ

$$= \frac{T^2_{\alpha A}}{4} + \frac{T^2_{\alpha B}}{4} + \frac{T^2_{\alpha C}}{4} + \frac{T^2_{\beta A}}{3} + \frac{T^2_{\beta B}}{3} + \frac{T^2_{\beta C}}{3}$$

$$+ \frac{T^2_{\gamma A}}{2} + \frac{T^2_{\gamma B}}{2} + \frac{T^2_{\gamma C}}{2} - C$$

and the $A\alpha$ cell-total df = number of cells − 1 = 9 − 1

The completely randomized matrix will be

α	$T_{\alpha 1}$	$T_{\alpha 2}$	$T_{\alpha 3}$	$T_{\alpha 4}$
β	$T_{\beta 5}$	$T_{\beta 6}$	$T_{\beta 7}$	
γ	$T_{\gamma 8}$	$T_{\gamma 9}$		

with the sum of squares being calculated from the squares of all these totals as before, and

cell total df = number of cells - 1 = 9 - 1 = 8

The LSD values for factor $\alpha\beta\gamma$ will be given by

$$t\sqrt{MS(BSE)\left(\frac{1}{n_\alpha}+\frac{1}{n_\beta}\right)}$$

or any of the other combinations of n_α, n_β, or n_γ.

The assumptions for the split-plot design are the same as those for other designs (see Section 10.4). The design should be used with caution, however, because statisticians warn that it is not as robust as the regular designs as far as breaking assumptions is concerned.

For further clarity, we will now consider a worked example.

14.4 WORKED EXAMPLE: SPLIT-PLOT ANALYSIS OF VARIANCE

Seven volunteer subjects were studied to determine the effect of a given disease on taste sensitivity, using three experimental procedures: *A*, *B*, and *C*. Four subjects had the disease while the remaining three other subjects did not. The first four were called the experimental group; the latter three formed the control group. This control group was small, but only these seven subjects were available for experimentation because of other considerations which will not be dealt with here. The sensitivity scores obtained from the three experimental procedures are given below; the higher the score, the more sensitive the subject.

Did the disease affect taste sensitivity (were sensitivities different for control and experimental groups), and if so, how? Did sensitivity values vary with the experimental procedure used, and if so, how? Were any differences noted between the experimental procedures consistent between the group with the disease and the group without?

We are interested only in this particular disease and the three tasting procedures *A*, *B*, *C*; we do not wish to extend our conclusions further. These factors are thus fixed effects. You may assume that the scores are drawn from populations of scores that are normally distributed with equal variance.

Before proceeding with the calculation, it is worth inspecting the data and noting the mean values obtained by the experimental and control groups (see Figure 14.1).

The data were as follows:

EXPERIMENTAL GROUP *E* WITH DISEASE

Subject	*A*	*B*	*C*		
S_1	2	3	4	$T_{E1} = 9$	$n_S = 3$ = number of scores per subject
S_2	3	3	4	$T_{E2} = 10$	
S_3	1	2	3	$T_{E3} = 6$	
S_4	1	2	4	$T_{E4} = 7$	
	$T_{EA} = 7$	$T_{EB} = 10$	$T_{EC} = 15$	$T_E = 32$	$n_E = 12$ = number of scores in experimental group
	$\bar{X}_{EA} = 1.75$	$\bar{X}_{EB} = 2.5$	$\bar{X}_{EC} = 3.75$	$\bar{X}_E = \frac{32}{12} = 2.67$	

CONTROL GROUP *C* WITHOUT DISEASE

Subject	*A*	*B*	*C*		
S_5	2	2	2	$T_{C5} = 6$	$n_C = 9$ = number of scores in control group
S_6	3	2	3	$T_{C6} = 8$	
S_7	3	3	3	$T_{C7} = 9$	
	$T_{CA} = 8$	$T_{CB} = 7$	$T_{CC} = 8$	$T_C = 23$	
	$\bar{X}_{CA} = 2.7$	$\bar{X}_{CB} = 2.3$	$\bar{X}_{CC} = 2.7$	$\bar{X}_C = \frac{23}{9} = 2.56$	
	$T_A = 15$	$T_B = 17$	$T_C = 23$	Grand total $T = 55$	$N = 21$

$$\bar{X}_A = \frac{15}{7} = 2.14 \qquad \bar{X}_C = \frac{23}{7} = 3.29$$

$$\bar{X}_B = \frac{17}{7} = 2.43$$

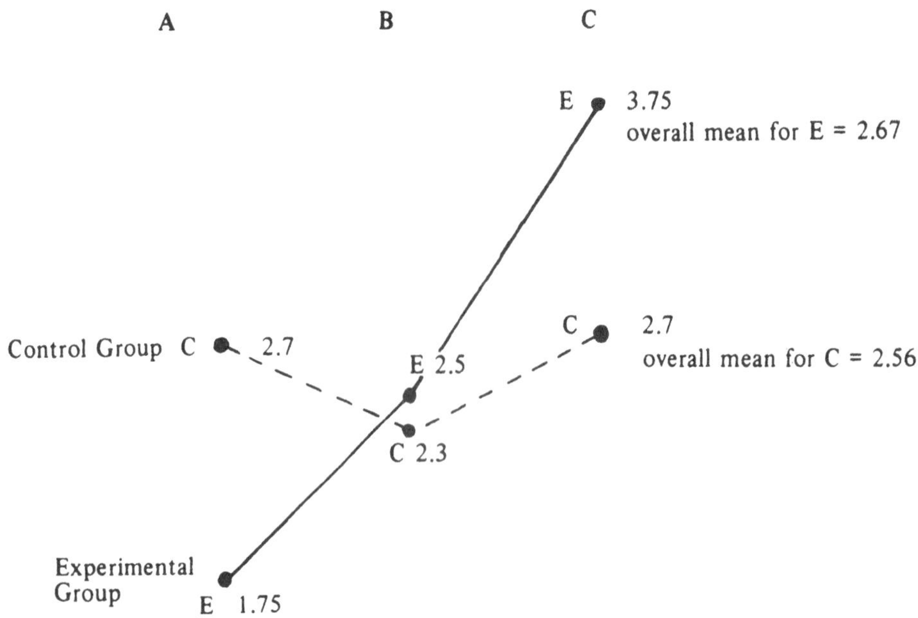

Overall means

A = 2.2 B = 2.4 C = 3.2

Figure 14.1 Mean values for experimental and control groups.

It can be seen that the experimental group means increase on passing from condition A to B to C. However, in the control group, the means do not vary much. Thus we would expect there to be an interaction; the ANOVA will indicate whether this is significant. The overall means increase on passing from A to B to C, while the mean for the experimental group is only slightly higher than for the control group. The ANOVA will indicate whether the treatment means are significantly different.

The calculation proceeds as outlined in the computation scheme (Section 14.3).

$$\text{Correction term, } C = \frac{T^2}{N} = \frac{55^2}{21} = 144.05$$

$$\text{Total, } SS_T = 2^2 + 3^2 + 4^2 + 3^2 + 3^2 + \cdots + 3^2 + 3^2 + 3^2 - C$$

$$= 159 - 144.05 = 14.95$$

$$\text{Total, } df_T = N - 1 = 21 - 1 = 20$$

Between-experimental procedures (A, B, C) effects:

$$SS_P = \frac{T_A^2 + T_B^2 + T_C^2}{n_A} - C \qquad \text{where } n_A = \text{number of scores making up totals } T_A,\ T_B, \text{ or } T_C$$

$$= \frac{15^2 + 17^2 + 23^2}{7} - 144.05 = 149 - 144.05 = 4.95$$

Between-experimental procedures df_P = number of procedures − 1 = 3 − 1 = 2

Between-groups (experimental, E vs. control, C) effects:

$$SS_G = \frac{T_E^2}{n_E} + \frac{T_C^2}{n_C} - C$$

$$= \frac{32^2}{12} + \frac{23^2}{9} - 144.05 = 85.33 + 58.77 - 144.05 = 0.05$$

Between groups, df_G = number of groups − 1 = 2 − 1 = 1

We now consider the experimental procedures × groups two-way matrix:

		Experimental Procedure		
		A	B	C
Group	E	$T_{EA} = 7$	$T_{EB} = 10$	$T_{EC} = 15$
	C	$T_{CA} = 8$	$T_{CB} = 7$	$T_{CC} = 8$

$$\text{Two-way cell-total } SS = \frac{T_{EA}^2}{n_{EA}} + \frac{T_{EB}^2}{n_{EB}} + \frac{T_{EC}^2}{n_{EC}} + \frac{T_{CA}^2}{n_{CA}} + \frac{T_{CB}^2}{n_{CB}} + \frac{T_{CC}^2}{n_{CC}} - C$$

$$= \frac{7^2}{4} + \frac{10^2}{4} + \frac{15^2}{4} + \frac{8^2}{3} + \frac{7^2}{3} + \frac{8^2}{3} - 144.05$$

$$= 152.05 - 144.05 = 8.45$$

Two-way cell-total df = number of cells − 1 = 6 − 1 = 5

Procedures × group interaction, $SS_{P \times G}$ = cell-total $SS - SS_P - SS_G$

= 845. − 4.95 − 0.05 = 3.45

Procedures × group interaction $df_{P \times G}$ = cell total − $df_P - df_G$

= 5 − 2 − 1 = 2

We now consider the completely randomized matrix from the second set of subtotals:

Group					
	E	$T_{E1} = 9$	$T_{E2} = 10$	$T_{E3} = 6$	$T_{E4} = 7$
	C	$T_{C5} = 6$	$T_{C6} = 8$	$T_{C7} = 9$	

$$\text{Cell-total } SS = \frac{T_{E1}^2 + T_{E2}^2 + T_{E3}^2 + T_{E4}^2 + T_{C5}^2 + T_{C6}^2 + T_{C7}^2}{n_{\text{cell}}} - C$$

$$= \frac{9^2 + 10^2 + 6^2 + 7^2 + 6^2 + 8^2 + 9^2}{3} - 144.05$$

$$= \frac{447}{3} - 144.05 = 4.95$$

$$\text{Cell-total } df = \text{number of cells} - 1 = 7 - 1 = 6$$

$$\text{Between-subjects error, } SS(BSE) = \text{cell-total } SS - SS_G = 4.95 - 0.05 = 4.90$$

$$\text{Between-subjects error, } df(BSE) = \text{cell-total } df - df_G = 6 - 1 = 5$$

Finally, the error term is obtained by subtraction:

$$SS_E = SS_T - SS_P - SS_G - SS_{P\times G} - SS(BSE)$$

$$= 14.95 - 4.95 - 0.05 - 3.45 - 4.90 = 1.60$$

$$df_E = df_T - df_P - df_G - df_{P\times G} - df(BSE)$$

$$= 20 - 2 - 1 - 2 - 5 = 10$$

We now have all SS and df terms to construct the ANOVA table.

Source	*SS*	*df*	*MS*	*F*
Total	14.95	20		
Between groups Experimental vs. Control	0.05	1	0.05	$\frac{0.05}{0.98} = 0.05$ *df*: 1, 5
Between-subjects error (*BSE*)	4.90	5	0.98	
Between procedures *A, B, C*	4.95	2	2.475	$\frac{2.475}{0.16} = 15.47$*** *df*: 2, 10
Procedures × groups interaction	3.45	2	1.725	$\frac{1.725}{0.16} = 10.78$** *df*: 2, 10
Error	1.60	10	0.16	

We now examine the values in Table G.11, to test the significance of the calculated F values. For $df = 1, 5, p = 0.05, F = 6.61$. This exceeds our between-groups F value of 0.05, so we do not reject H_0. Thus there is no difference between the experimental and the control group means (2.67 vs. 2.56), so we note no effects on taste sensitivity due to the disease.

For $df = 2, 10$,	$p = 0.05$	$F = 4.10$
	0.01	7.56
	0.001	14.91

The between-procedures F value (15.47) exceeds all these values, so we reject H_0 ($p < 0.001$). We deduce that the experimental procedures A, B, and C used for the taste testing produced differences in the taste sensitivity means (2.2, 2.4, 3.2). Multiple-comparison tests are needed to explore these differences further. The procedures × groups interaction F value (10.78) exceeds the value in Table G.11 at $p = 0.01$. Thus again, we reject H_0 and deduce that the suspected two-way interaction was indeed highly significant. The trends over the testing procedures are not the same for the diseased (experimental) group and the disease-free (control) group. The lack of difference between these two groups does not mean that the groups have similar scores in conditions A, B, and C; there are differences which cancel each other when the values are averaged over the three procedures.

The difference between the overall means for procedures *A*, *B*, and *C* can be investigated using LSD tests; at this extreme level of significance, LSD tests will be safe to use (see Section 9.3) with a low risk of Type I error. First, let us inspect the mean values and the differences between them.

$$\begin{array}{cccccc} A & & B & & C & \\ 2.2 & & 2.4 & & 3.2 & \text{means of seven values} \\ & \leftarrow 0.29 \rightarrow & & \leftarrow 0.86 \rightarrow & & \end{array}$$

We will do LSD comparisons at the level of significance of the ANOVA ($p <$ 0.001). The test will be two-tailed to match the two-tailed nature of the *F* test.

$$\text{LSD range} = t\sqrt{\frac{2MS_E}{n}} \qquad \begin{aligned} MS_E &= 0.16 \\ df_E &= 10 \end{aligned} \qquad n = 7$$

From Table G.8, t for $df_E = 10$ at 0.1% level = 4.587.

$$\text{LSD range} = 4.587\sqrt{\frac{2 \times 0.16}{7}} = 0.98$$

The means *A* and *C* differ significantly ($p < 0.001$); *B* did not differ significantly from *A* or *C*. This can be represented as follows:

$$\begin{array}{ccc} A & B & C \\ 2.2 & 2.4 & 3.2 \\ \cline{1-2} \\ \cline{2-3} \end{array}$$

Thus the different taste procedures elicited differences between means, procedure *C* giving higher taste-sensitivity scores than procedure *A*; procedure *B* elicited intermediate values which were not significantly different from either.

Thus, in summary, the different taste procedures elicit different taste-sensitivity scores, although the interaction indicated that these differences vary between the disease and nondisease groups. Averaged over all taste procedures, the disease and nondisease groups did not differ, but the interaction indicated that there were indeed differences between the two groups for individual tasting procedures.

15

Correlation and Regression

15.1 SOME PRELIMINARIES

Use of Correlation and Regression

So far, we have considered statistical tests that search for differences between sets of data. In sensory testing, we may be interested in whether scores change under different experimental conditions, whether groups of human subjects vary, or whether different sets of food samples give different flavor scores. We will now consider a different approach. We will consider a test that looks for similarities between two sets of data. It examines how well they relate to each other, or in other words, how they co-relate or correlate. Such a test may be useful to determine, for example, whether a set of intensity ratings, representing the perceived taste of a seasoning, varies directly with the concentration of that seasoning in the food. We may be interested in such perceived intensity relationships not only for taste but also for odor, sound, color, or pain. Perhaps we may not actually look at the correlation between the intensity and concentration scores; we may instead look for correlation between, say, the logarithms of these scores (such a relationship is called *Stevens' power law*). As far as food research is concerned, there are many uses for correlation, especially correlation between processing and sensory variables.

A further technique is of interest here. We may have some data and suspect that a graph of the data is best fitted by a straight line. We could draw the best-fitting straight line through the data by eye. Alternatively, we could use a mathematical technique to draw the line; one such mathematical technique is called *linear regression*.

In this chapter we consider correlation and linear regression. Before doing so, however, it is necessary to consider a little calculus.

Brief Recapitulation of Some Calculus

First, two definitions. We must explain the difference between dependent and independent variables. An *independent variable* is one which is deliberately manipulated by the experimenter. A *dependent variable* is the one being measured in an experiment; its values are the result of manipulation of the independent variable. For example, should subjects be rating the intensity of a light, the experimenter will choose given light intensities (independent variable) and see what ratings (dependent variable) are given by her subjects. Should she be investigating the effects of processing on the characteristics of the flavor of a food, the various processing treatments constitute the independent variable, and the flavor scores obtained from a panel of judges constitute the dependent variable.

When plotting the results of an experiment on a graph, it is traditional to plot the values for the dependent variable on the vertical axis (also called the Y axis or the ordinate) and to plot the values for the independent variable on the horizontal axis (also called the X axis or the abscissa).

Any graph of Y values plotted against X values (graph of a dependent variable plotted against an independent variable) can be described by an equation relating Y to X. For example, imagine a graph where every Y value is the same as the X value. When $X = 1$, $Y = 1$; when $X = 10$, $Y = 10$; when $X = 20$, $Y = 20$; etc. Such a graph would like Figure 15.1. It would be a straight line at an angle of 45° passing through the origin (the origin is the zero point for the graph, where $Y = X = 0$). Because the Y values are always equal to X values, we can write

$$Y = X$$

This is the equation representing this straight line.

Imagine now the case where every value of Y is twice the value of X. When $X = 1$, $Y = 2$, when $X = 10$, $Y = 20$; when $X = 20$, $Y = 40$; etc. We could write such a relationship as

$$Y = 2X$$

A graph of such a relationship would look like Figure 15.2. From the graph it can be seen that for any given increase in X the increase in Y will be twice

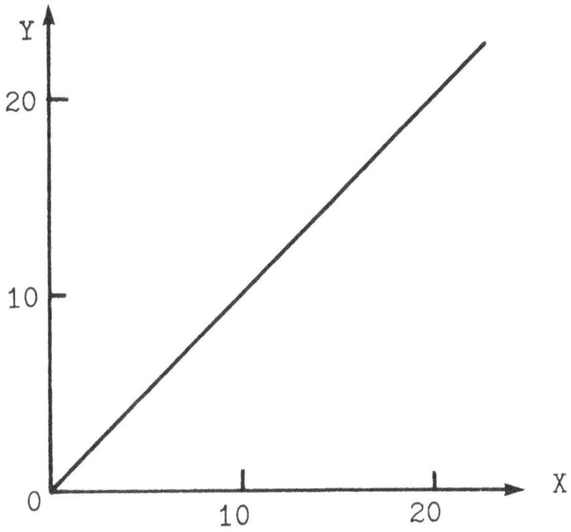

Figure 15.1 $Y = X$.

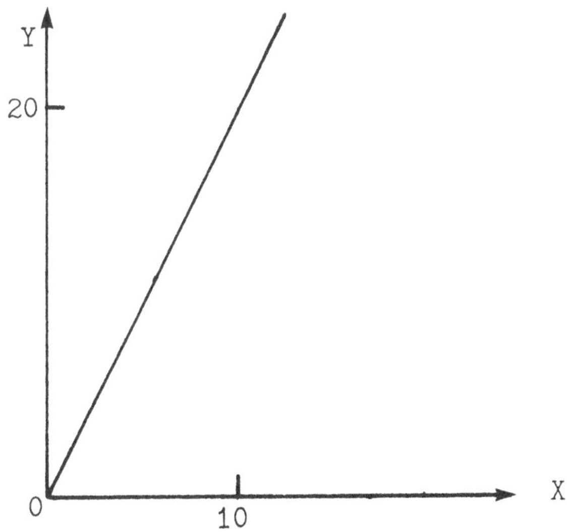

Figure 15.2 $Y = 2X$.

as great. The graph rises up the Y axis twice the distance it travels along the X axis; in other words, the slope of the graph = 2. In general, should the slope of the graph = a, we can write

$$Y = aX$$

A graph with a slope of 3 ($Y = 3X$) will be steeper than the one shown in Figure 15.2; a graph with a slope of 1/2 ($Y = 1/2X$) will be less steep.

Now, back to our graph of $Y = X$. This time imagine it is raised up the Y axis by a value of 6, as in Figure 15.3. Here, when $X = 1$, $Y = 7$; when $X = 10$, $Y = 16$; when $X = 20$, $Y = 26$. When $X = 0$, $Y = 6$; the graph crosses the Y axis at a point where $Y = 6$. This point is called the *intercept.* Y always has a value of 6 greater than X; this relationship can be represented by the equation

$$Y = X + 6$$

In general, we can write

$$Y = X + b$$

where b is the intercept. Combining these two arguments, a straight-line graph of slope a and intercept b can be represented by the equation

$$Y = aX + b$$

Such a graph is represented in Figure 15.4.

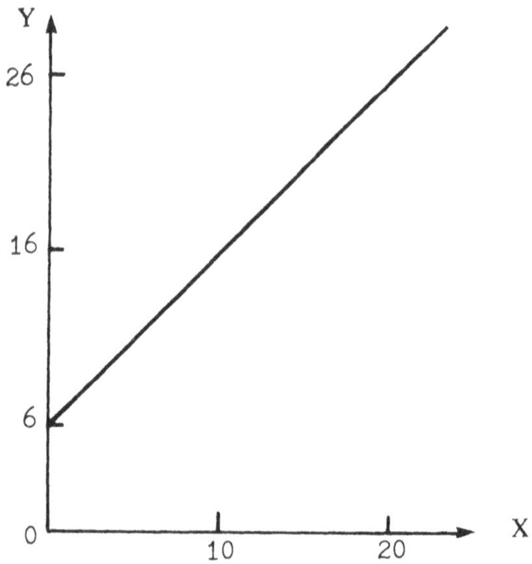

Figure 15.3 $Y = X + 6$.

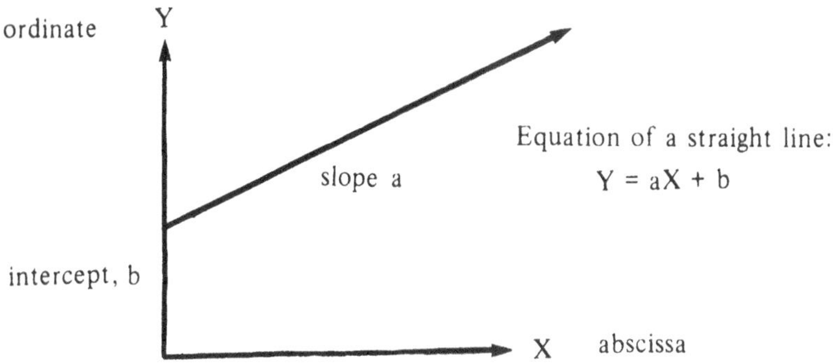

Figure 15.4 Equation of a straight line: $Y = aX + b$.

Note: $Y = aX + b$ is the general equation of a straight line [some authors write $Y = bX + a$, or $Y =$ (slope)$X +$ (intercept); the actual symbols themselves are not important]. Other types of lines (curves, circles, etc.) are represented by more complex equations.

Having considered some preliminary calculus, we will now go on to examine correlation.

15.2 WHAT ARE CORRELATION COEFFICIENTS?

The degree of correlation between two variables is represented by the *correlation coefficient*. The correlation coefficient range is as follows:

+1	for perfect correlation
0	for no correlation
−1	for perfect negative correlation

The correlation coefficient considered here is *Pearson's product-moment correlation coefficient*, denoted by r. If two variables, say, growth rate of a plant and the rainfall the plant receives, are perfectly correlated, then should growth rate be plotted against rainfall, the graph will be a straight line with a positive slope (Figure 15.5). All points will fall on this straight line, indicating perfect correlation ($r = 1$); the growth rate will always increase by an amount proportional to the rainfall, the proportion always being the same. The actual slope of the line is not important as long as all points lie on the line. The correlation coefficient will be positive as long as the slope is positive; if the slope is negative, the correlation coefficient will be negative.

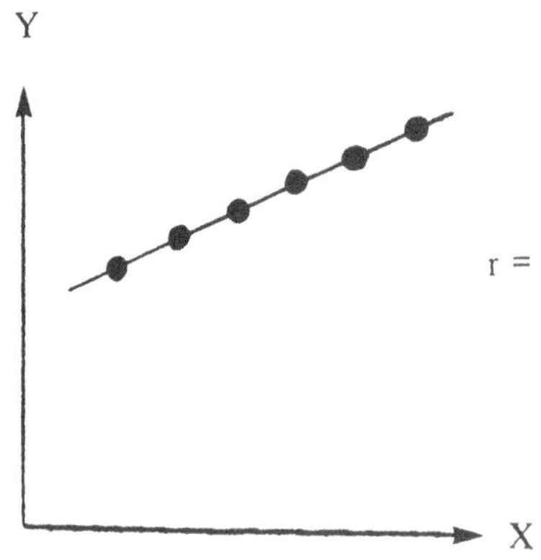

Figure 15.5 Perfect correlation.

In real life, correlations are rarely perfect (+1 or −1); it is very unusual for all points to fall on the line. They are usually scattered about the line and the correlation coefficient is reduced accordingly (Figure 15.6). The correlation is

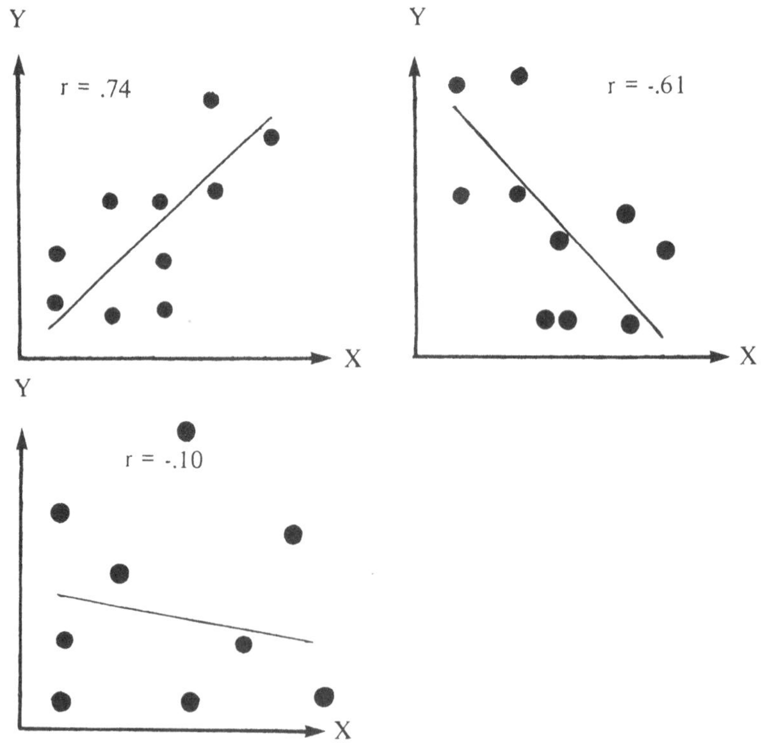

Figure 15.6 Imperfect correlation.

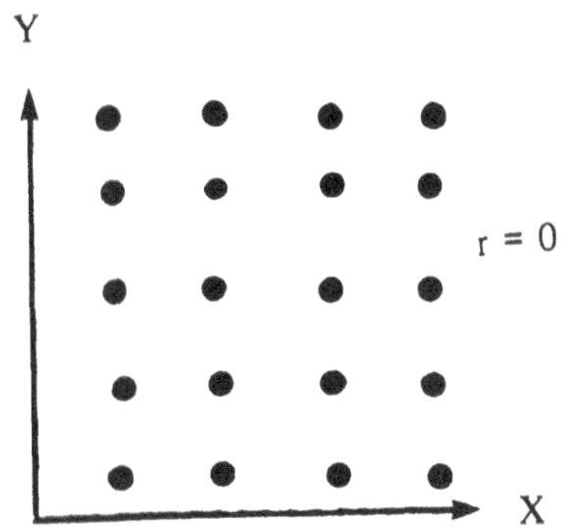

Figure 15.7 Example of zero correlation.

reduced as the points are scattered more, until the correlation becomes zero, when the points are scattered randomly. Zero correlation coefficients can also be obtained in other ways; consider the examples of nonrandom points shown in Figures 15.7 and 15.8. It is difficult to know where to draw a line in these figures.

Note: The correlation coefficient is a coefficient of linear correlation. It is high only when the points fall on a straight line. If the relationship between X and Y is curvilinear, as in Figure 15.8, the correlation will be zero. More complex coefficients are required to describe such a relationship.

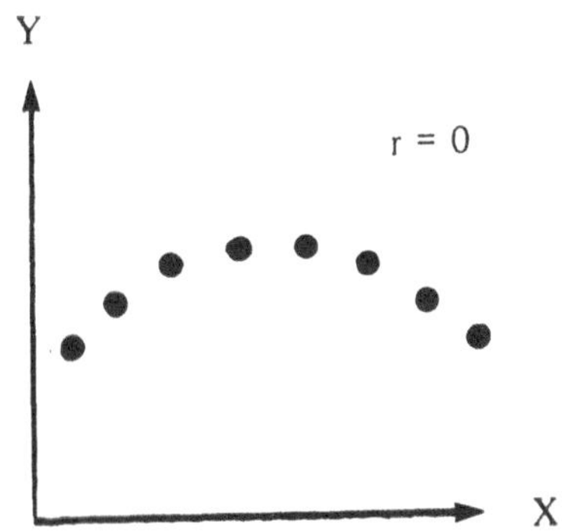

Figure 15.8 Curvilinear relationship yields zero correlation.

15.3 HOW TO COMPUTE PEARSON'S PRODUCT-MOMENT CORRELATION COEFFICIENT

To measure the correlation between a set of Y and X values, Pearson's product-moment correlation coefficient, developed by Karl Pearson, is used. We will simply give the formula for the correlation coefficient; we will not derive it formally. It is given by the formula

$$r = \frac{\Sigma (X - \bar{X})(Y - \bar{Y})}{NS_x S_y}$$

where $\bar{X}$ and $\bar{Y}$ are the mean of the X and Y values being correlated, S_x and S_y are their standard deviations, and N is the number of X scores (or the number of Y scores, but *not* the number of X scores + the number of Y scores), generally the number of subjects tested.

Note:

$$S_x = \sqrt{\frac{\Sigma (X - \bar{X})^2}{N - 1}} = \sqrt{\frac{\Sigma X^2 - \frac{(\Sigma X)^2}{N}}{N - 1}}$$

This formula is inconvenient to use. However, it can be rearranged into a more convenient, albeit longer form:

$$r = \frac{N \Sigma XY - \Sigma X \Sigma Y}{\sqrt{[N \Sigma X^2 - (\Sigma X)^2] [N \Sigma Y^2 - (\Sigma Y)^2}}$$

Although this formula is not derived here, we can see that it is of an intuitively sensible form. Looking at the numerator, we can see that the larger the value of ΣXY, the larger the value of the numerator and hence r. Should the two sets of scores (X and Y) be positively correlated, large X and large Y values will be associated, giving large XY values and ΣXY will be correspondingly large. This can be clarified by an example. If there are two X scores (10 and 2) and two Y scores (20 and 4), then if X and Y are correlated, one subject will get the high scores (10 and 20, so that $XY = 200$) while the other subject will get the low scores (2 and 4, so that $XY = 8$ and $\Sigma XY = 208$). If the X and Y are not correlated (each subject has a high and low score: 10 and 4, $XY = 40$; 20 and 2, $XY = 40$), ΣXY will not be so large ($40 + 40 = 80$). As the degree of correlation between the X and Y values decreases, ΣXY decreases and so will r. Should the X and Y values be negatively correlated, the largest Y values will be associated with the smallest X values, and vice versa. The smallest XY values will be obtained and ΣXY will have a minimum value. In fact, ΣXY will be so small that it will be smaller than $\Sigma X \, \Sigma Y$, producing a negative correlation.

15.4 WORKED EXAMPLE: PEARSON'S PRODUCT-MOMENT CORRELATION

An example will make the use of the correlation coefficient clear. Let us assume that we are investigating whether there is any correlation between the average yield (Y, in tons) of acre plots of genetically engineered, pest resistant, sweet-tasting corn and the rainfall (X in cm) it receives. We have a set of yield (Y) and rainfall (X) values. The formula for r requires ΣX, ΣY, ΣX^2, ΣY^2, and ΣXY values, so we tabulate our X and Y data with columns for X^2, Y^2, and XY as follows:

Rainfall, X	Yield, Y	X^2	Y^2	XY
13	10	169	100	130
12	8	144	64	96
12	7	144	49	84
11	14	121	196	154
11	6	121	36	66
10	10	100	100	100
8	7	64	49	56
6	8	36	64	48
5	4	25	16	20
5	11	25	121	55
$\Sigma X = 93$	$\Sigma Y = 85$	$\Sigma X^2 = 949$	$\Sigma Y^2 = 795$	$\Sigma XY = 809$

Having obtained the relevant column totals, it merely remains to fill in the formula.

$$r = \frac{N \Sigma XY - \Sigma X \Sigma Y}{\sqrt{[N \Sigma X^2 - (\Sigma X)^2][N \Sigma Y^2 - (\Sigma Y)^2]}}$$

$$= \frac{10 \times 809 - 93 \times 85}{\sqrt{(10 \times 949 - 93^2)(10 \times 795 - 85^2)}} = 0.24$$

Thus the correlation coefficient for yield per acre (Y) and rainfall (X) is 0.24. The next question is whether this represents a significantly high correlation.

15.5 IS THE CORRELATION SIGNIFICANT?

Is 0.24 a significantly high correlation? Does the correlation coefficient from the samples of X and Y data represent a significant correlation between the population of X and Y values. Can we reject H_0? The null hypothesis in this case is that there is no correlation between the population of X and Y values.

To establish whether the correlation is significant, we need to know the highest correlation coefficients that can be obtained by chance (on H_0). In fact, we use Table G.16. If our correlation coefficient, for a given number of degrees of freedom (given by $N - 2$), is greater than or equal to the value in this table, the correlation is significant at the level of significance given.

We also need to decide whether our test is one-tailed or two-tailed. The rule is as follows: The test is

One-tailed: if, should H_0 be rejected, the direction of the correlation (positive or negative) can be predicted beforehand

Two-tailed: if, should H_0 be rejected, the direction of the correlation cannot be predicted beforehand

Let us assume that we could not predict the direction of any correlation between rainfall (X) and yield per acre (Y), should H_0 be rejected. Higher rainfall might speed up the growth of the corn, but it might also slow it down by causing rot or mold. Thus we choose a two-tailed test.

Examining Table G.16 we find columns of values for the various one-tailed or two-tailed levels of significance. For example, the second column indicates correlation values that are just significant at $p = 0.05$, two-tailed. We now pick the row appropriate to the number of degrees of freedom (see left-hand column headed $df = N - 2$); here $df = 10 - 2 = 8$. The corresponding values in the tables are inspected. Our value (0.24) exceeds none of the values in the row, not even 0.5494 ($p = 0.1$). So we do not reject H_0, the hypothesis that states that there is no correlation between the population of X and Y values. Thus the data in our sample do not indicate any correlation between rainfall (X) and yield per acre (Y), in the population of cornfields.

So it can be seen that the calculation of a correlation coefficient is a simple matter of filling in a formula, while testing its significance merely involves comparing it with a value in Table G.16. Note that the table does not give r values for given numbers of subjects N; it gives them for given numbers of degrees of freedom (df). Here, $df = N - 2$ (number of subjects − 2). It is not given by $N - 1$. One point is subtracted for X values and one for Y values, hence $N - 2$.

15.6 THE COEFFICIENT OF DETERMINATION

The coefficient of correlation (ranging −1 through 0 to +1) is a measure of how well all the points (for X and Y values) fit on a straight-line graph. However, another coefficient is traditionally used to express goodness of fit; this is the coefficient of determination. It is simply the square of the correlation coefficient, r^2. It is used because it is conceptually easy. r^2 is the proportion of ıriation in Y explained by variation in X, or more precisely, the proportion of variance in Y attributable to variance in X. Thus

If $r = 0.5$, $r^2 = 0.25$, 25% of the variance in Y is attributable to variance in X.
If $r = 0.7$, $r^2 = 0.49$, 49% of the variance in Y is attributable to variance in X.
If $r = 0.8$, $r^2 = 0.64$, 64% of the variance in Y is attributable to variance in X.

So it is not until a correlation of greater than 0.7 is obtained that a majority of the variation in Y is explained by variation in X.

What are the uses of the coefficient of determination and of significance testing? The coefficient of determination provides an easily understood numerical measure of the degree of association between Y and X values. It tells us how much of the variation in Y is attributable to variation in X. On the other hand, the significance of a correlation coefficient is quite different; it tells us whether the correlation in our sample is indicative of a real correlation in the population, or whether it is just a chance effect of sampling. Quite low correlation coefficients (with even lower coefficients of determination) can be significant, especially in large samples (see the low values in Table G.16 for $df = 100$). They merely indicate that there is, in fact, a small correlation in the population, even though only a small proportion of the variance in Y is attributable to variance in X.

To determine whether r^2 is significant, merely see whether r is. A coefficient of determination is significant when the coefficient of correlation is significant.

There are additional coefficients. Just as the coefficient of correlation (r) and determination (r^2) measure the degree of association between X and Y values, the coefficient of alienation (k) and nondetermination (k^2) measure the lack of association between them. They are related as follows:

$$k^2 = 1 - r^2$$

15.7 ASSUMPTIONS AND WHEN TO USE THE CORRELATION COEFFICIENT

To calculate Pearson's product-moment correlation coefficient (merely filling in a formula), you do not need assumptions, except that pairs of X and Y values should be independent or else the coefficient will have little practical value. Should you merely convert the correlation coefficient to r^2, to determine the proportion of variation in Y attributable to variation in X, no more assumptions need be considered.

However, once you test the signficance of r, assumptions must be made. These are as follows:

1. *Each pair of X and Y values must be independent.* What is meant by independent? It means that the pairs of X and Y values should all be independent of each other to the same degree. For example, if a sample of subjects is being tested, it means that each pair of X and Y values must come from a separate person. You cannot test someone twice to boost the sample size; their two

sets of scores would not then be as independent of each other as the other sets of scores. The value of r would also be biased, relying overmuch on the scores from one subject.

There is another situation in which correlation can be used. It is possible that all the X and Y scores could come from just one judge. Once again, each pair of X and Y scores would all be equally independent of each other, although not to such a degree as before. We could, for example, be correlating the sweetness of a series of food samples with their sugar content, for one judge. Each food sample would be independent and would provide a sweetness and a sugar-content score. These pairs of scores would be all equally independent. But you could not try to boost your data by testing one or two of the samples twice; such scores would not now be as independent as those from separate samples.

2. *The data must come from a population in which the joint distribution of X and Y is normal*; this is called a *bivariate normal distribution*. A bivariate normal distribution needs more explanation. Imagine a set of X and Y values as in Figure 15.9. Now imagine the population of these values. Let us pile the values that are the same on top of each other, so that the height above the page indicates the frequency of the scores. The distribution will now look like a mound resting on the paper (Figure 15.10). Such a distribution will be a bivariate normal distribution if, when sliced in any direction, it gives a normally distributed cross section (Figure 15.11). Thus for each X value there is a subpopulation of normally distributed Y values, and vice versa. As the correlation between X and Y increases, the mound becomes more and more pinched in, along the direction of the straight line through the X and Y values (Figure 15.12).

3. *Generally, X and Y values should be randomly sampled.* This means that you should randomly sample subjects and then see what their X and Y values are.

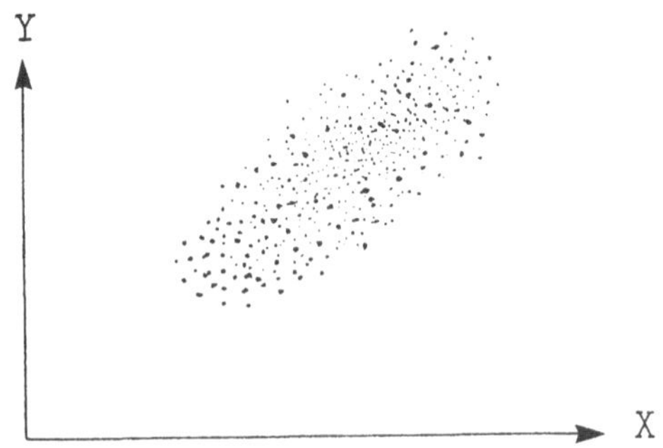

Figure 15.9 Set of values.

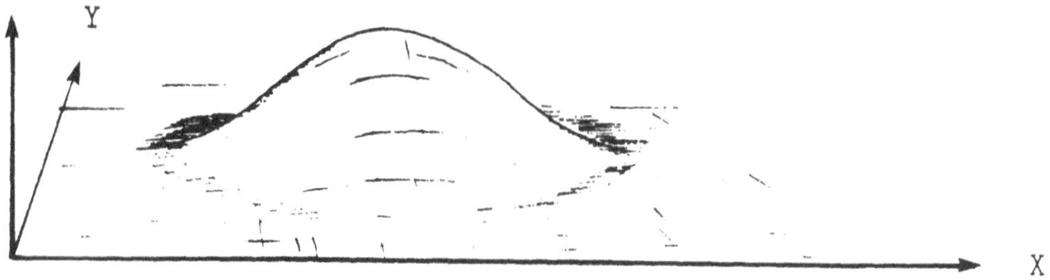

Figure 15.10 Set of values, with height representing frequency.

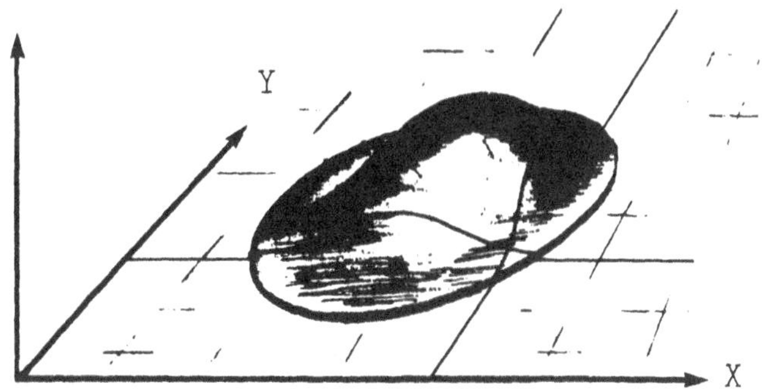

Figure 15.11 Cross section of Figure 15.10.

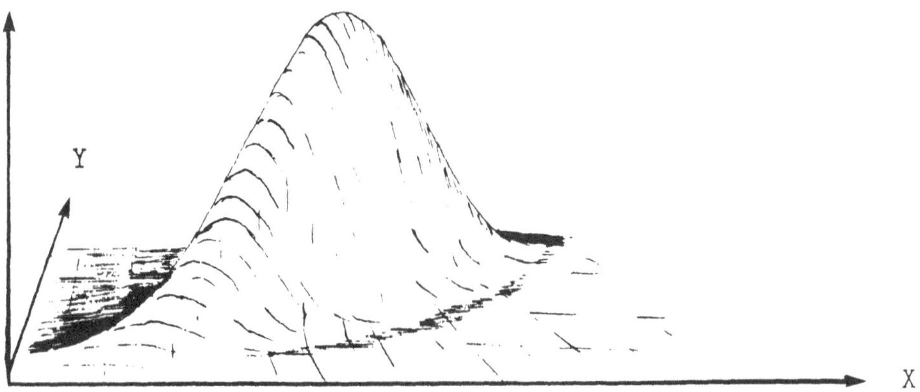

Figure 15.12 Set of values with increased correlation.

Sometimes an experimenter may wish to measure Y values for predetermined X values (or vice versa). In this case X values are not randomly sampled; they are predetermined. If you do this, you will tend to depress your value of r because you do not allow X the full range of variation with Y, to get as large a correlation as possible. Thus, fixing X makes it more difficult to get a significant r. Of course, should r still be significant, all well and good.

4. *The variances of Y values for each range of X values (Figure 15.13) should be approximately equal, and vice versa.* This, once again, is the assumption of homoscedasticity. This can often be seen by inspection of the data.

Should high correlations be obtained with small samples, the data should be inspected carefully to check that the correlation was not caused by one person who had excessively high (or low) X and Y values. This might be enough to cause r to be artificially high. In fact, it is generally a good idea to plot the data so as to inspect the relationship between X and Y values; it may reveal a curvilinear relationship when r (linear correlation) is low.

There are nonparametric alternatives to this correlation coefficient which are calculated from ranked data (the X and Y values are ranked in order): Spearman's ranked-correlation coefficient, ρ (Greek lowercase letter rho) and Kendall's ranked correlation coefficient, τ (Greek lowercase letter tau). Look at the Key to Statistical Tables to examine the relationship between r, ρ, and τ. ρ and τ are not as powerful as r if all the assumptions hold for r; if they do not, ρ and τ are preferable measures.

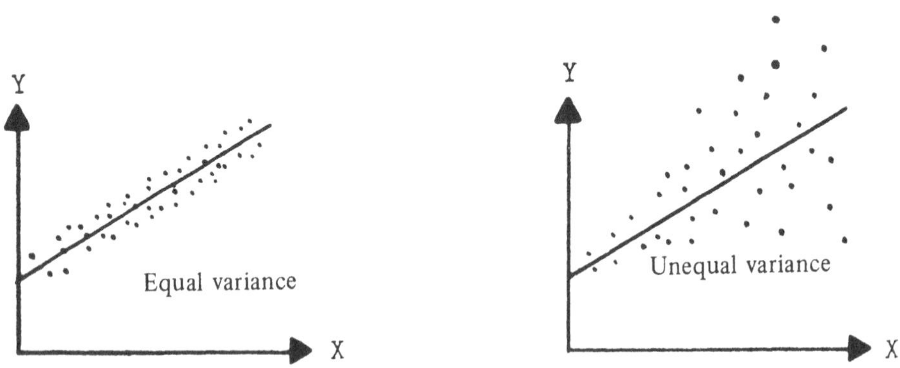

Figure 15.13 Equal (left) and unequal (right) variance.

15.8 CORRELATION DOES NOT IMPLY CAUSALITY

If two variables, X and Y, are correlated, it does not mean that X causes Y. Unfortunately, this faulty logic has been used all too often. If X causes Y, they will be correlated. But if X and Y are correlated, it does not mean that there is a causal relationship; X and Y may be causally related or they may not be. For instance, in one German town there was a high correlation between the number of storks' nests on the roof of a house and the number of children in that house. This does not mean that storks cause babies. It could just be that larger families with more children live in larger houses with more space, which are more likely to furnish nesting areas for storks.

Unfortunately, all too often a correlation is taken to indicate causality. However, a high and significant correlation between X and Y can mean:

1. X causes Y.
2. Y causes X.
3. X and Y are both caused by some other factor.

Of course, correlation can be a clue to causality. The science of epidemiology looks for correlations between diseases and possible causes. A high correlation indicates that there may be (but is not necessarily) a causal relationship. Research is then directed to investigate whether, in fact, this causal relationship actually exists. Here, correlation is used as a screening device to eliminate possible causes of a disease that are unlikely.

15.9 FURTHER TOPICS FOR CORRELATION: MEANS, DIFFERENCES, AND SIGNIFICANCE

There are a few further points that are worth noting as far as correlation is concerned.

Other Ways of Significance Testing

We tested the significance of r by merely comparing it with critical values in Table G.16; this is certainly the simplest way of doing so. However, there are alternative ways of testing r for significance. It can be established using a t test (Table G.8), where

$$t = r\sqrt{\frac{N-2}{1-r^2}} \qquad \text{where } df = N-2$$

And because $F = t^2$, we can also use Table G.11, where

$$F = r^2\left(\frac{N-2}{1-r^2}\right) \qquad \text{where } \frac{\text{numerator } df = 1}{\text{denominator } df = (N-2)}$$

These methods are older ways of testing the significance of r; not all texts supply the more convenient tables like Table G.16. You may often find research papers which test the significance of r using t or F tests.

Taking the Mean of Several Correlation Coefficients

We might want to take the mean of several correlation coefficients, but we cannot do this in the normal way. To calculate the mean of a set of r values, we first convert them to Fisher's Z values, take the mean of these, and then convert this back to an r value. A Fisher's Z value is given by

$$\text{Fisher's } Z = \frac{1}{2}\left[\log_e(1+r) - \log_e(1-r)\right]$$

or Table G.17 can be used to transform r values to Z values.

To find a mean of several r values, we use the following steps:

1. Transform all r values to Fisher's Z values (using given formula or Table G.17).
2. Multiply each Fisher's Z value by its appropriate $(N - 3)$ value [Z_1 by $(N_1 - 3)$, Z_2 by $(N_2 - 3)$, Z_3 by $(N_3 - 3)$, etc.].
3. Add all the $Z(N - 3)$ values.
4. Add all the $(N - 3)$ values.
5. Divide the sum of the $Z(N - 3)$ values by the sum of the $(N - 3)$ values. This gives a mean Fisher's Z.
6. Transform the Fisher's Z value back to a mean r value using Table G.17.
7. This gives the correct mean of all the original correlation coefficients.

Testing the Significance of a Difference Between Two Correlation Coefficients

Fisher's Z is also used when two correlation coefficients are being compared to see whether they are significantly different. The procedure is as follows:

1. Convert both r values to Fisher's Z values. r_1 is converted to Z_1; r_2 is converted to Z_2.
2. Now calculate a z score (a normalized z score, not a Fisher's Z) using the formula

$$\text{Normalized } z \text{ score} = \frac{Z_1 - Z_2}{\sqrt{\dfrac{1}{N_1 - 3} + \dfrac{1}{N_2 - 3}}}$$

3. Then use Table G.1 to look up the probability of getting a normalized z value greater than or equal to the calculated value. This gives the probability on H_0

of finding a difference between r_1 and r_2 as big as this by chance (see Section 4.3). Note that Table G.1 gives one-tailed values; for a two-tailed test, you need to double the value in the tables. For example, at the 5% level (two-tailed), the relevant probability value in Table G.1 is 0.025 because twice 0.025 equals 0.05. The normalized z score corresponding to this level is 1.96.

15.10 LOGIC OF LINEAR REGRESSION

We now consider linear regression, a procedure intimately tied up with correlation. Linear correlation measures the degree of association between two sets of data; it indicates to what degree a plot of the two sets of data fits on a straight line. Linear regression is a way of deciding where to draw that straight line; it is a way of fitting the best straight line through that data. As a reminder (see Section 15.1), the equation of a straight line (Figure 15.14) is given by

$$Y = aX + b$$

where a is the slope and b is the intercept.

If a line is to be fitted to some data (Figure 15.15), the question can be asked: What is the best line that can be fitted? A line could be fitted by eye, but there are better methods. Consider a line drawn through the data such that the sum of the vertical distances between data points and the line is a minimum (Figure 15.16). This would be a good line of best fit, but a better method would be to use the distance squared and draw a line such that the sum of the

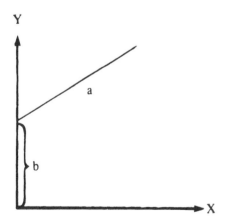

Figure 15.14 $Y = aX + b$.

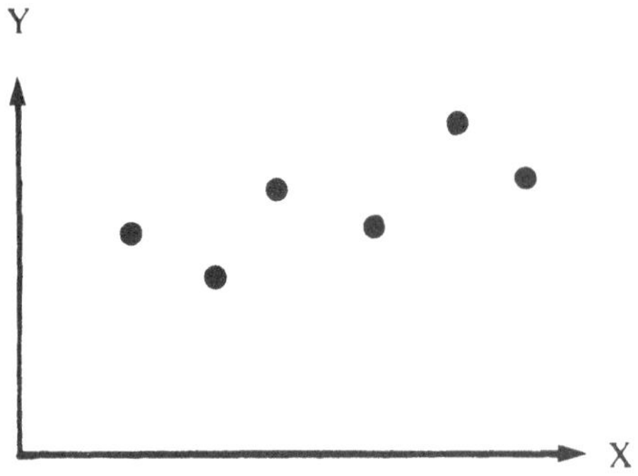

Figure 15.15 Set of data.

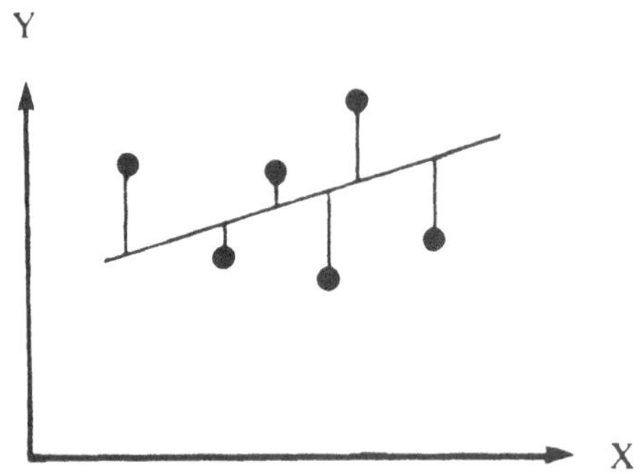

Figure 15.16 Line of best fit.

squares of the distances is a minimum; this makes the method more sensitive to large distances, that is, large deviations of the points from the line of best fit. This method is called the *method of least squares*, and the line drawn is called the *regression line of Y on X.*

The alternative regression line of X on Y would use the square of the horizontal distances between the data points for the line of best fit; the squares of these distances would be required to sum to a minimum value. However, this is rarely used because, in general, the independent variable is plotted on the X axis and such values are often fixed in an experiment. It is the Y values (dependent variable) that vary about the line of best fit and the task is to draw the best line through the means of the varying Y values.

15.11 COMPUTATION FOR LINEAR REGRESSION

Given that the equation of a straight line is $Y = aX + b$, then if the values of the slope (a) and intercept (b) are known, the straight line can be drawn (fit your ruler to the intercept b and tilt it so that the slope is a; then draw!). The least-squares method of obtaining the line of best fit supplies a value for the slope and for the intercept, so that the appropriate line can be drawn.

We will not derive the formulas for the slope and intercept values; we will merely provide them.

$$\text{Slope, } a = r\,\frac{S_y}{S_x}$$

where r is the correlation coefficient between X and Y values, and S_x and S_y are the standard deviations of the X and Y scores.

$$\text{Intercept, } b = \bar{Y} - r\,\frac{S_y}{S_x}\,\bar{X}$$

where $\bar{Y}$ and $\bar{X}$ are mean values for Y and X scores. However, there are more convenient versions of these formulas, especially if r and S values are unknown. These are:

$$\text{Slope, } a = \frac{\Sigma XY - \dfrac{\Sigma X\,\Sigma Y}{N}}{\Sigma X^2 - \dfrac{(\Sigma X)^2}{N}}$$

$$\text{Intercept, } b = \frac{\Sigma Y - a\Sigma X}{N}$$

The best line is fitted to the data. Values of Y, for values of X that have not been measured, can now be interpolated.

15.12 WORKED EXAMPLE: LINEAR REGRESSION

Let us assume that X and Y are positively correlated. X may be the concentration of a coloring agent in the food where Y may be its depth of color measured on some laboratory instrument. Say that we are given some X and Y values; we are to find the regression equation and also find the value of Y when $X = 30$. The X and Y values are as follows:

X	Y	X^2	XY
18	54	324	972
26	64	676	1,664
28	54	784	1,512
34	62	1,156	2,108
36	68	1,296	2,448
42	70	1,764	2,940
48	76	2,304	3,648
52	66	2,704	3,432
54	76	2,916	4,104
60	74	3,600	4,440
$\Sigma X = 398$	$\Sigma Y = 664$	$\Sigma X^2 = 17{,}524$	$\Sigma XY = 27{,}268$

$N = 10$

For the formulas for the slope and intercept, we need to calculate ΣX and ΣY. We also need ΣX^2 and ΣXY; we do not need ΣY^2 as we did for r. So we needed columns for X^2 and XY in the data. The slope and intercept are calculated simply by filling in the appropriate formulas.

$$\text{Slope, } a = \frac{\Sigma XY - \dfrac{\Sigma X \Sigma Y}{N}}{\Sigma X^2 - \dfrac{(\Sigma X)^2}{N}} = \frac{27{,}268 - \dfrac{398 \times 664}{10}}{17{,}524 - \dfrac{398^2}{10}}$$

$$= \frac{27{,}268 - 26{,}427}{17{,}524 - 15{,}840} = \frac{841}{1684} = 0.499$$

$$\text{Intercept, } b = \frac{\Sigma Y - a\Sigma X}{N} = \frac{644 - 0.499 \times 398}{10}$$

$$= \frac{664 - 199}{10} = 66.4 - 19.9 = 46.5$$

Thus the regression line is given by the equation $Y = 0.499X + 46.5$. To find a value of Y for a given value of X, we simply substitute the X value in the equation. When $X = 30$, $Y = 0.499 \times 30 + 46.5 = 61.47$.

15.13 ASSUMPTIONS AND WHEN TO USE LINEAR REGRESSION

The linear regression technique, using the least-squares technique to find the line of best fit through the data (regression line of Y and X), only makes the assumption that X and Y are linearly related (they are linearly correlated); if they were not linearly related, linear regression would be an inappropriate technique. Again the technique would only be appropriate, as with correlation, should the pairs of X and Y values be independent (see Section 15.7).

There are also techniques available for testing whether the regression line fitted to the data indicates a significant relationship in the data. In effect, the slope of the line is tested to determine whether it is significantly different from zero. If the slope is zero, there is no relationship between X and Y, the best predicted value of Y for any value of X is the mean value of Y (the best line is a horizontal line of height $\bar{Y}$). Should there be a linear relationship between Y and X, the line of best fit will have a slope which is not zero. To test whether there is a significant linear relationship between the X and Y scores, we test to see whether the slope is significantly different from zero. This procedure is beyond the scope of this book, but essentially it is done by comparing the variance in the Y scores accounted for by the regression line ($MS_{\text{regression}}$) with the variance due to error (MS_E); a significant F value ($MS_{\text{regression}}/MS_E$) indicates that the slope is significantly different from zero (we reject H_0, which states that the slope = 0). Such a test is really a test of whether X and Y are correlated. Should the slope be significantly greater than zero, the data will be significantly positively correlated (less than zero: negatively correlated).

To test whether the slope is significantly different from zero, we need the same first two assumptions required for resting the significance of a correlation coefficient: Each pair of X and Y values must be independent, while the data must be of the form of a bivariate normal distribution (see Section 15.7).

However, the third assumption is different. When testing for the significance of the slope, we assume that we measure Y values for fixed values of X. This is different from the assumptions for r, where the X values are not fixed (should you fix them, r will be reduced). The fixing of X values is necessary because of the mathematical model used to test the significance of the slope. This model considers the equation of the line of best fit to be

$$Y = aX + b + \epsilon$$

where X values are errorless and where ϵ is the error in Y; Y values for a given X value are distributed normally with equal variance for all such X values.

Finally, there is a technique that can be used after a regular analysis of variance instead of multiple comparisons. Instead of analyzing to determine which means are different from each other, the means are examined to see whether they are linearly or even curvilinearly related. This technique is called *trend analysis*.

15.14 WORKED EXAMPLE: CORRELATION AND LINEAR REGRESSION

For clarity, we will consider an example. The softness of canned plums was investigated to see whether it varied with the length of heat treatment during canning. Softness scores were evaluated by judges on a scale where higher scores indicated softer plums. The mean softness scores from a panel of judges were 3.0, 4.1, 4.9, 6.0 and 7.1 for lengths of heat treatment of 2, 4, 8, 10, and 12 minutes, respectively. Did the softness scores correlate with the length of the heat treatment? It was not known beforehand by the canning technologists whether a longer heat treatment would produce softer or harder plums for this particular canning process. Secondarily, what is the estimated softness score for a 9-minute heat treatment? You may assume that the pairs of softness and heat treatment scores come from a bivariate normal population and that the X and Y values exhibit homoescedasticity.

The problem requires, first, the calculation of a correlation coefficient. A two-tailed test of significance will be used because, should the correlation be significant, it could not be predicted by the canning technologists whether it would be significantly positive or negative. We can see that the pairs of softness scores and heat-treatment lengths are independent, so we proceed with the calculation. The independent variable (X) is length of heat treatment; the dependent variable (Y) is softness.

The data can be set out as follows ($N = 5$):

Heat-treatment time, X	Softness, Y	X^2	Y^2	XY
2	3.0	4	9.0	6.0
4	4.1	16	16.81	16.4
8	4.9	64	24.01	39.2
10	6.0	100	36.0	60.0
12	7.1	144	50.41	85.2
$\Sigma X = 36$	$\Sigma Y = 25.1$	$\Sigma X^2 = 328$	$\Sigma Y^2 = 136.23$	$\Sigma XY = 206.8$

From these totals, the correlation coefficient can be calculated.

$$r = \frac{N\Sigma XY - \Sigma X \Sigma Y}{\sqrt{[N\Sigma X^2 - (\Sigma X)^2][N\Sigma Y^2 - (\Sigma Y)^2]}}$$

$$= \frac{5 \times 206.8 - 36 \times 25.1}{\sqrt{(5 \times 328 - 36^2)(5 \times 136.23 - 25.1^2)}}$$

$$r = \frac{1034 - 903.6}{\sqrt{344 \times 51.14}} = \frac{130.4}{\sqrt{17{,}592.16}} = \frac{130.4}{132.63} = 0.98$$

The next question is whether or not the correlation is significant. This can be tested using Table G.16. It should be noted that the X values were fixed, reducing the r value. This should be taken into consideration when testing for significance. For a two-tailed test, for $df = N - 2 = 5 - 2 = 3$, the values in the table are 0.8783 ($p = 0.05$), 0.9587 ($p = 0.01$), and 0.9912 ($p = 0.001$). Our value exceeds the value of 0.9587, so our correlation is significant ($p < 0.01$). Thus the softness scores were correlated with the length of the heat treatment in the canning process. In fact, our depressed value of 0.98 might even have exceeded the value for $p = 0.001$ (0.9912) had the X value not been fixed. Our level of significance might even have been as low as $p < 0.001$.

Now we consider the second part of our problem. We need to estimate the softness score for a 9-minute heat treatment. For this, we will need to fit the best line through the data, so that we can read off softness values for given lengths of heat treatment. We go ahead and use the method of least squares to calculate the slope and intercept of this line.

$$\text{Slope, } a = \frac{\Sigma XY - \dfrac{\Sigma X \Sigma Y}{N}}{\Sigma X^2 - \dfrac{(\Sigma X)^2}{N}}$$

$$= \frac{206.8 - \dfrac{36 \times 25.1}{5}}{328 - \dfrac{36^2}{5}} = \frac{206.8 - 180.72}{328 - 259.2} = \frac{26.08}{68.8} = 0.379$$

$$\text{Intercept, } b = \frac{\Sigma Y - a\Sigma X}{N} = \frac{25.1 - 0.379 \times 36}{5} = \frac{11.46}{5} = 2.291$$

We can now write the equation of the best-fitted straight line for the data ($Y = aX + b$) as follows:

$$Y = 0.379X + 2.292$$

where Y is the softness score and X is the length of the heat treatment. To find the softness score for a heat treatment of 9 minutes, we substitute $X = 9$ in the equation. Thus

$$Y = 0.379 \times 9 + 2.292 = 5.7$$

Thus the mean softness scores from the panel of judges correlated with the length of the heat treatment ($p < 0.01$, maybe even $p < 0.001$). The estimated softness score for a 9-minute heat treatment is 5.7.

16
Additional Nonparametric Tests

16.1 WHEN TO USE NONPARAMETRIC TESTS

In this chapter we explore some nonparametric tests. Generally, other things being equal, these tests are not as powerful as the equivalent parametric tests, but they can be used when the usual assumptions about the data coming from normally distributed populations do not hold. Most important, included in this assumption of normality is the requirement that the data be on at least an interval or a ratio scale (see Section 2.6). Ranked (ordinal) or category (nominal) data are not normally distributed and are analyzed using nonparametric statistics. Also, a distribution of scores may be skewed away from the normal by, say, the distorting action of end effects on a rating scale (see Section 7.11); again, this would cause the assumptions for parametric statistical analysis to be broken.

An experimenter may decide that the scores from a rating scale cannot be realistically considered as having equal intervals due to the judges' lack of skill in generating numbers. He may decide that a score of 10 does not signify an intensity twice as great as one denoted by 5 and that the psychological distance between 4 and 5 is not the same as that between 1 and 2. However, a high score does at least indicate a higher intensity than a low value; the data may at least be ranked. After assigning ranks to his data, a nonparametric test will be needed for its analysis. The experimenter may even decide that rather than force judges to attempt the complex skill of rating, the simpler skill of ranking might be more appropriate; again a nonparametric analysis is required. Ranking would seem especially appropriate for consumer testing where little training is given and where there is not much time available for testing. Ranking has the advantage of simplicity and the fact that, in our culture, we are well practiced in this skill.

The nonparametric tests considered in this chapter will be ones suitable for analyzing ranked (ordinal) data. We will examine the Wilcoxon and Mann-Whitney U two-sample difference tests, as well as the rank sums test confusingly attributed to Wilcoxon, Mann and Whitney. For many-sample difference tests we will consider the Freidman, Kramer and Kruskal-Wallis tests as well as other tests which test for a ranking trend between treatments: the Page test and the Jonckheere test. We will also consider Spearman's rank correlation. These are nonparametric versions of t, F, and r; consult the Key to Statistical Tables to see the relationship between these and other tests. Also note their relationship to each other and to other nonparametric tests already considered: the binomial test, the McNemar test, the sign test, chi-square, the Cochran Q test, and the contingency coefficient.

We begin with two-sample difference tests.

16.2 THE WILCOXON TEST: A TWO-SAMPLE, RELATED-SAMPLES DIFFERENCE TEST

Computation of the Wilcoxon Test and Comparison with the Sign Test

This test is sometimes called the *Wilcoxon signed rank test*. The test is a related-samples difference test, the nonparametric equivalent of the related-samples t test. However, where the t test indicates whether means are significantly different, the Wilcoxon test indicates whether medians are significantly different. It can best be considered by first remembering the sign test. Consider plant growth rates (in cm per 2 weeks) before and after treatment by a gardener who claimed that talking to the plants made them grow more quickly. The rates are given on the opposite page. The sign test can be applied as discussed earlier (see Section 5.8). We compare growth rates before and after the talking treatment by the gardener. If they increase, we assign a + sign; if they decrease, a − sign; if they do not change, we assign 0.

There are eight plants ($N = 8$) and six increased their rate (six "+" values). The binomial test is used to determine whether this sample of data indicates that there are significantly more + values in the population than − values, whether significantly more plants increase their growth rate when talked to by the gardener. We use the binomial test (Table G.4.b, $N = 8$, $X = 6$). We use a two-

Before treatment	After treatment	Sign test
63	82	+
42	69	+
74	73	–
37	43	+
51	58	+
43	56	+
80	76	–
65	82	+

tailed test because we cannot predict, for a false H_0, whether the plants would increase or decrease their growth rate when spoken to by the gardener. From Table G.4.b we see that the probability of getting this result on H_0 is high, $p = 0.289$ (28.9%); thus we do not reject H_0. Thus there is not a significant majority of + signs; there was not a majority tendency for the growth rate to increase; the gardener's rhetoric did not change the plants' growth rate.

The sign test is all very well, but it does not consider the size of the changes taking place. An increase of 63 to 82 (denoted by +) is larger than a decrease of 74 to 73 (denoted by –). The sign test ignores this. On the other hand, the Wilcoxon test does not. Consider the same data, except that this time we give not only the direction of the change but also the difference between the scores (*d* values).

		Wilcoxon test	
Before treatment	After treatment	Difference in scores, *d*	Rank *d* (ignoring sign)
63	82	+19	7
42	69	+27	8
74	73	–1	(–)1
37	43	+6	3
51	58	+7	4
43	56	+13	5
80	76	–4	(–)2
65	82	+17	6
		$N = 8,\ W = 3\ (= 1 + 2)$	

The differences in scores (with due regard to sign) are noted and are then ranked (regardless of sign), the smallest difference being given rank 1, the next smallest rank 2, etc. The sum of the ranks from the sign of which there are fewer cases (here − because there were only two cases, as opposed to + which had six cases) equals Wilcoxon's W. The sign of W is always taken as positive, regardless of whether there are fewer positive or negative d values. Here $W = 3$, the sum of the two negative ranks 1 and 2.

If there were big increases in scores after the gardener's intervention, there would be very few negative d values (and vice versa), so the sum of their ranks ($= W$) would be small. If there was no significant increase or decrease in values, there would be equal numbers of positive and negative d values and the sum of their ranks would be approximately the same and W would not be small. Thus the smaller W becomes the more likely it is that there was a significant difference between the scores. Naturally, W could be small by chance, so what is needed is a table indicating how small such values can be by chance; a smaller value would indicate a significant difference. Table G.18 is such a table. If our W value is less than or equal to the critical value of W in Table G.18, a significant difference is indicated. We would reject H_0, which says that the two sets of scores are not significantly different, in favor of the alternative hypothesis (H_1), which states that the scores are different. Let us use a two-tailed test because we could not tell in advance the effect of the gardener, should we reject H_0.

Table G.18 gives critical values of W at various significance levels for various sample sizes (N). We identify our N value in the left-hand column ($N = 8$) and then read across to the column corresponding to various significance levels. For a two-tailed test, the critical W values are 3 ($p = 0.05$) and 0 ($p = 0.01$). The sample is not large enough for $p = 0.001$. Our value of W equals the critical value at the 5% level, so it is at this level that we reject H_0. Thus there was a significant increase in the growth rate of the plants when the gardener spoke to them ($p < 0.05$), but this was not significant at the 1% level.

It can be seen that the test is more powerful than the sign test; it is better at detecting differences because it uses more information, namely the magnitude of these differences.

Ties

There are special procedures for tied scores. There are two types of ties that can occur. What if you get the same score before and after the treatment? The d value will be zero:

Before	After	d
50	50	0

Discard these scores as in the sign test. This is a disadvantage of this test because information is lost. If there is a high proportion of ties, the test becomes misleading.

The second sort of tie is a tie in the d values. What if several d values are the same? Here, when ranking them, you take the mean rank of all the d scores that have that tied value. For example:

Before	After	d	Rank d
41	42	1	1st, 2nd, and 3rd tied, so give them the mean rank: 2nd
36	37	1	
51	50	-1	

Here the three lowest scores tie for first, second, and third places. The mean of 1, 2, and 3 is 2. Thus all three are awarded second place. This is contrary to the usual practice (in athletics, horse racing, etc.) of awarding them all first place.

Here are some more examples:

Ties for 1st, 2nd, 3rd, 4th, 5th: all given rank 3rd
Ties for 1st, 2nd: both given rank 1.5
Ties for 7th, 8th, 9th, 10th: all given rank 8.5

The more ties there are, the less accurate will Table G.18 be as a test of significance. However, as long as no more than 25% of the scores are involved in ties, Table G.18 will be reasonably accurate. Should there be more ties, the best procedure is to use an approximation to normality and convert W so that a z value can be used.

Approximation to Normality

When $N > 20$, an approximation to normality is good. Thus when $N > 50$ and Table G.18 cannot be used, the approximation to normality is used. Furthermore, should there be too many ties, an approximation to normality is necessary.

The following formula converts a Wilcoxon W value to a z value.

$$z = \frac{\left| W - \frac{N(N+1)}{4} \right| - \frac{1}{2}}{\sqrt{\frac{N(N+1)(2N+1)}{24}}}$$

This z value can be used with Table G.1 to calculate the probability of getting a W value as small as this or less, by chance. This will be a one-tailed probability; for a two-tailed test, double the probability value in Table G.1.

Assumptions for the Wilcoxon Test

As with parametric tests, the Wicoxon test requires a certain number of assumptions. These are as follows:

1. The *d* scale must be at least ordinal, or else it cannot be ranked.
2. The test is a related-samples test, so scores in each condition must be matched; they must be pairs of scores coming from the same subjects.
3. All *d* scores must be independent. They must all come from different subjects; you cannot use the same person twice.
4. Subjects must be chosen randomly.
5. Even though the data may only be ranked, the underlying variable must be continuous (i.e., theoretically capable of having any fractional value–1, 1.25, 1.5, 7.6, 3.9, etc.; see Section 2.6).
6. This is a more mathematical point. In fact, the test is not one that tests differences between two medians; it really tests whether the two distributions of scores are identical. If they are identical, the medians will be the same. If they are not, it could be because either the medians or the spread of the scores, or both, are different. It is generally assumed that the spread of the scores will be the same, so any differences will be due to differences in the medians. Samples of scores in the Wilcoxon test are generally assumed to be spread the same amount.

Look at the Key to Statistical Tables to see the relationship of the Wilcoxon test to other statistical procedures.

16.3 WORKED EXAMPLE: THE WILCOXON TEST

Eight subjects, who worked in the psychologically oriented marketing division of a nationwide fast-food chain, were given a psychoanalytically based personality test that was claimed to be related to decision-making efficiency. They were tested again under stress. Is there any evidence from the data given below that their personality scores changed under stress? The scores were known not to come from a normally distributed population.

Subject	Initial score	Score under stress	*d*	Rank *d*
1	40	30	-10	-5
2	29	32	+3	+2
3	60	40	-20	-7
4	12	10	-2	-1
5	25	20	-5	-3.5
6	15	0	-15	-6
7	54	49	-5	-3.5
8	23	23	0	Eliminated

$N = 7, \quad W = 2$

This is a two-tailed question; it was not known beforehand whether scores would increase or decrease should H_0 be rejected. H_0 states that stress had no effect on personality scores; the initial scores and scores under stress are no different. From Table G.18, for $N = 7$, $W = 2$ for $p = 0.05$, two-tailed; there are no values possible at the 1% and 0.1% levels for such a small sample size. Our value is *equal* to this; thus we can reject H_0 in favor of the alternative two-tailed hypothesis. Thus we reject the idea that the personality scores did not change under stress; they decreased ($p < 0.05$).

16.4 THE MANN-WHITNEY U TEST: A TWO-SAMPLE INDEPENDENT-SAMPLES DIFFERENCE TEST

The *Mann-Whitney U test* (devised by two statisticians, Mann and Whitney) is an independent-samples two-sample difference test; it is the independent-samples version of the Wilcoxon test. It uses ranked data and tests the medians of two sets of scores from separate groups of subjects to see whether they are significantly different.

Principle Behind the Test

Let us take two sets of scores, one from an experimental group, another from a control group. (We could be investigating the effect on a foodstuff of a seasoning that is claimed to have the physiological effect of making taste buds more sensitive; the experimental group may have eaten the foodstuff with the added seasoning, the control group without.)

Experimental group, E:	9	11	15	16		$n_1 = 4$	number of subjects per group
Control group, C:	$\underline{\underline{5}}$	$\underline{\underline{6}}$	$\underline{\underline{8}}$	$\underline{\underline{10}}$	$\underline{\underline{13}}$	$n_2 = 5$	

We can now arrange these scores in order of increasing size:

C	*C*	*C*	*E*	*C*	*E*	*C*	*E*	*E*
$\underline{\underline{5}}$	$\underline{\underline{6}}$	$\underline{\underline{8}}$	9	$\underline{\underline{10}}$	11	$\underline{\underline{13}}$	15	16

We now consider each score in one group and count the number of scores from the other groups which precede it in the rankings. For instance, scores $\underline{\underline{5}}$, $\underline{\underline{6}}$, and $\underline{\underline{8}}$ from the control group have no scores from the experimental group preceding them, while $\underline{\underline{10}}$ has one score (9 from experimental group), and $\underline{\underline{13}}$ has two (9, 11). In the experimental group, 9 has three control scores preceding it ($\underline{\underline{5}}$, $\underline{\underline{6}}$, $\underline{\underline{8}}$), 11 has four ($\underline{\underline{5}}$, $\underline{\underline{6}}$, $\underline{\underline{8}}$, $\underline{\underline{10}}$), and 15 and 16 have five ($\underline{\underline{5}}$, $\underline{\underline{6}}$, $\underline{\underline{8}}$, $\underline{\underline{10}}$, $\underline{\underline{13}}$) each.

The number of scores from the experimental group preceding each score in the control group may be added ($0 + 0 + 0 + 1 + 2 = 3$), while the same may be done for the experimental group (3 + 4 + 5 + 5 = 17). These sums are the Mann-Whitney U values for each group. Thus:

Control group, C:	0 + 0 + 0 + 1 + 2	= 3	U
Experimental group, E:	3 + 4 + 5 + 5	= 17	U′

If all the experimental scores were higher than the control scores, indicating a difference between the two groups, there would be no experimental scores preceding control scores. U would be zero and U′ would be large. But if the two groups were not significantly different, some control scores would precede experimental scores while some experimental scores preceded control scores. U and U′ would be approximately the same. Thus the smaller one of the U values, the larger the other value and the more likely it is that the two groups of scores are significantly different from each other (one median is significantly greater than the other). Of course, differences between the U values may occur by chance. What is needed is a table indicating how small (U) or how large (U′) the values can be by chance; Table G.19 is such a table.

Table G.19 gives both the large and the small U values. If our calculated smaller U value is equal to or smaller than the smaller U value in Table G.19, H_0 is rejected. It is also rejected if the calculated larger value of U is greater than or equal to the larger (underlined) value of U in Table G.19. The critical U values in the table are obtained by finding the appropriate row and column for the two sample sizes, n_1 and n_2. (Either sample can be n_1 or n_2; the same numbers will be found in the table.) Naturally, you do not really need to look up both U values in the table; if your large U is bigger than the large one in the table, your small U will be smaller than the small U in the table. Using both values in Table G.19 is merely a double check. Essentially, the values in Table G.19 are calculated using permutations.

We can now consider our example and test the U values (3 and 17) for significance. We will use a two-tailed test because we had no way of knowing in advance whether the control or experimental scores would be higher should H_0 be rejected. We locate the values for $n_1 = 4$ and $n_2 = 5$ (or vice versa). The values in the two-tailed section of Table G.19 are

1 for $p = 0.05$ no value for $p = 0.01$
$\underline{19}$

We now compare these with our calculated values (3, 17). Our value of 3 is not less than or equal to 1; 17 is not greater than or equal to 19. Thus we do not reject H_0. Even if we had worked at the 10% level (critical values: 2 and 8), we would not have rejected H_0. So we can say that the seasoning was not seen to have any effect on the flavor of the foodstuff in our samples ($p > 0.1$).

Ties

In the example above, none of the scores tied. Had they done so, the mean rank of the respective scores would have been taken. When computing U, had the tied rank been in the other group, it would have been taken as half preceding the score: for example, with two groups C and E, should their ranks be as follows:

C:	1	2		4.5		
E:			3	4.5	6	7

The number of scores preceding the E score 3 is 2, while the number preceding the score 4.5 is 2.5. Thus the U values for group E can be computed as usual:

$$U = 2 + 2.5 + 3 + 3 = 10.5$$

For group C:

$$U' = 0 + 0 + 1.5 = 1.5$$

These can be checked using the relationship

$$U = n_1 n_2 - U'$$

$$10.5 = 3 \times 4 - 1.5 = 10.5$$

This procedure for dealing with ties is all right as long as not too many scores (no more than 25%) are involved in ties. Should there be more ties, an alternative procedure (e.g., Wilcoxon-Mann-Whitney rank sums test, Section 16.6) provides a simple alternative.

Computational Formula for Convenience

The aforementioned way of calculating the Mann-Whitney U can become time consuming as the samples get larger; it can take time counting up, for each score in a given sample, the scores in the other sample that precede it. The process can be speeded up by ranking all the scores from the two samples in order of size. The ranks from each group are added and values of U calculated, using the formulas

$$U = n_1 n_2 + \frac{n_1 (n_1 + 1)}{2} - R_1$$

$$U' = n_1 n_2 + \frac{n_2 (n_2 + 1)}{2} - R_2$$

The use of the formulas will be shown for the data we have been considering. First the data are ranked in order of increasing value (they could be ranked in order of decreasing value; the same U values would be obtained). If any ranks are tied, the mean rank is assigned to each of the tied values.

	C	*C*	*C*	*E*	*C*	*E*	*C*	*E*	*E*
	$\underline{\underline{5}}$	$\underline{\underline{6}}$	$\underline{\underline{8}}$	9	$\underline{\underline{10}}$	11	$\underline{\underline{13}}$	15	16
Rank	$\underline{\underline{1}}$	$\underline{\underline{2}}$	$\underline{\underline{3}}$	4	$\underline{\underline{5}}$	6	$\underline{\underline{7}}$	8	9

$n_1 = 4, \quad n_2 = 5$

We now add up the rank values for each group (each rank total denoted by R_1 and R_2). Thus

$$R_1 = 4 + 6 + 8 + 9 = 27 \qquad n_1 = 4$$

$$R_2 = 1 + 2 + 3 + 5 + 7 = 18 \qquad n_2 = 5$$

Having obtained the rank totals, we now fill in the appropriate formulas:

$$U = n_1 n_2 + \frac{n_1(n_1 + 1)}{2} - R_1$$

$$= 4 \times 5 + \frac{4(4 + 1)}{2} - 27 = 3$$

check to see it is the same answer as before

The complementary formula is

$$U' = n_1 n_2 + \frac{n_2(n_2 + 1)}{2} - R_2$$

$$= 4 \times 5 + \frac{5 \times (5 + 1)}{2} - 18 = 17$$

As a check to make sure that you have made no mistakes, you can use the fact that the two U values are related to each other by

$$U = n_1 n_2 - U'$$

Substituting in our U values, it can be seen that the relationship holds and the U values are correct.

$$3 = 4 \times 5 - 17 = 20 - 17 = 3$$

Ties

For ties, mean ranks are taken and the calculation performed in the usual way. Again, this is all right as long as there are not too many (< 25%). Should there be more ties, it is simpler to use an alternative procedure (e.g., Wilcoxon-Mann-Whitney rank sums test, Section 16.6).

Assumptions for the Mann-Whitney U Test

These are similar to the Wilcoxon assumptions (see Section 16.2).

1. The data must be at least ordinal.
2. The test is an independent samples test, so the two sets of scores must be independent of each other (i.e., from different subjects).
3. All data within a given sample must be independent (i.e., all from different subjects; you cannot use someone twice).
4. Subjects must be chosen randomly from their respective populations.
5. Even though the data may only be ranked, the underlying variable must be continuous (see Section 2.6).
6. Like the Wilcoxon test, the Mann-Whitney U test really tests to see whether the two sets of scores have the same distribution. If the two distributions are of a similar type (and in the same experiment they most likely will be), the test becomes one of whether the two medians are the same. The test is generally more sensitive to differences in central tendency (or location: are one set of ranks ranked behind the other?) and for most practical purposes can be thought of as a test of central tendency or location; it is relatively insensitive to differences in distribution shape or spread.

Alternatives to the Mann-Whitney U Test

There are other unrelated-samples difference tests suitable for ranked data: the Kolmogorov-Smirnov test, the Wald-Wolfowitz test, the Moses test, etc. Although not dealt with in this text, these tests are sometimes used in behavioral research; they are given in the Key to Statistical Tables. Another alternative is the Wilcoxon-Mann-Whitney ranks sums test discussed in Sections 16.6 and 16.7.

Mann-Whitney U Test for Large Samples

Table G.19 gives critical values of U only for sample sizes no bigger than 20. For larger samples, an approximation to the normal distribution can be used. A z score can be obtained, and the probability of getting such a value, or greater, on H_0 can be found from Table G.1 (see Section 4.3). The z score is given by

$$z = \frac{U_{obs} - n_1 n_2/2}{\sqrt{\dfrac{n_1 n_2 (n_1 + n_2 + 1)}{12}}}$$

Note: Table G.1 gives one-tailed probabilities; for two-tailed tests, double the probability value.

16.5 WORKED EXAMPLE: THE MANN-WHITNEY U TEST

There is a chemical called phenylthiocarbamide (PTC) which some persons (tasters) taste as bitter and others (nontasters) find tasteless. In fact, the two groups, tasters and nontasters, overlap slightly so that it is more true to say that the two groups are differentially sensitive to PTC.

Five tasters and five nontasters of PTC had their sensitivities to a second chemical measured. The highest four sensitivities recorded belonged to nontasters while the lowest sensitivity to this second chemical also belonged to a nontaster. Are tasters and nontasters significantly different in their sensitivity to the second chemical? You do not know beforehand the direction of the difference, if any.

A two-tailed test is appropriate because the direction of the difference was not known beforehand. H_0 states that there is no difference in sensitivity to the second chemical between tasters and nontasters. The ranking of tasters (T) and nontasters (NT) can be deduced from the wording of the question. The four most sensitive and the least sensitive taster come from the NT group; the T group are sandwiched between them.

Greatest sensitivity	(NT)	(NT)	(NT)	(NT)	(T)	(T)	(T)	(T)	(T)	(NT)	Lowest sensitivity
Rank	1	2	3	4	$\underline{5}$	$\underline{6}$	$\underline{7}$	$\underline{8}$	$\underline{9}$	10	

number in the NT group, $n_1 = 5$

T group, $n_2 = 5$

We consider the nontaster group. We add the rank values to obtain R:

$$R_1 = 1 + 2 + 3 + 4 + 10 = 20$$

Thus

$$\text{U} = n_1 n_2 + \frac{n_1(n_1 + 1)}{2} - R_1$$

$$= 5 \times 5 + \frac{5 \times 6}{2} - 20$$

$$= 25 + 15 - 20 = 20$$

We could calculate the other value, U′, using

$$\text{U}' = n_1 n_2 - \text{U}$$

$$= 5 \times 5 - 20 = 5$$

Or we could calculate it directly:

$$R_2 = 5 + 6 + 7 + 8 + 9 = 35$$

and

$$U' = n_1 n_2 + \frac{n_2 (n_2 + 1)}{2} - R_2$$

$$= 5 \times 5 + \frac{5 \times 6}{2} - 35$$

$$= 25 + 15 - 35 = 5$$

The two U values (20 and 5) can now be compared with values in Table G.19. The two-tailed values for $n_1 = n_2 = 5$ are

2	for $p = 0.05$	0	for $p = 0.01$
<u>23</u>		<u>25</u>	

Our values of 20 and 5 are neither equal to, nor do they exceed the values in the table (e.g., $20 < 23$; $5 > 2$; etc.), so we cannot reject H_0 at the 5% or 1% levels of significance. In fact, we cannot even reject it at the 10% level (critical values: 4, 21). Thus tasters and nontasters of PTC are not significantly different in their sensitivities to the second chemical ($p > 0.1$).

16.6 THE WILCOXON-MANN-WHITNEY RANK SUMS TEST

Relationship to the Mann-Whitney U Test

This test, rather unfortunately, has a lot of different names. It was apparently devised by Frank Wilcoxon and has been called the *rank sums test* or the *Wilcoxon rank sums test* (not to be confused with the Wilcoxon test of Section 16.2). However, because it is equivalent to the U test of Mann and Whitney (Section 16.4), it is also sometimes called the Mann-Whitney test. But do not confuse it with the Mann-Whitney U test. It is not the same test; it is merely equivalent. To try and avoid confusion with these other tests, as well as diplomatically credit all possible authors, we will give this test the unwieldy name of the Wilcoxon-Mann-Whitney rank sums test. If that seems too long, it can be called the rank sums test. Consult the Key to Statistical Tests to see the relationship of this test to others.

This independent-samples, two-sample difference test is used as an alternative to the Mann-Whitney U test. It is also related to the Mann-Whitney U test;

in fact, it can be considered as a Mann-Whitney U test in disguise. The index for the rank sums test, S, is merely the difference between the two Mann-Whitney U values.

$$S = |U - U'|$$

where the vertical lines (signifying the term "modulus") indicate that the difference between U and U′ is arranged to be positive.

Should there be a significant difference between two unrelated samples it was seen (see Section 16.4) that the two U values would be different; one would be large, one would be small. Should there be no significant difference between the two samples, the two U values would be similar. The Mann-Whitney U test calculated how large or small these two values could be by chance (on H_0). This test uses an index, S, which is the difference between the two U values. The greater the difference between the two independent samples tested, the greater the difference between U and U′ and the greater the value of S. Should there be no difference between the samples tested (H_0 true), the U values would tend to be the same and S would be small. What is needed is a table giving the largest values of S that can occur by chance; should the calculated value of S exceed the chance value, H_0 can be rejected. Such a table is Table G.20. This table gives critical values of S such that should an experimental value of S be equal to or exceed the value in the table, H_0 is rejected. Critical values for various significance levels are located in rows indicated by the two sample sizes n_1 and n_2 (see the left-hand column).

Example 1: Experimental and Control Groups from the Mann-Whitney U Test

In the Mann-Whitney U test example (see Section 16.4), considering experimental and control groups,

$$U = 3 \qquad U' = 17 \qquad \text{for two groups of subjects: } n_1 = 4, \quad n_2 - 5$$

$$S = |U - U'| = 3 - 17 = 14$$

Using Table G.20, for sample sizes of 5 and 4 (5 is the largest sample denoted by n_1 in the table), two-tailed, the critical values are

$p = 0.5$: $S = 18$

$p = 0.01$: S has no value

Our value (14) is less than 18. Thus we do not reject H_0 at the 5% level. We would also not reject it at the 10% level but would at 20%. This result is compatible with that obtained by the Mann-Whitney U test (see Section 16.4).

Example 2: PTC Tasters from the Mann-Whitney U Test

In the worked example about PTC tasters and nontasters (Sec. 16.5) for the Mann-Whitney U test, we had two independent samples of five subjects each and two U values of 5 and 20. Here, $S = |U - U'| = 20 - 5 = 15$. According to Table G.20, two-tailed, $n_1 = n_2 = 5$, the critical values of S are:

For $p = 0.05$, $S = 21$.
For $p = 0.01$, S has no value.

Our value (15) is less than 21; thus we do not reject H_0 at the 5% level. We would also not reject it at the 10% level but would at 20%. This result is compatible with that obtained by the Mann-Whitney U test (see Section 16.4).

Alternative Ways of Calculating *S*

Just as there are a variety of procedures for calculating the Mann-Whitney U values (counting scores in the second sample preceding those in the first, or using a formula), there are a variety of procedures for calculating S. We will consider three procedures of which the third is the one generally used.

Let us take an example. Subjects from an experimental and control group are ranked along some dimension (e.g., liking for a given food, salivary salt level, smell sensitivity, or whatever). Judges from the experimental (E) group came 1st, 2nd, 3rd, and 7th. From the control (C) group they came 4th, 8th, 9th, and 10th. One subject from each group tied for 5th and 6th place and are thus given the mean rank of 5.5. The data can be displayed as follows:

E:	1	2	3		5.5	7			
C:				4	5.5		8	9	10

where $n_1 = n_2 = 5$.

Alternative 1: Calculation of S, Using the Computational Formulas to Obtain U and U′

To calculate S, the two U values can be calculated, in the normal way, using the formulas

$$U = n_1 n_2 + \frac{n_1(n_1 + 1)}{2} - R_1$$

$$U' = n_1 n_2 + \frac{n_2(n_2 + 1)}{2} - R_2$$

The rank totals for each group are as follows:

E:	1	2	3		5.5	7				$R_1 = 18.5$
C:				4	5.5		8	9	10	$R_2 = 36.5$

Thus for E,

$$U = n_1 n_2 + \frac{n_1(n_1+1)}{2} - R_1 = 5 \times 5 + \frac{5 \times 6}{2} - 18.5 = 21.5$$

and for C,

$$U' = n_1 n_2 + \frac{n_2(n_2+1)}{2} - R_2 = 5 \times 5 + \frac{5 \times 6}{2} - 36.5 = 3.5$$

Thus $S = |U - U'| = 21.5 - 3.5 = 18.$

Using Table G.20 for $n_1 = n_2 = 5$, two-tailed, our calculated S (18) only exceeds the tabular value for $p = 0.1$. It does not exceed the $p = 0.05$ value (21). Because the 10% level is customarily too high for most purposes, we will state that we do not reject H_0 at $p < 0.05$. On this basis we state that we do not have any evidence of a significant difference between experimental and control groups ($0.1 > p > 0.05$).

Alternative 2: *Calculation of S, Using the Method of Counting How Many Scores from One Sample Precede Those from the Second Sample, to Obtain U and U'*

The ranks are as follows:

E:	1	2	3		5.5	7			
C:				4	5.5		8	9	10

We count the number of judges in the E group that come in front of each judge in the C group (e.g., three came before 4, three and a half before 5.5, five before 8, etc.). This gives the Mann-Whitney U value. Doing the same for each judge in the E group gives the other Mann-Whitney U value. Note that with ties, one score in the tie is said to half-precede the tied score in the other group.

For group E:

Rank: (1) (2) (3) (5.5) (7)

Number preceding: $U = 0 + 0 + 0 + 1.5 + 2 = 3.5$

For group C:

Rank: (4) (5.5) (8) (9) (10)

Number preceding: $U' = 3 + 3.5 + 5 + 5 + 5 = 21.5$

$$S = |U - U'| = |3.5 - 21.5| = 18$$

This is the value obtained by the first method.

Alternative 3: *Calculation of S, Using the Traditional Rank Sums Method to Obtain u and u'*

There is a third alternative method which is similar to the second method, except that ties are ignored completely when the number of scores in one sample preceding those in the other sample are being counted. When this is done, another value, u, not the regular Mann-Whitney U, is obtained. This is the same as the S value obtained by other methods. The only difference is that the half-values derived from ties are not included in the u score. Thus for u, the 5.5 score in the E group is scored as having one C score in front of it, while for U it is scored as having one and half C scores in front of it. As these half-values are eliminated from both sets of u scores (u and u'), the difference $u - u'$ will be the same as the difference $U - U'$, where the half-scores are included. Thus S is unaffected. The procedure is merely a simplification, because awkward fractional values are omitted.

In fact, rather than counting the number of scores in one group that come *before* a given score in the second group, and vice versa, we count the number of scores that come before and then the number that come after scores in the same group. This is, in fact, an equivalent procedure. We count the number of C scores coming before each E score as usual. We then count the number of C scores coming after each E score; this is equivalent to counting the number of E scores coming before each C score.

E:	1	2	3		5.5	7			
C:				4	5.5		8	9	10

Consider group E. We count the number of scores in group C preceding and also coming after score in group E.

Rank in group E:		(1)	(2)	(3)	(5.5)	(7)	
Number of scores in group C preceding each value in group E:	u =	0 +	0 +	0 +	1 +	2	= 3
Number of scores in group C that are preceded by each value in group E:	u' =	5 +	5 +	5 +	3 +	3	= 21

$$S = |u - u'| = |3 - 21| = 18$$

It can be seen that the S value obtained here is exactly the same as that obtained using Mann-Whitney U values.

Alternative 3 is the commonly used procedure for the rank sums test.

Approximation to Normal Distribution

Should either sample size, n, be greater than 25, Table G.20 cannot be used. However, the S distribution approximates to normal and we can calculate a z value as follows:

$$z = \frac{S - 1}{\sqrt{\frac{n_1 n_2 (n_1 + n_2 + 1)}{3}}}$$

Substracting 1 from S is a correction for continuity. This value of z can be used with Table G.1 to determine the (one-tailed) probability of getting a z (or S) value this large by chance. For a two-tailed test, double the probability value in Table G.1.

Ties

For ties, mean ranks are taken, as before, and Table G.20 used in the normal manner, although this procedure will tend to make the test a little more conservative (harder to reject H_0). If more than 25% of the scores are involved in ties, an alternative procedure is required (see later).

Assumptions for the Wilcoxon-Mann-Whitney Rank Sums Test

These are the same as those for the Mann-Whitney U test.

1. The data must be at least ordinal.
2. The test is an independent samples test, so the two sets of scores must be independent of each other (i.e., from different subjects).
3. All data within a given sample must be independent (i.e., all from different subjects; you cannot use someone twice).
4. Subjects must be chosen randomly from their respective populations.
5. Even though the data may only be ranked, the underlying variable must be continuous (see Section 2.6).
6. The procedure really tests whether the distributions from the two samples are different. It can pick up differences in the shape or spread of the distribution but is particulary sensitive to difference in central tendency (median). It is rare that any differences are due to anything other than location or central tendency, so for most practical purposes, it can be seen as a test of differences between the medians of the two samples, in other words, their locations (one sample being ranked behind the other).

Example with Many Ties

When there are many ties (more than 25% of the data involved in ties) the procedure for testing the significance of *S* needs to be changed, because Table G.20 becomes inaccurate. In this case a *z* value is calculated, using the approximation of the *S* distribution to normality, although the calculation varies somewhat (a different formula for *z* is required).

Let us imagine an example with many ties. Imagine two groups of scores, one called *S* and the other called *N*. Imagine that the scores are categorized into four ranked classes: 1st, 2nd, 3rd, and 4th. Let there be 10 scores in each group. With only four categories, this will produce a lot of ties.

The scores could each come from different judges under two experimental conditions *S* and *N*; the responses could be on a four-point category scale rating taste intensity of a given food. They could be hedonic or intensity responses from separate groups of judges to two foods *S* and *N*. Naturally, a related-samples design would be preferable, so as to allow for individual differences between judges. As each response comes from a different judge, the test is an independent-samples difference test, each score being equally independent from every other score, coming from separate subjects. They could also all be responses from one single judge, because the degree of independence within a group would be the same as that between the two groups; it would still be an independent-samples design.

Let us imagine that in the *S* group, six scores were ranked in the 1st class, two were 2nd, and two were 3rd. In the *N* group, one was ranked 2nd, two were 3rd, and seven were 4th. Such data can be represented on a matrix as follows:

	1	2	3	4	
S	6	2	2	0	$n_1 = 10$
N	0	1	2	7	$n_2 = 10$

Note that this matrix represents a *one*-factor, *not* a *two*-factor design. There is only one factor with two treatments *S* and *N*. 1, 2, 3, and 4 do not represent treatments under a second factor; they are responses given by different subjects under treatments *S* or *N*. A response matrix like this can be a little confusing. It can look as though two factors (*S*, *N* and 1, 2, 3, 4) are operating; this is not so. This is discussed further in Section 16.20.

Again, there are three alternative approaches to finding *S*, of which the traditional rank sums method, the first outlined here, is the quickest and easiest.

Alternative 1: Traditional Rank Sums Method, Calculating u and u′

Consider the S group.

	Rank value				
	1	2	3	4	
The number in group N that are preceded by each value in the S group, u =	$6 \times (1+2+7)$ +	$2 \times (2+7)$ +	2×7 +	0	= 92
The number in group N that precede each value in the S group, u' =	6×0	+ 2×0	+ 2×1 +	0	= 2

Thus $S = |u - u'| = 92 - 2 = 90$.

This procedure is the fastest and easiest to use. The equivalent method of calculating the Mann-Whitney U values by counting the number of scores in each group preceding each score in the other group is almost as convenient. This first method, however, is readily applicable to the analysis of R-index matrices (see Appendix E).

Alternative 2: Method of Counting How Many Scores from One Sample Precede Those from the Second Sample, to Find U Values

	1	3	3	4
S	6	2	2	
N		1	2	7

$n_1 = 10$

$n_2 = 10$

Counting the number of scores in one group preceding each score in the other group is equivalent to the traditional rank sums method, except in the treatment of ties. Here tied scores are taken as half-preceding each other.

	Rank value 1	2	3	4	
Number of scores in group S preceding each score in group N, U =	0	$+ 1 \times \left(6 + \frac{2}{2}\right)$	$+ 2 \times \left(6 + 2 + \frac{2}{2}\right)$	$+ 7 \times (6 + 2 + 2)$	= 95
Number of scores in group N preceding each score in group S, U′ =	6×0	$+ 2 \times \frac{1}{2}$	$+ 2 \times \left(1 + \frac{2}{2}\right)$	+ 0	= 5

Thus $S = |U - U'| = 95 - 5 = 90$. This is the same as the value obtained using $u - u$!

Alternative 3. Method Using Computational Formulas to Find U Values

This method tends to take longer because all the scores have to be ranked in order, not left in their four categories. This is done as follows:

	1	2	3	4	
S	6	2	2	0	$n_1 = 10$
N		1	2	7	$n_2 = 10$

Range of ranks in this category:	1-6	7-9	10-13	14-20
Mean rank:	3.5	8	11.5	17

For S,

$$R_1 = 6 \times 3.5 + 2 \times 8 + 2 \times 11.5 + 0 = 21 + 16 + 23 = 60$$

For N,

$$R_2 = 0 + 1 \times 8 + 2 \times 11.5 + 7 \times 17 = 8 + 23 + 119 = 150$$

Using the formulas for U and U′, we have

$$U = n_1 n_2 + \frac{n_1(n_1 + 1)}{2} - R_1 = 10 \times 10 + \frac{10 \times 11}{2} - 60$$

$$= 100 + 55 - 60 = 95$$

$$U' = n_1 n_2 + \frac{n_2(n_2 + 1)}{2} - R_2 = 10 \times 10 + \frac{10 \times 11}{2} - 150$$

$$= 100 + 55 - 150 = 5$$

Thus $S = |U - U'| = 95 - 5 = 90$. These are the same as the S values obtained by the last method.

Testing an Example with Many Ties for Significance

With a high proportion of ties, we cannot use Table G.20 to test the significance of S. Instead, we must use the approximation of the S distribution to normal, and convert S values to z values. However, we do not use the regular formula to do this; the formula required is

$$z = \frac{S - K}{\text{SD}}$$

where S is the calculated S value, K is a correction for continuity, and SD is a standard deviation value.

Let us look at the example we have just examined, where $S = 90$. We test the significance of S (by computing z) using the formula above. To calculate values for the SD and for K, the response matrix must be reconsidered, with the column and row totals being computed.

	1	2	3	4
S	6	2	2	
N		1	2	7
Column totals, c =	6	3	4	7
c^3 =	216	27	64	343

number of response categories, $k = 4$

$n_1 = 10 =$ row total for S

$n_2 = 10 =$ row total for N

$N = n_1 + n_2 = 20$

$\Sigma c^3 = 650$

The value of c in the first column, $c_1 = 6$; the value of c in the last, kth column, $c_k = 7$.

We can now use these values to calculate the parts in the formula for z. The standard deviation term is given by

$$\mathrm{SD} = \sqrt{\frac{n_1 n_2 (N^3 - \Sigma c^3)}{3N(N-1)}}$$

$$= \sqrt{\frac{100(20^3 - 650)}{3 \times 20 \times 19}} = \sqrt{\frac{100(8000 - 650)}{1140}}$$

$$= \sqrt{\frac{100 \times 7350}{1140}} = 25.39$$

The correction for continuity is given by

$$K = \frac{2N - c_1 - c_k}{2(k-1)} = \frac{40 - 6 - 7}{2(4-1)} = \frac{27}{6} = 4.5$$

Thus

$$z = \frac{S - K}{\mathrm{SD}} = \frac{90 - 4.5}{25.39} = 3.37$$

Using Table G.1, the probability of obtaining a z score of 3.37 or larger by chance (on H_0) is between 0.0004 and 0.0003. This is the one-tailed probability. It is doubled to obtain the two-tailed probability; it thus lies between 0.0008 and 0.0006. This is small, so H_0 can be rejected. Thus we conclude that there is a significant difference between the S and N groups.

This analysis is particularly useful for the analysis of R-index data from a single judge obtained from signal detection difference testing (see Appendix E).

16.7 WORKED EXAMPLE: THE WILCOXON-MANN-WHITNEY RANK SUMS TEST

Consider a judge who is tasting a random order of 2 mM NaCl solutions and water samples and rating them as having salt present in the solution (rated as containing a salt signal, S) or being merely water (the mere background noise, N, of the taste of the water). He is also required to state whether he is sure of the judgment or not. He can thus respond "I am sure it is a salt solution" (S), "I think it is a salt solution but I am not sure" (S?), "I think it is only water but I am not sure" (N?), "I am sure it is only water" (N). This would be a typical procedure for measuring salt taste sensitivity using the shortcut signal detection R-index procedure (see Appendix E).

Let us assume that the judge rated six of the salt solutions as (S), three as (S?), and one as (N?), while rating six water solutions as (N), two as (N?), and two as (S?). These results are summarized in the following matrix:

Judge's response

		S	S?	N?	N		
Salt stimulus,	S	6	3	1	0	$n_1 = 10$	Total number of responses, $N = 20$
Water stimulus,	N	0	2	2	6	$n_2 = 10$	
Column totals,	$c =$	6	5	3	6		
	$c^3 =$	216	125	27	216	$\Sigma c^3 = 584$	

There are $k = 4$ response conditions

$c_1 = 6 \qquad c_k = 6$

We use the traditional rank sums method for calculating S, because it is the simplest method.

For the salt stimulus:

Number of water samples preceded by salt stimuli,

For stimuli rated: S, S?, N?

$$u = 6 \times (2 + 2 + 6) + 3 \times (2 + 6) + 1 \times 6 = 60 + 24 + 6 = 90$$

Number of water samples that precede salt stimuli,

For stimuli rated: S, S?, N?

$$u' = 6 \times 0 + 3 \times 0 + 1 \times 2 = 2$$

Thus $S = |u - u'| = 90 - 2 = 88$.

There are many ties; thus the significance of S will be tested using the approximation to normality. Thus z is computed:

$$z = \frac{S - K}{\text{SD}}$$

where

$$\text{SD} = \sqrt{\frac{n_1 n_2 (N^3 - \Sigma\, c^3)}{3N(N-1)}} = \sqrt{\frac{100(8000-584)}{3 \times 20 \times 19}}$$

$$= \sqrt{\frac{100 \times 7416}{1140}} = 25.51$$

$$K = \frac{2N - c_1 - c_k}{2(k-1)} = \frac{2 \times 20 - 6 - 6}{2(4-1)} = \frac{28}{6} = 4.67$$

and thus

$$z = \frac{S - K}{\text{SD}} = \frac{88 - 4.67}{25.51} = 3.27$$

From Table G.1, the one-tailed probability of getting a z value as great as 3.27 is between 0.0005 and 0.0006; the two-tailed probability is thus between 0.001 and 0.0012. This is low; thus H_0 is rejected. So the responses to the salt solutions are significantly different in rank from the responses to the water samples (they tend more to S); thus the judge distinguished significantly between salt and water.

16.8 SPEARMAN'S RANKED CORRELATION

Spearman's ranked correlation coefficient is the nonparametric version of Pearson's product-moment correlation (see Chapter 15). Like Pearson's r, ρ (Greek lowercase rho) can range from 1 (perfect correlation), through 0 (no correlation), to −1 (perfect negative correlation). Again, it should be stated that correlation does not imply causality.

The calculation proceeds as follows. X and Y values are obtained from the judges. The X values are ranked and so are the Y values. Alternatively, the original data obtained may be ranked. The differences between the ranks (d) are calculated, as are their squares (d^2) which are summed ($\Sigma\, d^2$). A typical calculation proceeds as follows:

Subject	X	Y	d	d^2
A	1	1	0	0
B	2	5	3	9
C	3	2	-1	1
D	4	4	0	0
E	5	3	-2	4
F	6	7	1	1
G	7	6	-1	1
H	8	8	0	0
I	9	10	1	1
J	10	9	-1	1
			$\Sigma d^2 =$	18

$N = 10$

The correlation coefficient, ρ, is calculated from the simple formula

$$\rho = 1 - \frac{6\Sigma d^2}{N(N^2 - 1)} = 1 - \frac{6 \times 18}{10(100 - 1)} = 1 - \frac{108}{990} = 0.891$$

Should the X and Y values be highly correlated, subjects ranked first for X will also be ranked first for Y, and subjects ranked last for X will also be ranked last for Y; there will be little difference in ranks for X and Y values, so d values will be small, as will Σd^2. It can be seen from the formula for ρ that the smaller Σd^2 the greater is ρ. With perfect correlation, the d values will all be zero, as will be Σd^2 and $\rho = 1$. On the other hand, should X and Y not be correlated, their rankings will not correspond and the differences between the d values will be large, with a resulting large Σd^2 value and a low value for ρ. Thus, although we have not derived the formula for ρ, it can be seen to be intuitively sensible.

The greater the correlation coefficient obtained from the sample of data, the more likely it is to represent a significant correlation in the population. Of course, when there is no correlation in the population, ρ will not always be zero in the sample; chance correlations can occur. A table is required to indicate how large a value of ρ can be obtained by chance, how large ρ can be on the null hypothesis, which states that there is no correlation between X and Y.

Table G.21 gives critical values of ρ such that calculated ρ values equal to or greater than values in the table will be significant. Values are given for various sample sizes (N in the left-hand column) at various significance levels, for one- and two-tailed tests. A one-tailed test is used when it can be predicted beforehand that the correlation will be positive (or negative) if H_0 is rejected. Otherwise, if the direction of any significant correlation cannot be predicted, a two-tailed test is used.

Turning to the example, there are ten subjects ($N = 10$). We will use a two-tailed test because we will assume that we could not predict the direction of the correlation, should H_0 be rejected. We find the row corresponding to $N = 10$ (see the left-hand column). The two-tailed critical values of ρ are

0.649 $p = 0.05$
0.794 $p = 0.01$

Our calculated value (0.891) exceeds both of these, so the correlation is significant at the 1% level, and also at the 0.2% level (0.879). Thus we reject H_0 and state that X and Y values are correlated ($p<0.002$).

Large Samples: Testing the Significance of ρ by a t Test

Table G.21 gives values of ρ for sample sizes up to $N = 30$. For larger samples, we can use a t test to test the significance of ρ using the relationship

$$t = \rho \sqrt{\frac{N-2}{1-\rho^2}} \qquad \text{with } df = N-2$$

Use Table G.8, the table of critical t values.

Ties

Should the ranked values of X and Y be the same, d values will be zero and are included in the calculation in the usual way. Should some of the X values tie with other X values (or the same happen for Y values), mean ranks are used and the formula used in the usual way (e.g., should there be a tie for 2nd, 3rd, and 4th place, all are given the rank of 3). However, if the proportion of ties is too great (25%), ρ tends to be overestimated. This opens the possibility of correlations appearing significant when they should not (Type I error). An alternative procedure is adopted.

Imagine a calculation in which some of the X data tie, as do some of the Y data. Let the data be as follows (for convenience not all the data will be shown):

X	Y	d	d^2
⋮	⋮	⋮	⋮
3	.	.	.
3	.	.	.
3	.	.	.
⋮	⋮	⋮	⋮
7.5	.	.	.
7.5	.	.	.
⋮	⋮	⋮	⋮

The calculation proceeds as follows. Let

t = number of X observations tied at a given rank

= 3 (at 3rd) and 2 (at 7.5th)

Then calculate

$$T_x = \frac{t^2(t-1)}{12} \qquad \text{for each value of } t$$

$$= 1.5, 0.33$$

Then calculate $\Sigma T_x = 1.5 + 0.33 = 1.83$.

Now do the same for the Y values and calculate ΣT_y. Then calculate

$$\Sigma x^2 = \frac{N(N^2-1)}{12} - \Sigma T_x$$

$$\Sigma y^2 = \frac{N(N^2-1)}{12} - \Sigma T_y$$

Then the correlation coefficient is given by

$$\rho = \frac{\Sigma x^2 + \Sigma y^2 - \Sigma d^2}{2\sqrt{\Sigma x^2 \, \Sigma y^2}}$$

The formula reduces to the original formula should there be no ties. Some authors recommend that the t-test method of testing for significance is preferable when there is a high proportion of ties.

As with Pearson's product-moment correlation coefficient (r) (see Section 15.7), the only assumption required for the calculation of ρ is:

Assumption for the Rank Sums Test

1. The pairs of ranked values of X and Y must be independent. They must each come from separate subjects; do not test subjects twice to boost the size of the sample.

Another assumption is also necessary for the calculation of ρ if d values are to be obtained.

2. The measurement must be at least ordinal (ranked) in character.

As with Pearson's r, there are further assumptions that should be made if the significance of Spearman's ρ is to be tested (Table G.21). These are:

3. Samples should be randomly selected.
4. Even though the data may only be ranked, the underlying variable must be continuous (see Section 2.6).

A cautionary note: Spearman's ρ is not a measure of linear association between X and Y; it is a measure of linear association between the ranks of X and Y. This is a completely different thing. X and Y could be related in a curvilinear fashion, producing a low r value, but as long as X and Y both increase monotonically, the ranks of X and Y will be linearly related and thus $\rho = 1$. Thus, when $r = 1$, the relationship between X and Y is linear; when $\rho = 1$, the relationship between X and Y may be anything as long as X and Y both increase monotonically.

A commonly used alternative to Spearman's correlation coefficient, ρ, is Kendall's τ (Greek lowercase letter tau). Look at the Key to Statistical Tables to see the relation of Kendall's and Spearman's tests to other statistical tests.

16.9 WORKED EXAMPLE: SPEARMAN'S RANKED CORRELATION

A geneticist wished to test whether the blueness of a man's eyes was related to his smell sensitivity. He took a sample of seven men and ranked them in order for the two variables. He did not know beforehand whether the relationship would be direct or inverse. The men were (in order of blueness of eyes) Heinz, Wilhelm, Adolf, Herman, Carl, Wolfgang, and poor old Sigmund with brown eyes. In order of smell sensitivity they were Herman, Wilhelm, Heinz, Adolf, Carl, Sigmund, and Wolfgang. Did his results support the notion of a correlation between blueness of eyes and smell sensitivity?

Because the geneticist could not predict whether the correlation would be positive or negative (direct or inverse relationship) should H_0 be rejected, we will use a two-tailed test. Using the given rank orders, the data can be set up as follows:

	Blueness of eyes	Smell sensitivity	d	d^2	
Heinz	1	3	-2	4	
Wilhelm	2	2	0	0	
Adolf	3	4	-1	1	
Herman	4	1	3	9	
Carl	5	5	0	0	
Wolfgang	6	7	-1	1	
Sigmund	7	6	1	1	
			$\Sigma d^2 =$	16	$N = 7$

Filling in the formula for ρ, we obtain

$$\rho = 1 - \frac{6 \Sigma d^2}{N(N^2 - 1)} = 1 - \frac{6 \times 16}{7(49 - 1)}$$

$$= 1 - \frac{96}{336} = 1 - 0.286$$

$$= 0.714$$

H_0 states that there is no correlation between blueness of eyes and smell sensitivity. The critical values in Table G.21 ($N = 7$, two-tailed) are

0.786 ($p = 0.05$) 0.929 ($p = 0.01$)

Both these values are larger than our calculated value (0.714), as is the value at $p = 0.1$ (0.715). Thus we do not have sufficient evidence to reject H_0 ($p > 0.1$). Thus the data do not indicate any ranked correlation between blueness of eyes and smell sensitivity.

16.10 THE FRIEDMAN TWO-FACTOR RANKED ANALYSIS OF VARIANCE (TWO-FACTOR REPEATED-MEASURES DESIGN)

Having dealt with correlation and two-sample difference tests, we now pass on to many sample difference tests, the nonparametric equivalent of ANOVA. The *Friedman test* (developed by Milton Friedman, the economist) is a related-samples many-sample difference test. Where the sign test or Wilcoxon test deal with two samples, the Friedman test deals with more than two samples. The Friedman test is the equivalent of the two-factor, repeated-measures ANOVA for ranked data. Look at the Key to Statistical Tables to see the relationship of the Friedman test to these other tests. In fact, theoretically, the Friedman test is an extension of the sign test (see Section 5.8).

The test is best demonstrated by an example. Consider scores for three judges S_1, S_2, and S_3 under four treatment conditions 1, 2, 3, and 4. The scores in each treatment are ranked in order of size for each subject or the original data itself might have been in ranked form. The ranked data can be presented as below. For each treatment, ranks are summed giving rank totals (e.g., 4 + 3 + 4 = 11).

		Condition			
		1	2	3	4
	S_1	4	2	1	3
Judge	S_2	3	1	2	4
	S_3	4	2	1	3
Rank total, R =		11	5	4	10

N = number of rows = 3 = number of subjects

k = number of columns = 4 = number of treatments or conditions

We now calculate a type of chi-square value using the following formula:

$$\chi_r^2 = \frac{12}{Nk(k+1)} \Sigma R^2 - 3N(k+1)$$

$$= \frac{12}{3 \times 4 \times (4+1)} (11^2 + 5^2 + 4^2 + 10^2) - 3 \times 3 \times (4+1)$$

$$= 7.4$$

Now consider the formula for χ_r^2. Should there be no difference between treatments 1, 2, 3, and 4, the order of ranks will vary; for one subject the conditions may be ranked in one way, and for the next subject the rankings may be quite the opposite:

	Treatment				
	1	2	3	4	
S_1	1	2	3	4	
S_2	4	3	2	1	
R =	5	5	5	5	
R^2 =	25	25	25	25	$\Sigma R^2 = 100$

The rank totals (R) will then tend to be the same, as will the R^2 values, and ΣR^2 will have a moderate value. On the other hand, should there be clear differences between the four conditions, the rankings will tend to correspond for each subject:

	Treatment				
	1	2	3	4	
S_1	1	2	3	4	
S_2	1	2	3	4	
R =	2	4	6	8	
R^2 =	4	16	32	64	$\Sigma R^2 = 120$

In this case some of the column totals (R) may be small but others will be large. Some R^2 values will be small while some will be very large, causing the ΣR^2 to be larger. This can be seen from the simple values given above. Squaring the R values has the effect of exaggerating the differences between the column totals and causing higher ΣR^2 values.

Thus the more different the experimental conditions, the more different the R and R^2 values and the greater ΣR^2. It can be seen from the formula for χ_r^2 that the larger ΣR^2, the larger χ_r^2. Thus the greater the differences between the experimental conditions, the greater χ_r^2. The question becomes one of how great χ_r^2 can be by chance; such information can be provided in a table.

Table G.22 gives probabilities for obtaining given χ_r^2 values under H_0 for given values of N and k. The table is divided into two sections, one for $k = 3$ and one for $k = 4$. Having chosen the appropriate k value, choose the section for the appropriate N value. Then go down the column to the χ_r^2 value obtained and identify the p value to the right of this; this is the probability of obtaining a χ_r^2 value as great as this, on the null hypothesis. Thus for our example, for $k = 4$ and $N = 3$, the probability of obtaining a χ_r^2 value of 7.4, on H_0, is given in the p column as 0.033. 3.3% is low and thus we can reject H_0. Thus the data indicate that there are significant differences between experimental treatment conditions 1, 2, 3, and 4 ($p = 0.033$).

For values of k or N greater than those shown in Table G.22, we can use the fact that for higher N and k values, the index for Friedman's test, χ_r^2, approximates to χ^2. The significance of this χ^2 value can be tested using Table G.7, in the normal way (see Section 6.2) with $df = k - 1$, two-tailed.

Multiple Comparisons

Just as multiple-comparison tests are used to identify which treatments differ from each other after finding a significant F value in a parametric ANOVA test (see Chapter 9), multiple comparisons are needed after finding that significant differences exist between treatments in the Friedman test. Because the Friedman test is an extension of the sign test, the appropriate multiple-comparison procedure to use is a set of sign tests.

However, the performance of a set of sign tests produces the same problem as performing a series of t tests for multiple comparisons (see Section 8.1); the level of significance creeps up. The more the comparisons that are performed, the greater the chance that the tests will identify a difference that is not there, at the supposed level of significance (commit a Type I error).

The exact increase in the significance level when a whole set of sign tests is performed is difficult to calculate exactly, but it can be estimated. The greater the number of comparisons, the lower the level of significance that must be chosen for each sign test to keep the overall level at the value chosen (5%, 1%, etc.). This can be achieved approximately by dividing the level of significance for each sign test by the total number of comparisons that can be made. In our example there were four treatments, allowing a total of six different paired comparisons to be made between them. Thus each sign test should be used at 1/6 the level of significance of that of the Friedman test to maintain a comparable level of significance over the six sign tests used.

The number of pair comparisons possible is given by

$$\text{Number of comparisons} = \frac{k(k-1)}{2}$$

where k is the number of treatments.

For our example, with $k = 4$,

$$\text{Number of comparisons} = \frac{4 \times 3}{2} = 6$$

Some authors state that if only a few planned comparisons are to be made, a regular Wilcoxon test can be used in the usual way.

It should be noted, however, that when the sample sizes are small, the treatments may not involve enough scores to allow rejection of H_0 for the multiple sign tests at the low significance levels required. For instance, for a Friedman test significant at $p = 0.01$ with four treatments (thus six comparisons), the significance level for each sign test should be $p = 0.01/6 = 0.0017$. In Table G.4.b you need a large-sample size ($N = 14$ or 15 at least) to establish such a significant level. Thus, for smaller samples, inspection of the data is all that is available.

Ties

Should any of the treatments tie in rank, mean ranks are assigned (tie for 1st and 2nd place, assign rank of 1.5). Should there be only a few ties, the computation procedure is not altered. On the other hand, should there be extensive ties (say, over 25% of the scores involved in ties), the formula for χ_r^2 is altered as follows:

$$\chi_r^2 = \frac{\dfrac{12}{Nk(k+1)} \Sigma R^2 - 3N(k+1)}{1 - \dfrac{\Sigma T}{N(k^2 - k)}}$$

All values in this formula are familiar, except for ΣT. ΣT is obtained as follows:

1. Consider the scores for each subject.
2. For a given subject, examine the number of ties that occur at a given rank.
3. For each set of tied scores, let t be the number of scores tied at that given rank.
4. For each t value, calculate $(t^3 - t)$.

5. Add up the $(t^3 - t)$ values for each subject; call this value T.
6. Thus $T = \Sigma\,(t^3 - t)$.
7. Add the T values for each subject to give $\Sigma\,T$.

Assumptions for the Friedman Test

The assumptions for the Friedman test are as follows:

1. The test is a related-samples test, so data from each treatment must be matched; each subject should be tested under each treatment.
2. Data within a treatment must be independent. Do not test a subject twice within treatments.
3. Subjects should be randomly selected.
4. The data should be at least on an ordinal scale.
5. Even though the data may only be ranked, the underlying variable must be continuous.
6. The test actually investigates whether distributions are different; however, it will generally be a good test of central tendency rather than shape or spread of the distribution.

It is worth noting that the Friedman test is, like the parametric ANOVA, inherently two-tailed. As far as power is concerned, Friedman himself compared his test with the parametric ANOVA over 56 appropriate data analyses; he concluded that the tests were roughly comparable.

The Friedman test is a repeated-measures (related-samples) design. It could also be used as a completely randomized design, where S_1, S_2, and S_3 would be different experimental treatment and each rank would come from a different judge. Although statistically possible, such a design would be of little use in sensory evaluation, where treatment differences would become confused with subject differences.

16.11 WORKED EXAMPLE: THE FRIEDMAN TEST

A psychiatrist believed that a certain type of kleptomania resulted in differential sensitivity to the four "basic" tastes. He believed that by giving taste tests to shoppers in department stores, he could identify potential shoplifters. As part of his investigations, he gave diagnostic taste tests to four kleptomaniac patients. They were each ranked in order of sensitivity over the four tastes. Did the patients show significant differences in sensitivity to these four tastes at $p < 0.05$? Use the Friedman two-factor ANOVA.

The subjects had sensitivities in the following orders for the basic tastes:

Subject A: salty > sour > sweet > bitter
Subject B: salty > sour > bitter > sweet
Subject C: bitter > sour > salty > sweet
Subject D: salty > sweet > sour > bitter

The subjects were all tested with each basic taste (i.e., under each condition), so a repeated-measures design was used. Thus the Friedman two-way ANOVA can be used.

The data can be displayed as follows:

	Salty	Sour	Sweet	Bitter
S_1	1	2	3	4
S_2	1	2	4	3
S_3	3	2	4	1
S_4	1	3	2	4
R	6	9	13	12

N, number of rows = 4 = number of subjects

k, number of columns = 4 = number of treatments

We now fill in the formula for χ_r^2.

$$\chi_r^2 = \frac{12}{Nk(k+1)} \Sigma R^2 - 3N(k+1)$$

$$= \frac{12}{(4)(4)(4+1)}(6^2 + 9^2 + 13^2 + 12^2) - (3)(4)(4+1)$$

$$= \frac{12}{80}(36 + 81 + 169 + 144) - 60$$

$$= 64.5 - 60 = 4.5$$

We now use Table G.22 to test whether this value of χ_r^2 indicates that there were differences in sensitivity to the four tastes. For $k = 4$ (treatments) and $N = 4$ (subjects), the probability of obtaining χ_r^2 as large at 4.5 on H_0 is 0.242. This value is high (24.2%); thus this result could easily have occurred under H_0 (sensitivity same in all conditions). Thus we cannot reject H_0. Thus the data do not indicate any significant differences in sensitivity of the patients to the four basic tastes.

Had H_0 been rejected, the sign test, with an adjusted significance level, would have been used to make multiple comparisons.

16.12 THE KRAMER TWO-FACTOR RANKED ANALYSIS OF VARIANCE (TWO-FACTOR REPEATED-MEASURES DESIGN)

The *Kramer two-factor ranked ANOVA* basically does what the Friedman two-factor ANOVA does; it is a related-samples design, testing whether there are any differences between two or more treatments. However, whereas the Friedman

**Note added in 3rd printing*: The Kramer test has recently been shown to be invalid.

test merely determines whether differences exist, so that sign tests are then needed to pinpoint these differences, the Kramer test combines these two procedures. In fact, the Kramer test does not go as far as pinpointing all the differences between treatments, it merely categorizes the treatments into: significantly higher, medium, significantly lower. Although the anlaysis may not be as thorough, it is sufficient for many needs and has the appeal of speed and great simplicity. The test is best understood by considering an example.

Imagine six subjects who each perform in four conditions. They could be ranking wines A, B, C, and D for sweetness. Let the ranks be as follows:

		Treatment			
		A	B	C	D
	S_1	4	2	3	1
	S_2	4	2	3	1
Subject	S_3	4	3	2	1
	S_4	2	4	3	1
	S_5	4	2	3	1
	S_6	4	1	3	2
	Total	22↑	14	17	7↓

The ranks are totaled and the totals compared with values in Table G.23. Should treatments A, B, C, and D have no effect (the wines all have the same sweetness), the totals will all be fairly similar. Should any treatments have an effect (some of the wines may be noticeably sweeter, having lower ranks, or others be noticeably less sweet, having higher ranks), the totals will vary; some will be high, some will be low. All that is needed is a table indicating how high or how low such rank totals may be by chance. Kramer published computer-generated tables, giving critical values of the rank totals that must be exceeded for H_0 to be rejected (A. Kramer, G. Kahan, D. Cooper, and Papavasiliou, "A non-parametric ranking method for the statistical evaluation of sensory data," *Chemical Senses and Flavor*, Vol. 1, 1974, pp. 121-123). These tables are given as Table G.23.

Table G.23 is rather large. Table G.23.a gives critical rank totals at the 5% level; Table G.23.b gives values at the 1% level. The table provides critical values for a given number of subjects or judges (left-hand column) up to 75, and a given number of treatments (column headings) up to 20. The critical values are selected for the appropriate number of judges (select appropriate row) and treatments (select appropriate column).

The critical rank totals are presented as a set of four numbers. For our example with six subjects and four treatments, the critical values are as follows:

5% level	*1% level*
9-21	8-22
11-19	9-21

Take the top two figures for the 1% level, 8-22. Should any of our rank totals (22, 14, 17, 7) be more extreme than the tabular values, that is, exceed 22 or be below 8 (not equal to), there are significant differences between treatments (differences in sweetness of wine). The total 7 is below the value 8, so differences exist.

We have now performed the first part of the Kramer test, the part equivalent to an ANOVA, establishing that differences exist. We now need to move on to the second part, the part equivalent to multiple comparisons. In this part, treatments are categorized as: significantly higher, medium, significantly lower.

We now go to the lower two numbers 9-21. Again, any totals that are more extreme than the values in Table G.23 indicate treatments where the differences exist. In this case, treatment A (22) is higher and treatment D is lower (7). We have indicated this with the rank totals on the original data, using vertical arrows.

The operation can be performed again at the 5% level to see whether any lesser differences exist. Total A (22) exceeds the tabular value 21, so differences exist. (Also, total D is less than the tabular value.) To establish where the differences are, the lower two tabular values (11-19) are examined. Once again, total A is higher, total D is lower. So the same differences exist at the 5% level as at the 1% level.

Of course, there may not be sufficient differences for a categorization into: significantly higher, medium, significantly lower. The categorization may merely be: significantly higher vs. significantly lower.

It might be asked why there are two sets of critical values, one for the first ANOVA step and one for the second multiple-comparison step, the second set being easier to exceed and thus establish significance. Kramer states that the lower set can be used directly in the special case where all the conditions give the same result except one, and this one is being tested to see whether the rank total is significantly higher or lower. However, when several comparisons are to be made, a stricter preliminary test (upper pair of figures) is needed as a screen to protect against detecting differences too readily (Type I error). It should be pointed out that some researchers do not feel happy with the strategy of making the second part of the test less conservative once the first step has established that differences exist.

As far as assumptions are concerned, Kramer does not state these specifically. However, it can be said that the test is a related-samples test so that each subject must be tested under each treatment. Data within a treatment must be independent; and the data should be at least ordinal. The test is regarded as two-tailed, testing rank totals that are both significantly higher and lower.

The Kramer test uses a two-factor design. Generally, it is used as a repeated-measures test, where one factor is subjects, who are each tested under a set of treatments. It is possible, however, to use the test as a completely randomized design whereby the two factors are experimental treatments. In this case, each score comes from a different judge. Although possible statistically, such a design would not be so desirable experimentally because differences between treatments will be due not only to the effect of the treatment but also to individual differences between judges. A related-samples design would be more desirable because it would separate out this between-subjects effect. (This is discussed further in Section 16.20.)

16.13 WORKED EXAMPLE: THE KRAMER TEST

Ten judges evaluated the firmness of samples of fresh peaches and peaches canned under four different treatments: A, B, C, and D. They evaluated samples of the peaches by ranking them for firmness, with the lowest rank (1) being given to the most firm and the highest rank (5) to the least firm. The results were as follows:

		Canning treatments			
Judge	Fresh	A	B	C	D
1	1	2	3	4	5
2	1	2	3	5	4
3	1	4	5	2	3
4	1	3	4	5	2
5	1	2	3	5	4
6	1	3	5	2	4
7	1	5	4	3	2
8	1	2	3	5	4
9	1	2	4	3	5
10	1	5	2	4	3
Rank total	10	30	36	38	36

Were there any differences in firmness for the peach treatments? Here the Kramer test can be used because of the repeated-measures design.

Table G.23 shows tabular values for 10 judges and five treatments as follows:

5%	*1%*
20-40	18-42
23-37	20-40

For the 1% level of significance, the top two tabular values are taken (18-42). The total for fresh peaches (10) is more extreme than these values (10 < 18), so differences exist. To locate these differences the lower tabular values (20-40) are taken. Only the total for fresh peaches is more extreme than the tabular values (10 < 20). Thus no differences exist except for the fresh peaches.

At the 1% level, the fresh peaches are significantly firmer than the canned peaches; the four separate canning procedures do not differ in their firmness. The same picture emerges at the 5% level.

16.14 THE KRUSKAL-WALLIS ONE-FACTOR RANKED ANALYSIS OF VARIANCE (ONE-FACTOR COMPLETELY RANDOMIZED DESIGN)

The *Kruskal-Wallis test* is an independent-samples many-samples differences test; it is the nonparametric equivalent of the completely randomized one-factor ANOVA. It is, in fact, an extension of the Wilcoxon-Mann-Whitney rank sums test.

The test is best demonstrated by an example. Consider 10 tasters trained by different consultancy firms (A, B, C), all of whom claim to have a superior psychophysical method for the sensory evaluation of flavor intensity. The 10 tasters were tested in their flavor intensity measurement skills and ranked accordingly (1 = best; 10 = worst).

These rankings are given below under the heading of the appropriate consultancy company:

	Consultancy Company		
	A	B	C
	1	5	6
	2	8	7
	3	9	10
	4		
Rank total, R:	10	22	23
Number in treatment, n:	4	3	3

Total number of scores, $N = 10$

We will now fill these values into the following formula to calculate an H value:

$$H = \frac{12}{N(N+1)} \Sigma \frac{R^2}{n} - 3(N+1)$$

$$= \frac{12}{10 \times 11} \left(\frac{10^2}{4} + \frac{22^2}{3} + \frac{23^2}{3} \right) - 3 \times 11$$

$$= \frac{12}{10 \times 11} \left(\frac{100}{4} + \frac{484 + 529}{3} \right) - 33$$

$$= \frac{12}{10 \times 11} \times 371 - 33 = 40.47 - 33 = 7.47$$

Mistake in book, real answer 6.56

Should there be no difference between treatments A, B, and C, the judges will be evenly spread between the treatments; lowly ranked and highly ranked subjects will appear in all three treatments and the mean ranks (R/n) will be approximately equal. However, should a treatment prove more effective, the lower ranks will occur in this treatment (given low R/n); higher ranks will occur in the less effective treatment (giving high R/n). Thus, should differences occur between treatments, some mean rank value (R/n) will be high and others will be low.

With no differences between the treatments (R/n values roughly the same), R^2/n values will not vary greatly and $\Sigma(R^2/n)$ will have some middle value. With differences between treatments (some R/n values small, others large), some R^2/n values will be small while others will be very large (squaring R causes a great increase for large R values). This will cause $\Sigma(R^2/n)$ to be much larger. From the formula for H it can be seen that as $\Sigma(R^2/n)$ increases, so does H. Thus the greater the differences between the treatments, the larger the value of H. Should H_0 be true (no differences between treatments), H will be smaller, although some moderately large values may occur by chance. What is needed is a table of values indicating how large H may be by chance.

Table G.24 gives, for various sample sizes (n_1, n_2, n_3; left-hand column), two-tailed probabilities on H_0 (in right-hand column headed "p") of obtaining given H values (in center column headed "H"). In our example, the sample sizes were 4, 3, and 3; we locate this section in Table G.24 (see below). Our H value of 7.47 exceeds all the H values in the table; it exceeds 6.7455, the value obtained on H_0 at $p = 0.01$. Thus the probability of obtaining a value as large as this is less than 1%, so we reject H_0.

n_1	n_2	n_3	H	p
4	3	3	6.7455	.010
			6.7091	.013
			5.7909	.046
		7.47	5.7273	.050
			4.7091	.092
			4.7000	.101

should be 6.56.

Thus, on rejecting H_0 we conclude that the different consultancy companies had differential effectiveness for training subjects in the sensory evaluation of flavor intensity.

Had we obtained an H value, say, of 5.75 it would have fallen between the tabular values of 5.7909 (with an associated probability on H_0, $p = 0.046$) and 5.7273 ($p = 0.050$). Thus the probability of obtaining an H value as large as 5.75 on H_0 would have fallen between 0.046 and 0.05; this would have been small enough to reject H_0.

Should there be more than three treatments or more than five subjects per treatment Table G.24 cannot be used. In such cases, H approximates to chi-square and Table G.7 can be used in the normal way ($df = k - 1$, here $= 3 - 1 = 2$) to test the significance of differences between A, B, and C (again a two-tailed test is used).

Having established that differences exist between treatments A, B, and C, multiple comparisons are required to determine where these differences occur. As the Kruskal-Wallis test is an extension of the Wilcoxon-Mann-Whitney rank sums test, the latter test is suitable for multiple comparisons (as with the Kruskal-Wallis test, the multiple-comparisons tests will be two-tailed). However, as with multiple comparisons after an ANOVA using t tests (see Section 8.1) or after the Friedman test using sign tests (see Section 16.10), the level of significance goes up with repeated testing. The more tests that are performed, the greater the chance of making a Type I error (rejecting H_0 when it is true). The overall level of significance for a set of paired comparison tests is higher than that for each single test. To maintain the overall level of significance chosen, each test must work at a lower level of significance. This is achieved by dividing by the total number of possible comparisons. Thus, for three comparisons, to maintain a level of significance of 5%, each comparison must have a significance level of 5/3%. This same strategy was adopted for the multiple comparisons conducted after a Friedman test, using sign tests (see Section 16.10). For k treatments, the total possible number of comparisons is given by

$$\text{Number of comparisons} = \frac{k(k-1)}{2}$$

In our example with three treatments, the possible number of comparisons is $(3 \times 2)/2 = 3$. Because the Kruskal-Wallis test was significant at the 1% level, the three Wilcoxon-Mann-Whitney rank sums tests should be performed at the 1/3% = 0.003 level of significance. Table G.20 showing critical values of S for the Wilcoxon-Mann-Whitney rank sums test does not present values at the 0.003 level; the nearest level suitable for use is 0.002. It should also be noted that should sample sizes be very small, there may be too few scores in a mere pair of treatments to allow rejection of H_0 at such significance levels. Inspection of Table G.20 indicates that samples need to be a certain size before significance can be attained at the 0.003 level (at least seven scores in one sample and six in the other, for the 0.002 level). If samples are too small, likely differences can only be determined by inspection.

Ties

Should any ranks tie, mean ranks are used. Should two subjects tie for 3rd and 4th place, both are given the rank of 3.5. Should there not be too many ties, the normal procedure is used.

Should there be excessive ties (more than 25% of the scores involved in ties), the formula for H is modified as follows:

$$H = \frac{\frac{12}{N(N+1)} \Sigma \frac{R^2}{n} - 3(N+1)}{1 - \frac{\Sigma T}{N^3 - N}}$$

This is basically the same formula for H as before except that a denominator has been added. All values in this formula are obtained as before except for ΣT. This is obtained as follows:

1. For each set of tied scores, note the number of scores involved in the tie = t.
2. For each t value calculate $t^3 - t = T$.
3. Calculate ΣT.

Assumptions for the Kruskal-Wallis Test

The assumptions for the Kruskal-Wallis test are as follows:

1. This is an independent-samples test, so the data in each treatment must be independent of each other; they must come from different subjects. Do not allow a subject to be tested under more than one treatment.
2. Data within a treatment must be independent; they must come from different subjects. Do not test a subject twice under a given treatment.
3. Subjects must be randomly selected.
4. Data must be on at least an ordinal scale.
5. Although the data may be ranked, the underlying variable must be continuous.
6. The test really tests for differences between the distributions, not just their medians. However, the test is relatively insensitive to the shape or breadth of the distributions; it is almost exclusively a test of differences in the medians.

Like ANOVA, the Kruskal-Wallis test is inherently two-tailed. The Kruskal-Wallis test is not quite as powerful as ANOVA; it would need more subjects to reject H_0.

The Kruskal-Wallis test is an unrelated-samples test. For sensory testing it is generally better to use a related-samples design so that each subject can be tested under each treatment; this prevents us confounding the effect of the experimental treatments and the effects due to differences between the subjects. This would mean that the Kruskal-Wallis test would be used only rarely. However, if all the data in the analysis come from only one subject, an unrelated-samples design is appropriate because the degree of independence between treatments would be the same as that within treatments (see Section 16.20). Thus the Kruskal-Wallis test may be used to analyze a set of data generated by only one judge.

16.15 WORKED EXAMPLE: THE KRUSKAL-WALLIS TEST

In an attempt to find ways for improving quality control for a hamburger restaurant chain, three sets of judges rated odor differences in hamburgers under different-colored lights (white, red, and yellow). Judges were ranked in order of their success at rating odor differences. Those under white light came 1st, 2nd, 4th, and 6th; those under red light came 7th, 8th, and 9th; those under yellow light came 3rd and 5th. Do the results indicate that the three groups of subjects rated odor differences significantly differently?

The design used different groups of subjects under different-colored lights. It was thus an independent-samples design. (A related-samples design would have been better.) We use the Kruskal-Wallis test.

The ranked data can be represented as follows:

	White light	Red light	Yellow light
	1	7	3
	2	8	5
	4	9	
	6		
Rank total, R:	13	24	8
Group size, n:	4	3	2

Total number of subjects, $N = 9$

We now use these values in the formula to calculate H.

$$H = \frac{12}{N(N+1)} \Sigma \frac{R^2}{n} - 3(N+1)$$

$$= \frac{12}{9(9+1)} \left(\frac{13^2}{4} + \frac{24^2}{3} + \frac{8^2}{2} \right) - 3(9+1)$$

$$= \frac{12}{90} (42.25 + 192 + 32) - 30 = \frac{12}{90} (266.25) - 30$$

$$= 35.5 - 30 = 5.5$$

Using Table G.24 for sample sizes 4, 3, and 2, the H value of 5.5 is between the H values in the table of 6.3000 and 5.444. Thus the probability of this result happening on H_0 is between 0.046 and 0.011 (nearer 0.046). This probability is sufficiently low to reject H_0. Thus we conclude that the judges under different lighting conditions had different success at rating odor. By inspection of mean ranks (white: 3.25; red: 8; yellow: 4) it can be seen that the most disparate condition is under red light. Generally, multiple-difference testing, using the Wilcoxon-Mann-Whitney rank sums test, at an adjusted level of significance (see Section 16.14) can be used. Here, with three possible comparisons, the significance level would be 0.046/3 = 0.015. However, with the small samples used here, it can be seen from Table G.20 that there would not be sufficient scores to attain such significance levels; thus inspection of the data is needed to determine which are the treatments that differ.

16.16 THE PAGE TEST: TESTING FOR A RANKED TENDENCY

The *Friedman test* is a repeated-measures test which indicates whether there are any differences at all between treatments. The null hypothesis (H_0) states that there are no differences between treatments, while the alternative hypothesis (H_1) states that some or all of the treatments differ in some unspecified way. Multiple comparisons are then used (sign tests at modified significance levels; see Section 16.10) to determine where these differences occur.

However, in some circumstances, it may well be that the only difference that is expected to occur between the treatments will be a specified ranked trend. With four treatments A, B, C, and D, it may be expected before the experiment is performed that if H_0 is rejected, the scores in treatment A will be higher than those in treatment B, which are higher than C, etc. A ranked trend A $>$ B $>$ C $>$ D is expected. The different treatments could be different degrees of training, where more training is expected to elicit better performance in judges and thus higher scores. The different treatments may be foods with different amounts of sugar added, so that sweetness scores would be expected to be ranked in a given order according to sugar content.

Thus rather than the alternative to H_0 (no differences between treatments) being "some unspecified differences," the alternative to H_0 is "a specified rank order of treatments." It is rather like replacing the multiple comparisons with just one specified comparison, a test to see whether the treatments fall in a specified order. So we test our sample data to determine whether it indicates either no differences between treatments (H_0) or that the treatments are ranked in the population according to the order specified. Because we are only testing for one alternative to H_0, we can alter our probability calculations accordingly (rather like using a one-tailed test); this is done using the *Page test*. Of course, the ranking trend over treatments that we are testing must be known before the experiment. We cannot perform the experiment and then, by inspection of the data, determine what ranking trend to test for; when we do not know what the trends will be beforehand, we must go through the regular Friedman procedure followed by multiple comparisons (or Kramer test).

So the Page test is useful under specific circumstances. Should we suspect that the treatments either have no effect or that their effect is ranked in a given order, we can shorten the procedure by using the Page test. The Page test is a related-samples test; the independent-samples version, the Jonckheere test, is given in Section 16.18.

The test is best understood by considering an example. Consider subjects tested under conditions A, B, C, D, and E. Each subject is ranked for his or her performance under each condition. We have theoretical reasons for expecting that if subjects perform differently under these conditions they will do better in condition A than B, than C, than D, than E (ranked A > B > C > D > E); we thus use a Page test to test whether our data indicate this ranked trend in the population.

Consider the following rankings for three subjects:

		Conditions				
		A	B	C	D	E
	S_1	2	1	4	5	3
Subject	S_2	1	2	3	5	4
	S_3	1	2	3	4	5
Rank total:		$R_1 = 4$	$R_2 = 5$	$R_3 = 10$	$R_4 = 14$	$R_5 = 12$

Number of treatments, $k = 5$ Number of judges, $n = 3$

We test for our expected trend ($A > B > C > D > E$) by using Page's L. This is given by the formula

$$L = R_1 + 2R_2 + 3R_3 + 4R_4 + 5R_5 + \cdots$$

For our data

$$L = 4 + (2 \times 5) + (3 \times 10) + (4 \times 14) + (5 \times 12)$$

$$= 160$$

Should there be a ranked trend, the rank totals from A to E should get steadily greater. Thus the highest rank total will be multiplied by the highest number (5), the next highest rank total by the next highest number (4), etc. The highest rank totals will be multiplied by the highest numbers, giving the highest possible total value of L. Should this ranked trend not be there, the rank totals for each treatment will be roughly equal and will not steadily increase to be multiplied by larger and larger numbers; the overall total, Page's L, will thus not be so large. Note also that we are testing only for the order $A > B > C > D > E$, not the reverse. In the reverse case, L would be low (110). Thus, in this way, the test is one-tailed (only ABCDE), not two-tailed (ABCDE or EDCBA). The greater the ranking trend, the greater Page's L. Should L be greater than a value that can occur by chance (on H_0, no differences between treatments), we can reject H_0 and say that the data are compatible with the ranked order of treatments in the population.

Table G.25 gives critical values of L such that should the calculated value of L be equal to or exceed the value in the tables, H_0 is rejected in favor of the alternative ranked hypothesis. The table gives critical values of L for a given number of treatments (select appropriate column) and a given number of subjects (select appropriate row). The critical values are given in sets of three; the upper value is for the 0.1% level ($p = 0.001$), the middle value is for the 1% level ($p = 0.01$), and the lower value is for the 5% level ($p = 0.05$). In our example (five treatments, three subjects) the critical values are:

For p =	0.001	0.01	0.05
L =	160	155	150

Our calculated value of L (160) equals the critical value at the 0.001 significance level. Thus we can say that the data from our sample indicate that H_0 can be rejected in favor of the specific ranking trend hypothesis, $A > B > C > D > E$; our data are compatible with a ranking of $A > B > C > D > E$ in the population. Note that although the overall ranking trend in the sample was $A > B > C > E > D$, it is still compatible with a trend in the population of $A > B > C > D > E$ (at $p < 0.001$).

The only alternative to H_0 is a specific ranking trend, so no multiple comparisons are needed should L be significant; the trend is known. When ties occur, mean ranks are used as usual.

Assumptions for the Page Test

The assumptions required for the use of the Page test are mainly those required for the Friedman test. They are, as follows:

1. The test is a related-samples test; the scores are matched between treatments. All subjects must be tested under each treatment.
2. Scores within a given treatment are independent; they must come from different subjects.
3. Subjects must be chosen randomly.
4. The data must be at least ordinal.
5. Although the data may be ranked, the underlying variable must be continuous.
6. The order of ranks for the treatments, should H_0 be rejected, must be predicted in advance.

It is worth remembering that the Page test is one-tailed; the only alternative to H_0 is the rank order "A > B > C > D > E," not "A > B > C > D > E or E > D > C > B > A." The Jonckheere test, on the other hand, is two-tailed; both A > B > C > D > E and A < B < C < D < E are alternatives to H_0 (see Section 16.18). If the treatments are ranked in order, it means that the scores increase as you pass from treatment to treatment. It does not mean that this increase is linear, however.

Just as with the Friedman test, the Page test is a two-factor test in which one factor is treatments and the other is subjects. It is also statistically possible that both factors might be treatments and that each score might come from a different subject; the design would then be a two-factor completely randomized design. This would not be a good experimental design for sensory work, however, because the experimental treatments effects would be confounded with the differences between judges per se. A related-samples design is preferable (see Section 16.20).

16.17 WORKED EXAMPLE: THE PAGE TEST

Four judges evaluated the rubbery texture of a set of new, entirely nonnutritive sausages to be used with synthetic nonnutritive bread buns in low-calorie "skinny" hot dogs. Four types of nonnutritive sausage were assessed: A, B, C, and D. It was expected from the ingredients of these sausages that should any differences be perceived, the sausages would be ranked A, B, C, D, from most to least rubbery. Do the data confirm this expectation? The data were as follows, listing each judge's rank ordering from most to least rubbery:

Judge 1:	A	B	C	D
Judge 2:	A	C	B	D
Judge 3:	A	B	C	D
Judge 4:	A	B	D	C

The question is not one of whether or not the sausages differ in rubbery texture. It is a question of whether, if differences do occur, they are ranked in order A > B > C > D. The Page test will thus be used. The data are tabulated as follows:

		Condition				
		A	B	C	D	$k = 4$
Subject	1	1	2	3	4	
	2	1	3	2	4	
	3	1	2	3	4	
	4	1	2	4	3	$n = 4$
Sum of ranks, R:		4	9	12	15	
		R_1	R_2	R_3	R_4	

$$L = R_1 + 2R_2 + 3R_3 + 4R_4$$
$$= 4 + 2(9) + 3(12) + 4(15)$$
$$= 118$$

From Table G.25, ($k = 4$, $n = 4$) the critical values of L are 111 (5%), 114 (1%), 117 (0.1%). Our value exceeds all these values. Thus the data indicate a significant tendency ($p < 0.001$) for the sausages to be ranked in order of decreasing rubbery texture: A, B, C, D. Another way of expressing this is to say that the data in the sample are consistent with a population where the sausages are ranked for rubbery texture: A, B, C, D.

16.18 THE JONCKHEERE TEST: TESTING FOR A RANKED TENDENCY

Relation Between the Jonckheere Test and the Rank Sums Test

The Page test (see Section 16.16) examines data to see whether H_0 is true (there is no difference between treatments) or whether, instead, the data indicate a specific ranked alternative hypothesis (the treatments have a specific ranking effect on the data, the order of ranking being determined before the experiment). The Page test examines data for a specific ranking trend, using a related-samples design; the Jonckheere test does the same, using an unrelated-samples design. Examine the Key to Statistical Tables to see the relationship between these tests.

The *Jonckheere test*, like the Kruskal-Wallis test, is an extension of the Wilcoxon-Mann-Whitney rank sums test. Its similarity to the Wilcoxon-Mann-Whitney rank sums test is apparent from its computation (see Section 16.6). Instead of calculating an S value from only two independent samples ($U - U'$ or $u - u'$), the same procedure is used to calculate U and U′ values (or u and u' values) from each possible pairing of (independent) treatments; this gives J value ($\Sigma U - \Sigma U'$ or $\Sigma u - \Sigma u'$). As with the Wilcoxon-Mann-Whitney rank sums test, the difference between the U or u values is arranged to be positive (take the modulus, indicated by two vertical lines). Thus:

Wilcoxon-Mann-Whitney rank sums test:

$$S = |U - U'| \quad \text{or} \quad |u - u'|$$

Jonckheere test, repeated application of the rank sums technique to various treatment pairs:

$$J = |\Sigma U - \Sigma U'| \quad \text{or} \quad |\Sigma u - \Sigma u'|$$

Just as the Wilcoxon-Mann-Whitney S value can be calculated in several different ways, so can Jonckheere's J. These will all be considered, but only the third method is recommended for use, being quick and easy. The first two methods are included only to demonstrate the test's relationshp to the rank sums and the Mann-Whitney U test.

Alternative 1: Calculation of J, Using the Computational Formulas to Obtain Σ U and Σ U′

This calculation of J from U and U′ values is included in this text to illustrate the theory behind Jonckheere's test. It is not recommended that J values be calculated using this procedure; the third procedure utilizing u and u' values is more satisfactory. U values are calculated for each pair of treatments using the standard formulas for U (see Section 16.4):

$$U = n_1 n_2 + \frac{n_1(n_1 + 1)}{2} - R_1$$

$$U' = n_1 n_2 + \frac{n_2(n_2 + 1)}{2} - R_2$$

The method is best illustrated by an example. Let us imagine that 10 judges are tested under one of three conditions A, B, and C. It is expected that should the treatment have an effect, the effect will be a ranking such that subjects will do best under treatment A and worst under treatment C. Thus the Jonckheere test is used to test where there is no difference between treatments (H_0) or, alternatively, whether the treatments are ranked in the specific order (ABC). Let the 10 subjects be ranked where the subject doing best is given the rank of 1st and the subject doing worst has the rank of 10th. Let the data be as follows:

A	B	C
1	3	6
2	5	9
4	7	10
	8	

Should these data be subjected to each pairwise comparison to calculate U and U′ values, the data will have to be ranked again for each pair comparison. Doing this, the reordered rankings for each of the possible pair comparisons are as follows:

	A	B	A	C	B	C
	1	3	1	4	1	3
	2	5	2	5	2	6
	4	6	3	6	4	7
		7			5	
Rank totals:	$R_1 = 7$	$R_2 = 21$	$R_1 = 6$	$R_2 = 15$	$R_1 = 12$	$R_2 = 16$
Sample size:	$n_1 = 3$	$n_2 = 4$	$n_1 = 3$	$n_2 = 3$	$n_1 = 4$	$n_2 = 3$

A — B

$$U = n_1 n_2 + \frac{n_1(n_1 + 1)}{2} - R_1$$

$$= 12 + 6 - 7 = 11$$

$$U' = n_1 n_2 + \frac{n_2(n_2 + 1)}{2} - R_2$$

$$= 12 + 10 - 21 = 1$$

A — C

$$U = 9 + 6 - 6 = 9$$
$$U' = 9 + 6 - 15 = 0$$

B — C

$$U = 12 + 10 - 12 = 10$$
$$U' = 12 + 6 - 16 = 2$$

Adding the U values, we have

$$\Sigma\,U = 11 + 9 + 10 = 30$$

$$\Sigma\,U' = 1 + 0 + 2 = 3$$

$$J = \Sigma\,U - \Sigma\,U' = 30 - 3 = 27$$

It was seen that for the Mann-Whitney U test (see Section 16.4), the greater the difference between two samples, the more discrepant the two U values. Subtract the two U values to give an S value (as for the Wilcoxon-Mann-Whitney ranks sums test; see Section 16.6) and the more different the two samples, the greater the S value (because U and U′ differ more). A similar logic can be applied to this test. Should the treatments be ranked in the order A > B > C, there will be a tendency for rank totals (R) to be lower in treatment A than in treatment B than in treatment C, allowing for sample sizes. Thus, if the treatments are arranged in pairs, with the treatment with the lower mean rank total (R) on the left (A to the left of B or C; B to the left of C), the U value of the treatment on the left will tend to be higher (subtract a smaller R to obtain a larger U). Similarly, the treatments on the right will tend to have higher R values and lower U′ values. Thus, overall, Σ U will be greater than Σ U′ and J will be large. The greater the differences between the treatments, the greater the discrepancy between U values and hence the greater will be J (Σ U − Σ U′). Should there be no differences between treatments, the discrepancy between U values and hence J will be small. Table G.26 indicates how large J can be by chance; should the calculated J value be equal to or greater than the value in the table, H_0 is rejected and the data are compatible with a ranking in the population of A > B > C.

For given sample sizes, Table G.26 provides critical values of J. If the calculated J value exceeds or is equal to the value in the tables, H_0 is rejected; the data in the sample are compatible with the treatments in the population being ranked in the order suspected. The sample sizes are selected in the left-hand column of the table: n_1 is made equal to the number of observations in the largest sample, n_3 is taken as the number in the smallest sample. In this example:

n_1 = 4, the largest sample, B.
n_3 = 3, the smallest sample, A or C.
The size of the third remaining sample, $n_2 = 3$.

The row of these sample sizes (4, 3, 3) is selected and critical J values inspected, each column representing a level of significance indicated at the top of the column.

It should be noted that critical values are given for one-tailed and for two-tailed tests. For a one-tailed test, the alternative to H_0 is only the ranked tendency A > B > C. This is the case we are considering in this example. We could also consider a two-tailed test, however, where the alternative to H_0 was the ranking "A > B > C or C > B > A." This would be used when there is an expected ordering of treatments but the order of magnitude of scores in the treatments cannot be predicted. For this example, the one-tailed critical values are as follows:

Significance level (one-tailed):	0.1	0.05	0.025	0.01	0.005
Critical values of J:	15	19	23	25	27

Our calculated value, 27, is equal to the highest value. Thus we reject H_0 in favor of the alternative one-tailed ranked hypothesis ($p < 0.005$). Thus our sample of data is compatible with a ranking in the population of A > B > C; those subjects in treatment A tended to come before those in B, before those in C.

Alternative 2: Calculation of J, Using the Method of Counting How Many Scores from One Sample Precede Those from the Second Sample, to Obtain Σ U and Σ U′

Like the first procedure, this calculation is included in this text to illustrate the theory behind Jonckheere's test. It is not recommended that J values be calculated using this procedure; the third procedure utilizing u and u' values is more satisfactory.

This approach uses the second method of obtaining U and U′ values for each possible pairing of the treatments. The treatments are paired and reranked as before. For the A-B pair the number of scores in sample B preceding each score in sample A (U) and the number in sample A preceding each score in Sample B (U′) is calculated as for the Mann-Whitney U test (see Section 16.4). The same is done for the A-C and the B-C pairs:

A	B	A	C	B	C
1	3	1	4	1	3
2	5	2	5	2	6
4	6	3	6	4	7
	7			5	

The number of scores in one sample preceded by those in the second sample are as follows:

0	2	0	3	0	2
0	3	0	3	0	4
1	3	0	3	1	4
	3			1	
U = 1	U′ = 11	U = 0	U′ = 9	U = 2	U′ = 10

For example, take the score 1st in sample A in the A-B pairing. No scores in sample B have a lower rank, so score this one 0. Now take the score in sample B which was ranked 3rd. Two scores in sample A (1st and 2nd) precede it, while the third one, 4th, does not; give this a score of 2. Thus

$$\Sigma U = 1 + 0 + 2 = 3$$

$$\Sigma U' = 11 + 9 + 10 = 30$$

Hence $J = |\Sigma U - \Sigma U'| = |3 - 30| = 27$.

Again, the Σ U values are the same as those calculated by the last procedure (their signs are swapped due to the particular ordering of the samples chosen; this does not matter because the sign is ignored); J is thus the same value. Should there be ties, the usual procedure used with U values is adopted; a tied score is taken as half-preceding the tied score in the other sample.

For samples as small as this, this second procedure is quicker and easier. The next procedure is even simpler.

Alternative 3: Calculation of J, Using the Traditional Rank Sums Method to Obtain Σu and $\Sigma u'$

The last two methods were outlined to demonstrate the theory behind the Jonckheere test. The method outlined here, however, is the one recommended for the calculation of J, because of its simplicity and speed. When the rank sums procedure is used, all treatments are considered together; they are not reranked in their respective pairs.

A	B	C
1	3	6
2	5	9
4	7	10
	8	

The procedure used is merely an extension of the rank sums approach used with the Wilcoxon-Mann-Whitney rank sums test (see Section 16.6). A given sample (say A) is chosen and for each score in this sample, the number of scores to the right (from samples B and C) preceding it, as well as the number preceded by it, are noted. This exercise is performed for samples A and B. It cannot be performed for sample C because there are no samples to its right.

For sample A:

Rank	Number of scores to right preceded by this score	Number of scores to right preceding this score
1	7	0
2	7	0
4	6	1
	total, u = 20	total, u' = 1

For sample B:

Rank	Number of scores to right preceded by this score	Number of scores to right preceding this score
3	3	0
5	3	0
7	2	1
8	2	1
	total, u = 10	total, u' = 2

Thus

$\Sigma u = 20 + 10 = 30$

$\Sigma u' = 1 + 2 = 3$

Remember to be consistent in the use of u (number of scores "preceded by") and u' (number of scores "preceding")

$$J = \Sigma u - \Sigma u' = 30 - 3 = 27$$

Again J is the same value as before. It is worth noting that the Σu and $\Sigma u'$ values are the same as the Σ U and Σ U$'$ values. This is because there were no ties. Should there have been ties, the U and u values would have differed because of the use of half-values in the former case to score them (see Section 16.6).

It can be seen that for the Jonckheere test, this rank sums procedure is simple and quick, because all the treatments do not have to be reranked for a series of paired comparisons.

Approximation to Normal

Should a sample have more than five scores or should there be more than three samples, Table G.26 cannot be used. In this case, an approximation of the J distribution to normal is used. A z score (see Section 4.2) is calculated and the probability of getting a score this large by chance (on H_0) can be calculated, using Table G.1 (see Section 4.3). The probability obtained in Table G.1 is a one-tailed probability, corresponding to the one-tailed Jonckheere test; the probability value is doubled for the two-tailed test.

The z value is given by

$$z = \frac{J - 1}{\text{SD}}$$

where J is the J value obtained from the data. A value of unity is subtracted from it as a correction for continuity. The SD value is given by

$$\text{SD} = \sqrt{\frac{2(N^3 - \Sigma n^3) + 3(N^2 - \Sigma n^2)}{18}}$$

where

N = total number of scores

n = number of scores in a given sample

Ties

Should there be any tied ranks, mean rank values are taken. Should there not be an excessive number of scores involved in ties (no more than 25%), the usual procedure using mean rank values can be used to calculate J. Should there be more ties, the procedure for excessive ties, outlined later, should be used.

Assumptions for the Jonckheere Test

The assumptions for the use of the Jonckheere test are similar to those for the Page test, except that the Jonckheere test is an unrelated-samples test, the Page test is a related-samples test. The assumptions are as follows:

1. The test is an unrelated-samples test; the scores in each treatment must be independent. No subject may be tested under more than one treatment.
2. Scores within a given treatment are independent; they must come from different subjects.
3. Subjects must be chosen randomly.
4. The data must be at least ordinal.
5. Although the data may be ranked, the underlying variable must be continuous.
6. The order of ranks for the treatments, should H_0 be rejected, must be predicted in advance.

It is worth remembering that the Jonckheere test can be one-tailed or two-tailed; in the former case the alternative to H_0 is the single ranking trend A > B > C; in the latter case the alternative is A > B > C or C > B > A. The

Page test, on the other hand, is only one-tailed (see Section 16.16). If the treatments are ranked in order, it means that the scores increase as you pass from one treatment to another. It does not mean that this increase is necessarily linear.

The Jonckheere test is an unrelated-samples test. For sensory testing, it is generally better to use a related-samples design so that each subject can be tested under each treatment; this prevents confounding the treatment differences with intersubject differences per se. This would mean that the Jonckheere test would be used only rarely. However, if all the data in the analysis come from only one subject, an unrelated-samples design is appropriate, because the degree of independence between treatments would be the same as that within treatments (see Section 16.20). Thus the Jonckheere test is appropriate for analysis of data from a single judge; it can be used to analyze the data obtained from a single judge for *R*-index multiple sensory difference testing (see Appendix E).

Procedure for Excessive Ties

Should there be more than 25% of the scores involved in ties, the procedure is altered somewhat. The new procedure is best illustrated using an example.

Imagine that subjects are ranked as being class 1, class 2, class 3, or class 4. However, imagine also that there are four treatments called N, S_1, S_2, and S_3 and that there are 10 subjects under each treatment. There being only four response categories (classes 1, 2, 3, and 4) will mean that there will be many ties. Let us also assume that we have good reason for assuming that if H_0 (no difference between treatments N, S_1, S_2, S_3) is not true, the only alternative is a ranking of treatments: $S_3 > S_2 > S_1 > N$; that is, S_3 will tend to have subjects ranked more toward class 1, S_2 less so, S_3 even less so, and N even less than S_3.

Let the number of subjects in each of the four treatments (N, S_1, S_2, S_3), ranked into each class (1, 2, 3, 4) be represented by the matrix below. For example, under treatment S_3, nine subjects are ranked as class 1 and one is class 2. In treatment N, seven are class 4, two are class 3, and one is class 2.

Response Class Assigned to Subject

		1	2	3	4	Row total:
Treatment	S_3	9	1			$n_1 = 10$
	S_2	6	3	1		$n_2 = 10$
	S_1	6	2	2		$n_3 = 10$
	N		1	2	7	$n_4 = 10$
Column total:		$c_1 = 21$	$c_2 = 7$	$c_3 = 5$	$c_4 = 7$	$N = 40$

c values are column totals = number of subjects in each response class
n values are row totals = number of subjects under each treatment; here all n values are the same (10), but this is not necessary for an independent samples design

N = total number of subjects = $n_1 + n_2 + n_3 + n_4$ or $c_1 + c_2 + c_3 + c_4$ = 40

A Jonckheere J value is obtained as before; the traditional rank sums method is used, being the simplest approach here. Each score in each treatment is compared to all the scores in treatments below to ascertain how many precede it and how many are preceded by it. A J value is calculated and from it a z value calculated, which can be used with Table G.1 to ascertain the chance probability (on H_0) of obtaining a J value so large. Table G.1 gives one-tailed probabilities; double the value to obtain a two-tailed value.

The z value is given by the formula

$$z = \frac{J}{\text{SD}}$$

where SD is a standard-deviation term given by $\sqrt{\text{variance}}$, and the variance is given by a horrifying formula, as follows:

$$\text{Variance} = \frac{2(N^3 - \Sigma n^3 - \Sigma c^3) + 3(N^2 - \Sigma n^2 - \Sigma c^2) + 5N}{18}$$

$$+ \frac{(\Sigma n^3 - 3\Sigma n^2 + 2N)(\Sigma c^3 - 3\Sigma c^2 + 2N)}{9N(N-1)(N-2)}$$

$$+ \frac{(\Sigma n^2 - N)(\Sigma c^2 - N)}{2N(N-1)}$$

For this formula, the following items are required:

N	N^2	N^3	$N-1$	$N-2$
Σn^2	Σn^3			
Σc^2	Σc^3			

We will now proceed with this calculation, first calculating J and then a z value.

We take each score in treatment S_1 and count those scores in treatment N that are preceded by it (higher ranks) and those that precede it (lower ranks). Tied ranks are ignored, neither preceding nor being preceded. Thus in treatment S_1, considering the six scores rated 1, all scores in treatment N (1 + 2 + 7) are preceded by them while none precede them. Thus:

Number preceded: $6 \times (1 + 2 + 7) = 60$
Number preceding: $6 \times 0 = 0$

For the two scores rated 2, the scores in treatment N rated 3 and 4 (2 + 7) are preceded while none precede them. Thus:

Number preceded: $2 \times (2 + 7) = 18$
Number preceding: $2 \times 0 = 0$

For the two scores rated 3, the scores in treatment N rated 4 (7) are preceded and the score rated 2 precedes.

Number preceded: $2 \times 7 = 14$
Number preceding: $2 \times 1 = 2$

Using the same approach for treatment S_2, it can be seen for the six scores rated 1, none precede them while they precede all scores in treatments S_1 and N rated 2, 3, and 4 (2 + 2 + 1 + 2 + 7 = 14); score 6×14 preceded = 84, and so on.

This procedure of comparing scores with those in treatments below is applied through the matrix starting with treatment S_1, then considering S_2, and finally S_3. There is no point considering treatment N, because there are no treatments below it. Note that this procedure is the same as that used in the example without ties; we have merely laid the treatments above each other rather than side by side.

Let us now consider the total analysis of the matrix; u values give the number of scores preceded and u' give the number of scores preceding.

Treatment S_1:

Number of scores below that S_1 values precede:

$$\begin{aligned} u &= 6 \times (1 + 2 + 7) + 2 \times (2 + 7) + 2 \times 7 = \\ &= 6 \times 10 + 2 \times 9 + 2 \times 7 = 92 \end{aligned}$$

Number of scores below that precede S_2 values:

$$u' = 6 \times 0 + 2 \times 0 + 2 \times 1 = 2$$

Treatment S_2:

Number of scores below that S_2 values precede:

$$\begin{aligned} u &= 6 \times (2 + 2 + 1 + 2 + 7) + 3 \times (2 + 2 + 7) + 1 \times 7 \\ &= 6 \times 14 + 3 \times 11 + 1 \times 7 = 124 \end{aligned}$$

Number of scores below that precede S_2 values:

$$\begin{aligned} u' &= 6 \times 0 + 3 \times 6 + 1 \times (6 + 2 + 1) \\ &= 0 + 3 \times 6 + 1 \times 9 = 27 \end{aligned}$$

Treatment S_3:

Number of scores below that S_3 values precede:

$$u = 9 \times (3+1+2+2+1+2+7) \quad + \; 1 \times (1+2+2+7)$$
$$= 9 \times 18 \quad + \; 1 \times 12 = 174$$

Number of scores below that precede S_3 values:

$$u' = 9 \times 0 \quad + 1 \times (6+6) \quad = 12$$

Thus

$$\Sigma u = 92 + 124 + 174 = 390$$

$$\Sigma u' = 2 + 27 + 12 = 41$$

$$J = |\Sigma u - \Sigma u'| = 390 - 41 = 349$$

We now need a z value to test the significance of J. For this we must tackle the horrific formula for the variance. For this we need the following values:

$$N = 40 \qquad N^2 = 1600 \qquad N^3 = 64{,}000$$
$$N - 1 = 39 \qquad N - 2 = 38$$

$$\Sigma c^2 = 21^2 + 7^2 + 5^2 + 7^2 = 441 + 49 + 25 + 49 = 564$$

$$\Sigma c^3 = 21^3 + 7^3 + 5^3 + 7^3 = 9261 + 343 + 125 + 343 = 10{,}072$$

$$\Sigma n^2 = 10^2 + 10^2 + 10^2 + 10^2 = 400$$

$$\Sigma n^3 = 10^3 + 10^3 + 10^3 + 10^3 = 4000$$

We now place these values in the formula for the variance, with a certain amount of care and double checking.

$$\text{Variance} = \frac{2 \times (64{,}000 - 4000 - 10{,}072) + 3(1600 - 400 - 564) + 200}{18}$$

$$+ \frac{(4000 - 3 \times 400 + 80)(10{,}072 - 3 \times 564 + 80)}{9 \times 40 \times 39 \times 38}$$

$$+ \frac{(400 - 40)(564 - 40)}{80 \times 39}$$

$$= \frac{99{,}856 + 1908 + 200}{18} + \frac{2880 \times 8460}{533{,}520} + \frac{360 \times 524}{80 \times 39}$$

$$= 5664.67 + 45.67 + 60.46 = 5770.80$$

$$\text{SD} = \sqrt{\text{variance}} = \sqrt{5770.80} = 75.97$$

Exhausted by this effort, we proceed shakily to the calculation of the z value:

$$z = \frac{J}{\text{SD}} = \frac{349}{75.97} = 4.59$$

Consulting Table G.1, the highest z value available is 4, with an associated one-tailed probability of 0.00003. Thus we reject H_0 in favor of a hypothesis that states that the treatments are ranked in the order $S_3 > S_2 > S_1 > N$; that is, subjects in S_3 tend more to class 1, down to those in N tending more to class 4 ($p < 0.00003$).

Such a Jonckheere analysis is applied when there are many ties. It is particularly useful in the analysis of R-index data obtained by sensory multiple difference testing (see Appendix E). Such testing is used to determine the degree of difference between the products S_1, S_2, and S_3 and a product N; it is expected that S_3 is most different, S_2 less so, and S_1 even less so. Being an unrelated-samples test, it is applicable to the analysis of R-index data from a single subject (see Section 16.20). In fact, the response matrix analyzed in this example is the one given in Appendix E.

16.19 WORKED EXAMPLE: THE JONCKHEERE TEST

A further example using the Jonckheere test will now be examined. If you are brave enough, dear reader, to tackle that frightening variance formula, we will consider an example with extensive ties.

A whisky was rated on a three-point quality scale as the result of an extensive assessment by expert whisky tasters. Tastings were made after 1, 10, and 12 years of maturing. Unfortunately, all tastings were made by different judges, owing to retirement and death of the tasters. However, all the tasters had been trained for at least 10 years to rate whiskies in the same way, with extensive use of standards, practice, and cross-checks. Newer judges had been shown to be compatible with the older judges who had died or retired during the study. Also, the 3-point grading scale was simple and clear-cut. So, despite the undesirable completely randomized nature of the design, a cautious analysis was performed.

After one year, both of two judges graded the whisky as grade 3. After 10 years, the whisky was given grades 1, 2, and 3 by each of three new judges. After 12 years, with three even newer judges, two gave the whisky a grade 1, while one gave it a grade 2. Do the data indicate that the whisky improved significantly over time?

The treatments vary in terms of maturing time; they are ordinal. We examine the data for a ranking trend; do the grades improve over time? The design is an independent-samples design, because each assessment was made by a different judge. The Jonckheere test for a ranked tendency will thus be used.

The number of judgements of the various grades over maturing time are represented in the following matrix:

		Whisky Grade 1	2	3	Number of tasters
Years maturing:	12	2	1		$n_1 = 3$
	10	1	1	1	$n_2 = 3$
	1			2	$n_3 = 2$
Column total:		$c_1 = 3$	$c_2 = 2$	$c_3 = 3$	$N = 8$

If the grade improved with time, the tendency would be for scores to become nearer to grade 1 as the years passed. There would be a ranked trend: 12 years > 10 years > 1 year. The Jonckheere test is used to examine whether this trend is significant or whether there are no differences between the maturing time (H_0). The rank sums technique will be used here.

For 10-year-whisky samples,

The number of younger whisky scores with lower grade, $u = 1 \times 2 + 1 \times 2 = 4$.
The number of younger whisky scores with higher grades, $u' = 0$.

For 12-year-whisky samples,

The number of younger whisky scores with lower grades, $2 \times (1 + 1 + 2) + 1 \times (1 + 2) = 11$.
The number of younger whisky scores with higher grades, $1 \times 1 = 1$.

$$J = |\Sigma u - \Sigma u'| = (4 + 11) - (0 + 1) = 14$$

Next, the significance of J is tested. An approximation to normality is required owing to the relatively high proportion of ties. The value for the appropriate standard z score is given by

$$z = \frac{J}{\text{SD}} = \frac{14}{\text{SD}}$$

where $\text{SD} = \sqrt{\text{variance}}$ and

$$\text{Variance} = \frac{2(N^3 - \Sigma n^3 - \Sigma c^3) + 3(N^2 - \Sigma n^2 - \Sigma c^2) + 5N}{18}$$

$$+ \frac{(\Sigma n^3 - 3\Sigma n^2 + 2N)(\Sigma c^3 - 3\Sigma c^2 + 2N)}{9N(N-1)(N-2)}$$

$$+ \frac{(\Sigma n^2 - N)(\Sigma c^2 - N)}{2N(N-1)}$$

For this formula we need the following values:

$$N = 8 \qquad N^2 = 64 \qquad N^3 = 512$$

$$N - 1 = 7 \qquad N - 2 = 6$$

$$\Sigma n^3 = 3^3 + 3^3 + 2^3 = 27 + 27 + 8 = 62$$

$$\Sigma n^2 = 3^2 + 3^2 + 2^2 = 9 + 9 + 4 = 22$$

$$\Sigma c^3 = 3^3 + 2^3 + 3^3 = 27 + 8 + 27 = 62$$

$$\Sigma c^2 = 3^2 + 2^2 + 3^2 = 9 + 4 + 9 = 22$$

$$\text{Variance} = \frac{2(512 - 62 - 62) + 3(64 - 22 - 22) + 40}{18}$$

$$+ \frac{(62 - 66 + 16)(62 - 66 + 16)}{9 \times 8 \times 7 \times 6}$$

$$+ \frac{(22 - 8)(22 - 8)}{16 \times 7}$$

$$= \frac{776 + 60 + 40}{18} + \frac{12 \times 12}{3024} + \frac{14 \times 14}{16 \times 7}$$

$$= 48.67 + 0.048 + 1.75 = 50.47$$

$$\text{SD} = \sqrt{\text{variance}} = \sqrt{50.47} = 7.10$$

Thus

$$z = \frac{J}{\text{SD}} = \frac{14}{7.10} = 1.97$$

It is reasonable to expect that if the maturing does have an effect on the whisky (reject H_0), the grades would get better (not worse) as time passed. Thus the direction of the change is known; a one-tailed test can be used. From Table G.1, the one-tailed probability on H_0 of getting a z score as high as 1.97 or higher is 0.0244. This is low; thus we reject H_0.

We conclude that there is a significant trend ($p = 0.0244$) whereby the whisky improves over time. More precisely, our judges indicate that the improvement in our samples of whisky indicates a significant improvement over time in the populations of whisky the samples were drawn from.

Of course, it would be far better to have the same judges assess 1- 10-, and 12-year-old whiskies, all in one session. This would avoid difficulties due to different judges possibly using the three-point scale in a different way. On the other hand, the whiskies would not have come from the same batch and so it would not have been possible to assess the maturing of a single brew. Problems such as these are encountered in long-term storage studies. See Appendix F for further discussion.

16.20 FINAL THEORETICAL NOTE REGARDING NONPARAMETRIC TESTS

Nonparametric Versus Parametric Tests for Scaled Data

This section draws together and expands on arguments that have been mentioned intermittently throughout the text (Sections 5.8, 7.11, 8.11, 10.5, Appendix F, etc.). The aim is to put in focus some of the design aspects of sensory experiments and accompanying statistical issues. For further discussion, see "Some assumptions and difficulties with common statistics," *Food Technology*, Vol. 36, 1982, pp. 75-82.

When data obtained from an experiment are only ordinal (ranks) or nominal (categories), nonparametric statistics are appropriate. When the data obtained are at least on an interval scale, parametric statistics are generally appropriate. With sensory measurement, however, there is a problem that occurs quite often; it is the problem of data obtained by scaling (e.g., 9-point category scale, placing a mark on a 10-cm line, etc.). Scaling is a common procedure for measuring flavor intensity or the degree of liking for a product. With numbers obtained from a scaling procedure there are doubts about the applicability of parametric statistics for their analysis. End effects and the fact that the scale may have unequal intervals, psychologically, can distort a distribution of scores from normality; with magnitude estimation, a favored number effect can also destroy assumptions of normality. End effects will further destroy the assumptions of homoscedasticity so often required. Thus parametric statistical analyses should be applied with caution to such data (see Section 7.11 for a discussion on the application of the t test and Section 8.11 for a discussion of the application of ANOVA to scaling data). In fact, although some authors state that ANOVA and t tests are robust to violations of their assumptions, these statistical tests are probably better thought of as approximate analyses. This does not mean that they should not be used; it merely means that as the tests are rather approximate, one should not be too rigid about using the 5% or 1% level as a cutoff for significance. As the assumptions are violated, the critical values of t and F given in the tables become more and more approximate.

Because data from a scaling procedure can break the assumptions required for parametric analysis, some authors prefer to use a nonparametric analysis. Even though the actual scaled values may not be on an interval scale, at least they can be ranked in order. Thus a nonparametric analysis using ranked data may be more appropriate (e.g., Wilcoxon, Mann-Whitney U, Wilcoxon-Mann-Whitney rank sums, Friedman, Kruskal-Wallis, Page, Jonckheere tests, etc.). The disadvantage of a nonparametric analysis is that some information is lost if actual scaled values are reduced to ranks. In fact, should it be difficult to decide between a parametric and nonparametric analysis, both might be used; they should give the same result as long as the 5% and 1% level is not rigidly adhered to.

There is one point of caution, however. Although nonparametric tests do not make assumptions about the data being distributed normally in the population and in this way are distribution free, the tests do make some assumptions. They are essentially tests of whether distributions are different or not. The differences may be due to variation in the shape or width of the distributions or they may be due to different medians (differences in medians are also called differences in central tendency or location). Generally, these nonparametric tests are more sensitive to differences in location, but sometimes shape differences may cause a nonparametric test to reach significance. Thus, such tests assume that there are no differences in the shape of the distribution or that such differences are unimportant. Herein lies the assumption. Should scaling data (say, from a category or straight-line scale) be ranked for nonparametric analysis, it is quite possible that factors that affect the spread of the distributions of scores (variations between judges in "end effects" or "favored numbers" on the scale) could affect the spread of the ranked distributions. This could cause the test to indicate a significant difference even if the central tendencies were the same. However, nonparametric tests vary in how much they are biased by this effect.

In the face of the possibilities of the assumptions for statistical tests being broken, with varying effects, the choice of a statistical test can be problematical. If trends are revealed by ANOVA, it is a good idea to examine the subjects to see whether a majority of them exhibit that trend (binomial test) or whether the trend is caused by one or two extreme cases. A ranked analysis can also be applied. Such an approach may well reveal different aspects of the trends being studied and lead to a better understanding of the data. The analogy of a set of data being like a statue is useful. It should be viewed from several different angles; to look at it from just one angle could give a distorted view.

Related- and Independent-Samples Designs

The nonparametric difference tests can be seen to be divided simply into related-samples (repeated-measures) designs and independent-samples (completely randomized) designs. It is worth reexamining the various ANOVA designs once again in terms of whether they are related- or independent-samples designs.

The parametric one-factor ANOVA is applied to an independent-samples (completely randomized) design when different subjects are assigned to different sets of treatments (along one factor) for testing; different treatments (A, B, C, etc.) have assigned to them quite different subjects (S_1, S_2, S_3, S_4, etc.).

Such a design can be represented schematically as follows:

A	B	C
S_1	S_6	S_8
S_2	S_7	S_9
S_3		S_{10}
S_4		S_{11}
S_5		

Note that there are not necessarily the same number of subjects (hence scores) under each treatment. The design is an independent-samples design; there is the same degree of independence between scores in different treatments A, B, and C (different subjects) as there is for scores within the treatments (different subjects). It is this which makes the design an independent-samples design. This same degree of independence is attained using a different subject for each score in the data. Independent-samples tests considered in this text are the independent-samples t test and the completely randomized ANOVA (parametric) and for ranked data (nonparametric): the Mann-Whitney U test, the Wilcoxon-Mann-Whitney rank sums tests, the Kruskal-Wallis ranked ANOVA, and the Jonckheere test.

However, part of the differences between treatments may be due to the very fact that the subjects tested under each treatment are themselves different. Thus it would be more sensible, where possible, to use a related-samples design, whereby each subject was tested under each treatment. The differences themselves would then be due to the effects of the treatments per se rather than any effects due to differences between subjects. Such a related-samples design can be represented schematically as follows:

A	B	C
S_1	S_1	S_1
S_2	S_2	S_2
S_3	S_3	S_3
S_4	S_4	S_4

Here there must be the same number of scores under each treatment because each subject is tested once under each treatment. The scores are matched between the treatments A, B, and C; thus matching is obtained by having the same subjects tested under each treatment. It is possible that such matching could be obtained by using clones or identical twins, but in sensory measurement such refinements are not used. Matched samples mean that the same subjects are used in each condition, while unmatched or independent samples mean that different subjects are used in each condition. It is easy to tell the difference between the two types of designs. In other disciplines, deciding whether or not samples are matched is not as simple. Are animals from the same litter close enough to be matched? Are seeds from the same plant, plants from the same forest, crops from the same country?

The related-samples t test and the two-factor repeated-measures ANOVA are parametric related-samples tests while the Wilcoxon, Friedman, Kramer, and Page tests are nonparametric versions. All involve testing every subject under each treatment. The degree of independence of the scores between treatments is less (same subjects) than the degree of independence of the scores within treatments (different subjects). It is this difference in the degrees of independence within and between treatments that gives definition to the related-samples design.

When the repeated-measures ANOVA is used it comes under the general heading of a two-factor design. One factor provides the different experimental treatments experienced by each subject, while the second factor provides the subjects who experience each treatment. Each treatment is tested with every subject; each subject is tested under every treatment. In this way, effects due to subjects and treatments can be compared separately. In fact, in more complex designs, more factors and even their interactions can be examined.

In fact, it is not necessary that one of the factors in a two-factor design be "subjects"; both factors could be sets of experimental treatments (this is discussed in Section 11.1). Three wines A, B, and C (one factor) may be tested under the matched conditions of being tasted at low, medium, and high temperatures. In this case a form of matching between sets of scores for A, B, and C is obtained by testing all the wines at given temperatures rather than with given subjects. Different subjects are thus tested in each combination of conditions (each wine at each temperature); each score comes from a different subject. This can be represented schematically as follows:

	A	B	C
High	S_1	S_4	S_7
Medium	S_2	S_5	S_8
Low	S_3	S_6	S_9

This two-factor design is called a completely randomized design to reflect the fact that one of the factors is not "subjects." All the repeated-measures designs

(ANOVA, Friedman, Kramer, Page, etc.) could be utilized in this way, but for sensory measurement, this would not be a good strategy. For instance, should wines be tested at different temperatures, it is far better that each judge tastes each wine at each temperature, so as to allow a more realistic comparison between them. In sensory science, two factor designs are nearly always repeated-measures designs.

We have seen that for a completely randomized (unrelated-samples) design, each score under each treatment comes from a different judge; the degree of independence of the scores between treatments is the same as that within treatments. For a related-samples design where each judge is tested under every treatment, the degree of independence of the scores between treatments (same judges) is less than that of the scores within treatments (different judges). What about the case where all the data came from only one judge? In this case, an independent-samples design would be used because, once again, the degree of independence of scores between treatments (same judge) is no different from the degree of independence of the scores within treatments (same judge). Admittedly, both degrees of independence are less than if different judges were used in each combination of conditions, but what is important is that the relative degree of independence between and within treatments is the same. Thus it is the Wilcoxon-Mann-Whitney rank sums test and the Jonckheere test (both independent-samples tests) which are used to look for trends in the response matrices for single judges, obtained using R-index measures (see Appendix E).

Thus, in summary, a related-samples design requires each subject to be tested under every treatment. The independent-samples design requires different subjects to be tested under each treatment or it may be used when only one judge is used to produce all the data. This is represented by the following diagrams.

Independent Samples, Completely Randomized Design

A	B	C
	Many subjects	
S_1	S_4	S_7
S_2	S_5	S_8
S_3	S_6	S_9
	One subject	
S_1	S_1	S_1
S_1	S_1	S_1
S_1	S_1	S_1

Related Samples, Repeated-Measures Design

A	B	C
	Many subjects	
S_1	S_1	S_1
S_2	S_2	S_2
S_3	S_3	S_3

There is one final point that sometimes causes confusion. The matrices obtained for R-index measurement (see Appendix E) are one-factor unrelated-samples matrices. Hence the Wilcoxon-Mann-Whitney rank sums test and the Jonckheere test are applied to them. Consider the matrix

	S	$S?$	$N?$	N
S_3	9	1		
S_2	6	3	1	
S_1	6	2	2	
N		1	2	7

This matrix does not represent data under treatments from two factors (S_3, S_2, S_1, N and S, $S?$, $N?$, N); it represents data under treatments from only one factor (S_3, S_2, S_1, N). The dimension S, $S?$, $N?$, N merely represents the responses given by a judge under these treatments. Thus the data represent an unrelated-samples design for treatments S_3, S_2, S_1, and N, all scores coming from one subject.

Appendix A

Proof That $\Sigma(X - \bar{X})^2 = \Sigma X^2 - \frac{(\Sigma X)^2}{N}$

This relationship was mentioned first in Section 2.4. It is also the basis for the procedures used in analysis of variance. In the text, the reader is asked to accept this relationship; here it will be demonstrated more formally.

The proof is as follows: Taking the expression $\Sigma (X - \bar{X})^2$, expand it in the manner

$$(a - b)^2 = a^2 + b^2 - 2ab$$

The expression now becomes

$$\Sigma (X^2 + \bar{X}^2 - 2X\bar{X})$$

The sum of several terms taken together is the sum of each of the separate terms. Thus

$$\Sigma (a + b + c) = \Sigma a + \Sigma b + \Sigma c$$

Therefore, the expression becomes

$$\Sigma X^2 + \Sigma \bar{X}^2 - \Sigma 2X\bar{X}$$

The sum of a constant taken N times is N times the constant. Thus Σk (sum of N k-values) $= Nk$. $\bar{X}$ is a constant, so $\Sigma \bar{X}^2 = N\bar{X}^2$. Substituting this into the expression, we get

$$\Sigma X^2 + N\bar{X}^2 - \Sigma 2X\bar{X}$$

The sum of a constant times a variable equals a constant times the sum of the variable. $\Sigma\, kX = k\Sigma\, X$. $2\bar{X}$ is a constant, so $\Sigma\, 2X\bar{X} = 2\bar{X}\Sigma\, X$. Substituting this into the expression, we get

$$\Sigma\, X^2 + N\bar{X}^2 - 2\bar{X}\Sigma\, X$$

By definition, the mean, $\bar{X} = \Sigma\, X/N$. So substituting $\Sigma\, X/N$ for $\bar{X}$, we get

$$\Sigma\, X^2 + N\left(\frac{\Sigma\, X}{N}\right)^2 - 2\left(\frac{\Sigma\, X}{N}\right)\Sigma\, X$$

Canceling N in the second term and rearranging the third term, we get

$$\Sigma\, X^2 + \frac{(\Sigma\, X)^2}{N} - 2\,\frac{(\Sigma\, X)^2}{N}$$

Subtracting the two last terms,

$$\frac{(\Sigma\, X)^2}{N} - \frac{2(\Sigma\, X)^2}{N} = -\frac{(\Sigma\, X)^2}{N}$$

we get

$$\Sigma\, X^2 - \frac{(\Sigma\, X)^2}{N}$$

This is what we set out to prove. We have shown that

$$\Sigma\,(X - \bar{X})^2 = \Sigma\, X^2 - \frac{(\Sigma\, X)^2}{N}$$

This term is the numerator for the variance and standard deviation formulas, S^2 and S. Hence

$$S^2 = \frac{\Sigma\,(X - \bar{X})^2}{N - 1} = \frac{\Sigma\, X^2 - \dfrac{(\Sigma\, X)^2}{N}}{N - 1}$$

$$S = \sqrt{\frac{\Sigma\,(X - \bar{X})^2}{N - 1}} = \sqrt{\frac{\Sigma\, X^2 - \dfrac{(\Sigma\, X)^2}{N}}{N - 1}}$$

Appendix B

Binomial Expansion

In Chapter 5 we deal with the binomial test. This is based on a method of calculating probability values using a mathematical technique called the binomial expansion (see Section 5.1). It is assumed in Chapter 5 that the reader has the mathematical knowledge to be able to perform the binomial expansion. In case this knowledge is absent or forgotten, we will now outline this procedure.

The binomial expansion or distribution is the expansion of $(p + q)^n$. It is a straightforward piece of algebra. We will see here what the values of $(p + q)^n$ are for various values of n.

B.1 HOW TO CALCULATE THE EXPANSION: FIRST PRINCIPLES

We will look at values of $(p + q)^n$ for various values of n. (Incidentally, n is called an *index* or an *exponent* and numbers coming in front of p or q, defining how many units there are, are called *coefficients*. The plural of "index" is "indices.")

$$(p+q)^2 = (p+q)(p+q) = p^2 + pq + qp + q^2$$

$$= p^2 + 2pq + q^2 \qquad \text{where } p^2 \qquad\qquad 2\,pq$$

where in p^2 the 2 is an index or exponent, and in $2\,pq$ the 2 is a coefficient.

$$(p+q)^3 = (p+q)(p+q)^2 = (p+q)(p^2+2pq+q^2)$$
$$= p^3+2p^2q+pq^2+qp^2+2pq^2+q^3$$
$$= p^3+3p^2q+3pq^2+q^3$$

$$(p+q)^4 = (p+q)(p+q)^3 = (p+q)(p^3+3p^2q+3pq^2+q^3)$$
$$= p^4+3p^3q+3p^2q^2+pq^3+qp^3+3p^2q^2+3pq^3+q^4$$
$$= p^4+4p^3q+6p^2q^2+4pq^3+q^4$$

In the same way,

$$(p+q)^5 = (p+q)(p+q)^4 = p^5+5p^4q+10p^3q^2+10p^2q^3+5pq^4+q$$

A Shortcut

Now these calculations are getting a bit tedious. Just imagine the trouble involved in working out $(p+q)^{10}$. So we look for a shortcut. We do this by examining the expansion to see whether there are any trends. To help us do this, we write out the expansions once again.

$$(p+q)^1 = p+q$$
$$(p+q)^2 = p^2+2pq+q^2$$
$$(p+q)^3 = p^3+3p^2q+3pq^2+q^3$$
$$(p+q)^4 = p^4+4p^3q+6p^2q^2+4pq^3+q^4$$
$$(p+q)^5 = p^5+5p^4q+10p^3q^2+10p^2q^3+5pq^4+q^5$$

One trend emerges if we ignore all the coefficients and just consider the indices. Each expansion starts with the highest power of p [for $(p+q)^n$ it starts with p^n] and then each succeeding term has a p value with an index reduced by one, combined with a q value that increases its index by one. So the sequence for $(p+q)^5$ goes

$$p^5, p^4q, p^3q^2, p^2q^3, pq^4, q^5$$

and the expansion $(p+q)^6$ would go

$$p^6, p^5q, p^4q^2, p^3q^3, p^2q^4, pq^5, q^6$$

Note the indices all add up to the same value for each term, that is, 5 for $(p+q)^5$, 6 for $(p+q)^6$, and will add up to 7 for $(p+q)^7$.

Now that we know how the indices are sequenced, the next problem is to work out what the coefficients are. Let's just look at how they are sequenced.

For	$(p+q)^1$	they go	1									1
	$(p+q)^2$		1				2					1
	$(p+q)^3$		1			3		3				1
	$(p+q)^4$		1		4		6		4			1
	$(p+q)^5$		1	5	10				10		5	1

There is a trend, but it is not so easy to see. We go to the work of the French physicist-mathematician Blaise Pascal, an infant prodigy born in 1623. When he was only 16 he published his first book on conic sections, an when he was 19 he invented a calculating machine. He went on to be one of the founders of modern probability theory. He also gave us Pascal's triangle, which is a clever way of calculating the coefficients for the binomial distribution. It goes like this.

For the top line of the triangle, take two 1s:

1 1

For the second line put two more 1s at the end of the line, giving

  1 1
1   1

and then to get a number in the middle, add up the two numbers on the line above, 1 + 1 = 2. So now we have

line 1: 1 1
↘ ↙
line 2: 1 2 1

For the third line, put in the two ones at the ends again, giving

  1 1
 1 2 1
1   1

And fill in the rest of the numbers by adding together the two numbers above:

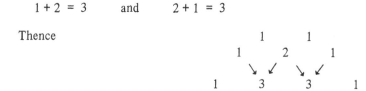

1 + 2 = 3 and 2 + 1 = 3

Thence

  1 1
 1 2 1
 ↘ ↙ ↘ ↙
1 3 3 1

Add the fourth line of the triangle and we get

```
1st line:              1       1
2nd line:          1       2       1
3rd line:      1       3       3       1
                 ↘   ↙   ↘   ↙   ↘   ↙
4th line:  1       4       6       4       1
```

We got the 4 by adding the two numbers above it: 1 + 3. Then 6 was obtained by adding the two numbers above it: 3 + 3. Now we complete the fifth line and get

```
1st line:                  1       1
2nd line:              1       2       1
3rd line:          1       3       3       1
4th line:      1       4       6       4       1
                 ↘   ↙   ↘   ↙   ↘   ↙   ↘   ↙
5th line:  1       5      10      10       5       1
```

If you turn back, you will see that the sequence of coefficients follows Pascal's triangle.

$(p+q)^1$, first line: $p+q$ (i.e., $1p$, $1q$)

$(p+q)^2$, second line: $p^2 + \underline{2}pq + q^2$ (i.e., $1p^2$, $1q^2$)

$(p+q)^3$, third line: $p^3 + \underline{3}p^2q + \underline{3}pq^3 + q^3$

$(p+q)^4$, fourth line: $p^4 + \underline{4}p^3q + \underline{6}p^2q^2 + \underline{4}pq^3 + q^4$

$(p+q)^5$, fifth line: $p^5 + \underline{5}p^4q + \underline{10}p^3q^2 + \underline{10}p^2q^3 + \underline{5}pq^4 + q^5$

B.2 HOW TO CALCULATE THE EXPANSION WITHOUT EFFORT

So now we can write out any expansion of $(p+q)^n$ by first putting down the squence of pq's with their appropriate indices and using "boy wonder" Pascal's triangle to do the rest: for example: $(p+q)^6$. The sequence of indices goes as follows, starting at p^6:

$$p^6,\ p^5q,\ p^4q^2,\ p^3q^3,\ p^2q^4,\ pq^5,\ q^6$$

and we take the sixth line of Pascal's triangle for the coefficients

1 1
1 2 1
1 3 3 1
1 4 6 4 1
1 5 10 10 5 1
1 6 15 20 15 6 1

And now just match them up to get

$$p^6 + 6p^5q + 15p^4q^2 + 20p^3q^3 + 15p^2q^4 + 6pq^5 + q^6$$

We can speed things up a little by looking up these coefficients in Table G.3. For the calculation above we merely look down the n column to 6 and read along the row to get the coefficients: 1, 6, 15, 20, 15, 6, 1. One more example: $(p + q)^{10}$:

$$p^{10}, p^9q, p^8q^2, p^7q^3, p^6q^4, p^5q^5, p^4q^6, p^3q^7, p^2q^8, pq^9, q^{10}$$

Now use Table G.3 for the coefficients. For $n = 10$, we get 1, 10, 45, 120, 210, 252, 210, 120, 45, 10, 1. Combine the two and we get:

$$p^{10} + 10p^9q + 45p^8q^2 + 120p^7q^3 + 210p^6q^4 + 252p^5q^5 + 210p^4q^6 +$$

$$120p^3q^7 + 45p^2q^8 + 10pq^9 + q^{10}$$

Should we want any one term in the expansion, we can use a further relationship which makes use of the fact that the coefficients in the expansion are really combinations. Remember, combinations are discussed in Section 3.8 and are the number of separate samples of a given size n that can be drawn from a population of size N. This is denoted by

$$_NC_n \quad \text{or} \quad \binom{N}{n} = \frac{N!}{(N-n)!n!}$$

Thus the number of separate samples of four items that can be drawn from a sample of six items is given by

$$_6C_4 \quad \text{or} \quad \binom{6}{4} = \frac{6!}{(6-4)!4!} = \frac{6 \times 5 \times 4 \times 3 \times 2 \times 1}{2 \times 1 \times 4 \times 3 \times 2 \times 1} = 15$$

The coefficients for the expansion

$$(p+q)^6 = p^6 + \underline{6}p^5q + \underline{15}p^4q^2 + \underline{20}p^3q^3 + \underline{15}p^2q^4 + \underline{6}pq^5 + q^6$$

are given by the number of combinations obtained for various sample sizes from a population of 6. Hence

For p^0 or q^0: ${}_6C_0 = 1$ by definition for q^6 or p^6

For p^1 or q^1: ${}_6C_1 = \frac{6!}{(6-1)!1!} = 6$ for pq^5 or p^5q

For p^2 or q^2: ${}_6C_2 = \frac{6!}{(6-2)!2!} = 15$ for p^2q^4 or p^4q^2

For p^3 or q^3: ${}_6C_3 = \frac{6!}{(6-3)!3!} = 20$ for p^3q^3 or p^3q^3

For p^4 or q^4: ${}_6C_4 = \frac{6!}{(6-4)!4!} = 15$ for p^4q^2 or p^2q^4

For p^5 or q^5: ${}_6C_5 = \frac{6!}{(6-5)!5!} = 6$ for p^5q or pq^5

For p^6 or q^6: ${}_6C_6 = 1$ by definition for p^6 or q^6

Note that these values are symmetrical. ${}_6C_4$ gives the coefficient for p^2q^4 or p^4q^2. So does ${}_6C_2$. In general, the coefficients for $(p + q)^n$ are given by ${}_nC_0$, ${}_nC_1$, ${}_nC_2$, ${}_nC_3$, . . . , ${}_nC_n$.

For the term p^aq^{n-a} or $p^{n-a}q^a$, the coefficient is ${}_nC_a$. For the term p^4q^2, the coefficient is ${}_6C_4 = 15$. For the term p^2q^4 the coefficient is the same. It is now possible for us to write the coefficient for any term.

What is the coefficient of p^9q^3 in the expansion $(p + q)^{12}$? It is

$$ {}_{12}C_9 = \frac{12!}{(12-9)!9!} = 220 $$

This can be checked in Table G.3.

Appendix C

Proof That $SS_T = SS_B + SS_E$

In the development of the procedure for calculating F values in analysis of variance (see Section 8.3), the relationship between the total sum of squares (SS_T), the between-treatments sum of squares (SS_B), and the error sum of squares (SS_E) was given as

$$SS_T = SS_B + SS_E$$

This relationship was merely given. To avoid distraction from the main argument, it was not derived in the text. It will be derived here.

Imagine a completely randomized ANOVA design with k treatments, A, B, C, etc. There are n scores in each treatment and thus the total number of scores $N = nk$. Each score comes from a different subject; there are thus N subjects participating, each providing one score. The means for treatments A, B, C, etc. are $\bar{X}_A$, $\bar{X}_B$, $\bar{X}_C$, etc. The totals of the X values for these treatments are T_A, T_B, T_C, etc. (where T_k is the general term for T_A, T_B, T_C, etc.). The grand mean of all the scores is denoted by m, the grand total by T. This is shown below:

	A	B	C	. . . k treatments
	X	X	X	
	X	X	X	
	|	|	|	
	X	X	X	
Total	T_A	T_B	T_C	T (grand total)
Mean	$\bar{X}_A$	$\bar{X}_B$	$\bar{X}_C$	m (grand mean)

There are n scores per treatment and k treatments

Total scores, $N = kn$

T_k is a general term for T_A, T_B, T_C, etc.

We now consider the various sum-of-squares values, the numerators in the variance formulas $\Sigma (X - \bar{X})^2 = \Sigma X^2 - (\Sigma X)^2/N$, denoted by SS.

C.1 TOTAL VARIANCE: SUM OF SQUARES

The total variance of all scores around the grand mean can be calculated. The numerator, SS_T, is given by

$$SS_T = \Sigma (X - m)^2 = \Sigma X^2 - \frac{(\Sigma X)^2}{N}$$

where m, the grand mean, is subtracted from every score. But we denote ΣX by T, for convenience. Therefore,

$$SS_T = \Sigma X^2 - \frac{T^2}{N}$$

Because T^2/N occurs so frequently, it is called the correction term and denoted by C.

$$SS_T = \Sigma X^2 - C$$

C.2 BETWEEN-TREATMENTS ESTIMATE OF THE TOTAL VARIANCE: SUM OF SQUARES

Here we examine how the treatment means $\bar{X}_A$, $\bar{X}_B$, $\bar{X}_C$, etc. vary about the grand mean. We estimate the total variance as if it were caused only by between-treatment effects. It is as though every score in a given treatment were equal to the treatment mean, the effects of the treatment being the only thing that determines the score. The numerator for the between-treatments effect, SS_B is given by

$$SS_B = n \Sigma (\bar{X} - m)^2$$

Here we look at the total variance as though all values in a given treatment were equal to the treatment mean. We subtract the grand mean from each of the scores within a treatment (each score has the value of the respective treatment mean).

We now use the relationship $\Sigma (X - \bar{X})^2 = \Sigma X^2 - (\Sigma X)^2/N$. For $\Sigma (\bar{X} - m)^2$, the appropriate N value is k, because there are k treatments and thus a total of k means. Hence

$$SS_B = n \Sigma (\bar{X} - m)^2 = n \left[\Sigma \bar{X}^2 - \frac{(\Sigma \bar{X})^2}{k} \right]$$

Now $\bar{X} = T_k/n$ by definition; for example, $\bar{X}_A = T_A/n$, $\bar{X}_B = T_B/n$, $\bar{X}_C = T_c/n$. Thus

$$\Sigma \bar{X}^2 = \Sigma \left(\frac{T_k}{n}\right)^2$$

But if a set of values each divided by n^2 are added together, the total is the same as the total for all those values then divided by n^2. For example

$$\frac{3}{n^2} + \frac{2}{n^2} = \frac{3+2}{n^2}$$

So not only does

$$\Sigma \bar{X}^2 = \Sigma \left(\frac{T_k}{n}\right)^2 \qquad \text{but it also equals} \qquad \frac{\Sigma T_k^2}{n^2}$$

Therefore,

$$SS_B = n\left[\Sigma \bar{X}^2 - \frac{(\Sigma \bar{X})^2}{k}\right] = n\left[\frac{\Sigma T_k^2}{n^2} - \frac{(\Sigma \bar{X})^2}{k}\right]$$

Now

$$\Sigma \bar{X} = \bar{X}_A + \bar{X}_B + \bar{X}_C = \frac{T_A}{n} + \frac{T_B}{n} + \frac{T_C}{n} = \frac{T_A + T_B + T_C}{n} = \frac{T}{n}$$

Thus

$$(\Sigma \bar{X})^2 = \frac{T^2}{n^2}$$

$$SS_B = n\left[\frac{\Sigma T_k^2}{n^2} - \frac{T^2}{kn^2}\right] = \frac{\Sigma T_k^2}{n} - \frac{T^2}{kn}$$

But $kn = N$, so

$$\frac{T^2}{kn} = \frac{T^2}{N} = C$$

Therefore,

$$SS_B = \frac{\Sigma T_k^2}{n} - C \qquad \text{or} \qquad \frac{T_A^2 + T_B^2 + T_C^2}{n} - C$$

C.3 WITHIN-TREATMENTS ESTIMATE OF THE TOTAL VARIANCE: SUM OF SQUARES

In this case we examine variation within a given treatment: how X values vary about their treatment mean. Here we calculate the total variance as though it were made up only of variation of the scores about their respective treatment means. The numerator for the within-treatments (error) effect, SS_E, in this case is given by

$$SS_E = \Sigma (X_A - \bar{X}_A)^2 + \Sigma (X_B - \bar{X}_B)^2 + \Sigma (X_C - \bar{X}_C) + \cdots$$

Using the relationship $\Sigma (X - \bar{X})^2 = \Sigma X^2 - (\Sigma X)^2/N$, where N here is the number of scores in a given treatment = n, we obtain:

$$SS_E = \Sigma X_A^2 - \frac{(\Sigma X_A)^2}{n} + \Sigma X_B^2 - \frac{(\Sigma X_B)^2}{n} + \Sigma X_C^2 - \frac{(\Sigma X_C)^2}{n} + \cdots$$

But $(\Sigma X_A)^2 = T_A^2$, $(\Sigma X_B)^2 = T_B^2$, etc. Therefore,

$$SS_E = \Sigma X_A^2 - \frac{T_A^2}{n} + \Sigma X_B^2 - \frac{T_B^2}{n} + \Sigma X_C^2 - \frac{T_C^2}{n} + \cdots$$

$$= \Sigma X_A^2 + \Sigma X_B^2 + \Sigma X_C^2 - \left(\frac{T_A^2}{n} + \frac{T_B^2}{n} + \frac{T_C^2}{n}\right)$$

But $\Sigma X_A^2 + \Sigma X_B^2 + \Sigma X_C^2 + \cdots = \Sigma X^2$. Should the sum of the X^2 values for each treatment be added together, it will come to the sum of all the X^2 values. Also,

$$\frac{T_A^2}{n} + \frac{T_B^2}{n} + \frac{T_C^2}{n} + \cdots = \frac{\Sigma T_k^2}{n}$$

Therefore,

$$SS_E = \Sigma X^2 - \frac{\Sigma T_k^2}{n} \quad \text{or} \quad \Sigma X^2 - \frac{T_A^2 + T_B^2 + T_C^2}{n}$$

C.4 FINAL EXPRESSION

Values have now been calculated for SS_T, SS_B, and SS_E.

$$SS_T = \Sigma X^2 - C$$

$$SS_B = \frac{\Sigma T_k^2}{n} - C$$

$$SS_E = \Sigma X^2 - \frac{\Sigma T_k^2}{n}$$

Thus it can be seen that

$$SS_T = SS_B + SS_E$$

$$\Sigma X^2 - C = \frac{\Sigma T_k^2}{n} - C + \Sigma X^2 - \frac{\Sigma T_k^2}{n}$$

or

$$\Sigma X^2 - C = \frac{T_A^2 + T_B^2 + T_C^2}{n} - C + \Sigma X^2 - \frac{T_A^2 + T_B^2 + T_C^2}{n}$$

Appendix D
A Note on Confidence Intervals

The sampling distribution has been discussed earlier. It is a distribution of sample means around the population mean whose standard deviation ($\sigma/\sqrt{N}$) is called the *standard error*. This implies that sample means are not always the same as the population mean; they will vary and their variation is described by the sampling distribution. Thus any single random sample mean will usually not be the same as the population mean, no matter how carefully the sample is taken. It would be useful, then, to have some sort of range value around the sample mean within which the population mean is likely to fall. The width of this range value will depend on how certain we wish to be that the population mean is to be included. Such a range value is given by what is called a *confidence interval.*

Looking at the sampling distribution, we can take a range of means such that 95% of all sample means fall within this range. Then any sample mean has a 95% chance of falling within this range at the center of which, of course, is the population mean. To look at it another way, if we take a sample, the population mean will have a 95% chance of being within that range and only a 5% chance of falling outside. Thus it is sometimes useful when quoting a sample mean to quote such a range. This gives, for any given sample mean, the range or confidence interval within which the population mean will fall for a given probability level.

We know that for large samples the sampling distribution is always normal; this is called the *central limit theorem*. The mean of the sampling distribution will be the same as the population mean, μ, and the standard deviation of this sampling distribution, called the standard error, will be $\sigma/\sqrt{N}$ (σ = standard deviation of population, N = sample size). This is all very well but we generally do not know σ so we have to estimate it from the sample; the estimate is called S. For small samples S will not be near σ, so it is better to use the t distribution for probability calculations, than the normal distribution. In essence, the t distribution becomes a better model for the sampling distribution and t tests are used. For large samples it does not matter which distribution is used; they become pretty well identical, so t tests and z tests give the same answer. This can be seen as S and σ becoming identical as the sample size is increased.

To calculate confidence limits for small samples, then, we turn to the t distribution as our model for the sampling distribution. For large samples either the t or normal distributions can be used; the procedure is the same for both.

If we want a confidence interval within which 95% of the sample means will fall, and we have a large sample, we can go to the normal distribution table (Table G.1). This tells us within how many standard deviations from the population mean 95% of all the sample means must fall; it gives us our confidence interval in terms of z values. Should we want the confidence interval in terms of actual raw scores, we merely multiply by the size of the standard deviation, which in this case is the standard deviation of the sampling distribution, the standard error ($\sigma/\sqrt{N}$, which approximates to $S/\sqrt{N}$ for large samples). Thus the confidence limits will be $z\sigma/\sqrt{N}$ or $zS/\sqrt{N}$ as long as the sample is large. Thus we can say that the population mean (μ) has a 95% chance of falling within $zS/\sqrt{N}$ of our sample mean ($\bar{X}$) for this particular z value.

For small (or large) samples the argument is the same except that we now use t values and the t distribution (Table G.8) rather than z values and the normal distribution. Again, we say that the population mean (μ) has a 95% chance of falling with $tS/\sqrt{N}$ of our sample mean ($\bar{X}$). Of course, we need not pick 95%, we could pick 99% or any level; we merely vary our t value accordingly.

As an example, let us assume that we took a sample of 61 consumers and gave them a questionnaire regarding a new food product. Let us assume that the mean score on this questionnaire ($\bar{X}$) was 100 and the estimated standard deviation for the population (S) = 78.10. The estimated standard error will thus be $78.10/\sqrt{61} = 10$. We wish to know the confidence limits whereby we have a 95% chance of the population mean falling within this range of our sample mean. In other words, a range whereby we have a 5% chance of the population mean not falling within this range. We thus look up t for $df = 60$ ($df = N - 1 = 61 - 1$) at the 5% level for a two-tailed test (the population mean could be above or below sample value, thus use a two-tailed value). The value of t from Table G.8 is 2.000. The confidence limits are thus a distance of $tS/\sqrt{N} = 2.000 \times 78.10/\sqrt{61} = 2.000 \times 10 = 20.00$ from the sample mean.

Thus we know that the upper limit is 100 + 20.00 and the lower limit is 100 - 20.00. Alternatively, we may say that the population mean has a 95% chance of falling in the range 100 ± 20.

For a 99% confidence interval, the appropriate t value is 2.660, so the confidence interval is 100 ± 26.60. This range is larger than the 95% confidence interval, as one would expect if there is more chance of the population mean falling there (less chance, 1%, of it falling outside).

The larger our sample the surer we are of where the population mean will be. Thus the smaller will be the confidence interval at a given probability level. This can be seen mathematically because the larger N, the smaller the standard error $S/\sqrt{N}$ and the smaller the interval $(tS/\sqrt{N})$.

Confidence intervals have not been emphasized in this text because they are rarely used in sensory evaluation or psychophysics; they are more common in survey data. While workers in sensory evaluation and psychophysics describe the variation in their data by standard deviations, a confidence interval also takes into account how the sample size affects the degree of precision of the data. Thus confidence intervals provide more information than the more traditional standard deviations and variances; this could be an argument for making more use of them.

Appendix E
Sensory Multiple-Difference Testing, Using the *R*-Index to Ascertain Degrees of Difference

To measure whether differences in flavor occur between two foods, the commonly used difference tests, paired-comparison, triangle, and duo-trio are suitable (see Section 5.11). Whether any observed differences are significant or merely due to chance is tested statistically using nonparametric binomial tests (see Sections 5.11 and 5.12).

Should differences occur, it is sometimes useful to determine the degree of difference. For instance, should several products (reformulations perhaps) be tested against a standard product, it might be useful to know the degree of difference between these products and the standard (so that the most similar reformulations can be determined). The traditional method of doing this is to use some sort of scaling procedure (9-point category scale, magnitude estimation, a 10-cm line scale) to scale the degree of difference. However, a scaling procedure is best used when the items to be scaled are easily perceived as different. Judges tend not to be skilled at generating scale numbers. The numbers are best thought of only as approximate values; they tend to have a high degree of error (large ± value). This may not be so important when the differences between the items to be scaled are large, but when they are barely perceptible, the error in scaling may well obscure the small differences being scaled. Another approach might seem better.

One alternative procedure is derived from an approach to sensory measurement called *signal detection theory*. This analyzes the perceptual system in terms of a communication system (rather like a set of telephone lines) and provides a set of more sophisticated techniques for sensory measurement. Basically, it provides an improved approach to sensitivity (threshold) measurement but also provides an alternative approach to sensory difference testing. This alternative approach would seem especially useful for avoiding the need for judges to use scaling procedures when the degree of difference between several items is required (multiple-difference testing). The approach requires the measurement of *R*-indices. An *R-index* is a probability value; it is the probability of a given judge distinguishing correctly between two items. The probability of a judge distinguishing between two products is a useful measure of their difference; the greater the degree of difference, the higher the probability of distinguishing between them.

We will develop the theory behind the use of *R*-indices to measure degrees of difference by first considering its use as a measure of difference between two products using a rating or categorization procedure. We will then apply this method to measuring degrees of difference between several products and will finally approach the latter task using an alternative procedure: ranking.

E.1 DEGREE OF DIFFERENCE BETWEEN TWO PRODUCTS USING RATING

Let us assume that a judge is required to distinguish by flavor between two products *S* and *N*. *N* could be a regular product while *S* could be a reformulated version of this product. (In signal detection jargon, *S* is a signal and *N* the noise. The task is to distinguish any flavor change signal, due to reformulation, from the background noise of the flavor of the regularly formulated product.) We give the judge a given number of *S* samples and *N* samples in random order and require her to say whether each sample is *S* or *N*. She could make these judgments based on a session of practice at distinguishing between the two; she could make the judgments on the basis of "same as *N*," "different from *N*" with a standard *N*-sample provided. She could even make the judgments in terms of "same as *S*." Should there be a predictable specific difference in flavor between *S* and *N*, this could also be used as the basis for the judgment. For example, should *S* have added sugar, the judge could be required to distinguish the samples in terms of "sweet" vs. "not sweet," or "sweeter" vs. "less sweet."

Thus the judge is required to distinguish which of the randomly presented samples are *S* and which are *N*. She is also required to say whether or not she is sure of this judgment. Thus each sample can be responded to as "definitely *S*" (*S*), "perhaps *S* but not sure" (*S*?), "definitely *N*" (*N*), "perhaps *N* but not sure" (*N*?). This procedure is called the rating procedure but is not nearly as complex as the regular scaling procedures (e.g., generating numbers on a 9-point category scale, magnitude estimation, etc.). To state whether one is sure or not of one's categorization is a simple task. Differences in the level of sureness of the judgment have little effect on the index obtained.

How many samples of S and N should the judge taste? This is a matter of convenience for the specific test being performed. We will choose 10 samples of each, merely for the sake of simple mathematics.

When presented with the random selection of 10 S and 10 N samples, let us assume that the judge rates six of the S samples as S, two as S?, and two as N?. Let us also assume that she rates seven of the N samples as N, two as N?, and one as S?. Her performance can be summarized by the following matrix:

		Judge's Rating				
		S	S?	N?	N	
Sample	S	6	2	2		Total, n_S = 10
Presented	N		1	2	7	Total, n_N = 10

The question now becomes: Can we estimate from this performance matrix the probability of the judge distinguishing between S and N? The answer is yes! The estimation can be made in the following manner.

If there were 10 S and 10 N samples, how many possible paired comparisons could we get? With each S sample being compared to each of the 10 N samples (10 pair comparisons) and there being 10 S samples for such comparison, the total number of possible pair comparisons is 100 ($n_S \times n_N = 10 \times 10$). Can we predict from the matrix how many of these 100 paired comparisons would be correct? How many times would the S or N samples be identified correctly? This percentage is a good estimate of the probability of correctly distinguishing between the two. (73% correct paired comparisons would mean that there is an estimated 73% chance, 0.73, of distinguishing between the two.) We can now go on to predict the number of correct pair comparisons.

Let us consider the 6 S samples identified as "definitely S" (S). When paired with any of the N samples identified as N, N? or S? (7N, 2 N?, 1S?), they would be correctly identified as S samples. Even the N sample rated as S? would not be chosen as S because faced with a choice between a sample with a flavor rated S and one rated S?, the judge should sensibly choose the one rated S as the S sample. So this gives $6 \times (1 + 2 + 7) = 60$ correct identifications, so far. In the same way, the 2 S samples rated S? would be identified correctly in pair comparison with the 2 N samples rated N? or the 7 N samples rated N. This gives a further $2 \times (2 + 7) = 18$ correct identifications. However, when these 2 S samples are compared to the N sample rated S?, the judge would not know which to choose as the S sample, because they were both rated as exactly the same (S?); so these two comparisons are scored as "don't know." The 2 S samples rated N? will be identified correctly when compared to the 7 N samples rated N (score a further $2 \times 7 = 14$ correct), while comparison with the 2 N samples rated N? would leave the judge undecided (score another $2 \times 2 = 4$

don't knows). Thus the predicted final tally of paired comparisons is 92 (= 60 + 18 + 14) correct identifications of S and 6 (= 2 + 4) "don't knows." Incidentally, the two S samples rated N? would be identified incorrectly (as N samples) when compared with the N sample rated S? (score 2 × 1 incorrect responses so that the total tally comes, as it should, to 100:92 correct + 6 "don't knows" + 2 incorrect).

The test, however, is *forced choice*. Judges are not allowed a "don't know" response. Thus it is assumed for this calculation that when the judge is undecided, she guesses correctly by chance half of the time. Thus 6/2 = 3 of the "don't knows" will be correct. The final tally of correct scores is thus 92 + 3 = 95.

Thus the judge will correctly distinguish S (or N) on an estimated 95 out of 100 paired comparisons; the estimated probability of distinguishing S (or N) is thus 95% or 0.95. This estimated probability is called the R-index and is a useful measure of the degree of difference between S and N for this judge.

In the example, there were 10 replications given for both S and N samples; this number was chosen simply to make the illustrative example simple. Any convenient number of replicate samples may be chosen, bearing in mind that the larger the number chosen, the more representative they are as a sample of the food under consideration. In general, the R-index computation may be summarized by the formula below. For a given response matrix for two products:

		Judge's Response				
		S	$S?$	$N?$	N	
Samples	S	a	b	c	d	Total = $(a + b + c + d) = n_S$
Presented	N	e	f	g	h	Total = $(e + f + g + h) = n_N$

$$R = \frac{a(f+g+h) + b(g+h) + ch + \tfrac{1}{2}(ae + bf + cg + dh)}{n_S n_N}$$

This gives a fractional R. Often a percentage is conceptually easier (multiply by 100).

Of course, there is no reason why there need be only four rating points (S, S?, N?, N). More (not less) can be used to obtain greater resolution (e.g., S, S?, S??, N??, N?, N). There must be an even number, however; a middle "don't know" category often becomes the refuge of those who lack the confidence to commit themselves. Judges should be forced to choose, so as to stretch them to the best of their ability. For most purposes, four categories are usually satisfactory and carry the great bulk of the necessary information.

This procedure for measuring a degree of difference between two foods (probability of distinguishing between the two) may require more time and samples than the simpler paired-comparison, duo-trio, and triangle tests. This can be a distinct disadvantage. However, it does allow a more powerful parametric statistical analysis. Also, if judges are not being used as a representative sample of the population, but merely as measuring instruments, for a sample of food portions (replicate tastings), a large number of judges is not required. What is required is sensitive and accurate judges with a large sample of food tastings (replications), so as to sample the food as thoroughly as possible. Here, the *R*-index with its replicate tastings is well suited.

Given a series of choices in paired-comparisons tests between *S* and *N*, it is expected that half the choices would be correct by chance. Thus the chance level for the *R*-index is 50%. An *R*-index of 100% indicates perfect distinguishing ability. Intermediate values indicate less than perfect distinguishing ability.

Should you wish to inspect the matrix to determine whether the judge showed a significant distinguishing ability (did the judge tend to score *S* samples more as *S* and the *N* samples more as *N*; was her *R*-index significantly greater than the chance 50% level), the Wilcoxon-Mann-Whitney rank sums test (see Section 16.6) is suitable. In fact, a suitable analysis of this very matrix is given. It produces a Wilcoxon-Mann-Whitney *S* value of 90, which is highly significant. Thus the *R*-index of 95% is significantly greater than the chance 50% value; this is not surprising for such a large *R*-index value. Another approach to determining whether the *R*-index is greater than chance is to arrange a control experiment whereby the *S* and *N* samples are exactly the same. This experimentally obtained *R*-index thus gives the a posteriori chance value of *R*. Should the value obtained in the main experiment exceed this control value, the judge would be showing evidence of distinguishing between *S* and *N* at more than chance levels. Whether this is true for a whole panel of judges can be determined by comparing the means of the two sets of *R*-indices (main experiment and control) using a related-samples *t* test (see Section 7.4).

E.2 DEGREES OF DIFFERENCE BETWEEN SEVERAL PRODUCTS USING RATING

The *R*-index procedure may have some advantages for sensory difference testing of two foods but, in general, the more traditional tests will be quite adequate and have the advantage of brevity. Where the *R*-index procedure really shows some advantage is in the realm of multiple-difference testing, where several items are to be compared to determine their respective degrees of difference. This would generally involve difference testing between all pairs of items and scaling the degree of difference between them. For instance, following our example, instead of merely distinguishing between two samples *S* and *N* the degree of difference might be required between *N* and several *S* samples: S_1, S_2, and S_3. *N* might be a regular product and S_1, S_2, and S_3 three reformulations which are being tested to determine the degree of flavor difference between them and the regular product.

Such data could be used to determine which reformulations should be used for further research. Instead of following the usual procedure of difference testing between these products, followed by a scaling procedure to determine the degree of difference, the shorter R-index method could be used.

For multiple-difference testing, we would present 10 samples (we choose 10 for simplicity) of each product (40 samples total) in random order and require the judge to rate them as before. We could represent her results in a matrix as before, except this time the matrix would have four rows (for S_1, S_2, S_3, and N) rather than just two (S and N). Let us assume that the matrix is as follows:

		Judge's Rating			
		S	$S?$	$N?$	N
	S_3	9	1		
Stimulus	S_2	6	3	1	
Presented	S_1	6	2	2	
	N		1	2	7

R-index values can be calculated for the probability of distinguishing S_1 from N, S_2 from N, and S_3 from N, in the same way as before. (In fact, comparisons can also be made between S_1, S_2, and S_3, if required.) For S_1 (6, 2, 2) and N (1, 2, 7) the R-index is 95% (these are the same values as in the two-sample example). For S_2 and N, $R = 96.5$, and for S_3 and N, $R = 99.5\%$.

Thus the judge distinguishes between the reformulations and the regular product very well; the reformulations certainly need some more work! We know that S_3 is the most different from N and S_1 the least different; we also know the degree of difference between them. To test whether these differences in the degree of difference between the S samples and the N sample is significant over a whole panel of judges, a repeated-measures ANOVA (see Chapter 10) with multiple comparisons (see Chapter 9) can be applied to the R-index scores from all the judges, the three S samples S_1, S_2, S_3 providing the three treatments. This would determine whether S_3 is significantly more different from N than S_2 and S_1 and whether S_1 is significantly less different than S_3 or S_2, etc.

Another statistical analysis is possible. It may be that the treatments S_1, S_2, S_3, and N fall along an ordinal continuum. S_1, S_2, S_3, and N may be numerically ordered in some way that enables us to predict beforehand a ranked order for the differences, should they occur. For example, the samples may have had different amounts of sugar added or different processing times or storage times. In this case, the *R*-indices for each reformulation could be given a single overall rank order for each judge. Any tendency for a ranked order of difference (e.g., $S_3 > S_2 > S_1$) to occur over the whole panel of judges could be tested using the Page test (see Section 16.16). Whether any differences actually occur at all, rather than whether they occur in ranked order, could be tested using the Friedman two-factor ranked ANOVA (see Section 16.14) followed by sign tests (see Section 5.8) for multiple comparison. A Kramer two-factor ranked ANOVA (see Section 16.12) is also an alternative here.

There is a further question. For a single matrix from one given judge (not a whole panel of judges), was there a tendency for the degree of difference to be ordered $S_3 > S_2 > S_1$? Again this can only be tested statistically, should there be a prior reason for expecting them to be ordered in this way before the experiment. For instance, the reformulations could be ordered $S_3 > S_2 > S_1$ in terms of processing time or flavor additive content. To test for such a trend from a single judge, the Jonckheere test (see Section 16.18) is suitable. In fact, the example given for the Jonckheere test in this book is exactly the same as our *R*-index example; it yields a Jonckheere J value of 349, which is highly significant. The test actually tests whether the given sample of judge's behavior indicates a ranking trend in her whole population of behavior. An alternative to testing for a ranking trend is merely to test differences between S_1, S_2, and S_3 using the Kruskal-Wallis one-factor ranked ANOVA (see Section 16.14) followed by Wilcoxon-Mann-Whitney rank sums tests (see Section 16.6) for multiple comparisons. This merely tests for any differences between S_1, S_2, and S_3 when rejecting H_0, not a specific ordering; the procedure is more equivalent to the standard ANOVA. These tests are for a single judge, remember.

This *R*-index approach to multiple-difference testing has been described using a rating or categorization response (*S*, *S?*, *N?*, *N*) from the judge for a random presentation of samples. However, it can also be used when the judge is required to rank the samples to be compared. Such a procedure would involve the presentation of one sample each of S_1, S_2, S_3, and N to the judge, while asking her to rank them along some continuum such as degree of off-flavor, rancid odor, or any single dimension; here let us choose degree of sweet taste. First place is given to the sample that is sweetest and fourth place to the sample least sweet. The four samples are presented to the judge (in random order) and ranked accordingly. Then there are replicate rankings and a response matrix can be constructed, just as before, giving the number of occasions that a given product was given a particular rank. Such a response matrix can be constructed for a given judge from the following responses (for mathematical convenience we consider ten rankings):

Ranked responses:

1	2	3	4
S_3	S_2	S_1	N
S_3	S_1	S_2	N
S_3	N	S_2	S_1
S_3	S_2	S_1	N
S_3	S_1	S_2	N
S_3	S_2	N	S_1
S_3	S_2	S_1	N
S_3	S_2	S_1	N
S_3	S_1	N	S_2
S_3	S_2	N	S_1

Response matrix:

		Judges's Total Ranked Response			
		1st	2nd	3rd	4th
Sample Presented	S_3	10			
	S_2		6	3	1
	S_1		3	4	3
	N		1	3	6

For instance, S_3 comes first all 10 times, so it has a score of 10 in the 1st column of the matrix. S_2 is 2nd six times, 3rd three times, and 4th once; these values (6, 3, 1) are thus inserted in the appropriate columns of the matrix for S_2. The same is done for S_1 and N. The R-indices indicating the degree of difference in terms of the dimension along which the judge was ranking (sweetness) can be calculated in the usual way. They are as follows:

For S_3, R = 100%.
For S_2, R = 82.5%.
For S_1, R = 67.5%.

All the statistical analyses conducted with the previous matrix can be conducted with this ranked matrix. The only difference is that the former matrix uses responses whereby the judge is required to categorize whether stimuli are S or N (with an added sureness response) while the latter uses ranking responses. In both cases a graded response is required (S, S?, N?, N versus 1st, 2nd, 3rd, 4th).

It is worth noting a behavioral point. In the ranking procedure, judges are forced to spread their responses over the matrix. Each set of four samples to be ranked must have a response placed in each one of the response categories (1st, 2nd, 3rd, 4th). In the rating procedure, this forced spreading of responses across the matrix is not required; in theory, all samples could be rated identically. This tendency for ranked responses to be spread across the response matrix causes there to be fewer ties (due to identical ratings); ties only contribute by half to the R-index. This results in a tendency to obtain slightly higher R-index values.

When rating is used, it is important to use an even number of response categories (4: S, S?, N?, N or 6: S, S?, S??, N??, N?, N) to force the judge to distinguish between S and N samples; a middle "don't know" category would allow unsure judges not to try to distinguish. With ranking, choices are forced anyway; the number of categories equals the number of samples to be ranked and thus may well be an uneven number. How many samples can be ranked? For visual inspection, a judge can rank many samples. For taste or smell judgments, three, four, or sometimes five samples are possible, depending on the circumstances.

The R-index procedure provides a useful, simple, and economical approach to multiple-difference testing, susceptible to parametric statistical analysis. R-indices have the advantage of being calculated by more than one behavioral technique: rating and ranking.

Appendix F

Do We Have the Wrong Tools?

F.1 RELATED VERSUS INDEPENDENT SAMPLES DESIGNS AND STORAGE STUDIES

The statistical tests we have studied have been designed for certain purposes and these are often at odds with the aims of sensory analysis. For instance, for difference tests we can either use a related-samples (repeated-measures) design, where each judge tastes food samples under each treatment, or an independent-samples (completely randomized) design, where each treatment is tested by different sets of judges. Either the judges are all the same over the treatments or they are all different. You cannot have a little of each; you cannot substitute new judges to replace those who drop out. However, in a storage study where a panel tastes food samples over a period of a year or so, judges are very likely to drop out; it would be convenient to have a statistical analysis which enabled this to be done, provided that the psychological problems inherent in such an approach could be solved.

F.2 REJECTING H_0 AND QUALITY ASSURANCE

Statistical tests are set up to reject a null hypothesis. If H_0 is rejected, we know that there is a difference. If H_0 is not rejected, it does not mean that there is no difference, it merely means that H_0 has not been rejected. H_0 may be true or it may be false, with there being insufficient data to reject it; we have no way of knowing which is the case. Showing that H_0 is true is something that the regular statistical tests are not set up to do; they are only set up to show that H_0 is false. However, in quality assurance work, where the intention is to show that reformulated or differently treated food samples are no different from the regular food samples, the aim is completely opposite. The aim is to show that H_0 is true (demonstrate no difference). Before examining this problem further, it is worth recapitulating a little.

Should H_0 be true and we reject it, we commit a Type I error; we have said that a difference (or correlation) existed when it did not. The probability of doing this is called α, the level of significance. We choose a level of significance appropriate to our needs, although traditionally levels of 0.1%, 1%, and 5% are common. The lower the level of significance, the lower the probability of our finding a difference that is not there. The lower the level of significance, the harder it is to reject H_0 when it is true.

Should H_0 be false and we fail to reject it, we commit a Type II error. We miss rejecting a false H_0; we say that there is no difference (or correlation) where there is one. If the probability of doing this is β, the probability of rejecting H_0 when it is false is $1 - \beta$. Rejecting a false H_0 is desirable, so the larger $1 - \beta$ (hence the smaller β), the better. $1 - \beta$ is called the *power* of a test. The more powerful a test, the better it is at rejecting a false H_0.

As far as basic research is concerned, we formulate our problems in such a way that should we show differences in our data (reject H_0), we get the effect we are trying to demonstrate. But what if we publish our results and we are wrong? What if H_0 was not false? Obviously, we do not want this to happen. We do not want to say the effect was there when it was not; we do not want to commit a Type I error. We can minimize this chance by using a low level of significance (small α). The smaller the level of significance, the less our chance of thinking we have found an effect that is not there (rejecting a true H_0).

As far as quality assurance is concerned, however, the position is rather different. Typically, a quality assurance section is asked to test whether a reformulated product, or a product processed in a different way, is the same as the regular product. In this case, should the product be the same, it will be put on the market. But what if it was not the same? The consumers could reject it and the company would lose a great deal of money. Obviously, a quality assurance department wants to avoid saying that the products are the same when they are different; they want to avoid failing to reject a false H_0, committing a Type II error. Thus the interest here is in minimizing β, the probability of committing a Type II error. The aim is to maximize $1 - \beta$, the power of the test. Thus the more powerful the test, the better it is for quality assurance.

For a given population when H_0 is not true, the chances of rejecting H_0 are greater:

1. For greater differences between population means, $\mu_1 - \mu_2$.

2. For larger sample sizes; the sample size can be chosen by the experimenter, although limits are set by resources available.
3. The greater the power of the test (power = $1 - \beta$). Probability levels can be adjusted by changing α.
4. For tests that utilize more information (e.g., ratio data rather than categories).

Unfortunately, the differences that a quality assurance section may be interested in are sometimes very small (small $\mu_1 - \mu_2$). In fact, they may be so small that some of the available tests (chi-square, binomial for sensory difference testing) may have insufficient power to reject a false H_0 unless very large sample sizes ($\cong$ 500–1000) are used. If such tests are not able to reject false null hypotheses, effectively they will indicate that slightly different samples are the same, so defeating the purpose of quality assurance.

As a strategy, here, everything needs to be done to try and reject H_0, so that if no differences are found, the result can be believed. Thus the experiment should be designed so as to allow as powerful a statistical test as possible. Let's use the more powerful ANOVA even if we doubt its validity. As large a sample as possible needs to be chosen. The level of significance could be raised (say 10%, 20%, or 30%) because deducing a difference exists when it does not (Type I error) is not such a terrible thing in this case. In fact, we break all our cautious rules so as to try as hard as possible to reject H_0. However, all these strategies are merely measures to try to get our statistical tests to cope with analyses for which they were not designed. What would be far better would be to have tests especially designed for demonstrating that H_0 is true. Although statistical tests provide the right tools for basic psychophysical research, they are not ideally suited for some of the tasks encountered in sensory analysis. Perhaps the statisticians need to invent a few more tools for us sensory scientists.

Appendix G

Statistical Tables

Table G.1 Areas Under the Standard Normal Curve[a]

z	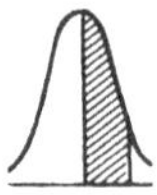 0 z	 0 z	z	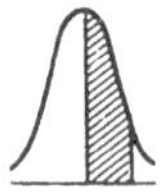 0 z	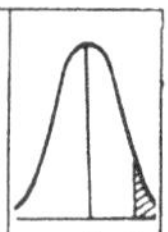 0 z	z	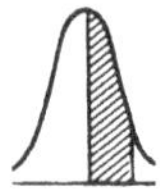 0 z	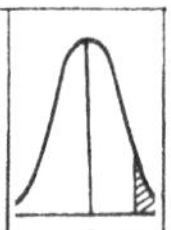 0 z
0.00	.0000	.5000	0.55	.2088	.2912	1.10	.3643	.1357
0.01	.0040	.4960	0.56	.2123	.2877	1.11	.3665	.1335
0.02	.0080	.4920	0.57	.2157	.2843	1.12	.3686	.1314
0.03	.0120	.4880	0.58	.2190	.2810	1.13	.3708	.1292
0.04	.0160	.4840	0.59	.2224	.2776	1.14	.3729	.1271
0.05	.0199	.4801	0.60	.2257	.2743	1.15	.3749	.1251
0.06	.0239	.4761	0.61	.2291	.2709	1.16	.3770	.1230
0.07	.0279	.4721	0.62	.2324	.2676	1.17	.3790	.1210
0.08	.0319	.4681	0.63	.2357	.2643	1.18	.3810	.1190
0.09	.0359	.4641	0.64	.2389	.2611	1.19	.3830	.1170
0.10	.0398	.4602	0.65	.2422	.2578	1.20	.3849	.1151
0.11	.0438	.4562	0.66	.2454	.2546	1.21	.3869	.1131
0.12	.0478	.4522	0.67	.2486	.2514	1.22	.3888	.1112
0.13	.0517	.4483	0.68	.2517	.2483	1.23	.3907	.1093
0.14	.0557	.4443	0.69	.2549	.2451	1.24	.3925	.1075
0.15	.0596	.4404	0.70	.2580	.2420	1.25	.3944	.1056
0.16	.0636	.4364	0.71	.2611	.2389	1.26	.3962	.1038
0.17	.0675	.4325	0.72	.2642	.2358	1.27	.3980	.1020
0.18	.0714	.4286	0.73	.2673	.2327	1.28	.3997	.1003
0.19	.0753	.4247	0.74	.2704	.2296	1.29	.4015	.0985
0.20	.0793	.4207	0.75	.2734	.2266	1.30	.4032	.0968
0.21	.0832	.4168	0.76	.2764	.2236	1.31	.4049	.0951
0.22	.0871	.4129	0.77	.2794	.2206	1.32	.4066	.0934
0.23	.0910	.4090	0.78	.2823	.2177	1.33	.4082	.0918
0.24	.0948	.4052	0.79	.2852	.2148	1.34	.4099	.0901
0.25	.0987	.4013	0.80	.2881	.2119	1.35	.4115	.0885
0.26	.1026	.3974	0.81	.2910	.2090	1.36	.4131	.0869
0.27	.1064	.3936	0.82	.2939	.2061	1.37	.4147	.0853
0.28	.1103	.3897	0.83	.2967	.2033	1.38	.4162	.0838
0.29	.1141	.3859	0.84	.2995	.2005	1.39	.4177	.0823
0.30	.1179	.3821	0.85	.3023	.1977	1.40	.4192	.0808
0.31	.1217	.3783	0.86	.3051	.1949	1.41	.4207	.0793
0.32	.1255	.3745	0.87	.3078	.1922	1.42	.4222	.0778
0.33	.1293	.3707	0.88	.3106	.1894	1.43	.4236	.0764
0.34	.1331	.3669	0.89	.3133	.1867	1.44	.4251	.0749
0.35	.1368	.3632	0.90	.3159	.1841	1.45	.4265	.0735
0.36	.1406	.3594	0.91	.3186	.1814	1.46	.4279	.0721
0.37	.1443	.3557	0.92	.3212	.1788	1.47	.4292	.0708
0.38	.1480	.3520	0.93	.3238	.1762	1.48	.4306	.0694
0.39	.1517	.3483	0.94	.3264	.1736	1.49	.4319	.0681
0.40	.1554	.3446	0.95	.3289	.1711	1.50	.4332	.0668
0.41	.1591	.3409	0.96	.3315	.1685	1.51	.4345	.0655
0.42	.1628	.3372	0.97	.3340	.1660	1.52	.4357	.0643
0.43	.1664	.3336	0.98	.3365	.1635	1.53	.4370	.0630
0.44	.1700	.3300	0.99	.3389	.1611	1.54	.4382	.0618
0.45	.1736	.3264	1.00	.3413	.1587	1.55	.4394	.0606
0.46	.1772	.3228	1.01	.3438	.1562	1.56	.4406	.0594
0.47	.1808	.3192	1.02	.3461	.1539	1.57	.4418	.0582
0.48	.1844	.3156	1.03	.3485	.1515	1.58	.4429	.0571
0.49	.1879	.3121	1.04	.3508	.1492	1.59	.4441	.0559
0.50	.1915	.3085	1.05	.3531	.1469	1.60	.4452	.0548
0.51	.1950	.3050	1.06	.3554	.1446	1.61	.4463	.0537
0.52	.1985	.3015	1.07	.3577	.1423	1.62	.4474	.0526
0.53	.2019	.2981	1.08	.3599	.1401	1.63	.4484	.0516
0.54	.2054	.2946	1.09	.3621	.1379	1.64	.4495	.0505

[a]In the second and third columns, probability associated with z is one-tailed; for two-tailed probability double value in column. For example, $z = 1.96$ is associated with $p = 0.025$ (one-tailed), $p = 0.05$ (two-tailed).

Table G.1 *(continued)*

z	0 z	0 z	z	0 z	0 z	z	0 z	0 z
1.65	.4505	.0495	2.22	.4868	.0132	2.79	.4974	.0026
1.66	.4515	.0485	2.23	.4871	.0129	2.80	.4974	.0026
1.67	.4525	.0475	2.24	.4875	.0125	2.81	.4975	.0025
1.68	.4535	.0465	2.25	.4878	.0122	2.82	.4976	.0024
1.69	.4545	.0455	2.26	.4881	.0119	2.83	.4977	.0023
1.70	.4554	.0446	2.27	.4884	.0116	2.84	.4977	.0023
1.71	.4564	.0436	2.28	.4887	.0113	2.85	.4978	.0022
1.72	.4573	.0427	2.29	.4890	.0110	2.86	.4979	.0021
1.73	.4582	.0418	2.30	.4893	.0107	2.87	.4979	.0021
1.74	.4591	.0409	2.31	.4896	.0104	2.88	.4980	.0020
1.75	.4599	.0401	2.32	.4898	.0102	2.89	.4981	.0019
1.76	.4608	.0392	2.33	.4901	.0099	2.90	.4981	.0019
1.77	.4616	.0384	2.34	.4904	.0096	2.91	.4982	.0018
1.78	.4625	.0375	2.35	.4906	.0094	2.92	.4982	.0018
1.79	.4633	.0367	2.36	.4909	.0091	2.93	.4983	.0017
1.80	.4641	.0359	2.37	.4911	.0089	2.94	.4984	.0016
1.81	.4649	.0351	2.38	.4913	.0087	2.95	.4984	.0016
1.82	.4656	.0344	2.39	.4916	.0084	2.96	.4985	.0015
1.83	.4664	.0336	2.40	.4918	.0082	2.97	.4985	.0015
1.84	.4671	.0329	2.41	.4920	.0080	2.98	.4986	.0014
1.85	.4678	.0322	2.42	.4922	.0078	2.99	.4986	.0014
1.86	.4686	.0314	2.43	.4925	.0075	3.00	.4987	.0013
1.87	.4693	.0307	2.44	.4927	.0073	3.01	.4987	.0013
1.88	.4699	.0301	2.45	.4929	.0071	3.02	.4987	.0013
1.89	.4706	.0294	2.46	.4931	.0069	3.03	.4988	.0012
1.90	.4713	.0287	2.47	.4932	.0068	3.04	.4988	.0012
1.91	.4719	.0281	2.48	.4934	.0066	3.05	.4989	.0011
1.92	.4726	.0274	2.49	.4936	.0064	3.06	.4989	.0011
1.93	.4732	.0268	2.50	.4938	.0062	3.07	.4989	.0011
1.94	.4738	.0262	2.51	.4940	.0060	3.08	.4990	.0010
1.95	.4744	.0256	2.52	.4941	.0059	3.09	.4990	.0010
1.96	.4750	.0250	2.53	.4943	.0057	3.10	.4990	.0010
1.97	.4756	.0244	2.54	.4945	.0055	3.11	.4991	.0009
1.98	.4761	.0239	2.55	.4946	.0054	3.12	.4991	.0009
1.99	.4767	.0233	2.56	.4948	.0052	3.13	.4991	.0009
2.00	.4772	.0228	2.57	.4949	.0051	3.14	.4992	.0008
2.01	.4778	.0222	2.58	.4951	.0049	3.15	.4992	.0008
2.02	.4783	.0217	2.59	.4952	.0048	3.16	.4992	.0008
2.03	.4788	.0212	2.60	.4953	.0047	3.17	.4992	.0008
2.04	.4793	.0207	2.61	.4955	.0045	3.18	.4993	.0007
2.05	.4798	.0202	2.62	.4956	.0044	3.19	.4993	.0007
2.06	.4803	.0197	2.63	.4957	.0043	3.20	.4993	.0007
2.07	.4808	.0192	2.64	.4959	.0041	3.21	.4993	.0007
2.08	.4812	.0188	2.65	.4960	.0040	3.22	.4994	.0006
2.09	.4817	.0183	2.66	.4961	.0039	3.23	.4994	.0006
2.10	.4821	.0179	2.67	.4962	.0038	3.24	.4994	.0006
2.11	.4826	.0174	2.68	.4963	.0037	3.25	.4994	.0006
2.12	.4830	.0170	2.69	.4964	.0036	3.30	.4995	.0005
2.13	.4834	.0166	2.70	.4965	.0035	3.35	.4996	.0004
2.14	.4838	.0162	2.71	.4966	.0034	3.40	.4997	.0003
2.15	.4842	.0158	2.72	.4967	.0033	3.45	.4997	.0003
2.16	.4846	.0154	2.73	.4968	.0032	3.50	.4998	.0002
2.17	.4850	.0150	2.74	.4969	.0031	3.60	.4998	.0002
2.18	.4854	.0146	2.75	.4970	.0030	3.70	.4999	.0001
2.19	.4857	.0143	2.76	.4971	.0029	3.80	.4999	.0001
2.20	.4861	.0139	2.77	.4972	.0028	3.90	.49995	.00005
2.21	.4864	.0136	2.78	.4973	.0027	4.00	.49997	.00003

Source: R. P. Runyon and A. Haber, *Fundamentals of Behavioral Statistics*, 1967, Addison-Wesley, Reading, Mass. Reprinted with permission.

Table G.2 Factorials

N	$N!$
0	1
1	1
2	2
3	6
4	24
5	120
6	720
7	5040
8	40320
9	362880
10	3628800
11	39916800
12	479001600
13	6227020800
14	87178291200
15	1307674368000
16	20922789888000
17	355687428096000
18	6402373705728000
19	121645100408832000
20	2432902008176640000

Source: Sidney Siegel, *Nonparametric Statistics for the Behavioral Sciences*, 1956, McGraw-Hill, New York. Reprinted with permission.

Table G.3 Binomial Coefficients (Pascal's Triangle)

n											
0	1										
1	1	1									
2	1	2	1								
3	1	3	3	1							
4	1	4	6	4	1						
5	1	5	10	10	5	1					
6	1	6	15	20	15	6	1				
7	1	7	21	35	35	21	7	1			
8	1	8	28	56	70	56	28	8	1		
9	1	9	36	84	126	126	84	36	9	1	
10	1	10	45	120	210	252	210	120	45	10	1
11	1	11	55	165	330	462	462	330	165	55	11
12	1	12	66	220	495	792	924	792	495	220	66
13	1	13	78	286	715	1287	1716	1716	1287	715	286
14	1	14	91	364	1001	2002	3003	3432	3003	2002	1001
15	1	15	105	455	1365	3003	5005	6435	6435	5005	3003
16	1	16	120	560	1820	4368	8008	11440	12870	11440	8008
17	1	17	136	680	2380	6188	12376	19448	24310	24310	19448
18	1	18	153	816	3060	8568	18564	31824	43758	48620	43758
19	1	19	171	969	3876	11628	27132	50388	75582	92378	92378
20	1	20	190	1140	4845	15504	38760	77520	125970	167960	184756

Source: Sidney Siegel, *Nonparametric Statistics for the Behavioral Sciences*, 1956, McGraw-Hill, New York. Reprinted with permission.

Table G.4.a Probability of X or More Correct Judgments in n Trials (one-tailed, p = ½)[a]

n \ X	0	1	2	3	4	5	6	7	8	9	10	11	12	13	14	15	16	17	18	19	20	21	22	23	24	25	26	27	28	29	30	31	32	33	34	35	36
5		969	812	500	188	031																															
6		984	891	656	344	109	016																														
7		992	938	773	500	227	062	008																													
8		996	965	855	637	363	145	035	004																												
9		998	980	910	746	500	254	090	020	002																											
10		999	989	945	828	623	377	172	055	011	001																										
11			994	967	887	726	500	274	113	033	006																										
12			997	981	927	806	613	387	194	073	019	003																									
13			998	989	954	867	709	500	291	133	046	011	002																								
14			999	994	971	910	788	605	395	212	090	029	006	001																							
15				996	982	941	849	696	500	304	151	059	018	004																							
16				998	989	962	895	773	598	402	227	105	038	011	002																						
17				999	994	975	928	834	685	500	315	166	072	025	006	001																					
18				999	996	985	952	881	760	593	407	240	119	048	015	004	001																				
19					998	990	968	916	820	676	500	324	180	084	032	010	002																				
20					999	994	979	942	868	748	588	412	252	132	058	021	006	001																			
21					999	996	987	961	905	808	668	500	332	192	095	039	013	004	001																		
22						998	992	974	933	857	738	584	416	262	143	067	026	008	002																		
23						999	995	983	953	895	798	661	500	339	202	105	047	017	005	001																	
24						999	997	989	968	924	846	729	581	419	271	154	076	032	011	003	001																
25							998	993	978	946	885	788	655	500	345	212	015	054	022	007	002																
26							999	995	986	962	916	837	721	577	423	279	163	084	038	014	005	001															
27							999	997	990	974	939	876	779	649	500	351	221	124	061	026	010	003	001														
28								998	994	982	956	908	828	714	575	425	286	172	092	044	018	006	002														
29								999	996	988	969	932	868	771	644	500	356	229	132	068	031	012	004	001													
30								999	997	992	979	951	900	819	708	572	428	292	181	100	049	021	008	003	001												
31									998	995	985	965	925	859	763	640	500	360	237	141	075	035	015	005	002												
32									999	997	990	975	945	892	811	702	570	430	298	189	108	055	025	010	004	001											
33									999	998	993	982	960	919	852	757	636	500	364	243	148	081	040	018	007	002											
34										999	995	988	971	939	885	804	696	568	432	304	196	115	061	029	012	005	002										
35										999	997	992	980	955	912	845	750	632	500	368	250	155	088	045	020	008	003	001									
36										999	998	994	986	967	934	879	797	691	566	434	309	203	121	066	033	014	006	002	001								
37											999	996	990	976	951	906	838	744	629	500	371	256	162	094	049	024	010	004	001								
38											999	997	993	983	964	928	872	791	686	564	436	314	209	128	072	036	017	007	003	001							
39											999	998	995	988	973	946	900	832	739	625	500	375	261	168	100	054	027	012	005	002	001						
40												999	997	992	981	960	923	866	785	682	563	437	318	215	134	077	040	019	008	003	001						
41												999	998	994	986	970	941	894	826	734	622	500	378	266	174	106	059	030	014	006	002	001					
42													999	996	990	978	956	918	860	780	678	561	439	322	220	140	082	044	022	010	004	001					
43													999	997	993	984	967	937	889	820	729	620	500	380	271	180	111	063	033	016	007	003	001				
44													999	998	995	989	976	952	913	854	774	674	560	440	326	226	146	087	048	024	011	005	002	001			
45														999	997	992	982	964	932	884	814	724	617	500	383	276	186	116	068	036	018	008	003	001			
46														999	998	994	987	973	948	908	849	769	671	558	442	329	231	151	092	052	027	013	006	002	001		
47														999	998	996	991	980	961	928	879	809	720	615	500	385	280	191	121	072	039	020	009	004	002	001	
48															999	997	993	985	970	944	903	844	765	667	557	443	333	235	156	097	056	030	015	007	003	001	
49															999	998	995	989	978	957	924	874	804	716	612	500	388	284	196	126	076	043	022	012	005	002	001
50																999	997	992	984	968	941	899	839	760	664	556	444	336	240	161	101	059	032	016	008	003	001

[a]Initial decimal point has been omitted.

Source: E. B. Roessler et al., *Journal of Food Science*, 1978, *43*, 940-947.

Table G.4.b Probability of X or More Agreeing Judgments in n Trials (two-tailed, p = ½)[a]

n \ X	3	4	5	6	7	8	9	10	11	12	13	14	15	16	17	18	19	20	21	22	23	24	25	26	27	28	29	30	31	32	33	34	35	36	37
5	625	312	062																																
6		688	219	031																															
7			453	125	016																														
8			727	289	070	008																													
9				508	180	039	004																												
10				754	344	109	021	002																											
11					549	227	065	011	001																										
12					774	388	146	039	006																										
13						581	267	092	022	003																									
14						791	424	180	057	013	002																								
15							607	302	118	035	007	001																							
16							804	454	210	077	021	004	001																						
17								629	332	143	049	013	002																						
18								815	481	238	096	031	008	001																					
19									648	359	167	064	019	004	001																				
20									824	503	263	115	041	012	003																				
21										664	383	189	078	027	007	001																			
22										832	523	286	134	052	017	004	001																		
23											678	405	210	093	035	011	003																		
24											839	541	307	152	064	023	007	002																	
25												690	424	230	108	043	015	004	001																
26												845	557	327	169	076	029	009	002	001															
27													701	442	248	122	052	019	006	002															
28													851	572	345	185	087	036	013	004	001														
29														711	458	265	136	061	024	008	002	001													
30														856	585	362	200	099	043	016	005	001													
31															720	473	281	150	071	030	011	003	001												
32															860	597	377	215	100	050	020	007	002	001											
33																728	487	296	163	080	035	014	005	001											
34																864	608	392	229	121	058	024	009	003	001										
35																	736	500	310	175	090	041	017	006	002										
36																	868	681	405	243	132	065	029	011	004	001									
37																		743	511	324	188	099	047	020	008	003	001								
38																		871	627	418	256	143	073	034	014	005	002								
39																			749	522	337	200	108	053	024	009	003	001							
40																			875	636	430	268	154	081	038	017	006	002	001						
41																				755	533	349	211	117	060	028	012	004	001						
42																				878	644	441	280	164	088	044	020	008	003	001					
43																					761	542	360	222	126	066	032	014	005	002	001				
44																					880	652	451	291	174	096	049	023	010	004	001				
45																						766	551	371	233	135	072	036	016	007	002	001			
46																						883	659	461	302	184	104	054	026	011	005	002	001		
47																							771	560	382	243	144	079	040	019	008	003	001		
48																							885	665	471	312	193	111	059	029	013	006	002	001	
49																								775	568	392	253	152	085	044	021	009	004	001	
50																								888	672	480	322	203	119	065	033	015	007	003	001

Table G.4.c Probability of X or More Correct Judgments in n Trials (one-tailed, p = 1/3)[a]

n \ X	0	1	2	3	4	5	6	7	8	9	10	11	12	13	14	15	16	17	18	19	20	21	22	23	24	25	26	27	28
5		868	539	210	045	004																							
6		912	649	320	100	018	001																						
7		941	737	429	173	045	007																						
8		961	805	532	259	088	020	003																					
9		974	857	623	350	145	042	008	001																				
10		983	896	701	441	213	077	020	003																				
11		988	925	766	527	289	122	039	009	001																			
12		992	946	819	607	368	178	066	019	004	001																		
13		995	961	861	678	448	241	104	035	009	002																		
14		997	973	895	739	524	310	149	058	017	004	001																	
15		998	981	921	791	596	382	203	088	031	008	002																	
16		998	986	941	834	661	453	263	126	050	016	004	001																
17		999	990	956	870	719	522	326	172	075	027	008	002																
18		999	993	967	898	769	588	391	223	108	043	014	004	001															
19			995	976	921	812	648	457	279	146	065	024	007	002															
20			997	982	940	848	703	521	339	191	092	038	013	004	001														
21			998	987	954	879	751	581	399	240	125	056	021	007	002														
22			998	991	965	904	794	638	460	293	163	079	033	012	003	001													
23			999	993	974	924	831	690	519	349	206	107	048	019	006	002													
24			999	995	980	941	862	737	576	406	254	140	068	028	010	003	001												
25			999	996	985	954	888	778	630	462	304	178	092	042	016	006	002												
26				997	989	964	910	815	679	518	357	220	121	058	025	009	003	001											
27				998	992	972	928	847	725	572	411	266	154	079	036	014	005	002											
28				999	994	979	943	874	765	623	464	314	191	104	050	022	008	003	001										
29				999	996	984	955	897	801	670	517	364	232	133	068	031	013	005	001										
30				999	997	988	965	916	833	714	568	415	276	166	090	043	019	007	002	001									
31					998	991	972	932	861	754	617	466	322	203	115	059	027	011	004	001									
32					998	993	978	946	885	789	662	516	370	243	144	078	038	016	006	002	001								
33					999	995	983	957	905	821	705	565	419	285	177	100	051	023	010	004	001								
34					999	996	987	965	922	849	744	612	468	330	213	126	067	033	014	006	002	001							
35					999	997	990	973	937	873	779	656	516	376	252	155	087	044	020	009	003	001							
36						998	992	978	949	895	810	697	562	422	293	187	109	058	028	012	005	002	001						
37						998	994	963	959	913	838	735	607	469	336	223	135	075	038	018	007	003	001						
38						999	996	987	967	928	863	769	650	515	381	261	164	095	051	025	011	004	002	001					
39						999	997	990	973	941	885	800	689	560	425	301	196	118	066	033	016	007	003	001					
40						999	997	992	979	952	903	829	726	603	470	342	231	144	083	044	021	010	004	001					
41							998	994	983	961	920	854	761	644	515	385	268	173	104	057	029	014	006	002	001				
42							999	995	987	968	933	876	791	683	558	428	307	205	127	073	038	019	008	003	001				
43							999	996	990	974	945	895	820	719	600	471	347	239	153	091	050	025	012	005	002	001			
44							999	997	992	980	955	912	845	753	639	514	389	275	182	111	063	033	016	007	003	001			
45							999	998	994	984	963	926	867	783	677	556	430	313	213	135	079	043	022	010	004	002	001		
46								998	995	987	970	938	887	811	713	596	472	352	246	161	098	055	029	014	006	003	001		
47								999	996	990	976	949	904	836	745	635	514	392	282	189	119	070	038	019	009	004	002	001	
48								999	997	992	980	958	919	859	776	672	554	433	318	220	142	086	048	025	012	006	002	001	
49								999	998	994	984	965	932	879	803	706	593	473	356	253	168	105	061	033	017	008	003	001	
50								999	998	995	987	972	943	896	829	739	631	513	395	287	196	126	076	042	022	011	005	002	001

Table G.5.a. Minimum Numbers of Correct Judgments to Establish Significance at Various Probability Levels for Paired-Comparison and Duo-Trio Tests (one-tailed, p = ½)

No. of trials (n)	Probability levels 0.05	0.04	0.03	0.02	0.01	0.005	0.001
7	7	7	7	7	7		
8	7	7	8	8	8	8	
9	8	8	8	8	9	9	
10	9	9	9	9	10	10	10
11	9	9	10	10	10	11	11
12	10	10	10	10	11	11	12
13	10	11	11	11	12	12	13
14	11	11	11	12	12	13	13
15	12	12	12	12	13	13	14
16	12	12	13	13	14	14	15
17	13	13	13	14	14	15	16
18	13	14	14	14	15	15	16
19	14	14	15	15	15	16	17
20	15	15	15	16	16	17	18
21	15	15	16	16	17	17	18
22	16	16	16	17	17	18	19
23	16	17	17	17	18	19	20
24	17	17	18	18	19	19	20
25	18	18	18	19	19	20	21
26	18	18	19	19	20	20	22
27	19	19	19	20	20	21	22
28	19	20	20	20	21	22	23
29	20	20	21	21	22	22	24
30	20	21	21	22	22	23	24
31	21	21	22	22	23	24	25
32	22	22	22	23	24	24	26
33	22	23	23	23	24	25	26
34	23	23	23	24	25	25	27
35	23	24	24	25	25	26	27
36	24	24	25	25	26	27	28
37	24	25	25	26	26	27	29
38	25	25	26	26	27	28	29
39	26	26	26	27	28	28	30
40	26	27	27	27	28	29	30
41	27	27	27	28	29	30	31
42	27	28	28	29	29	30	32
43	28	28	29	29	30	31	32
44	28	29	29	30	31	31	33
45	29	29	30	30	31	32	34
46	30	30	30	31	32	33	34
47	30	30	31	31	32	33	35
48	31	31	31	32	33	34	36
49	31	32	32	33	34	34	36
50	32	32	33	33	34	35	37
60	37	38	38	39	40	41	43
70	43	43	44	45	46	47	49
80	48	49	49	50	51	52	55
90	54	54	55	56	57	58	61
100	59	60	60	61	63	64	66

Source: E. B. Roessler et al., *Journal of Food Science*, 1978, *43*, 940-947. Copyright © by Institute of Food Technologists. Reprinted with permission of author and publisher.

Table G.5.b Minimum Numbers of Agreeing Judgments Necessary to Establish Significance at Various Probability Levels for the Paired-Preference Tests and Difference (two tailed, p = ½)

No. of trials (n)	Probability levels						
	0.05	0.04	0.03	0.02	0.01	0.005	0.001
7	7	7	7	7			
8	8	8	8	8	8		
9	8	8	9	9	9	9	
10	9	9	9	10	10	10	
11	10	10	10	10	11	11	11
12	10	10	11	11	11	12	12
13	11	11	11	12	12	12	13
14	12	12	12	12	13	13	14
15	12	12	13	13	13	14	14
16	13	13	13	14	14	14	15
17	13	14	14	14	15	15	16
18	14	14	15	15	15	16	17
19	15	15	15	15	16	16	17
20	15	16	16	16	17	17	18
21	16	16	16	17	17	18	19
22	17	17	17	17	18	18	19
23	17	17	18	18	19	19	20
24	18	18	18	19	19	20	21
25	18	19	19	19	20	20	21
26	19	19	19	20	20	21	22
27	20	20	20	20	21	22	23
28	20	20	21	21	22	22	23
29	21	21	21	22	22	23	24
30	21	22	22	22	23	24	25
31	22	22	22	23	24	24	25
32	23	23	23	23	24	25	26
33	23	23	24	24	25	25	27
34	24	24	24	25	25	26	27
35	24	25	25	25	26	27	28
36	25	25	25	26	27	27	29
37	25	26	26	26	27	28	29
38	26	26	27	27	28	29	30
39	27	27	27	28	28	29	31
40	27	27	28	28	29	30	31
41	28	28	28	29	30	30	32
42	28	29	29	29	30	31	32
43	29	29	30	30	31	32	33
44	29	30	30	30	31	32	34
45	30	30	31	31	32	33	34
46	31	31	31	32	33	33	35
47	31	31	32	32	33	34	36
48	32	32	32	33	34	35	36
49	32	33	33	34	34	35	37
50	33	33	34	34	35	36	37
60	39	39	39	40	41	42	44
70	44	45	45	46	47	48	50
80	50	50	51	51	52	53	56
90	55	56	56	57	58	59	61
100	61	61	62	63	64	65	67

Table G.5.c Minimum Numbers of Correct Judgments to Establish Significance at Various Probability Levels for the Triangle Tests (one tailed, p = 1/3)

No. of trials (n)	Probability levels 0.05	0.04	0.03	0.02	0.01	0.005	0.001
5	4	5	5	5	5	5	
6	5	5	5	5	6	6	
7	5	6	6	6	6	7	7
8	6	6	6	6	7	7	8
9	6	7	7	7	7	8	8
10	7	7	7	7	8	8	9
11	7	7	8	8	8	9	10
12	8	8	8	8	9	9	10
13	8	8	9	9	9	10	11
14	9	9	9	9	10	10	11
15	9	9	10	10	10	11	12
16	9	10	10	10	11	11	12
17	10	10	10	11	11	12	13
18	10	11	11	11	12	12	13
19	11	11	11	12	12	13	14
20	11	11	12	12	13	13	14
21	12	12	12	13	13	14	15
22	12	12	13	13	14	14	15
23	12	13	13	13	14	15	16
24	13	13	13	14	15	15	16
25	13	14	14	14	15	16	17
26	14	14	14	15	15	16	17
27	14	14	15	15	16	17	18
28	15	15	15	16	16	17	18
29	15	15	16	16	17	17	19
30	15	16	16	16	17	18	19
31	16	16	16	17	18	18	20
32	16	16	17	17	18	19	20
33	17	17	17	18	18	19	21
34	17	17	18	18	19	20	21
35	17	18	18	19	19	20	22
36	18	18	18	19	20	20	22
37	18	18	19	19	20	21	22
38	19	19	19	20	21	21	23
39	19	19	20	20	21	22	23
40	19	20	20	21	21	22	24
41	20	20	20	21	22	23	24
42	20	20	21	21	22	23	25
43	20	21	21	22	23	24	25
44	21	21	22	22	23	24	26
45	21	22	22	23	24	24	26
46	22	22	22	23	24	25	27
47	22	22	23	23	24	25	27
48	22	23	23	24	25	26	27
49	23	23	24	24	25	26	28
50	23	24	24	25	26	26	28
60	27	27	28	29	30	31	33
70	31	31	32	33	34	35	37
80	35	35	36	36	38	39	41
90	38	39	40	40	42	43	45
100	42	43	43	44	45	47	49

Table G.6 Some Comparisons Between Sensory Difference and Preference Tests

Test	Binomial Statistical Analysis	Advantages and disadvantages
	Difference tests[a]	
Paired comparison	One-tailed, if known which sample has greater amount of attribute being judged. p = q = 1/2 (Tables G.4.a and G.5.a) or Two-tailed, if not known which sample has greater amount of attribute being judged. p = q = 1/2 (Tables G.4.b. and G.5.b)	Must define attribute on which the samples differ; this can be difficult. Do not ask whether there is a difference or not; this introduces psychological bias. Need fewer samples (two per trial) than other tests. Traditionally regarded as the most sensitive difference test Reversals unlikely but possible; subjects may indicate by mistake the sample with a lesser degree of the attribute.
Duo trio	One-tailed p = q = 1/2 (Tables G.4.a and G.5.a)	Do not have to define attribute on which the samples differ. Can get reversals; the judge may mistakenly indicate the sample which is different from the standard. More samples need to be tasted (three per trial). Traditionally regarded as not as sensitive as the paired comparison.
Triangle	One-tailed p = 1/3, q = 2/3 (Table G.4.c and G.5.c)	Do not have to define the attribute on which the samples differ. Reversals very unlikely. More samples need to be tasted (three per trial). Traditionally regarded as not as sensitive as the paired comparison.
	Preference test	
Paired comparison	Two-tailed p = q = 1/2 (Table G.4.b and G.5.b)	Preference is not a difficult attribute to define. Other advantages and disadvantages: as for paired-comparison difference test.

[a]It is worth noting that there are many other forms of difference test; these are merely the most commonly used at present.

Table G.7 Critical Values of Chi-Square[a]

	Level of significance for one-tailed test					
	.10	.05	.025	.01	.005	.0005
	Level of significance for two-tailed test					
df	.20	.10	.05	.02	.01	.001
1	1.64	2.71	3.84	5.41	6.64	10.83
2	3.22	4.60	5.99	7.82	9.21	13.82
3	4.64	6.25	7.82	9.84	11.34	16.27
4	5.99	7.78	9.49	11.67	13.28	18.46
5	7.29	9.24	11.07	13.39	15.09	20.52
6	8.56	10.64	12.59	15.03	16.81	22.46
7	9.80	12.02	14.07	16.62	18.48	24.32
8	11.03	13.36	15.51	18.17	20.09	26.12
9	12.24	14.68	16.92	19.68	21.67	27.88
10	13.44	15.99	18.31	21.16	23.21	29.59
11	14.63	17.28	19.68	22.62	24.72	31.26
12	15.81	18.55	21.03	24.05	26.22	32.91
13	16.98	19.81	22.36	25.47	27.69	34.53
14	18.15	21.06	23.68	26.87	29.14	36.12
15	19.31	22.31	25.00	28.26	30.58	37.70
16	20.46	23.54	26.30	29.63	32.00	39.29
17	21.62	24.77	27.59	31.00	33.41	40.75
18	22.76	25.99	28.87	32.35	34.80	42.31
19	23.90	27.20	30.14	33.69	36.19	43.82
20	25.04	28.41	31.41	35.02	37.57	45.32
21	26.17	29.62	32.67	36.34	38.93	46.80
22	27.30	30.81	33.92	37.66	40.29	48.27
23	28.43	32.01	35.17	38.97	41.64	49.73
24	29.55	33.20	36.42	40.27	42.98	51.18
25	30.68	34.38	37.65	41.57	44.31	52.62
26	31.80	35.56	38.88	42.86	45.64	54.05
27	32.91	36.74	40.11	44.14	46.96	55.48
28	34.03	37.92	41.34	45.42	48.28	56.89
29	35.14	39.09	42.69	46.69	49.59	58.30
30	36.25	40.26	43.77	47.96	50.89	59.70
32	38.47	42.59	46.19	50.49	53.49	62.49
34	40.68	44.90	48.60	53.00	56.06	65.25
36	42.88	47.21	51.00	55.49	58.62	67.99
38	45.08	49.51	53.38	57.97	61.16	70.70
40	47.27	51.81	55.76	60.44	63.69	73.40
44	51.64	56.37	60.48	65.34	68.71	78.75
48	55.99	60.91	65.17	70.20	73.68	84.04
52	60.33	65.42	69.83	75.02	78.62	89.27
56	64.66	69.92	74.47	79.82	83.51	94.46
60	68.97	74.40	79.08	84.58	88.38	99.61

[a]The table lists the critical values of chi square for the degrees of freedom shown at the left for tests corresponding to those significance levels heading each column. If the observed value of χ^2_{obs} is *greater than or equal to* the tabled value, reject H_0.

Source: Table IV of Fisher and Yates, *Statistical Tables for Biological, Agricultural and Medical Research*, published by Longman Group Ltd., London (previously published by Oliver and Boyd Ltd., Edinburgh) and by permission of the authors and publishers.

Table G.8 Critical Values of t[a]

	Level of significance for one-tailed test					
	.10	.05	.025	.01	.005	.0005
	Level of significance for two-tailed test					
df	.20	.10	.05	.02	.01	.001
1	3.078	6.314	12.706	31.821	63.657	636.619
2	1.886	2.920	4.303	6.965	9.925	31.598
3	1.638	2.353	3.182	4.541	5.841	12.941
4	1.533	2.132	2.776	3.747	4.604	8.610
5	1.476	2.015	2.571	3.365	4.032	6.859
6	1.440	1.943	2.447	3.143	3.707	5.959
7	1.415	1.895	2.365	2.998	3.499	5.405
8	1.397	1.860	2.306	2.896	3.355	5.041
9	1.383	1.833	2.262	2.821	3.250	4.781
10	1.372	1.812	2.228	2.764	3.169	4.587
11	1.363	1.796	2.201	2.718	3.106	4.437
12	1.356	1.782	2.179	2.681	3.055	4.318
13	1.350	1.771	2.160	2.650	3.012	4.221
14	1.345	1.761	2.145	2.624	2.977	4.140
15	1.341	1.753	2.131	2.602	2.947	4.073
16	1.337	1.746	2.120	2.583	2.921	4.015
17	1.333	1.740	2.110	2.567	2.898	3.965
18	1.330	1.734	2.101	2.552	2.878	3.922
19	1.328	1.729	2.093	2.539	2.861	3.883
20	1.325	1.725	2.086	2.528	2.845	3.850
21	1.323	1.721	2.080	2.518	2.831	3.819
22	1.321	1.717	2.074	2.508	2.819	3.792
23	1.319	1.714	2.069	2.500	2.807	3.767
24	1.318	1.711	2.064	2.492	2.797	3.745
25	1.316	1.708	2.060	2.485	2.787	3.725
26	1.315	1.706	2.056	2.479	2.779	3.707
27	1.314	1.703	2.052	2.473	2.771	3.690
28	1.313	1.701	2.048	2.467	2.763	3.674
29	1.311	1.699	2.045	2.462	2.756	3.659
30	1.310	1.697	2.042	2.457	2.750	3.646
40	1.303	1.684	2.021	2.423	2.704	3.551
60	1.296	1.671	2.000	2.390	2.660	3.460
120	1.289	1.658	1.980	2.358	2.617	3.373
∞	1.282	1.645	1.960	2.326	2.576	3.291

[a]The value listed in the table is the critical value of t for the number of degrees of freedom listed in the left column for a one- or two-tailed test at the significance level indicated at the top of each column. If the observed t is *greater than or equal to* the tables value, reject H_0.

Source: Table III of Fisher and Yates, *Statistical Tables for Biological, Agricultural and Medical Research*, published by Longman Group Ltd., London (previously published by Oliver and Boyd Ltd., Edinburgh) and by permission of the authors and publishers.

Table G.9 $\sqrt{N_1 \cdot N_2(N_1 + N_2 - 2)/(N_1 + N_2)}$

	10	11	12	13	14	15	16	17	18	19
10	9·49	9·98	10·44	10·89	11·33	11·75	12·15	12·55	12·93	13·30
11	9·98	10·49	10·98	11·45	11·90	12·34	12·77	13·18	13·58	13·97
12	10·44	10·98	11·49	11·98	12·45	12·91	13·35	13·78	14·20	14·60
13	10·89	11·45	11·98	12·49	12·98	13·46	13·92	14·36	14·80	15·22
14	11·33	11·90	12·45	12·98	13·49	13·98	14·46	14·92	15·37	15·81
15	11·75	12·34	12·91	13·46	13·98	14·49	14·98	15·46	15·93	16·38
16	12·15	12·77	13·35	13·92	14·46	14·98	15·49	15·98	16·46	16·93
17	12·55	13·18	13·78	14·36	14·92	15·46	15·98	16·49	16·99	17·47
18	12·93	13·58	14·20	14·80	15·37	15·93	16·46	16·99	17·49	17·99
19	13·30	13·97	14·60	15·22	15·81	16·38	16·93	17·47	17·99	18·49
20	13·66	14·35	15·00	15·63	16·23	16·82	17·38	17·93	18·47	18·99
21	14·02	14·72	15·39	16·03	16·65	17·25	17·83	18·39	18·94	19·47
22	14·36	15·08	15·76	16·42	17·06	17·67	18·26	18·84	19·40	19·94
23	14·70	15·43	16·13	16·80	17·45	18·08	18·69	19·27	19·84	20·40
24	15·03	15·78	16·49	17·18	17·84	18·48	19·10	19·70	20·28	20·85
25	15·35	16·12	16·85	17·55	18·22	18·87	19·51	20·12	20·71	21·29
26	15·67	16·45	17·19	17·91	18·60	19·26	19·90	20·53	21·14	21·73
27	15·98	16·77	17·53	18·26	18·96	19·64	20·30	20·93	21·55	22·15
28	16·29	17·09	17·87	18·61	19·32	20·01	20·68	21·33	21·96	22·57
29	16·59	17·41	18·19	18·95	19·68	20·38	21·06	21·72	22·36	22·98
30	16·88	17·72	18·52	19·28	20·02	20·74	21·43	22·10	22·75	23·38
31	17·17	18·02	18·83	19·66	20·36	21·09	21·79	22·47	23·13	23·78
32	17·46	18·32	19·15	19·94	20·70	21·44	22·15	22·84	23·52	24·17
33	17·74	18·61	19·45	20·26	21·03	21·78	22·50	23·21	23·89	24·55
34	18·02	18·90	19·76	20·57	21·37	22·12	22·85	23·57	24·26	24·93
35	18·29	19·19	20·05	20·88	21·68	22·45	23·20	23·92	24·62	25·31
36	18·56	19·47	20·35	21·19	22·00	22·78	23·53	24·27	24·98	25·67
37	18·82	19·75	20·64	21·49	22·31	23·10	23·87	24·61	25·33	26·04
38	19·08	20·02	20·92	21·79	22·62	23·42	24·20	24·95	25·68	26·39
39	19·34	20·29	21·20	22·08	22·92	23·73	24·52	25·28	26·03	26·75
40	19·60	20·56	21·48	22·37	23·22	24·05	24·84	25·62	26·37	27·10
41	19·85	20·82	21·76	22·66	23·52	24·35	25·16	25·94	26·70	27·44
42	20·10	21·08	22·03	22·94	23·81	24·66	25·47	26·26	27·03	27·78
43	20·34	21·34	22·30	23·22	24·10	24·96	25·78	26·58	27·36	28·12
44	20·58	21·60	22·56	23·49	24·39	25·25	26·09	26·90	27·68	28·45
45	20·82	21·85	22·83	23·77	24·67	25·54	26·39	27·21	28·01	28·78
46	21·06	22·10	23·08	24·04	24·95	25·83	26·69	27·52	28·32	29·11
47	21·30	22·34	23·34	24·30	25·23	26·12	26·98	27·82	28·63	29·43
48	21·53	22·59	23·60	24·57	25·50	26·40	27·28	28·12	28·94	29·75
49	21·76	22·83	23·85	24·83	25·77	26·68	27·57	28·42	29·25	30·06
50	21·98	23·06	24·10	25·09	26·04	26·96	27·85	28·72	29·56	30·37

Source: E. G. Chambers, *Statistical Calculation for Beginners*, 1958, Cambridge University Press, Cambridge.

Table G.9 *(continued)*

	20	21	22	23	24	25	26	27	28	29
10	13·66	14·02	14·36	14·70	15·03	15·35	15·67	15·98	16·29	16·59
11	14·35	14·72	15·08	15·43	15·78	16·12	16·45	16·77	17·09	17·41
12	15·00	15·39	15·76	16·13	16·49	16·85	17·19	17·53	17·87	18·19
13	15·63	16·03	16·42	16·80	17·18	17·55	17·91	18·26	18·61	18·95
14	16·23	16·65	17·06	17·45	17·84	18·22	18·60	18·96	19·32	19·68
15	16·82	17·25	17·67	18·08	18·48	18·87	19·26	19·64	20·01	20·38
16	17·38	17·83	18·26	18·69	19·10	19·51	19·90	20·30	20·68	21·06
17	17·93	18·39	18·84	19·27	19·70	20·12	20·53	20·93	21·33	21·72
18	18·47	18·94	19·40	19·84	20·28	20·71	21·14	21·55	21·96	22·36
19	18·99	19·47	19·94	20·40	20·85	21·29	21·73	22·15	22·57	22·98
20	19·49	19·99	20·47	20·94	21·41	21·86	22·30	22·74	23·17	23·59
21	19·99	20·49	20·99	21·47	21·95	22·41	22·86	23·31	23·75	24·18
22	20·47	20·99	21·49	21·99	22·47	22·95	23·41	23·87	24·32	24·76
23	20·94	21·47	21·99	22·49	22·99	23·47	23·95	24·42	24·87	25·32
24	21·41	21·95	22·47	22·99	23·49	23·99	24·48	24·95	25·42	25·88
25	21·86	22·41	22·95	23·47	23·99	24·49	24·99	25·48	25·95	26·42
26	22·30	22·86	23·41	23·95	24·48	24·99	25·50	25·99	26·48	26·96
27	22·74	23·31	23·87	24·42	24·95	25·48	25·99	26·50	26·99	27·48
28	23·17	23·75	24·32	24·87	25·42	25·95	26·48	26·99	27·50	27·99
29	23·59	24·18	24·76	25·32	25·88	26·42	26·96	27·48	27·99	28·50
30	24·00	24·60	25·19	25·77	26·33	26·88	27·43	27·96	28·48	28·99
31	24·41	25·02	25·62	26·20	26·78	27·34	27·89	28·43	28·96	29·48
32	24·81	25·43	26·04	26·63	27·21	27·78	28·34	28·89	29·43	29·96
33	25·20	25·83	26·45	27·05	27·64	28·22	28·79	29·35	29·89	30·43
34	25·59	26·23	26·86	27·47	28·07	28·66	29·23	29·80	30·35	30·90
35	25·97	26·62	27·26	27·88	28·49	29·08	29·67	30·24	30·80	31·36
36	26·35	27·01	27·65	28·28	28·90	29·50	30·10	30·68	31·25	31·81
37	26·72	27·39	28·04	28·68	29·31	29·92	30·52	31·11	31·69	32·26
38	27·09	27·77	28·43	29·07	29·71	30·33	30·94	31·53	32·12	32·70
39	27·45	28·14	28·81	29·46	30·10	30·73	31·35	31·95	32·55	33·13
40	27·81	28·50	29·18	29·85	30·50	31·13	31·76	32·37	32·97	33·56
41	28·16	28·87	29·55	30·22	30·88	31·53	32·16	32·78	33·39	33·99
42	28·51	29·22	29·92	30·60	31·26	31·92	32·56	33·18	33·80	34·40
43	28·86	29·58	30·28	30·97	31·64	32·30	32·95	33·58	34·21	34·82
44	29·20	29·93	30·64	31·33	32·01	32·68	33·34	33·98	34·61	35·23
45	29·53	30·27	30·99	31·69	32·38	33·06	33·72	34·37	35·01	35·63
46	29·87	30·61	31·34	32·05	32·75	33·43	34·10	34·76	35·40	36·03
47	30·20	30·95	31·69	32·41	33·11	33·80	34·47	35·14	35·79	36·43
48	30·52	31·29	32·03	32·76	33·47	34·16	34·85	35·52	36·17	36·82
49	30·85	31·62	32·37	33·10	33·82	34·52	35·21	35·89	36·56	37·21
50	31·17	31·94	32·70	33·44	34·17	34·88	35·58	36·26	36·93	37·59

Table G.9 *(continued)*

	30	31	32	33	34	35	36	37	38	39
10	16·88	17·17	17·46	17·74	18·02	18·29	18·56	18·82	19·08	19·34
11	17·72	18·02	18·32	18·61	18·90	19·19	19·47	19·75	20·02	20·29
12	18·52	18·83	19·15	19·45	19·76	20·05	20·35	20·64	20·92	21·20
13	19·28	19·66	19·94	20·26	20·57	20·88	21·19	21·49	21·79	22·08
14	20·02	20·36	20·70	21·03	21·37	21·68	22·00	22·31	22·62	22·92
15	20·74	21·09	21·44	21·78	22·12	22·45	22·78	23·10	23·42	23·73
16	21·43	21·79	22·15	22·50	22·85	23·20	23·53	23·87	24·20	24·52
17	22·10	22·47	22·84	23·21	23·57	23·92	24·27	24·61	24·95	25·28
18	22·75	23·13	23·52	23·89	24·26	24·62	24·98	25·33	25·68	26·03
19	23·38	23·78	24·17	24·55	24·93	25·31	25·67	26·04	26·39	26·75
20	24·00	24·41	24·81	25·20	25·59	25·97	26·35	26·72	27·09	27·45
21	24·60	25·02	25·43	25·83	26·23	26·62	27·01	27·39	27·77	28·14
22	25·19	25·62	26·04	26·45	26·86	27·26	27·35	28·04	28·43	28·81
23	25·77	26·20	26·63	27·05	27·47	27·88	28·28	28·68	29·07	29·46
24	26·33	26·78	27·21	27·64	28·07	28·49	28·90	29·31	29·71	30·10
25	26·88	27·34	27·78	28·22	28·66	29·08	29·50	29·92	30·33	30·73
26	27·43	27·89	28·34	28·79	29·23	29·67	30·10	30·52	30·94	31·35
27	27·96	28·43	28·89	29·35	29·80	30·24	30·68	31·11	31·53	31·95
28	28·48	28·96	29·43	29·89	30·35	30·80	31·25	31·69	32·12	32·55
29	28·99	29·48	29·96	30·43	30·90	31·36	31·81	32·26	32·70	33·13
30	29·50	29·99	30·48	30·96	31·43	31·90	32·36	32·82	33·26	33·71
31	29·99	30·50	30·99	31·48	31·96	32·44	32·90	33·37	33·82	34·27
32	30·48	30·99	31·50	31·99	32·48	32·96	33·44	33·91	34·37	34·83
33	30·96	31·48	31·99	32·50	32·99	33·48	33·96	34·44	34·91	35·37
34	31·43	31·96	32·48	32·99	33·50	33·99	34·48	34·97	35·44	35·91
35	31·90	32·44	32·96	33·48	33·99	34·50	34·99	35·48	35·97	36·44
36	32·36	32·90	33·44	33·96	34·48	34·99	35·50	35·99	36·48	36·97
37	32·82	33·37	33·91	34·44	34·97	35·48	35·99	36·50	36·99	37·48
38	33·26	33·82	34·37	34·91	35·44	35·97	36·48	36·99	37·50	37·99
39	33·71	34·27	34·83	35·37	35·91	36·44	36·97	37·48	37·99	38·50
40	34·14	34·71	35·28	35·83	36·38	36·91	37·44	37·97	38·48	38·99
41	34·57	35·15	35·72	36·28	36·84	37·38	37·92	38·45	38·97	39·48
42	35·00	35·59	36·16	36·73	37·29	37·84	38·38	38·92	39·45	39·97
43	35·42	36·01	36·60	37·17	37·74	38·29	38·84	39·39	39·92	40·45
44	35·84	36·44	37·03	37·61	38·18	38·74	39·30	39·85	40·39	40·92
45	36·25	36·86	37·45	38·04	38·62	39·19	39·75	40·30	40·85	41·39
46	36·66	37·27	37·87	38·47	39·05	39·63	40·19	40·76	41·31	41·85
47	37·06	37·68	38·29	38·89	39·48	40·06	40·64	41·20	41·76	42·31
48	37·46	38·08	38·70	39·31	39·90	40·49	41·07	41·64	42·21	42·77
49	37·85	38·48	39·11	39·72	40·32	40·92	41·50	42·08	42·65	43·22
50	38·24	83·88	39·51	40·13	40·74	41·34	41·93	42·51	43·09	43·66

Table G.9 *(continued)*

	40	41	42	43	44	45	46	47	48	49
10	19·60	19·85	20·10	20·34	20·58	20·82	21·06	21·30	21·53	21·76
11	20·56	20·82	21·08	21·34	21·60	21·85	22·10	22·34	22·59	22·83
12	21·48	21·76	22·03	22·30	22·56	22·83	23·08	23·34	23·60	23·85
13	22·37	22·66	22·94	23·22	23·49	23·77	24·04	24·30	24·57	24·83
14	23·22	23·52	23·81	24·10	24·39	24·67	24·95	25·23	25·50	25·77
15	24·05	24·35	24·66	24·96	25·25	25·54	25·83	26·12	26·40	26·68
16	24·84	25·16	25·47	25·78	26·09	26·39	26·69	26·98	27·28	27·57
17	25·62	25·94	26·26	26·58	26·90	27·21	27·52	27·82	28·12	28·42
18	26·37	26·70	27·03	27·36	27·68	28·01	28·32	28·63	28·94	29·25
19	27·10	27·44	27·78	28·12	28·45	28·78	29·11	29·43	29·75	30·06
20	27·81	28·16	28·51	28·86	29·20	29·53	29·87	30·20	30·52	30·85
21	28·50	28·87	29·22	29·58	29·93	30·27	30·61	30·95	31·29	31·62
22	29·18	29·55	29·92	30·28	30·64	30·99	31·34	31·69	32·03	32·37
23	29·85	30·22	30·60	30·97	31·33	31·69	32·05	32·41	32·76	33·10
24	30·50	30·88	31·26	31·64	32·01	32·38	32·75	33·11	33·47	33·82
25	31·13	31·53	31·92	32·30	32·68	33·06	33·43	33·80	34·16	34·52
26	31·76	32·16	32·56	32·95	33·34	33·72	34·10	34·47	34·85	35·21
27	32·37	32·78	33·18	33·58	33·98	34·37	34·76	35·14	35·52	35·89
28	32·97	33·39	33·80	34·21	34·61	35·01	35·40	35·79	36·17	36·56
29	33·56	33·99	34·40	34·82	35·23	35·63	36·03	36·43	36·82	37·21
30	34·14	34·57	35·00	35·42	35·84	36·25	36·66	37·06	37·46	37·85
31	34·71	35·15	35·59	36·01	36·44	36·86	37·27	37·68	38·08	38·48
32	35·28	35·72	36·16	36·60	37·03	37·45	37·87	38·29	38·70	39·11
33	35·83	36·28	36·73	37·17	37·61	38·04	38·47	38·89	39·31	39·72
34	36·38	36·84	37·29	37·74	38·18	38·62	39·05	39·46	39·90	40·32
35	36·91	37·38	37·84	38·29	38·74	39·19	39·63	40·06	40·49	40·92
36	37·44	37·92	38·38	38·84	39·30	39·75	40·19	40·64	41·07	41·50
37	37·97	38·45	38·92	39·39	39·85	40·30	40·76	41·20	41·64	42·08
38	38·48	38·97	39·45	39·92	40·39	40·85	41·31	41·76	42·21	42·65
39	38·99	39·48	39·97	40·45	40·92	41·39	41·85	42·31	42·77	43·22
40	39·50	39·99	40·48	40·97	41·45	41·92	42·39	42·86	43·32	43·77
41	39·99	40·50	40·99	41·49	41·97	42·45	42·92	43·40	43·86	44·32
42	40·48	40·99	41·50	41·99	42·49	42·97	43·45	43·92	44·40	44·86
43	40·97	41·49	41·99	42·50	42·99	43·49	43·97	44·45	44·93	45·40
44	41·45	41·97	42·49	42·99	43·50	43·99	44·49	44·97	45·45	45·93
45	41·92	42·45	42·97	43·49	43·99	44·50	44·99	45·49	45·97	46·46
46	42·39	42·92	43·45	43·97	44·49	44·99	45·50	45·99	46·49	46·97
47	42·86	43·40	43·92	44·45	44·97	45·49	45·99	46·50	46·99	47·49
48	43·32	43·86	44·40	44·93	45·45	45·97	46·49	46·99	47·50	47·99
49	43·77	44·32	44·86	45·40	45·93	46·46	46·97	47·49	47·99	48·50
50	44·42	44·78	45·32	45·87	46·40	46·93	47·46	47·97	48·49	48·99

Table G.10 ANOVA Designs

The following is a chart of some common designs used for analysis of variance, together with their names given in the following leading textbooks of behavioral statistics:

Bruning, J. L., and Kintz, B. L., *Computational Handbook of Statistics*, 2nd ed. Scott, Foresman, Glenview, Ill., 1977.
Edwards, A. L., *Experimental Design in Psychological Research*, 4th ed. Holt, Rinehart and Winston, New York, 1972.
Keppel, G., *Design and Analysis: A Researcher's Handbook*. Prentice-Hall, Englewood Cliffs, N.J., 1973.
Kirk, R., *Experimental Design: Procedures for the Behavioral Sciences*. Brooks/Cole, Belmont, Calif., 1968.
Lindquist, E. F., *Design and Analysis of Experiments in Psychology and Education*. Houghton Mifflin, Boston, 1953.
McNemar, Q., *Psychological Statistics*, 4th ed. Wiley, New York, 1969.
Winer, B. J., *Statistical Principles in Experimental Design*, 2nd ed. McGraw-Hill Kogakusha, Tokyo, 1971.

One-Factor Completely Randomized Design

A	B	C
S_1	S_2	S_3
S_4	S_5	S_6
S_7	S_8	S_9

Here, S_1, S_2, S_3, etc., are all different subjects. It can be seen that each treatment uses different subjects: treatment A has S_1, S_4, and S_7; treatment B has S_2, S_5, S_8; and so on. This independent samples design is called a completely randomized design.

Bruning and Kintz	Completely randomized design
Edwards	Randomized group design
Keppel	Single-factor experiment or single-factor analysis of variance
Kirk	Completely randomized design; CR-k design (k = number of treatments)
Lindquist	Simple randomized design
McNemar	Analysis of variance: simple
Winer	Single-factor experiments

Table G.10 *(continued)*

Two-Factor Design Without Interaction, Repeated Measures

A	B	C
S_1	S_1	S_1
S_2	S_2	S_2
S_3	S_3	S_3

Here, subject S_1 is tested under treatments A, B, and C, so is listed under all three. The same is true for S_2 and S_3. This related samples design is called a repeated-measures design.

Bruning and Kintz	Treatments-by-subjects or repeated-measures design
Edwards	Randomized block designs: repeated measures
Keppel	Single-factor repeated-measures design or (A × S) design
Kirk	Randomized block design: RB-k design (k = number of treatments)
Lindquist	Treatments × subjects design
McNemar	Double or two-way classification
Winer	Single-factor experiments having repeated measures on the same elements

Two-Factor Design with Interaction, Completely Randomized

	A			B			C		
α	S_1	S_2	S_3	S_4	S_5	S_6	S_7	S_8	S_9
β	S_{10}	S_{11}	S_{12}	S_{13}	S_{14}	S_{15}	S_{16}	S_{17}	S_{18}
γ	S_{19}	S_{20}	S_{21}	S_{22}	S_{23}	S_{24}	S_{25}	S_{26}	S_{27}

(In the following texts, factors ABC and $\alpha\beta\gamma$ are considered as fixed effects except in McNemar, where fixed, random, and mixed models are considered.)

Bruning and Kintz	Factorial design: Two factors
Edwards	The n × n factorial experiment (n = number of treatments)
Keppel	Factorial experiment with two factors
Kirk	Completely randomized factorial design, CRF-pq design (p, q = number of treatments)
Lindquist	Factorial design (two factors) or two-factor (A × B) design
McNemar	Double classification with more than one score per cell
Winer	Factorial experiments: p × q factorial experiment (p, q = number of treatments)

Table G.10 *(continued)*

Three-Factor Design, One-Factor Repeated Measures

	A									B									C							
	α			β			γ			α			β			γ			α			β			γ	
S_1	S_1	S_1	S_1	S_1	S_1	S_1	S_1	S_1	S_1	S_1	S_1	S_1	S_1	S_1	S_1	S_1	S_1	S_1	S_1	S_1	S_1	S_1	S_1	S_1	S_1	S_1
S_2	S_2	S_2	S_2	S_2	S_2	S_2	S_2	S_2	S_2	S_2	S_2	S_2	S_2	S_2	S_2	S_2	S_2	S_2	S_2	S_2	S_2	S_2	S_2	S_2	S_2	S_2
S_3	S_3	S_3	S_3	S_3	S_3	S_3	S_3	S_3	S_3	S_3	S_3	S_3	S_3	S_3	S_3	S_3	S_3	S_3	S_3	S_3	S_3	S_3	S_3	S_3	S_3	S_3

(In the following texts, factors ABC and $\alpha\beta\gamma$ are considered as fixed effects while the "subjects" factor is a random effect, except in Kirk, where "subjects" is a fixed or random effect.)

Bruning and Kintz	Treatments-by-treatments-by-subjects or repeated-measures: Two-factors design
Edwards	Factorial experiment with repeated measures on all treatment combinations
Keppel	(A X B X S) design
Kirk	Randomized block factorial design: RBF-pq design (p, q = number of treatments)
Lindquist	Three-dimensional design: Treatments X treatments X subjects (A X B X S) design
McNemar	Three-way classification-special case where the rows stand for persons or matched individuals
Winer	Not listed

Three-Factor Design, Completely Randomized

	A									B									C								
		α			β			γ			α			β			γ			α			β			γ	
a	S_1	S_2	S_3	S_4	S_5	S_6	S_7	S_8	S_9	S_{10}	S_{11}	S_{12}	S_{13}	S_{14}	S_{15}	S_{16}	S_{17}	S_{18}	S_{19}	S_{20}	S_{21}	S_{22}	S_{23}	S_{24}	S_{25}	S_{26}	S_{27}
b	S_{28}	S_{29}	S_{30}	S_{31}	S_{32}	S_{33}	S_{34}	S_{35}	S_{36}	S_{37}	S_{38}	S_{39}	S_{40}	S_{41}	S_{42}	S_{43}	S_{44}	S_{45}	S_{46}	S_{47}	S_{48}	S_{49}	S_{50}	S_{51}	S_{52}	S_{53}	S_{54}
c	S_{55}	S_{56}	S_{57}	S_{58}	S_{59}	S_{60}	S_{61}	S_{62}	S_{63}	S_{64}	S_{65}	S_{66}	S_{67}	S_{68}	S_{69}	S_{70}	S_{71}	S_{72}	S_{73}	S_{74}	S_{75}	S_{76}	S_{77}	S_{78}	S_{79}	S_{80}	S_{81}

(In the following texts, factors ABC, $\alpha\beta\gamma$, and abc are considered as fixed effects.)

Bruning and Kintz	Factorial design: Three factors
Edwards	n X n X n factorial experiment (n = number of treatments)
Keppel	Three-factor case
Kirk	Completely randomized factorial design, CRF-pqr design (p, q, r = number of treatments)
Lindquist	Three-dimensional design, three-factor (A X B X C) design
McNemar	Three-way classification with m cases per cubicle
Winer	Factorial experiments: p X q X r factorial experiment

Table G.10 *(continued)*

Split Plot

	A	B	C
α	S_1	S_1	S_1
	S_2	S_2	S_2
β	S_3	S_3	S_3
	S_4	S_4	S_4
γ	S_5	S_5	S_5
	S_6	S_6	S_6

(In the following texts, ABC and $\alpha\beta\gamma$ are considered as fixed effects; the "subjects" factor is not considered.)

Bruning and Kintz	Two-factor mixed design: Repeated measures on one factor
Edwards	Split-plot design
Keppel	A × (B × S) design
Kirk	Split-plot design, SPF-pq design (p, q = number of treatments)
Lindquist	Mixed design, Type I design
McNemar	Split-plot design, three-way classification mixed model [$a_r A_b A_c$]
Winer	Two-factor experiment with repeated measures on one factor

Table G.10 *(continued)*

Multiple Comparisons

$\bar{X}_1 \quad \bar{X}_2 \quad \bar{X}_3 \quad \bar{X}_4 \quad \bar{X}_5 \quad \bar{X}_6$

Bruning and Kintz	Multiple comparisons (post hoc analyses)
Edwards	Multiple comparisons
Keppel	Multiple comparisons
Kirk	Multiple comparisons
Lindquist	Testing the significance of the difference in means for individual pairs of treatments
	Tests of significance applied to individual differences
	Testing differences in individual pairs of treatment means
McNemar	Selected contrasts
Winer	A posteriori tests

Table G.11 Critical Values of F[a]

a. 0.05 level in roman type, 0.01 level in boldface

Degrees of freedom for lesser mean square [denominator]	Degrees of freedom for greater mean square [numerator] 1	2	3	4	5	6	7	8	9	10	11	12	14	16	20	24	30	40	50	75	100	200	500	∞	
1	161 **4,052**	200 **4,999**	216 **5,403**	225 **5,625**	230 **5,764**	234 **5,859**	237 **5,928**	239 **5,981**	241 **6,022**	242 **6,056**	243 **6,082**	244 **6,106**	245 **6,142**	246 **6,169**	248 **6,208**	249 **6,234**	250 **6,261**	251 **6,286**	252 **6,302**	253 **6,323**	253 **6,334**	254 **6,352**	254 **6,361**	254 **6,366**	1
2	18.51 **98.49**	19.00 **99.00**	19.16 **99.17**	19.25 **99.25**	19.30 **99.30**	19.33 **99.33**	19.36 **99.36**	19.37 **99.37**	19.38 **99.39**	19.39 **99.40**	19.40 **99.41**	19.41 **99.42**	19.42 **99.43**	19.43 **99.44**	19.44 **99.45**	19.45 **99.46**	19.46 **99.47**	19.47 **99.48**	19.47 **99.48**	19.48 **99.49**	19.49 **99.49**	19.49 **99.49**	19.50 **99.50**	19.50 **99.50**	2
3	10.13 **34.12**	9.55 **30.82**	9.28 **29.46**	9.12 **28.71**	9.01 **28.24**	8.94 **27.91**	8.88 **27.67**	8.84 **27.49**	8.81 **27.34**	8.78 **27.23**	8.76 **27.13**	8.74 **27.05**	8.71 **26.92**	8.69 **26.83**	8.66 **26.69**	8.64 **26.60**	8.62 **26.50**	8.60 **26.41**	8.58 **26.35**	8.57 **26.27**	8.56 **26.23**	8.54 **26.18**	8.54 **26.14**	8.53 **26.12**	3
4	7.71 **21.20**	6.94 **18.00**	6.59 **16.69**	6.39 **15.98**	6.26 **15.52**	6.16 **15.21**	6.09 **14.98**	6.04 **14.80**	6.00 **14.66**	5.96 **14.54**	5.93 **14.45**	5.91 **14.37**	5.87 **14.24**	5.84 **14.15**	5.80 **14.02**	5.77 **13.93**	5.74 **13.83**	5.71 **13.74**	5.70 **13.69**	5.68 **13.61**	5.66 **13.57**	5.65 **13.52**	5.64 **13.48**	5.63 **13.46**	4
5	6.61 **16.26**	5.79 **13.27**	5.41 **12.06**	5.19 **11.39**	5.05 **10.97**	4.95 **10.67**	4.88 **10.45**	4.82 **10.29**	4.78 **10.15**	4.74 **10.05**	4.70 **9.96**	4.68 **9.89**	4.64 **9.77**	4.60 **9.68**	4.56 **9.55**	4.53 **9.47**	4.50 **9.38**	4.46 **9.29**	4.44 **9.24**	4.42 **9.17**	4.40 **9.13**	4.38 **9.07**	4.37 **9.04**	4.36 **9.02**	5
6	5.99 **13.74**	5.14 **10.92**	4.76 **9.78**	4.53 **9.15**	4.39 **8.75**	4.28 **8.47**	4.21 **8.26**	4.15 **8.10**	4.10 **7.98**	4.06 **7.87**	4.03 **7.79**	4.00 **7.72**	3.96 **7.60**	3.92 **7.52**	3.87 **7.39**	3.84 **7.31**	3.81 **7.23**	3.77 **7.14**	3.75 **7.09**	3.72 **7.02**	3.71 **6.99**	3.69 **6.94**	3.68 **6.90**	3.67 **6.88**	6
7	5.59 **12.25**	4.74 **9.55**	4.35 **8.45**	4.12 **7.85**	3.97 **7.46**	3.87 **7.19**	3.79 **7.00**	3.73 **6.84**	3.68 **6.71**	3.63 **6.62**	3.60 **6.54**	3.57 **6.47**	3.52 **6.35**	3.49 **6.27**	3.44 **6.15**	3.41 **6.07**	3.38 **5.98**	3.34 **5.90**	3.32 **5.85**	3.29 **5.78**	3.28 **5.75**	3.25 **5.70**	3.24 **5.67**	3.23 **5.65**	7
8	5.32 **11.26**	4.46 **8.65**	4.07 **7.59**	3.84 **7.01**	3.69 **6.63**	3.58 **6.37**	3.50 **6.19**	3.44 **6.03**	3.39 **5.91**	3.34 **5.82**	3.31 **5.74**	3.28 **5.67**	3.23 **5.56**	3.20 **5.48**	3.15 **5.36**	3.12 **5.28**	3.08 **5.20**	3.05 **5.11**	3.03 **5.06**	3.00 **5.00**	2.98 **4.96**	2.96 **4.91**	2.94 **4.88**	2.93 **4.86**	8
9	5.12 **10.56**	4.26 **8.02**	3.86 **6.99**	3.63 **6.42**	3.48 **6.06**	3.37 **5.80**	3.29 **5.62**	3.23 **5.47**	3.18 **5.35**	3.13 **5.26**	3.10 **5.18**	3.07 **5.11**	3.02 **5.00**	2.98 **4.92**	2.93 **4.80**	2.90 **4.73**	2.86 **4.64**	2.82 **4.56**	2.80 **4.51**	2.77 **4.45**	2.76 **4.41**	2.73 **4.36**	2.72 **4.33**	2.71 **4.31**	9
10	4.96 **10.04**	4.10 **7.56**	3.71 **6.55**	3.48 **5.99**	3.33 **5.64**	3.22 **5.39**	3.14 **5.21**	3.07 **5.06**	3.02 **4.95**	2.97 **4.85**	2.94 **4.78**	2.91 **4.71**	2.86 **4.60**	2.82 **4.52**	2.77 **4.41**	2.74 **4.33**	2.70 **4.25**	2.67 **4.17**	2.64 **4.12**	2.61 **4.05**	2.59 **4.01**	2.56 **3.96**	2.55 **3.93**	2.54 **3.91**	10
11	4.84 **9.65**	3.98 **7.20**	3.59 **6.22**	3.36 **5.67**	3.20 **5.32**	3.09 **5.07**	3.01 **4.88**	2.95 **4.74**	2.90 **4.63**	2.86 **4.54**	2.82 **4.46**	2.79 **4.40**	2.74 **4.29**	2.70 **4.21**	2.65 **4.10**	2.61 **4.02**	2.57 **3.94**	2.53 **3.86**	2.50 **3.80**	2.47 **3.74**	2.45 **3.70**	2.42 **3.66**	2.41 **3.62**	2.40 **3.60**	11
12	4.75 **9.33**	3.88 **6.93**	3.49 **5.95**	3.26 **5.41**	3.11 **5.06**	3.00 **4.82**	2.92 **4.65**	2.85 **4.50**	2.80 **4.39**	2.76 **4.30**	2.72 **4.22**	2.69 **4.16**	2.64 **4.05**	2.60 **3.98**	2.54 **3.86**	2.50 **3.78**	2.46 **3.70**	2.42 **3.61**	2.40 **3.56**	2.36 **3.49**	2.35 **3.46**	2.32 **3.41**	2.31 **3.38**	2.30 **3.36**	12
13	4.67 **9.07**	3.80 **6.70**	3.41 **5.74**	3.18 **5.20**	3.02 **4.86**	2.92 **4.62**	2.84 **4.44**	2.77 **4.30**	2.72 **4.19**	2.67 **4.10**	2.63 **4.02**	2.60 **3.96**	2.55 **3.85**	2.51 **3.78**	2.46 **3.67**	2.42 **3.59**	2.38 **3.51**	2.34 **3.42**	2.32 **3.37**	2.28 **3.30**	2.26 **3.27**	2.24 **3.21**	2.22 **3.18**	2.21 **3.16**	13

[a]The values in the table are the critical values of F for the degrees of freedom listed over the columns (the degrees of freedom for the greater mean square or numerator of the F ratio) and the degrees of freedom listed for the rows (the degrees of freedom for the lesser mean square for the denominator of the F ratio): The critical value for the 0.05 level of significance is presented first (roman type) followed by the critical value at the 0.01 level (boldface). If the observed value is *greater than or equal to* the tabled value, reject H_0.

Source: Reprinted by permission from *Statistical Methods*, Seventh Edition, by George W. Snedecor and William G. Cochran, copyright 1980 by the Iowa State University Press, Ames, Iowa 50010.

Table G.11 *(continued)*

Degrees of freedom for lesser mean square [denominator] (rows) × Degrees of freedom for greater mean square [numerator] (columns)

	1	2	3	4	5	6	7	8	9	10	11	12	14	16	20	24	30	40	50	75	100	200	500	x	
14	4.60 **8.86**	3.74 **6.51**	3.34 **5.56**	3.11 **5.03**	2.96 **4.69**	2.85 **4.46**	2.77 **4.28**	2.70 **4.14**	2.65 **4.03**	2.60 **3.94**	2.56 **3.86**	2.53 **3.80**	2.48 **3.70**	2.44 **3.62**	2.39 **3.51**	2.35 **3.43**	2.31 **3.34**	2.27 **3.26**	2.24 **3.21**	2.21 **3.14**	2.19 **3.11**	2.16 **3.06**	2.14 **3.02**	2.13 **3.00**	14
15	4.54 **8.68**	3.68 **6.36**	3.29 **5.42**	3.06 **4.89**	2.90 **4.56**	2.79 **4.32**	2.70 **4.14**	2.64 **4.00**	2.59 **3.89**	2.55 **3.80**	2.51 **3.73**	2.48 **3.67**	2.43 **3.56**	2.39 **3.48**	2.33 **3.36**	2.29 **3.29**	2.25 **3.20**	2.21 **3.12**	2.18 **3.07**	2.15 **3.00**	2.12 **2.97**	2.10 **2.92**	2.08 **2.89**	2.07 **2.87**	15
16	4.49 **8.53**	3.63 **6.23**	3.24 **5.29**	3.01 **4.77**	2.85 **4.44**	2.74 **4.20**	2.66 **4.03**	2.59 **3.89**	2.54 **3.78**	2.49 **3.69**	2.45 **3.61**	2.42 **3.55**	2.37 **3.45**	2.33 **3.37**	2.28 **3.25**	2.24 **3.18**	2.20 **3.10**	2.16 **3.01**	2.13 **2.96**	2.09 **2.98**	2.07 **2.86**	2.04 **2.80**	2.02 **2.77**	2.01 **2.75**	16
17	4.45 **8.40**	3.59 **6.11**	3.20 **5.18**	2.96 **4.67**	2.81 **4.34**	2.70 **4.10**	2.62 **3.93**	2.55 **3.79**	2.50 **3.68**	2.45 **3.59**	2.41 **3.52**	2.38 **3.45**	2.33 **3.35**	2.29 **3.27**	2.23 **3.16**	2.19 **3.08**	2.15 **3.00**	2.11 **2.92**	2.08 **2.86**	2.04 **2.79**	2.02 **2.76**	1.99 **2.70**	1.97 **2.67**	1.96 **2.65**	17
18	4.41 **8.28**	3.55 **6.01**	3.16 **5.09**	2.93 **4.58**	2.77 **4.25**	3.66 **4.01**	2.58 **3.85**	2.51 **3.71**	2.46 **3.60**	2.41 **3.51**	2.37 **3.44**	2.34 **3.37**	2.29 **3.27**	2.25 **3.19**	2.19 **3.07**	2.15 **3.00**	2.11 **2.91**	2.07 **2.83**	2.04 **2.78**	2.00 **2.71**	1.98 **2.68**	1.95 **2.62**	1.93 **2.59**	1.92 **2.57**	18
19	4.38 **8.18**	3.52 **5.93**	3.13 **5.01**	2.90 **4.50**	2.74 **4.17**	2.63 **3.94**	2.55 **3.77**	2.48 **3.63**	2.43 **3.52**	2.38 **3.43**	2.34 **3.36**	2.31 **3.30**	2.26 **3.19**	2.21 **3.12**	2.15 **3.00**	2.11 **2.92**	2.07 **2.84**	2.02 **2.76**	2.00 **2.70**	1.96 **2.63**	1.94 **2.60**	1.91 **2.54**	1.90 **2.51**	1.88 **2.49**	19
20	4.35 **8.10**	3.49 **5.85**	3.10 **4.94**	2.87 **4.43**	2.71 **4.10**	2.60 **3.87**	2.52 **3.71**	2.45 **3.56**	2.40 **3.45**	2.35 **3.37**	2.31 **3.30**	2.28 **3.23**	2.23 **3.13**	2.18 **3.05**	2.12 **2.94**	2.08 **2.86**	2.04 **2.77**	1.99 **2.69**	1.96 **2.63**	1.92 **2.56**	1.90 **2.53**	1.87 **2.47**	1.85 **2.44**	1.84 **2.42**	20
21	4.32 **8.02**	3.47 **5.78**	3.07 **4.87**	2.84 **4.37**	2.68 **4.04**	2.57 **3.81**	2.49 **3.65**	2.42 **3.51**	2.37 **3.40**	2.32 **3.31**	2.28 **3.24**	2.25 **3.17**	2.20 **3.07**	2.15 **2.99**	2.09 **2.88**	2.05 **2.80**	2.00 **2.72**	1.96 **2.63**	1.93 **2.58**	1.89 **2.51**	1.87 **2.47**	1.84 **2.42**	1.82 **2.38**	1.81 **2.36**	21
22	4.30 **7.94**	3.44 **5.72**	3.05 **4.82**	2.82 **4.31**	2.66 **3.99**	2.55 **3.76**	2.47 **3.59**	2.40 **3.45**	2.35 **3.35**	2.30 **3.26**	2.26 **3.18**	2.23 **3.12**	2.18 **3.02**	2.13 **2.94**	2.07 **2.83**	2.03 **2.75**	1.98 **2.67**	1.93 **2.58**	1.91 **2.53**	1.87 **2.46**	1.84 **2.42**	1.81 **2.37**	1.80 **2.33**	1.78 **2.31**	22
23	4.28 **7.88**	3.42 **5.66**	3.03 **4.76**	2.80 **4.26**	2.64 **3.94**	2.53 **3.71**	2.45 **3.54**	2.38 **3.41**	2.32 **3.30**	2.28 **3.21**	2.24 **3.14**	2.20 **3.07**	2.14 **2.97**	2.10 **2.89**	2.04 **2.78**	2.00 **2.70**	1.96 **2.62**	1.91 **2.53**	1.88 **2.48**	1.84 **2.41**	1.82 **2.37**	1.79 **2.32**	1.77 **2.28**	1.76 **2.26**	23
24	4.26 **7.82**	3.40 **5.61**	3.01 **4.72**	2.78 **4.22**	2.62 **3.90**	2.51 **3.67**	2.43 **3.50**	2.36 **3.36**	2.30 **3.25**	2.26 **3.17**	2.22 **3.09**	2.18 **3.03**	2.13 **2.93**	2.09 **2.85**	2.02 **2.74**	1.98 **2.66**	1.94 **2.58**	1.89 **2.49**	1.86 **2.44**	1.82 **2.36**	1.80 **2.33**	1.76 **2.27**	1.74 **2.23**	1.73 **2.21**	24
25	4.24 **7.77**	3.38 **5.57**	2.99 **4.68**	2.76 **4.18**	2.60 **3.86**	2.49 **3.63**	2.41 **3.46**	2.34 **3.32**	2.28 **3.21**	2.24 **3.13**	2.20 **3.05**	2.16 **2.99**	2.11 **2.89**	2.06 **2.81**	2.00 **2.70**	1.96 **2.62**	1.92 **2.54**	1.87 **2.45**	1.84 **2.40**	1.80 **2.32**	1.77 **2.29**	1.74 **2.23**	1.72 **2.19**	1.71 **2.17**	25
26	4.22 **7.72**	3.37 **5.53**	2.98 **4.64**	2.74 **4.14**	2.59 **3.82**	2.47 **3.59**	2.39 **3.42**	2.32 **3.29**	2.27 **3.17**	2.22 **3.09**	2.18 **3.02**	2.15 **2.96**	2.10 **2.86**	2.05 **2.77**	1.99 **2.66**	1.95 **2.58**	1.90 **2.50**	1.85 **2.41**	1.82 **2.36**	1.78 **2.28**	1.76 **2.25**	1.72 **2.19**	1.70 **2.15**	1.69 **2.13**	26

Table G.11 *(continued)*

Degrees of freedom for lesser mean square [denominator]	Degrees of freedom for greater mean square [numerator]																								
	1	2	3	4	5	6	7	8	9	10	11	12	14	16	20	24	30	40	50	75	100	200	500	∞	
27	4.21 **7.68**	3.35 **5.49**	2.96 **4.60**	2.73 **4.11**	2.57 **3.79**	2.46 **3.56**	2.37 **3.39**	2.30 **3.26**	2.25 **3.14**	2.20 **3.06**	2.16 **2.98**	2.13 **2.93**	2.08 **2.83**	2.03 **2.74**	1.97 **2.63**	1.93 **2.55**	1.88 **2.47**	1.84 **2.38**	1.80 **2.33**	1.76 **2.25**	1.74 **2.21**	1.71 **2.16**	1.68 **2.12**	1.67 **2.10**	27
28	4.20 **7.64**	3.34 **5.45**	2.95 **4.57**	2.71 **4.07**	2.56 **3.76**	2.44 **3.53**	2.36 **3.36**	2.29 **3.23**	2.24 **3.11**	2.19 **3.03**	2.15 **2.95**	2.12 **2.90**	2.06 **2.80**	2.02 **2.71**	1.96 **2.60**	1.91 **2.52**	1.87 **2.44**	1.81 **2.35**	1.78 **2.30**	1.75 **2.22**	1.72 **2.18**	1.69 **2.13**	1.67 **2.09**	1.65 **2.06**	28
29	4.18 **7.60**	3.33 **5.42**	2.93 **4.54**	2.70 **4.04**	2.54 **3.73**	2.43 **3.50**	2.35 **3.33**	2.28 **3.20**	2.22 **3.08**	2.18 **3.00**	2.14 **2.92**	2.10 **2.87**	2.05 **2.77**	2.00 **2.68**	1.94 **2.57**	1.90 **2.49**	1.85 **2.41**	1.80 **2.32**	1.77 **2.27**	1.73 **2.19**	1.71 **2.15**	1.68 **2.10**	1.65 **2.06**	1.64 **2.03**	29
30	4.17 **7.56**	3.32 **5.39**	2.92 **4.51**	2.69 **4.02**	2.53 **3.70**	2.42 **3.47**	2.34 **3.30**	2.27 **3.17**	2.21 **3.06**	2.16 **2.98**	2.12 **2.90**	2.09 **2.84**	2.04 **2.74**	1.99 **2.66**	1.93 **2.55**	1.89 **2.47**	1.84 **2.38**	1.79 **2.29**	1.76 **2.24**	1.72 **2.16**	1.69 **2.13**	1.66 **2.07**	1.64 **2.03**	1.62 **2.01**	30
32	4.15 **7.50**	3.30 **5.34**	2.90 **4.46**	2.67 **3.97**	2.51 **3.66**	2.40 **3.42**	2.32 **3.25**	2.25 **3.12**	2.19 **3.01**	2.14 **2.94**	2.10 **2.86**	2.07 **2.80**	2.02 **2.70**	1.97 **2.62**	1.91 **2.51**	1.86 **2.42**	1.82 **2.34**	1.76 **2.25**	1.74 **2.20**	1.69 **2.12**	1.67 **2.08**	1.64 **2.02**	1.61 **1.98**	1.59 **1.96**	32
34	4.13 **7.44**	3.28 **5.29**	2.88 **4.42**	2.65 **3.93**	2.49 **3.61**	2.38 **3.38**	2.30 **3.21**	2.23 **3.08**	2.17 **2.97**	2.12 **2.89**	2.08 **2.82**	2.05 **2.76**	2.00 **2.66**	1.95 **2.58**	1.89 **2.47**	1.84 **2.38**	1.80 **2.30**	1.74 **2.21**	1.71 **2.15**	1.67 **2.08**	1.64 **2.04**	1.61 **1.98**	1.59 **1.94**	1.57 **1.91**	34
36	4.11 **7.39**	3.26 **5.25**	2.86 **4.38**	2.63 **3.89**	2.48 **3.58**	2.36 **3.35**	2.28 **3.18**	2.21 **3.04**	2.15 **2.94**	2.10 **2.86**	2.06 **2.78**	2.03 **2.72**	1.98 **2.62**	1.93 **2.54**	1.87 **2.43**	1.82 **2.35**	1.78 **2,26**	1.72 **2.17**	1.69 **2.12**	1.65 **2.04**	1.62 **2.00**	1.59 **1.94**	1.56 **1.90**	1.55 **1.87**	36
38	4.10 **7.35**	3.25 **5.21**	2.85 **4.34**	2.62 **3.86**	2.46 **3.54**	2.35 **3.32**	2.26 **3.15**	2.19 **3.02**	2.14 **2.91**	2.09 **2.82**	2.05 **2.75**	2.02 **2.69**	1.96 **2.59**	1.92 **2.51**	1.85 **2.40**	1.80 **2.32**	1.76 **2.22**	1.71 **2.14**	1.67 **2.08**	1.63 **2.00**	1.60 **1.97**	1.57 **1.90**	1.54 **1.86**	1.53 **1.84**	38
40	4.08 **7.31**	3.23 **5.18**	2.84 **4.31**	2.61 **3.83**	2.45 **3.51**	2.34 **3.29**	2.25 **3.12**	2.18 **2.99**	2.12 **2.88**	2.07 **2.80**	2.04 **2.73**	2.00 **2.66**	1.95 **2.56**	1.90 **2.49**	1.84 **2.37**	1.79 **2.29**	1.74 **2.20**	1.69 **2.11**	1.66 **2.05**	1.61 **1.97**	1.59 **1.94**	1.55 **1.88**	1.53 **1.84**	1.51 **1.81**	40
42	4.07 **7.27**	3.22 **5.15**	2.83 **4.29**	2.59 **3.80**	2.44 **3.49**	2,32 **3.26**	2.24 **3.10**	2.17 **2.96**	2.11 **2.86**	2.06 **2.77**	2.02 **2.70**	1.99 **2.64**	1.94 **2.54**	1.89 **2.46**	1.82 **2.35**	1.78 **2.26**	1.73 **2.17**	1.68 **2.08**	1.64 **2.02**	1.60 **1.94**	1.57 **1.91**	1.54 **1.85**	1.51 **1.80**	1.49 **1.78**	42
44	4.06 **7.24**	3.21 **5.12**	2.82 **4.26**	2.58 **3.78**	2.43 **3.46**	2.31 **3.24**	2.23 **3.07**	2.16 **2.94**	2.10 **2.84**	2.05 **2.75**	2.01 **2.68**	1.98 **2.62**	1.92 **2.52**	1.88 **2.44**	1.81 **2.32**	1.76 **2.24**	1.72 **2.15**	1.66 **2.06**	1.63 **2.00**	1.58 **1.92**	1.56 **1.88**	1.52 **1.82**	1.50 **1.78**	1.48 **1.75**	44
46	4.05 **7.21**	3.20 **5.10**	2.81 **4.24**	2.57 **3.76**	2.42 **3.44**	2.30 **3.22**	2.22 **3.05**	2.14 **2.92**	2.09 **2.82**	2.04 **2.73**	2.00 **2.66**	1.97 **2.60**	1.91 **2.50**	1.87 **2.42**	1.80 **2.30**	1.75 **2.22**	1.71 **2.13**	1.65 **2.04**	1.62 **1.98**	1.57 **1.90**	1.54 **1.86**	1.51 **1.80**	1.48 **1.76**	1.46 **1.72**	46
48	4.04 **7.19**	3.19 **5.08**	2.80 **4.22**	2.56 **3.74**	2.41 **3.42**	2.30 **3.20**	2.21 **3.04**	2.14 **2.90**	2.08 **2.80**	2.03 **2.71**	1.99 **2.64**	1.96 **2.58**	1.90 **2.48**	1.86 **2.40**	1.79 **2.28**	1.74 **2.20**	1.70 **2.11**	1.64 **2.02**	1.61 **1.96**	1.56 **1.88**	1.53 **1.84**	1.50 **1.78**	1.47 **1.73**	1.45 **1.70**	48

Table G.11 *(continued)*

Degrees of freedom for greater mean square [numerator] (columns); Degrees of freedom for lesser mean square [denominator] (rows)

	1	2	3	4	5	6	7	8	9	10	11	12	14	16	20	24	30	40	50	75	100	200	500	∞	
50	4.03 **7.17**	3.18 **5.06**	2.79 **4.20**	2.56 **3.72**	2.40 **3.41**	2.29 **3.18**	2.20 **3.02**	2.13 **2.88**	2.07 **2.78**	2.02 **2.70**	1.98 **2.62**	1.95 **2.56**	1.90 **2.46**	1.85 **2.39**	1.78 **2.26**	1.74 **2.18**	1.69 **2.10**	1.63 **2.00**	1.60 **1.94**	1.55 **1.86**	1.52 **1.82**	1.48 **1.76**	1.46 **1.71**	1.44 **1.68**	50
55	4.02 **7.12**	3.17 **5.01**	2.78 **4.16**	2.54 **3.68**	2.38 **3.37**	2.27 **3.15**	2.18 **2.98**	2.11 **2.85**	2.05 **2.75**	2.00 **2.66**	1.97 **2.59**	1.93 **2.53**	1.88 **2.43**	1.83 **2.35**	1.76 **2.23**	1.72 **2.15**	1.67 **2.06**	1.61 **1.96**	1.58 **1.90**	1.52 **1.82**	1.50 **1.78**	1.46 **1.71**	1.43 **1.66**	1.41 **1.64**	55
60	4.00 **7.08**	3.15 **4.98**	2.76 **4.13**	2.52 **3.65**	2.37 **3.34**	2.25 **3.12**	2.17 **2.95**	2.10 **2.82**	2.04 **2.72**	1.99 **2.63**	1.95 **2.56**	1.92 **2.50**	1.86 **2.40**	1.81 **2.32**	1.75 **2.20**	1.70 **2.12**	1.65 **2.03**	1.59 **1.93**	1.56 **1.87**	1.50 **1.79**	1.48 **1.74**	1.44 **1.68**	1.41 **1.63**	1.39 **1.60**	60
65	3.99 **7.04**	3.14 **4.95**	2.75 **4.10**	2.51 **3.62**	2.36 **3.31**	2.24 **3.09**	2.15 **2.93**	2.08 **2.79**	2.02 **2.70**	1.98 **2.61**	1.94 **2.54**	1.90 **2.47**	1.85 **2.37**	1.80 **2.30**	1.73 **2.18**	1.68 **2.09**	1.63 **2.00**	1.57 **1.90**	1.54 **1.84**	1.49 **1.76**	1.46 **1.71**	1.42 **1.64**	1.39 **1.60**	1.37 **1.56**	65
70	3.98 **7.01**	3.13 **4.92**	2.74 **4.08**	2.50 **3.60**	2.35 **3.29**	2.23 **3.07**	2.14 **2.91**	2.07 **2.77**	2.01 **2.67**	1.97 **2.59**	1.93 **2.51**	1.89 **2.45**	1.84 **2.35**	1.79 **2.28**	1.72 **2.15**	1.67 **2.07**	1.62 **1.98**	1.56 **1.88**	1.53 **1.82**	1.47 **1.74**	1.45 **1.69**	1.40 **1.62**	1.37 **1.56**	1.35 **1.53**	70
80	3.96 **6.96**	3.11 **4.88**	2.72 **4.04**	2.48 **3.56**	2.33 **3.25**	2.21 **3.04**	2.12 **2.87**	2.05 **2.74**	1.99 **2.64**	1.95 **2.55**	1.91 **2.48**	1.88 **2.41**	1.82 **2.32**	1.77 **2.24**	1.70 **2.11**	1.65 **2.03**	1.60 **1.94**	1.54 **1.84**	1.51 **1.78**	1.45 **1.70**	1.42 **1.65**	1.38 **1.57**	1.35 **1.52**	1.32 **1.49**	80
100	3.94 **6.90**	3.09 **4.82**	2.70 **3.98**	2.46 **3.51**	2.30 **3.20**	2.19 **2.99**	2.10 **2.82**	2.03 **2.69**	1.97 **2.59**	1.92 **2.51**	1.88 **2.43**	1.85 **2.36**	1.79 **2.26**	1.75 **2.19**	1.68 **2.06**	1.63 **1.98**	1.57 **1.89**	1.51 **1.79**	1.48 **1.73**	1.42 **1.64**	1.39 **1.59**	1.34 **1.51**	1.30 **1.46**	1.28 **1.43**	100
125	3.92 **6.84**	3.07 **4.78**	2.68 **3.94**	2.44 **3.47**	2.29 **3.17**	2.17 **2.95**	2.08 **2.79**	2.01 **2.65**	1.95 **2.56**	1.90 **2.47**	1.86 **2.40**	1.83 **2.33**	1.77 **2.23**	1.72 **2.15**	1.65 **2.03**	1.60 **1.94**	1.55 **1.85**	1.49 **1.75**	1.45 **1.68**	1.39 **1.59**	1.36 **1.54**	1.31 **1.46**	1.27 **1.40**	1.25 **1.37**	125
150	3.91 **6.81**	3.06 **4.75**	2.67 **3.91**	2.43 **3.44**	2.27 **3.14**	2.16 **2.92**	2.07 **2.76**	2.00 **2.62**	1.94 **2.53**	1.89 **2.44**	1.85 **2.37**	1.82 **2.30**	1.76 **2.20**	1.71 **2.12**	1.64 **2.00**	1.59 **1.91**	1.54 **1.83**	1.47 **1.72**	1.44 **1.66**	1.37 **1.56**	1.34 **1.51**	1.29 **1.43**	1.25 **1.37**	1.22 **1.33**	150
200	3.89 **6.76**	3.04 **4.71**	2.65 **3.88**	2.41 **3.41**	2.26 **3.11**	2.14 **2.90**	2.05 **2.73**	1.98 **2.60**	1.92 **2.50**	1.87 **2.41**	1.83 **2.34**	1.80 **2.28**	1.74 **2.17**	1.69 **2.09**	1.62 **1.97**	1.57 **1.88**	1.52 **1.79**	1.45 **1.69**	1.42 **1.62**	1.35 **1.53**	1.32 **1.48**	1.26 **1.39**	1.22 **1.33**	1.19 **1.28**	200
400	3.86 **6.70**	3.02 **4.66**	2.62 **3.83**	2.39 **3.36**	2.23 **3.06**	2.12 **2.85**	2.03 **2.69**	1.96 **2.55**	1.90 **2.46**	1.85 **2.37**	1.81 **2.29**	1.78 **2.23**	1.72 **2.12**	1.67 **2.04**	1.60 **1.92**	1.54 **1.84**	1.49 **1.74**	1.42 **1.64**	1.38 **1.57**	1.32 **1.47**	1.28 **1.42**	1.22 **1.32**	1.16 **1.24**	1.13 **1.19**	400
1000	3.85 **6.66**	3.00 **4.62**	2.61 **3.80**	2.38 **3.34**	2.22 **3.04**	2.10 **2.82**	2.02 **2.66**	1.95 **2.53**	1.89 **2.43**	1.84 **2.34**	1.80 **2.26**	1.76 **2.20**	1.70 **2.09**	1.65 **2.01**	1.58 **1.89**	1.53 **1.81**	1.47 **1.71**	1.41 **1.61**	1.36 **1.54**	1.30 **1.44**	1.26 **1.38**	1.19 **1.28**	1.13 **1.19**	1.08 **1.11**	1000
∞	3.84 **6.64**	2.99 **4.60**	2.60 **3.78**	2.37 **3.32**	2.21 **3.02**	2.09 **2.80**	2.01 **2.64**	1.94 **2.51**	1.88 **2.41**	1.83 **2.32**	1.79 **2.24**	1.75 **2.18**	1.69 **2.07**	1.64 **1.99**	1.57 **1.87**	1.52 **1.79**	1.46 **1.69**	1.40 **1.59**	1.35 **1.52**	1.28 **1.41**	1.24 **1.36**	1.17 **1.25**	1.11 **1.15**	1.00 **1.00**	∞

Table G.11 *(continued)*

b. 0.001 level

Deg of freedom for denom.	Degrees of freedom for numerator 1	2	3	4	5	6	8	12	24	∞
1	405284	500000	540379	562500	576405	585937	598144	610667	623497	636619
2	998.5	999.0	999.2	999.2	999.3	999.3	999.4	999.4	999.5	999.5
3	167.0	148.5	141.1	137.1	134.6	132.8	130.6	128.3	125.9	123.5
4	74.14	61.25	56.18	53.44	51.71	50.53	49.00	47.41	45.77	44.05
5	47.18	37.12	33.20	31.09	29.75	28.84	27.64	26.42	25.14	23.78
6	35.51	27.00	23.70	21.92	20.81	20.03	19.03	17.99	16.89	15.75
7	29.25	21.69	18.77	17.19	16.21	15.52	14.63	13.71	12.73	11.69
8	25.42	18.49	15.83	14.39	13.49	12.86	12.04	11.19	10.30	9.34
9	22.86	16.39	13.90	12.56	11.71	11.13	10.37	9.57	8.72	7.81
10	21.04	14.91	12.55	11.28	10.48	9.92	9.20	8.45	7.64	6.76
11	19.69	13.81	11.56	10.35	9.58	9.05	8.35	7.63	6.85	6.00
12	18.64	12.97	10.80	9.63	8.89	8.38	7.71	7.00	6.25	5.42
13	17.81	12.31	10.21	9.07	8.35	7.86	7.21	6.52	5.78	4.97
14	17.14	11.78	9.73	8.62	7.92	7.43	6.80	6.13	5.41	4.60
15	16.59	11.34	9.34	8.25	7.57	7.09	6.47	5.81	5.10	4.31
16	16.12	10.97	9.00	7.94	7.27	6.81	6.19	5.55	4.85	4.06
17	15.72	10.66	8.73	7.68	7.02	6.56	5.96	5.32	4.63	3.85
18	15.38	10.39	8.49	7.46	6.81	6.35	5.76	5.13	4.45	3.67
19	15.08	10.16	8.28	7.26	6.62	6.18	5.59	4.97	4.29	3.52
20	14.82	9.95	8.10	7.10	6.46	6.02	5.44	4.82	4.15	3.38
21	14.59	9.77	7.94	6.95	6.32	5.88	5.31	4.70	4.03	3.26
22	14.38	9.61	7.80	6.81	6.19	5.76	5.19	4.58	3.92	3.15
23	14.19	9.47	7.67	6.69	6.08	5.65	5.09	4.48	3.82	3.05
24	14.03	9.34	7.55	6.59	5.98	5.55	4.99	4.39	3.74	2.97
25	13.88	9.22	7.45	6.49	5.88	5.46	4.91	4.31	3.66	2.89
26	13.74	9.12	7.36	6.41	5.80	5.38	4.83	4.24	3.59	2.82
27	13.61	9.02	7.27	6.33	5.73	5.31	4.76	4.17	3.52	2.75
28	13.50	8.93	7.19	6.25	5.66	5.24	4.69	4.11	3.46	2.70
29	13.39	8.85	7.12	6.19	5.59	5.18	4.64	4.05	3.41	2.64
30	13.29	8.77	7.05	6.12	5.53	5.12	4.58	4.00	3.36	2.59
40	12.61	8.25	6.60	5.70	5.13	4.73	4.21	3.64	3.01	2.23
60	11.97	7.76	6.17	5.31	4.76	4.37	3.87	3.31	2.69	1.90
120	11.38	7.32	5.79	4.95	4.42	4.04	3.55	3.02	2.40	1.54
∞	10.83	6.91	5.42	4.62	4.10	3.74	3.27	2.74	2.13	1.00

Source: Table V of Fisher and Yates, *Statistical Tables for Biological, Agricultural and Medical Research*, published by Longman Group Ltd., London (previously published by Oliver and Boyd Ltd., Edinburgh) and by permission of the authors and publishers.

Table G.12 Significant Studentized Ranges (critical Q values) for Newman-Keuls' and Tukey Multiple-Comparison Tests

Significant Studentized Ranges (two-tailed): Q

Error *df*	α	r = number of means or number of steps between ordered means									
		2	3	4	5	6	7	8	9	10	11
5	.05	3.64	4.60	5.22	5.67	6.03	6.33	6.58	6.80	6.99	7.17
	.01	5.70	6.98	7.80	8.42	8.91	9.32	9.67	9.97	10.24	10.48
6	.05	3.46	4.34	4.90	5.30	5.63	5.90	6.12	6.32	6.49	6.65
	.01	5.24	6.33	7.03	7.56	7.97	8.32	8.61	8.87	9.10	9.30
7	.05	3.34	4.16	4.68	5.06	5.36	5.61	5.82	6.00	6.16	6.30
	.01	4.95	5.92	6.54	7.01	7.37	7.68	7.94	8.17	8.37	8.55
8	.05	3.26	4.04	4.53	4.89	5.17	5.40	5.60	5.77	5.92	6.05
	.01	4.75	5.64	6.20	6.62	6.96	7.24	7.47	7.68	7.86	8.03
9	.05	3.20	3.95	4.41	4.76	5.02	5.24	5.43	5.59	5.74	5.87
	.01	4.60	5.43	5.96	6.35	6.66	6.91	7.13	7.33	7.49	7.65
10	.05	3.15	3.88	4.33	4.65	4.91	5.12	5.30	5.46	5.60	5.72
	.01	4.48	5.27	5.77	6.14	6.43	6.67	6.87	7.05	7.21	7.36
11	.05	3.11	3.82	4.26	4.57	4.82	5.03	5.20	5.35	5.49	5.61
	.01	4.39	5.15	5.62	5.97	6.25	6.48	6.67	6.84	6.99	7.13
12	.05	3.08	3.77	4.20	4.51	4.75	4.95	5.12	5.27	5.39	5.51
	.01	4.32	5.05	5.50	5.84	6.10	6.32	6.51	6.67	6.81	6.94
13	.05	3.06	3.73	4.15	4.45	4.69	4.88	5.05	5.19	5.32	5.43
	.01	4.26	4.96	5.40	5.73	5.98	6.19	6.37	6.53	6.67	6.79
14	.05	3.03	3.70	4.11	4.41	4.64	4.83	4.99	5.13	5.25	5.36
	.01	4.21	4.89	5.32	5.63	5.88	6.08	6.26	6.41	6.54	6.66
15	.05	3.01	3.67	4.08	4.37	4.59	4.78	4.94	5.08	5.20	5.31
	.01	4.17	4.84	5.25	5.56	5.80	5.99	6.16	6.31	6.44	6.55
16	.05	3.00	3.65	4.05	4.33	4.56	4.74	4.90	5.03	5.15	5.26
	.01	4.13	4.79	5.19	5.49	5.72	5.92	6.08	6.22	6.35	6.46
17	.05	2.98	3.63	4.02	4.30	4.52	4.70	4.86	4.99	5.11	5.21
	.01	4.10	4.74	5.14	5.43	5.66	5.85	6.01	6.15	6.27	6.38
18	.05	2.97	3.61	4.00	4.28	4.49	4.67	4.82	4.96	5.07	5.17
	.01	4.07	4.70	5.09	5.38	5.60	5.79	5.94	6.08	6.20	6.31
19	.05	2.96	3.59	3.98	4.25	4.47	4.65	4.79	4.92	5.04	5.14
	.01	4.05	4.67	5.05	5.33	5.55	5.73	5.89	6.02	6.14	6.25
20	.05	2.95	3.58	3.96	4.23	4.45	4.62	4.77	4.90	5.01	5.11
	.01	4.02	4.64	5.02	5.29	5.51	5.69	5.84	5.97	6.09	6.19
24	.05	2.92	3.53	3.90	4.17	4.37	4.54	4.68	4.81	4.92	5.01
	.01	3.96	4.55	4.91	5.17	5.37	5.54	5.69	5.81	5.92	6.02
30	.05	2.89	3.49	3.85	4.10	4.30	4.46	4.60	4.72	4.82	4.92
	.01	3.89	4.45	4.80	5.05	5.24	5.40	5.54	5.65	5.76	5.85
40	.05	2.86	3.44	3.79	4.04	4.23	4.39	4.52	4.63	4.73	4.82
	.01	3.82	4.37	4.70	4.93	5.11	5.26	5.39	5.50	5.60	5.69
60	.05	2.83	3.40	3.74	3.98	4.16	4.31	4.44	4.55	4.65	4.73
	.01	3.76	4.28	4.59	4.82	4.99	5.13	5.25	5.36	5.45	5.53
120	.05	2.80	3.36	3.68	3.92	4.10	4.24	4.36	4.47	4.56	4.64
	.01	3.70	4.20	4.50	4.71	4.87	5.01	5.12	5.21	5.30	5.37
∞	.05	2.77	3.31	3.63	3.86	4.03	4.17	4.29	4.39	4.47	4.55
	.01	3.64	4.12	4.40	4.60	4.76	4.88	4.99	5.08	5.16	5.23

To use table, locate the critical Q value for the appropriate error *df* and *r* values at the level of signficance desired. For the Tukey test, only the value for the largest range is needed.

Table G.12 *(continued)*

r = number of means or number of steps between ordered means									α	Error *df*
12	13	14	15	16	17	18	19	20		
7.32	7.47	7.60	7.72	7.83	7.93	8.03	8.12	8.21	.05	5
10.70	10.89	11.08	11.24	11.40	11.55	11.68	11.81	11.93	.01	
6.79	6.92	7.03	7.14	7.24	7.34	7.43	7.51	7.59	.05	6
9.48	9.65	9.81	9.95	10.08	10.21	10.32	10.43	10.54	.01	
6.43	6.55	6.66	6.76	6.85	6.94	7.02	7.10	7.17	.05	7
8.71	8.86	9.00	9.12	9.24	9.35	9.46	9.55	9.65	.01	
6.18	6.29	6.39	6.48	6.57	6.65	6.73	6.80	6.87	.05	8
8.18	8.31	8.44	8.55	8.66	8.76	8.85	8.94	9.03	.01	
5.98	6.09	6.19	6.28	6.36	6.44	6.51	6.58	6.64	.05	9
7.78	7.91	8.03	8.13	8.23	8.33	8.41	8.49	8.57	.01	
5.83	5.93	6.03	6.11	6.19	6.27	6.34	6.40	6.47	.05	10
7.49	7.60	7.71	7.81	7.91	7.99	8.08	8.15	8.23	.01	
5.71	5.81	5.90	5.98	6.06	6.13	6.20	6.27	6.33	.05	11
7.25	7.36	7.46	7.56	7.65	7.73	7.81	7.88	7.95	.01	
5.61	5.71	5.80	5.88	5.95	6.02	6.09	6.15	6.21	.05	12
7.06	7.17	7.26	7.36	7.44	7.52	7.59	7.66	7.73	.01	
5.53	5.63	5.71	5.79	5.86	5.93	5.99	6.05	6.11	.05	13
6.90	7.01	7.10	7.19	7.27	7.35	7.42	7.48	7.55	.01	
5.46	5.55	5.64	5.71	5.79	5.85	5.91	5.97	6.03	.05	14
6.77	6.87	6.96	7.05	7.13	7.20	7.27	7.33	7.39	.01	
5.40	5.49	5.57	5.65	5.72	5.78	5.85	5.90	5.96	.05	15
6.66	6.76	6.84	6.93	7.00	7.07	7.14	7.20	7.26	.01	
5.35	5.44	5.52	5.59	5.66	5.73	5.79	5.84	5.90	.05	16
6.56	6.66	6.74	6.82	6.90	6.97	7.03	7.09	7.15	.01	
5.31	5.39	5.47	5.54	5.61	5.67	5.73	5.79	5.84	.05	17
6.48	6.57	6.66	6.73	6.81	6.87	6.94	7.00	7.05	.01	
5.27	5.35	5.43	5.50	5.57	5.63	5.69	5.74	5.79	.05	18
6.41	6.50	6.58	6.65	6.73	6.79	6.85	6.91	6.97	.01	
5.23	5.31	5.39	5.46	5.53	5.59	5.65	5.70	5.75	.05	19
6.34	6.43	6.51	6.58	6.65	6.72	6.78	6.84	6.89	.01	
5.20	5.28	5.36	5.43	5.49	5.55	5.61	5.66	5.71	.05	20
6.28	6.32	6.45	6.52	6.59	6.65	6.71	6.77	6.82	.01	
5.10	5.18	5.25	5.32	5.38	5.44	5.49	5.55	5.59	.05	24
6.11	6.19	6.26	6.33	6.39	6.45	6.51	6.56	6.61	.01	
5.00	5.08	5.15	5.21	5.27	5.33	5.38	5.43	5.47	.05	30
5.93	6.01	6.08	6.14	6.20	6.26	6.31	6.36	6.41	.01	
4.90	4.98	5.04	5.11	5.16	5.22	5.27	5.31	5.36	.05	40
5.76	5.83	5.90	5.96	6.02	6.07	6.12	6.16	6.21	.01	
4.81	4.88	4.94	5.00	5.06	5.11	5.15	5.20	5.24	.05	60
5.60	5.67	5.73	5.78	5.84	5.89	5.93	5.97	6.01	.01	
4.71	4.78	4.84	4.90	4.95	5.00	5.04	5.09	5.13	.05	120
5.44	5.50	5.56	5.61	5.66	5.71	5.75	5.79	5.83	.01	
4.62	4.68	4.74	4.80	4.85	4.89	4.93	4.97	5.01	.05	∞
5.29	5.35	5.40	5.45	5.49	5.54	5.57	5.61	5.65	.01	

Source: Table 29 of E. S. Pearson and H. O. Hartley, *Biometrika Tables for Statisticians*, Vol. 1, 3rd ed., 1966. Used by permission of the Biometrika Trustees.

Table G.12 *(continued)*

b. Distribution of the Studentized Range Statistic : Supplementary Table

df_E	α	r = number of steps between ordered means													
		2	3	4	5	6	7	8	9	10	11	12	13	14	15
1	.05	18.0	27.0	32.8	37.1	40.4	43.1	45.4	47.4	49.1	50.6	52.0	53.2	54.3	55.4
	.01	90.0	135	164	186	202	216	227	237	246	253	260	266	272	277
2	.05	6.09	8.3	9.8	10.9	11.7	12.4	13.0	13.5	14.0	14.4	14.7	15.1	15.4	15.7
	.01	14.0	19.0	22.3	24.7	26.6	28.2	29.5	30.7	31.7	32.6	33.4	34.1	34.8	35.4
3	.05	4.50	5.91	6.82	7.50	8.04	8.48	8.85	9.18	9.46	9.72	9.95	10.2	10.4	10.5
	.01	8.26	10.6	12.2	13.3	14.2	15.0	15.6	16.2	16.7	17.1	17.5	17.9	18.2	18.5
4	.05	3.93	5.04	5.76	6.29	6.71	7.05	7.35	7.60	7.83	8.03	8.21	8.37	8.52	8.66
	.01	6.51	8.12	9.17	9.96	10.6	11.1	11.5	11.9	12.3	12.6	12.8	13.1	13.3	13.5

ource: From WADC Tech. Rep. 58-484, vol. 2, 1959, Wright Air Development Center.

Table G.13 Significant Studentized Ranges (Critical Q_D Values) for Duncan's New Multiple-Range Test

a. $\alpha = 0.05$

df_E \ k	2	3	4	5	6	7	8	9	10	11	12	13	14	15	16	17	18	19
2	6.085																	
3	4.501	4.516																
4	3.927	4.013	4.033															
5	3.635	3.749	3.797	3.814														
6	3.461	3.587	3.649	3.680	3.694													
7	3.344	3.477	3.548	3.588	3.611	3.622												
8	3.261	3.399	3.475	3.521	3.549	3.566	3.575											
9	3.199	3.339	3.420	3.470	3.502	3.523	3.536	3.544										
10	3.151	3.293	3.376	3.430	3.465	3.489	3.505	3.516	3.522									
11	3.113	3.256	3.342	3.397	3.435	3.462	3.480	3.493	3.501	3.506								
12	3.082	3.225	3.313	3.370	3.410	3.439	3.459	3.474	3.484	3.491	3.496							
13	3.055	3.200	3.289	3.348	3.389	3.419	3.442	3.458	3.470	3.478	3.484	3.488						
14	3.033	3.178	3.268	3.329	3.372	3.403	3.426	3.444	3.457	3.467	3.474	3.479	3.482					
15	3.014	3.160	3.250	3.312	3.356	3.389	3.413	3.432	3.446	3.457	3.465	3.471	3.476	3.478				
16	2.998	3.144	3.235	3.298	3.343	3.376	3.402	3.422	3.437	3.449	3.458	3.465	3.470	3.473	3.477			
17	2.984	3.130	3.222	3.285	3.331	3.366	3.392	3.412	3.429	3.441	3.451	3.459	3.465	3.469	3.473	3.475		
18	2.971	3.118	3.210	3.274	3.321	3.356	3.383	3.405	3.421	3.435	3.445	3.454	3.460	3.465	3.470	3.472	3.474	
19	2.960	3.107	3.199	3.264	3.311	3.347	3.375	3.397	3.415	3.429	3.440	3.449	3.456	3.462	3.467	3.470	3.472	3.473
20	2.950	3.097	3.190	3.255	3.303	3.339	3.368	3.391	3.409	3.424	3.436	3.445	3.453	3.459	3.464	3.467	3.470	3.472
24	2.919	3.066	3.160	3.226	3.276	3.315	3.345	3.370	3.390	3.406	3.420	3.432	3.441	3.449	3.456	3.461	3.465	3.469
30	2.888	3.035	3.131	3.199	3.250	3.290	3.322	3.349	3.371	3.389	3.405	3.418	3.430	3.439	3.447	3.454	3.460	3.466
40	2.858	3.006	3.102	3.171	3.224	3.266	3.300	3.328	3.352	3.373	3.390	3.405	3.418	3.429	3.439	3.448	3.456	3.463
60	2.829	2.976	3.073	3.143	3.198	3.241	3.277	3.307	3.333	3.355	3.374	3.391	3.406	3.419	3.431	3.442	3.451	3.460
120	2.800	2.947	3.045	3.116	3.172	3.217	3.254	3.287	3.314	3.337	3.359	3.377	5.394	3.409	3.423	3.435	3.446	3.457
∞	2.772	2.918	3.017	3.089	3.146	3.193	3.232	3.265	3.294	3.320	3.343	3.363	3.382	3.399	3.414	3.428	3.442	3.454

Source: Reproduced from H. L. Harter, "Critical Values for Duncan's New Multiple Range Test," *Biometrics, 16*, 671-685 (1960). With permission from The Biometric Society.

Table G.13 used in same way as Table G.12.

Table G.13 *(continued)*

b. $\alpha = 0.01$

df_E \ k	2	3	4	5	6	7	8	9	10	11	12	13	14	15	16	17	18	19
2	14.04																	
3	8.261	8.321																
4	6.512	6.677	6.740															
5	5.702	5.893	5.989	6.040														
6	5.243	5.439	5.549	5.614	5.655													
7	4.949	5.145	5.260	5.334	5.383	5.416												
8	4.746	4.939	5.057	5.135	5.189	5.227	5.256											
9	4.596	4.787	4.906	4.986	5.043	5.086	5.118	5.142										
10	4.482	4.671	4.790	4.871	4.931	4.975	5.010	5.037	5.058									
11	4.392	4.579	4.697	4.780	4.841	4.887	4.924	4.952	4.975	4.994								
12	4.320	4.504	4.622	4.706	4.767	4.815	4.852	4.883	4.907	4.927	4.944							
13	4.260	4.442	4.560	4.644	4.706	4.755	4.793	4.824	4.850	4.872	4.889	4.904						
14	4.210	4.391	4.508	4.591	4.654	4.704	4.743	4.775	4.802	4.824	4.843	4.859	4.872					
15	4.168	4.347	4.463	4.547	4.610	4.660	4.700	4.733	4.760	4.783	4.803	4.820	4.834	4.846				
16	4.131	4.309	4.425	4.509	4.572	4.622	4.663	4.696	4.724	4.748	4.768	4.786	4.800	4.813	4.825			
17	4.099	4.275	4.391	4.475	4.539	4.589	4.630	4.664	4.693	4.717	4.738	4.756	4.771	4.785	4.797	4.807		
18	4.071	4.246	4.362	4.445	4.509	4.560	4.601	4.635	4.664	4.689	4.711	4.729	4.745	4.759	4.772	4.783	4.792	
19	4.046	4.220	4.335	4.419	4.483	4.534	4.575	4.610	4.639	4.665	4.686	4.705	4.722	4.736	4.749	4.761	4.771	4.780
20	4.024	4.197	4.312	4.395	4.459	4.510	4.552	4.587	4.617	4.642	4.664	4.684	4.701	4.716	4.729	4.741	4.751	4.761
24	3.956	4.126	4.239	4.322	4.386	4.437	4.480	4.516	4.546	4.573	4.596	4.616	4.634	4.651	4.665	4.678	4.690	4.700
30	3.889	4.056	4.168	4.250	4.314	4.366	4.409	4.445	4.477	4.504	4.528	4.550	4.569	4.586	4.601	4.615	4.628	4.640
40	3.825	3.988	4.098	4.180	4.244	4.296	4.339	4.376	4.408	4.436	4.461	4.483	4.503	4.521	4.537	4.553	4.566	4.579
60	3.762	3.922	4.031	4.111	4.174	4.226	4.270	4.307	4.340	4.368	4.394	4.417	4.438	4.456	4.474	4.490	4.504	4.518
120	3.702	3.858	3.965	4.044	4.107	4.158	4.202	4.239	4.272	4.301	4.327	4.351	4.372	4.392	4.410	4.426	4.442	4.456
∞	3.643	3.796	3.900	3.978	4.040	4.091	4.135	4.172	4.205	4.235	4.261	4.285	4.307	4.327	4.345	4.363	4.379	4.394

Table G.13 *(continued)*

c. $\alpha = 0.001$

df_E \ k	2	3	4	5	6	7	8	9	10	11	12	13	14	15	16	17	18	19
2	44.69																	
3	18.28	18.45																
4	12.18	12.52	12.67															
5	9.714	10.05	10.24	10.35														
6	8.427	8.743	8.932	9.055	9.139													
7	7.648	7.943	8.127	8.252	8.342	8.409												
8	7.130	7.407	7.584	7.708	7.799	7.869	7.924											
9	6.762	7.024	7.195	7.316	7.407	7.478	7.535	7.582										
10	6.487	6.738	6.902	7.021	7.111	7.182	7.240	7.287	7.327									
11	6.275	6.516	6.676	6.791	6.880	6.950	7.008	7.056	7.097	7.132								
12	6.106	6.340	6.494	6.607	6.695	6.765	6.822	6.870	6.911	6.947	6.978							
13	5.970	6.195	6.346	6.457	6.543	6.612	6.670	6.718	6.759	6.795	6.826	6.854						
14	5.856	6.075	6.223	6.332	6.416	6.485	6.542	6.590	6.631	6.667	6.699	6.727	6.752					
15	5.760	5.974	6.119	6.225	6.309	6.377	6.433	6.481	6.522	6.558	6.590	6.619	6.644	6.666				
16	5.678	5.888	6.030	6.135	6.217	6.284	6.340	6.388	6.429	6.465	6.497	6.525	6.551	6.574	6.595			
17	5.608	5.813	5.953	6.056	6.138	6.204	6.260	6.307	6.348	6.384	6.416	6.444	6.470	6.493	6.514	6.533		
18	5.546	5.748	5.886	5.988	6.068	6.134	6.189	6.236	6.277	6.313	6.345	6.373	6.399	6.422	6.443	6.462	6.480	
19	5.492	5.691	5.826	5.927	6.007	6.072	6.127	6.174	6.214	6.250	6.281	6.310	6.336	6.359	6.380	6.400	6.418	6.434
20	5.444	5.640	5.774	5.873	5.952	6.017	6.071	6.117	6.158	6.193	6.225	6.254	6.279	6.303	6.324	6.344	6.362	6.379
24	5.297	5.484	5.612	5.708	5.784	5.846	5.899	5.945	5.984	6.020	6.051	6.079	6.105	6.129	6.150	6.170	6.188	6.205
30	5.156	5.335	5.457	5.549	5.622	5.682	5.734	5.778	5.817	5.851	5.882	5.910	5.935	5.958	5.980	6.000	6.018	6.036
40	5.022	5.191	5.308	5.396	5.466	5.524	5.574	5.617	5.654	5.688	5.718	5.745	5.770	5.793	5.814	5.834	5.852	5.869
60	4.894	5.055	5.166	5.249	5.317	5.372	5.420	5.461	5.498	5.530	5.559	5.586	5.610	5.632	5.653	5.672	5.690	5.707
120	4.771	4.924	5.029	5.109	5.173	5.226	5.271	5.311	5.346	5.377	5.405	5.431	5.454	5.476	5.496	5.515	5.532	5.549
∞	4.654	4.798	4.898	4.974	5.034	5.085	5.128	5.166	5.199	5.229	5.256	5.280	5.303	5.324	5.343	5.361	5.378	5.394

Table G.14 Critical Values of the Dunnett Test for Comparing Treatment Means with a Control

ONE-TAILED COMPARISONS

df_{error}	α	k = number of treatment means, including control: 2	3	4	5	6	7	8	9	10
5	.05	2.02	2.44	2.68	2.85	2.98	3.08	3.16	3.24	3.30
	.01	3.37	3.90	4.21	4.43	4.60	4.73	4.85	4.94	5.03
6	.05	1.94	2.34	2.56	2.71	2.83	2.92	3.00	3.07	3.12
	.01	3.14	3.61	3.88	4.07	4.21	4.33	4.43	4.51	4.59
7	.05	1.89	2.27	2.48	2.62	2.73	2.82	2.89	2.95	3.01
	.01	3.00	3.42	3.66	3.83	3.96	4.07	4.15	4.23	4.30
8	.05	1.86	2.22	2.42	2.55	2.66	2.74	2.81	2.87	2.92
	.01	2.90	3.29	3.51	3.67	3.79	3.88	3.96	4.03	4.09
9	.05	1.83	2.18	2.37	2.50	2.60	2.68	2.75	2.81	2.86
	.01	2.82	3.19	3.40	3.55	3.66	3.75	3.82	3.89	3.94
10	.05	1.81	2.15	2.34	2.47	2.56	2.64	2.70	2.76	2.81
	.01	2.76	3.11	3.31	3.45	3.56	3.64	3.71	3.78	3.83
11	.05	1.80	2.13	2.31	2.44	2.53	2.60	2.67	2.72	2.77
	.01	2.72	3.06	3.25	3.38	3.48	3.56	3.63	3.69	3.74
12	.05	1.78	2.11	2.29	2.41	2.50	2.58	2.64	2.69	2.74
	.01	2.68	3.01	3.19	3.32	3.42	3.50	3.56	3.62	3.67
13	.05	1.77	2.09	2.27	2.39	2.48	2.55	2.61	2.66	2.71
	.01	2.65	2.97	3.15	3.27	3.37	3.44	3.51	3.56	3.61
14	.05	1.76	2.08	2.25	2.37	2.46	2.53	2.59	2.64	2.69
	.01	2.62	2.94	3.11	3.23	3.32	3.40	3.46	3.51	3.56
15	.05	1.75	2.07	2.24	2.36	2.44	2.51	2.57	2.62	2.67
	.01	2.60	2.91	3.08	3.20	3.29	3.36	3.42	3.47	3.52
16	.05	1.75	2.06	2.23	2.34	2.43	2.50	2.56	2.61	2.65
	.01	2.58	2.88	3.05	3.17	3.26	3.33	3.39	3.44	3.48
17	.05	1.74	2.05	2.22	2.33	2.42	2.49	2.54	2.59	2.64
	.01	2.57	2.86	3.03	3.14	3.23	3.30	3.36	3.41	3.45
18	.05	1.73	2.04	2.21	2.32	2.41	2.48	2.53	2.58	2.62
	.01	2.55	2.84	3.01	3.12	3.21	3.27	3.33	3.38	3.42
19	.05	1.73	2.03	2.20	2.31	2.40	2.47	2.52	2.57	2.61
	.01	2.54	2.83	2.99	3.10	3.18	3.25	3.31	3.36	3.40
20	.05	1.72	2.03	2.19	2.30	2.39	2.46	2.51	2.56	2.60
	.01	2.53	2.81	2.97	3.08	3.17	3.23	3.29	3.34	3.38
24	.05	1.71	2.01	2.17	2.28	2.36	2.43	2.48	2.53	2.57
	.01	2.49	2.77	2.92	3.03	3.11	3.17	3.22	3.27	3.31
30	.05	1.70	1.99	2.15	2.25	2.33	2.40	2.45	2.50	2.54
	.01	2.46	2.72	2.87	2.97	3.05	3.11	3.16	3.21	3.24

Table G.14 *(continued)*

df_{error}	α	k = number of treatment means, including control: 2	3	4	5	6	7	8	9	10
40	.05	1.68	1.97	2.13	2.23	2.31	2.37	2.42	2.47	2.51
	.01	2.42	2.68	2.82	2.92	2.99	3.05	3.10	3.14	3.18
60	.05	1.67	1.95	2.10	2.21	2.28	2.35	2.39	2.44	2.48
	.01	2.39	2.64	2.78	2.87	2.94	3.00	3.04	3.08	3.12
120	.05	1.66	1.93	2.08	2.18	2.26	2.32	2.37	2.41	2.45
	.01	2.36	2.60	2.73	2.82	2.89	2.94	2.99	3.03	3.06
∞	.05	1.64	1.92	2.06	2.16	2.23	2.29	2.34	2.38	2.42
	.01	2.33	2.56	2.68	2.77	2.84	2.89	2.93	2.97	3.00

TWO-TAILED COMPARISONS

df_{error}	α	k = number of treatment means, including control: 2	3	4	5	6	7	8	9	10
5	.05	2.57	3.03	3.29	3.48	3.62	3.73	3.82	3.90	3.97
	.01	4.03	4.63	4.98	5.22	5.41	5.56	5.69	5.80	5.89
6	.05	2.45	2.86	3.10	3.26	3.39	3.49	3.57	3.64	3.71
	.01	3.71	4.21	4.51	4.71	4.87	5.00	5.10	5.20	5.28
7	.05	2.36	2.75	2.97	3.12	3.24	3.33	3.41	3.47	3.53
	.01	3.50	3.95	4.21	4.39	4.53	4.64	4.74	4.82	4.89
8	.05	2.31	2.67	2.88	3.02	3.13	3.22	3.29	3.35	3.41
	.01	3.36	3.77	4.00	4.17	4.29	4.40	4.48	4.56	4.62
9	.05	2.26	2.61	2.81	2.95	3.05	3.14	3.20	3.26	3.32
	.01	3.25	3.63	3.85	4.01	4.12	4.22	4.30	4.37	4.43
10	.05	2.23	2.57	2.76	2.89	2.99	3.07	3.14	3.19	3.24
	.01	3.17	3.53	3.74	3.88	3.99	4.08	4.16	4.22	4.28
11	.05	2.20	2.53	2.72	2.84	2.94	3.02	3.08	3.14	3.19
	.01	3.11	3.45	3.65	3.79	3.89	3.98	4.05	4.11	4.16
12	.05	2.18	2.50	2.68	2.81	2.90	2.98	3.04	3.09	3.14
	.01	3.05	3.39	3.58	3.71	3.81	3.89	3.96	4.02	4.07
13	.05	2.16	2.48	2.65	2.78	2.87	2.94	3.00	3.06	3.10
	.01	3.01	3.33	3.52	3.65	3.74	3.82	3.89	3.94	3.99
14	.05	2.14	2.46	2.63	2.75	2.84	2.91	2.97	3.02	3.07
	.01	2.98	3.29	3.47	3.59	3.69	3.76	3.83	3.88	3.93
15	.05	2.13	2.44	2.61	2.73	2.82	2.89	2.95	3.00	3.04
	.01	2.95	3.25	3.43	3.55	3.64	3.71	3.78	3.83	3.88

Table G.14 *(continued)*

		k = number of treatment means, including control								
df_{error}	α	2	3	4	5	6	7	8	9	10
16	.05	2.12	2.42	2.59	2.71	2.80	2.87	2.92	2.97	3.02
	.01	2.92	3.22	3.39	3.51	3.60	3.67	3.73	3.78	3.83
17	.05	2.11	2.41	2.58	2.69	2.78	2.85	2.90	2.95	3.00
	.01	2.90	3.19	3.36	3.47	3.56	3.63	3.69	3.74	3.79
18	.05	2.10	2.40	2.56	2.68	2.76	2.83	2.89	2.94	2.98
	.01	2.88	3.17	3.33	3.44	3.53	3.60	3.66	3.71	3.75
19	.05	2.09	2.39	2.55	2.66	2.75	2.81	2.87	2.92	2.96
	.01	2.86	3.15	3.31	3.42	3.50	3.57	3.63	3.68	3.72
20	.05	2.09	2.38	2.54	2.65	2.73	2.80	2.86	2.90	2.95
	.01	2.85	3.13	3.29	3.40	3.48	3.55	3.60	3.65	3.69
24	.05	2.06	2.35	2.51	2.61	2.70	2.76	2.81	2.86	2.90
	.01	2.80	3.07	3.22	3.32	3.40	3.47	3.52	3.57	3.61
30	.05	2.04	2.32	2.47	2.58	2.66	2.72	2.77	2.82	2.86
	.01	2.75	3.01	3.15	3.25	3.33	3.39	3.44	3.49	3.52
40	.05	2.02	2.29	2.44	2.54	2.62	2.68	2.73	2.77	2.81
	.01	2.70	2.95	3.09	3.19	3.26	3.32	3.37	3.41	3.44
60	.05	2.00	2.27	2.41	2.51	2.58	2.64	2.69	2.73	2.77
	.01	2.66	2.90	3.03	3.12	3.19	3.25	3.29	3.33	3.37
120	.05	1.98	2.24	2.38	2.47	2.55	2.60	2.65	2.69	2.73
	.01	2.62	2.85	2.97	3.06	3.12	3.18	3.22	3.26	3.29
∞	.05	1.96	2.21	2.35	2.44	2.51	2.57	2.61	2.65	2.69
	.01	2.58	2.79	2.92	3.00	3.06	3.11	3.15	3.19	3.22

Sources: One-tailed comparisons: C. W. Dunnett, "A multiple comparison procedure for comparing several treatments with a control," *Journal of the American Statistical Association, 50*, 1096-1121 (1955); two-tailed comparisons: C. W. Dunnett, "New tables for multiple comparisons with a control," *Biometrics, 20*, 482-491 (1964); by permission of the author and the editors.

Table G.15 Critical Values of the Dunn Multiple-Comparison Test[a]

Number of Comparisons		error *df*											
(*C*)	α	5	7	10	12	15	20	24	30	40	60	120	∞
2	.05	3.17	2.84	2.64	2.56	2.49	2.43	2.39	2.36	2.33	2.30	2.27	2.24
	.01	4.78	4.03	3.58	3.43	3.29	3.16	3.09	3.30	2.97	2.92	2.86	2.81
3	.05	3.54	3.13	2.87	2.78	2.69	2.61	2.58	2.54	2.50	2.47	2.43	2.39
	.01	5.25	4.36	3.83	3.65	3.48	3.33	3.26	3.19	3.12	3.06	2.99	2.94
4	.05	3.81	3.34	3.04	2.94	2.84	2.75	2.70	2.66	2.62	2.58	2.54	2.50
	.01	5.60	4.59	4.01	3.80	3.62	3.46	3.38	3.30	3.23	3.16	3.09	3.02
5	.05	4.04	3.50	3.17	3.06	2.95	2.85	2.80	2.75	2.71	2.66	2.62	2.58
	.01	5.89	4.78	4.15	3.93	3.74	3.55	3.47	3.39	3.31	3.24	3.16	3.09
6	.05	4.22	3.64	3.28	3.15	3.04	2.93	2.88	2.83	2.78	2.73	2.68	2.64
	.01	6.15	4.95	4.27	4.04	3.82	3.63	3.54	3.46	3.38	3.30	3.22	3.15
7	.05	4.38	3.76	3.37	3.24	3.11	3.00	2.94	2.89	2.84	2.79	2.74	2.69
	.01	6.36	5.09	4.37	4.13	3.90	3.70	3.61	3.52	3.43	3.34	3.27	3.19
8	.05	4.53	3.86	3.45	3.31	3.18	3.06	3.00	2.94	2.89	2.84	2.79	2.74
	.01	6.56	5.21	4.45	4.20	3.97	3.76	3.66	3.57	3.48	3.39	3.31	3.23
9	.05	4.66	3.95	3.52	3.37	3.24	3.11	3.05	2.99	2.93	2.88	2.83	2.77
	.01	6.70	5.31	4.53	4.26	4.02	3.80	3.70	3.61	3.51	3.42	3.34	3.26
10	.05	4.78	4.03	3.58	3.43	3.29	3.16	3.09	3.303	2.97	2.92	2.86	2.81
	.01	6.86	5.40	4.59	4.32	4.07	3.85	3.74	3.65	3.55	3.46	2.37	3.29

15	.05	5.25	4.36	3.83	3.65	3.48	3.33	3.26	3.19	3.12	3.06	2.99	2.94
	.01	7.51	5.79	4.86	4.56	4.29	4.03	3.91	3.80	3.70	3.59	3.50	3.40
20	.05	5.60	4.59	4.01	3.80	3.62	3.46	3.38	3.30	3.23	3.16	3.09	3.02
	.01	8.00	6.08	5.06	4.73	4.42	4.15	4.04	3.90	3.79	3.69	3.58	3.48
25	.05	5.89	4.78	4.15	3.93	3.74	3.55	3.47	3.39	3.31	3.24	3.16	3.09
	.01	8.37	6.30	5.20	4.86	4.53	4.25	4.1*	3.98	3.88	3.76	3.64	3.54
30	.05	6.15	4.95	4.27	4.04	3.82	3.63	3.54	3.46	3.38	3.30	3.22	3.15
	.01	8.68	6.49	5.33	4.95	4.61	4.33	4.2*	4.13	3.93	3.81	3.69	3.59
35	.05	6.36	5.09	4.37	4.13	3.90	3.70	3.61	3.52	3.43	3.34	3.27	3.19
	.01	8.95	6.67	5.44	5.04	4.71	4.39	4.3*	4.26	3.975	3.84	3.73	3.63
40	.05	6.56	5.21	4.45	4.20	3.97	3.76	3.66	3.57	3.48	3.39	3.31	3.23
	.01	9.19	6.83	5.52	5.12	4.78	4.46	4.3*	4.1*	4.01	3.89	3.77	3.66
45	.05	6.70	5.31	4.53	4.26	4.02	3.80	3.70	3.61	3.51	3.42	3.34	3.26
	.01	9.41	6.93	5.60	5.20	4.84	4.52	4.3*	4.2*	4.1*	3.93	3.80	3.69
50	.05	6.86	5.40	4.59	4.32	4.07	3.85	3.74	3.65	3.55	3.46	3.37	3.29
	.01	9.68	7.06	5.70	5.27	4.90	4.56	4.4*	4.2*	4.1*	3.97	3.83	3.72
100	.05	8.00	6.08	5.06	4.73	4.42	4.15	4.04	3.90	3.79	3.68	3.58	3.48
	.01	11.04	7.80	6.20	5.70	5.20	4.80	4.7*	4.4*	4.5*		4.00	3.89
250	.05	9.68	7.06	5.70	5.27	4.90	4.56	4.4*	4.2*	4.1*	3.97	3.83	3.72
	.01	13.26	8.83	6.9*	6.3*	5.8*	5.2*[a]	5.0*	4.9*	4.8*			4.11

[a] Values followed by asterisks were obtained by graphical interpolation.

Source: O. J. Dunn, "Multiple comparisons among means," *Journal of the American Statistical Association, 56*, 52-64 (1961); by permission of the author and the editor.

Table G.16 Critical Values of the Pearson Product-Moment Correlation Cofficient[a]

	Level of significance for one-tailed test				
	0.05	0.025	0.01	0.005	0.0005
	Level of significance for two-tailed test				
df = N - 2	0.10	0.05	0.02	0.01	0.001
1	0.9877	0.9969	0.9995	0.9999	1.0000
2	0.9000	0.9500	0.9800	0.9900	0.9990
3	0.8054	0.8783	0.9343	0.9587	0.9912
4	0.7293	0.8114	0.8822	0.9172	0.9741
5	0.6694	0.7545	0.8329	0.8745	0.9507
6	0.6215	0.7067	0.7887	0.8343	0.9249
7	0.5822	0.6664	0.7498	0.7977	0.8982
8	0.5494	0.6319	0.7155	0.7646	0.8721
9	0.5214	0.6021	0.6851	0.7348	0.8471
10	0.4973	0.5760	0.6581	0.7079	0.8233
11	0.4762	0.5529	0.6339	0.6835	0.8010
12	0.4575	0.5324	0.6120	0.6614	0.7800
13	0.4409	0.5139	0.5923	0.6411	0.7603
14	0.4259	0.4973	0.5742	0.6226	0.7420
15	0.4124	0.4821	0.5577	0.6055	0.7246
16	0.4000	0.4683	0.5425	0.5897	0.7084
17	0.3887	0.4555	0.5285	0.5751	0.6932
18	0.3783	0.4438	0.5155	0.5614	0.6787
19	0.3687	0.4329	0.5034	0.5487	0.6652
20	0.3598	0.4227	0.4921	0.5368	0.6524
25	0.3233	0.3809	0.4451	0.4869	0.5974
30	0.2960	0.3494	0.4093	0.4487	0.5541
35	0.2746	0.3246	0.3810	0.4182	0.5189
40	0.2573	0.3044	0.3578	0.3932	0.4896
45	0.2428	0.2875	0.3384	0.3721	0.4648
50	0.2306	0.2732	0.3218	0.3541	0.4433
60	0.2108	0.2500	0.2948	0.3248	0.4078
70	0.1954	0.2319	0.2737	0.3017	0.3799
80	0.1829	0.2172	0.2565	0.2830	0.3568
90	0.1726	0.2050	0.2422	0.2673	0.3375
100	0.1638	0.1946	0.2301	0.2540	0.3211

[a] If the observed value of r is *greater than or equal to* the tabled value for the appropriate level of significance (columns) and degrees of freedom (rows), then reject H_0. The degrees of freedom are the number of pairs of scores minus two, or N − 2.

Source: Table VII of Fisher and Yates, *Statistical Tables for Biological, Agricultural and Medical Research*, published by Longman Group Ltd., London (previously published by Oliver and Boyd Ltd., Edinburgh) and by permission of the authors and publishers.

Table G.17 Transformation of r to Z (Fisher's Z)

r	Z	r	Z	r	Z	r	Z	r	Z
.000	.000	.200	.203	.400	.424	.600	.693	.800	1.099
.005	.005	.205	.208	.405	.430	.605	.701	.805	1.113
.010	.010	.210	.213	.410	.436	.610	.709	.810	1.127
.015	.015	.215	.218	.415	.442	.615	.717	.815	1.142
.020	.020	.220	.224	.420	.448	.620	.725	.820	1.157
.025	.025	.225	.229	.425	.454	.625	.733	.825	1.172
.030	.030	.230	.234	.430	.460	.630	.741	.830	1.188
.035	.035	.235	.239	.435	.466	.635	.750	.835	1.204
.040	.040	.240	.245	.440	.472	.640	.758	.840	1.221
.045	.045	.245	.250	.445	.478	.645	.767	.845	1.238
.050	.050	.250	.255	.450	.485	.650	.775	.850	1.256
.055	.055	.255	.261	.455	.491	.655	.784	.855	1.274
.060	.060	.260	.266	.460	.497	.660	.793	.860	1.293
.065	.065	.265	.271	.465	.504	.665	.802	.865	1.313
.070	.070	.270	.277	.470	.510	.670	.811	.870	1.333
.075	.075	.275	.282	.475	.517	.675	.820	.875	1.354
.080	.080	.280	.288	.480	.523	.680	.829	.880	1.376
.085	.085	.285	.293	.485	.530	.685	.838	.885	1.398
.090	.090	.290	.299	.490	.536	.690	.848	.890	1.422
.095	.095	.295	.304	.495	.543	.695	.858	.895	1.447
.100	.100	.300	.310	.500	.549	.700	.867	.900	1.472
.105	.105	.305	.315	.505	.556	.705	.877	.905	1.499
.110	.110	.310	.321	.510	.563	.710	.887	.910	1.528
.115	.116	.315	.326	.515	.570	.715	.897	.915	1.557
.120	.121	.320	.332	.520	.576	.720	.908	.920	1.589
.125	.126	.325	.337	.525	.583	.725	.918	.925	1.623
.130	.131	.330	.343	.530	.590	.730	.929	.930	1.658
.135	.136	.335	.348	.535	.597	.735	.940	.935	1.697
.140	.141	.340	.354	.540	.604	.740	.950	.940	1.738
.145	.146	.345	.360	.545	.611	.745	.962	.945	1.783
.150	.151	.350	.365	.550	.618	.750	.973	.950	1.832
.155	.156	.355	.371	.555	.626	.755	.984	.955	1.886
.160	.161	.360	.377	.560	.633	.760	.996	.960	1.946
.165	.167	.365	.383	.565	.640	.765	1.008	.965	2.014
.170	.172	.370	.388	.570	.648	.770	1.020	.970	2.092
.175	.177	.375	.394	.575	.655	.775	1.033	.975	2.185
.180	.182	.380	.400	.580	.662	.780	1.045	.980	2.298
.185	.187	.385	.406	.585	.670	.785	1.058	.985	2.443
.190	.192	.390	.412	.590	.678	.790	1.071	.990	2.647
.195	.198	.395	.418	.595	.685	.795	1.085	.995	2.994

Source: A. L. Edwards, *Experimental Design in Psychological Research*, 1968, 3rd edition, Holt, Rinehart and Winston, New York. By permission of the author.

Table G.18 Critical Values of W for the Wilcoxon Test

	One-tailed significance level, α						
	0.100	0.050	0.025	0.010	0.005	0.0025	0.0005
	Two-tailed significance level, α						
n	0.200	0.100	0.050	0.020	0.010	0.005	0.001
4	0						
5	2	0					
6	3	2	0				
7	5	3	2	0			
8	8	5	3	1	0		
9	10	8	5	3	1	0	
10	14	10	8	5	3	1	
11	17	13	10	7	5	3	0
12	21	17	13	9	7	5	1
13	26	21	17	12	9	7	2
14	31	25	21	15	12	9	4
15	36	30	25	19	15	12	6
16	42	35	29	23	19	15	8
17	48	41	34	27	23	19	11
18	55	47	40	32	27	23	14
19	62	53	46	37	32	27	18
20	69	60	52	43	37	32	21
21	77	67	58	49	42	37	25
22	86	75	65	55	48	42	30
23	94	83	73	62	54	48	35
24	104	91	81	69	61	54	40
25	113	100	89	76	68	60	45
26	124	110	98	84	75	67	51
27	134	119	107	92	83	74	57
28	145	130	116	101	91	82	64
29	157	140	126	110	100	90	71
30	169	151	137	120	109	98	78
31	181	163	147	130	118	107	86
32	194	175	159	140	128	116	94
33	207	187	170	151	138	126	102
34	221	200	182	162	148	136	111

Table G.18 *(continued)*

	One-tailed significance level, α						
	0.100	0.050	0.025	0.010	0.005	0.0025	0.0005
	Two-tailed significance level, α						
n	0.200	0.100	0.050	0.020	0.010	0.005	0.001
35	235	213	195	173	159	146	120
36	250	227	208	185	171	157	130
37	265	241	221	198	182	168	140
38	281	256	235	211	194	180	150
39	297	271	249	224	207	192	161
40	313	286	264	238	220	204	172
41	330	302	279	252	233	217	183
42	348	319	294	266	247	230	195
43	365	336	310	281	261	244	207
44	384	353	327	296	278	258	220
45	402	371	343	312	291	272	233
46	422	389	361	328	307	287	246
47	441	407	378	348	322	302	260
48	462	426	396	362	339	318	274
49	482	446	415	379	355	334	289
50	503	466	434	397	373	350	304

For a given N (the number of pairs of scores), if the observed value is less than or equal to the value in the table for the appropriate level of significance, then reject H_0.

Source: R. L. McCormack, "Extended tables of the Wilcoxon matched pair signed rank statistic," *Journal of the American Statistical Association, 60*, 864-871 (1965); by permission of the author and the editor.

Table G.19 Critical Values of the Mann-Whitney U Test[a,b]

a. One-Tailed Test at 0.005 or Two-Tailed Test at 0.01

n_2 \ n_1	1	2	3	4	5	6	7	8	9	10	11	12	13	14	15	16	17	18	19	20
1																				
2																			0	0
																			38	40
3									0	0	0	1	1	1	2	2	2	2	3	3
									27	30	33	35	38	41	43	46	49	52	54	57
4						0	0	1	1	2	2	3	3	4	5	5	6	6	7	8
						24	28	31	35	38	42	45	49	52	55	59	62	66	69	72
5					0	1	1	2	3	4	5	6	7	7	8	9	10	11	12	13
					25	29	34	38	42	46	50	54	58	63	67	71	75	79	83	87
6				0	1	2	3	4	5	6	7	9	10	11	12	13	15	16	17	18
				24	29	34	39	44	49	54	59	63	68	73	78	83	87	92	97	102
7				0	1	3	4	6	7	9	10	12	13	15	16	18	19	21	22	24
				28	34	39	45	50	56	61	67	72	78	83	89	94	100	105	111	116
8				1	2	4	6	7	9	11	13	15	17	18	20	22	24	26	28	30
				31	38	44	50	57	63	69	75	81	87	94	100	106	112	118	124	130
9			0	1	3	5	7	9	11	13	16	18	20	22	24	27	29	31	33	36
			27	35	42	49	56	63	70	77	83	90	97	104	111	117	124	131	138	144

10		0	2	4	6	9	11	13	16	18	21	24	26	29	31	34	37	39	42
		30	38	46	54	61	69	77	84	92	99	106	114	121	129	136	143	151	158
11		0	2	5	7	10	13	16	18	21	24	27	30	33	36	39	42	45	48
		33	42	50	59	67	75	83	92	100	108	116	124	132	140	148	156	164	172
12		1	3	6	9	12	15	18	21	24	27	31	34	37	41	44	47	51	54
		35	45	54	63	72	81	90	99	108	117	125	134	143	151	160	169	177	186
13		1	3	7	10	13	17	20	24	27	31	34	38	42	45	49	53	56	60
		38	49	58	68	78	87	97	106	116	125	125	144	153	163	172	181	191	200
14		1	4	7	11	15	18	22	26	30	34	38	42	46	50	54	58	63	67
		41	52	63	73	83	94	104	114	124	134	144	154	164	174	184	194	203	213
15		2	5	8	12	16	20	24	29	33	37	42	46	51	55	60	64	69	73
		43	55	67	78	89	100	111	121	132	143	153	164	174	185	195	206	216	227
16		2	5	9	13	18	22	27	31	36	41	45	50	55	60	65	70	74	79
		46	59	71	83	94	106	117	129	140	151	163	174	185	196	207	218	230	241
17		2	6	10	15	19	24	29	34	39	44	49	54	60	65	70	75	81	86
		49	62	75	87	100	112	124	148	148	160	172	184	195	207	219	231	242	254
18		2	6	11	16	21	26	31	37	42	47	53	58	64	70	75	81	87	92
		52	66	79	92	105	118	131	143	156	169	181	194	206	218	231	243	255	268
19	0	3	7	12	17	22	28	33	39	45	51	56	63	69	74	81	87	93	99
	38	54	69	83	97	111	124	138	151	164	177	191	203	216	230	242	255	268	281
20	0	3	8	13	18	24	30	36	42	48	54	60	67	73	79	86	92	99	105
	40	57	72	87	102	116	130	144	158	172	186	200	213	227	241	254	268	281	295

Table G.19 *(continued)*

b. One-Tailed Test at 0.01 or Two-Tailed Test at 0.02

n_2 \ n_1	1	2	3	4	5	6	7	8	9	10	11	12	13	14	15	16	17	18	19	20
1																				
2													0 26	0 28	0 30	0 32	0 34	0 36	1 37	1 39
3							0 21	0 24	1 26	1 29	1 32	2 34	2 37	2 40	3 42	3 45	4 47	4 50	4 52	5 55
4					0 20	1 23	1 27	2 30	3 33	3 37	4 40	5 43	5 47	6 50	7 53	7 57	8 60	9 63	9 67	10 70
5				0 20	1 24	2 28	3 32	4 36	5 40	6 44	7 48	8 52	9 56	10 60	11 64	12 68	13 72	14 76	15 80	16 84
6				1 23	2 28	3 33	4 38	6 42	7 47	8 52	9 57	11 61	12 66	13 71	15 75	16 80	18 84	19 89	20 94	22 93
7			0 21	1 27	3 32	4 38	6 43	7 49	9 54	11 59	12 65	14 70	16 75	17 81	19 86	21 91	23 96	24 102	26 107	28 112
8			0 24	2 30	4 36	6 42	7 49	9 55	11 61	13 67	15 73	17 79	20 84	22 90	24 96	26 102	28 108	30 114	32 120	34 126
9			1 26	3 33	5 40	7 47	9 54	11 61	14 67	16 74	18 81	21 87	23 94	26 100	28 107	31 113	33 120	36 126	38 133	40 140

10		1	3	6	8	11	13	16	19	22	24	27	30	33	36	38	41	44	47
		29	37	44	52	59	67	74	81	88	96	103	110	117	124	132	139	146	153
11		1	4	7	9	12	15	18	22	25	28	31	34	37	41	44	47	50	53
		32	40	48	57	65	73	81	88	96	104	112	120	128	135	143	151	159	167
12		2	5	8	11	14	17	21	24	28	31	35	38	42	46	49	53	56	60
		34	43	52	61	70	79	87	96	104	113	121	130	138	146	155	163	172	180
13	0	2	5	9	12	16	20	23	27	31	35	39	43	47	51	55	59	63	67
	26	37	47	56	66	76	84	94	103	112	121	130	139	148	157	166	175	184	193
14	0	2	6	10	13	17	22	26	30	34	38	43	47	51	56	60	65	69	73
	28	40	50	60	71	81	90	100	110	120	130	139	149	159	168	178	187	197	207
15	0	3	7	11	15	19	24	28	33	37	42	47	51	56	61	66	70	75	80
	30	42	53	64	75	86	96	107	117	128	138	148	159	169	179	189	200	210	220
16	0	3	7	12	16	21	26	31	36	41	46	51	56	61	66	71	76	82	87
	32	45	57	68	80	91	102	113	124	135	146	157	168	179	190	201	212	222	233
17	0	4	9	13	18	23	28	33	38	44	49	55	60	66	71	77	82	88	93
	34	47	60	72	84	96	108	120	132	143	155	166	178	189	201	212	224	234	247
18	0	4	9	14	19	24	30	36	41	47	53	59	65	70	76	82	88	94	100
	36	50	63	76	89	102	114	126	139	151	163	175	187	200	212	224	236	248	260
19	1	4	9	15	20	26	32	38	44	50	46	63	69	75	82	88	94	101	107
	37	53	67	80	94	107	120	133	146	159	172	184	197	210	272	235	248	260	273
20	1	5	10	16	22	28	34	40	47	53	60	67	73	80	87	93	100	107	114
	39	55	70	84	98	112	126	140	153	167	180	193	207	220	233	247	260	273	286

Table G.19 *(continued)*

c. One-Tailed Test at 0.025 or Two-Tailed Test at 0.05

n_2 \ n_1	1	2	3	4	5	6	7	8	9	10	11	12	13	14	15	16	17	18	19	20
1																				
2								0	0	0	0	1	1	1	1	1	2	2	2	2
								16	18	20	22	23	25	27	29	31	32	34	36	38
3					0	1	1	2	2	3	3	4	4	5	5	6	5	7	7	8
					15	17	20	22	25	27	30	32	35	37	40	42	45	47	50	52
4				0	1	2	3	4	4	5	6	7	8	9	10	11	11	12	13	13
				16	19	22	25	28	32	35	38	41	44	47	50	53	57	60	63	67
5			0	1	2	3	5	6	7	8	9	11	12	13	14	15	17	18	19	20
			15	19	23	27	30	34	38	42	46	49	53	57	61	65	68	72	76	80
6			1	2	3	5	6	8	10	11	13	14	16	17	19	21	22	24	25	27
			17	22	27	31	36	40	44	49	53	58	62	67	71	75	80	84	89	93
7			1	3	5	6	8	10	12	14	16	18	20	22	24	26	28	30	32	34
			20	25	30	36	41	46	51	56	61	66	71	76	81	86	91	96	101	106
8		0	2	4	6	8	10	13	15	17	19	22	24	26	29	31	34	36	38	41
		16	22	28	34	40	46	51	57	63	69	74	80	86	91	97	102	108	111	119
9		0	2	4	7	10	12	15	17	20	23	26	28	31	34	37	39	42	45	48
		18	25	32	38	44	51	57	64	70	76	82	89	95	101	107	114	120	126	132

10	0	3	5	8	11	14	17	20	23	26	29	33	36	39	42	45	48	52	55
	20	27	35	42	49	56	63	70	77	84	91	97	104	111	118	125	132	138	145
11	0	3	6	9	13	16	19	23	26	30	33	37	40	44	47	51	55	58	62
	22	30	38	46	53	61	69	76	84	91	99	106	114	121	129	136	143	151	158
12	1	4	7	11	14	18	22	26	29	33	37	41	45	49	53	57	61	65	69
	23	32	41	49	58	66	74	82	91	99	107	115	123	131	139	147	155	163	171
13	1	4	8	12	16	20	24	28	33	37	41	45	50	54	59	63	67	72	76
	25	35	44	53	62	71	80	89	97	106	115	124	132	141	149	158	167	175	184
14	1	5	9	13	17	22	26	31	36	40	45	50	55	59	64	67	74	78	83
	27	37	47	51	67	76	86	95	104	114	123	132	141	151	160	171	178	188	197
15	1	5	10	14	19	24	29	34	39	44	49	54	59	64	70	75	80	85	90
	29	40	50	61	71	81	91	101	111	121	131	141	151	161	170	180	190	200	210
16	1	6	11	15	21	26	31	37	42	47	53	59	64	70	75	81	86	92	98
	31	42	53	65	75	86	97	107	118	129	139	149	160	170	181	191	202	212	222
17	2	6	11	17	22	28	34	39	45	51	57	63	67	75	81	87	93	99	105
	32	45	57	68	80	91	102	114	125	136	147	158	171	180	191	202	213	224	235
18	2	7	12	18	24	30	36	42	48	55	61	67	74	80	86	93	99	106	112
	34	47	60	72	84	96	108	120	132	143	155	167	178	190	202	213	225	236	248
19	2	7	13	19	25	32	38	45	52	58	65	72	78	85	92	99	106	113	119
	36	50	63	76	89	101	114	126	138	151	163	175	188	200	212	224	236	248	261
20	2	8	13	20	27	34	41	48	55	62	69	76	83	90	98	105	112	119	127
	38	52	67	80	93	106	119	132	145	158	171	184	197	210	222	235	248	261	273

Table G.19 *(continued)*

d. One-Tailed Test at 0.05 or Two-Tailed Test at 0.10

n_2 \ n_1	1	2	3	4	5	6	7	8	9	10	11	12	13	14	15	16	17	18	19	20
1																			0	0
																			19	20
2					0	0	0	1	1	1	1	2	2	2	3	3	3	4	4	4
					10	12	14	15	17	19	21	22	24	26	27	29	31	32	34	36
3			0	0	1	2	2	3	3	4	5	5	6	7	7	8	9	9	10	11
			9	12	14	16	19	21	24	26	28	31	33	35	38	40	42	45	47	49
4			0	1	2	3	4	5	6	7	8	9	10	11	12	14	15	16	17	18
			12	15	18	21	24	27	30	33	36	39	42	45	48	50	53	56	59	62
5		0	1	2	4	5	6	8	9	11	12	13	15	16	18	19	20	22	23	25
		10	14	18	21	25	29	32	36	39	43	47	50	54	57	61	65	68	72	75
6		0	2	3	5	7	8	10	12	14	16	17	19	21	23	25	26	28	30	32
		12	16	21	25	29	34	38	42	46	50	55	59	63	67	71	76	80	84	88
7		0	2	4	6	8	11	13	15	17	19	21	24	26	28	30	33	35	37	39
		14	19	24	29	34	38	43	48	53	58	63	67	72	77	82	86	91	96	101
8		1	3	5	8	10	13	15	18	20	23	26	28	31	33	36	39	41	44	47
		15	21	27	32	38	43	49	54	60	65	70	76	81	87	92	97	103	108	113
9		1	3	6	9	12	15	18	21	24	27	30	33	36	39	42	45	48	51	54
		17	24	30	36	42	48	54	60	66	72	78	84	90	96	102	108	114	120	126
10		1	4	7	11	14	17	20	24	27	31	34	37	41	44	48	51	55	58	62
		19	26	33	39	46	53	60	66	73	79	86	93	99	106	112	119	125	132	136

11		1	5	8	12	16	19	23	27	31	34	38	42	46	50	54	57	61	65	69
		21	28	36	43	50	58	65	72	79	87	94	101	108	115	122	130	137	144	151
12		2	5	9	13	17	21	26	30	34	38	42	47	51	55	60	64	68	72	77
		22	31	39	47	55	63	70	78	86	94	102	109	117	125	132	140	148	156	163
13		2	6	10	15	19	24	28	33	37	42	47	51	56	61	65	70	75	80	84
		24	33	42	50	59	67	76	84	93	101	109	118	126	134	143	151	159	167	176
14		2	7	11	16	21	26	31	36	41	46	51	56	61	66	71	77	82	87	92
		26	35	45	54	63	72	81	90	99	108	117	126	135	144	153	161	100	100	100
15		3	7	12	18	23	28	33	39	44	50	55	61	66	72	77	83	00	94	100
		27	38	48	57	67	77	87	96	106	115	125	134	144	153	163	100	100	100	100
16		3	8	14	19	25	30	36	42	48	54	60	65	71	77	83	89	95	101	107
		29	40	50	61	71	82	92	102	112	122	132	143	153	163	173	183	193	203	213
17		3	9	15	20	26	33	39	45	51	57	64	70	77	83	89	96	102	109	115
		31	42	53	65	76	86	97	108	119	130	140	151	161	172	183	193	204	214	225
18		4	9	16	22	28	35	41	48	55	61	68	75	82	88	95	102	109	116	123
		32	45	56	68	80	91	103	114	123	137	148	159	170	182	193	204	215	226	237
19	0	4	10	17	23	30	37	44	51	58	65	72	80	87	94	101	109	116	123	130
	19	34	47	59	72	84	96	108	120	132	144	156	167	179	191	203	214	226	238	250
20	0	4	11	18	25	32	39	47	54	62	69	77	84	92	100	107	115	123	130	138
	20	36	49	62	75	88	101	113	126	138	151	163	176	188	200	213	225	237	250	262

[a] The absence of values in the body of the table indicate that no decision is possible at the stated level of significance.
[b] If the observed value of U falls between the two values presented in the table for n_1 and n_2, *accept* H_0. Otherwise, reject H_0.

Source: H. B. Mann and D. R. Whitney, "On a test of whether one or two random variables is a stochastically larger than the other," *Annals of Mathematical Statistics*, *18*, 50-60 (1947); and D. Auble, "Extended tables for the Mann-Whitney statistic," *Bulletin of the Institute of Educational Research at Indiana University*, 1953, *1*, No. 2, as used in R. P. Runyon and A. Haber, *Fundamentals of Behavioral Statistics*, 1967, Addison-Wesley, Reading, Mass.

Table G.20 Critical Values of S for the Wilcoxon Rank-Sum Test[a]

		One-tailed significance level, α					
		0.100	0.050	0.025	0.010	0.005	0.001
		Two-tailed significance level, α					
n_1	n_2	0.200	0.100	0.050	0.020	0.010	0.002
3	2	6					
3	3	7	9				
4	2	8					
4	3	10	12				
4	4	10	14	16			
5	2	8	10				
5	3	11	13	15			
5	4	12	16	18	20		
5	5	15	17	21	23		
6	2	10	12				
6	3	12	14	16			
6	4	14	18	20	22	24	
6	5	16	20	24	26	28	
6	6	18	22	26	30	32	
7	2	12	14				
7	3	13	17	19	21		
7	4	16	20	22	26	28	
7	5	19	23	25	29	33	
7	6	20	26	30	34	36	42
7	7	23	27	33	37	41	47
8	2	12	14	16			
8	3	14	18	20	24		
8	4	18	22	24	28	30	
8	5	20	24	28	32	36	40
8	6	22	28	32	36	40	46
8	7	24	30	36	42	44	52
8	8	26	34	38	46	50	56
9	1	9					
9	2	14	16	18			
9	3	17	19	23	25	27	
9	4	18	24	28	30	34	
9	5	21	27	31	35	39	43
9	6	24	30	34	40	44	50
9	7	27	33	39	45	49	57
9	8	28	36	42	50	54	62
9	9	31	39	47	53	59	67
10	1	10					
10	2	14	18	20			
10	3	18	22	24	28	30	
10	4	20	26	30	34	36	40
10	5	24	28	34	38	42	48
10	6	26	32	38	44	48	54
10	7	28	36	42	48	52	60
10	8	32	40	46	54	58	68
10	9	34	42	50	58	64	74
10	10	36	46	54	62	68	80
11	1	11					
11	2	16	20	22			
11	3	19	23	27	31	33	
11	4	22	28	32	36	40	44
11	5	25	31	37	41	45	51
11	6	28	34	40	48	52	58
11	7	31	39	45	53	57	65
11	8	34	42	50	58	62	72
11	9	37	45	53	63	67	79
11	10	38	48	58	66	74	86
11	11	41	53	61	71	79	91
12	1	12					
12	2	16	20	22			
12	3	20	26	28	32	34	
12	4	24	30	34	38	42	48
12	5	26	34	38	44	48	56
12	6	30	38	44	50	54	64
12	7	32	42	48	56	60	70
12	8	36	44	52	62	66	78
12	9	38	48	56	66	72	84
12	10	42	52	62	72	78	92
12	11	44	56	66	76	84	98
12	12	46	60	70	82	90	104
13	1	13					
13	2	18	22	24	26		
13	3	21	27	31	35	37	
13	4	26	32	36	42	46	50
13	5	29	35	41	47	51	59
13	6	32	40	46	54	58	68
13	7	35	43	51	59	65	75

[a] n_1 is the number of scores in the larger sample, n_2 the number in the smaller sample. Reject H_0 if calculated value of S is equal to or greater than the value in the table. If calculated S value is negative, ignore the negative sign when using the table.

Table G.20 *(continued)*

n_1	n_2	One-tailed significance level, α: 0.100 / Two-tailed significance level, α: 0.200	0.050 / 0.100	0.025 / 0.050	0.010 / 0.020	0.005 / 0.010	0.001 / 0.002
13	8	38	48	56	64	70	82
13	9	41	51	61	71	77	89
13	10	44	56	64	76	82	96
13	11	47	59	69	81	89	103
13	12	50	62	74	86	94	110
13	13	53	67	79	91	101	117
14	1	14					
14	2	18	22	26	28		
14	3	22	28	32	38	40	
14	4	26	34	38	44	48	54
14	5	30	38	44	50	56	64
14	6	34	42	50	58	62	72
14	7	36	46	54	64	68	80
14	8	40	50	60	68	76	88
14	9	44	54	64	74	82	96
14	10	46	58	68	80	88	102
14	11	50	62	74	86	94	110
14	12	52	66	78	92	100	118
14	13	56	70	82	96	106	124
14	14	58	74	86	102	112	132
15	1	15					
15	2	20	24	28	30		
15	3	25	31	35	39	41	
15	4	28	36	40	46	50	58
15	5	31	39	47	53	59	67
15	6	36	44	52	60	66	76
15	7	39	49	57	67	73	85
15	8	42	54	62	72	80	92
15	9	45	57	67	79	87	101
15	10	48	62	72	84	92	108
15	11	51	65	77	91	99	117
15	12	54	70	82	96	106	124
15	13	59	73	87	101	111	131
15	14	62	78	92	108	118	138
15	15	65	81	97	113	123	145
16	1	16					
16	2	22	26	30	32		
16	3	26	32	36	42	44	
16	4	30	36	42	50	54	60
16	5	34	42	50	56	62	70
16	6	38	46	54	64	70	80
16	7	40	52	60	70	76	90
16	8	44	56	66	76	84	98
16	9	48	60	70	82	90	106
16	10	52	64	76	88	98	114
16	11	54	68	82	94	104	122
16	12	58	72	86	100	110	130
16	13	60	78	90	106	118	138
16	14	64	82	96	112	124	146
16	15	68	86	100	118	130	154
16	16	70	90	106	124	136	160
17	1	17					
17	2	22	28	30	34		
17	3	27	33	39	43	47	51
17	4	32	38	46	52	56	64
17	5	35	45	51	59	65	75
17	6	40	50	58	66	72	84
17	7	43	53	63	73	81	93
17	8	46	58	68	80	88	102
17	9	49	63	75	87	95	111
17	10	54	68	80	94	102	120
17	11	57	73	85	99	109	129
17	12	60	76	90	106	116	136
17	13	63	81	95	111	123	145
17	14	68	84	100	118	130	152
17	15	71	89	105	123	135	161
17	16	74	94	110	130	142	168
17	17	77	97	115	135	149	175
18	1	18					
18	2	24	28	32	36		
18	3	28	36	40	46	50	54
18	4	32	40	48	54	60	66
18	5	36	46	54	62	68	78
18	6	40	52	60	70	76	88
18	7	44	56	66	78	84	98
18	8	48	62	72	84	92	108
18	9	52	66	78	90	100	116
18	10	56	70	84	98	106	126
18	11	60	76	88	104	114	134
18	12	62	80	94	110	122	142
18	13	66	84	100	116	128	150
18	14	70	88	104	122	136	160
18	15	74	94	110	130	142	168
18	16	76	98	116	136	148	176
18	17	80	102	120	142	156	184
18	18	84	106	126	148	162	192
19	1	17	19				
19	2	24	30	34	36	38	
19	3	29	37	43	49	51	57
19	4	34	42	50	58	62	70

Table G.20 *(continued)*

		One-tailed significance level, α					
		0.100	0.050	0.025	0.010	0.005	0.001
		Two-tailed significance level, α					
n_1	n_2	0.200	0.100	0.050	0.020	0.010	0.002
19	5	39	49	57	65	71	81
19	6	42	54	64	74	80	92
19	7	47	59	69	81	89	103
19	8	50	64	76	88	96	112
19	9	55	69	81	95	105	121
19	10	58	74	86	102	112	132
19	11	63	79	93	109	119	141
19	12	66	84	98	116	126	148
19	13	69	87	103	121	133	157
19	14	72	92	110	128	140	166
19	15	77	97	115	135	147	175
19	16	80	102	120	140	156	184
19	17	83	105	125	147	161	191
19	18	86	110	130	154	168	200
19	19	91	115	135	159	175	207
20	1	18	20				
20	2	26	32	36	38	40	
20	3	30	38	44	50	54	60
20	4	36	44	52	60	64	74
20	5	40	50	60	68	74	86
20	6	44	56	66	76	84	96
20	7	48	62	72	84	92	108
20	8	52	66	78	92	100	118
20	9	56	72	84	100	108	128
20	10	60	76	90	106	116	136
20	11	64	82	96	114	124	146
20	12	68	86	102	120	132	156
20	13	72	92	108	126	140	164
20	14	76	96	114	134	146	172
20	15	80	100	120	140	154	182
20	16	82	106	124	146	162	190
20	17	86	110	130	154	168	200
20	18	90	114	136	160	176	208
20	19	94	120	142	166	182	216
20	20	98	124	146	172	190	224
21	1	19	21				
21	2	26	32	36	40	42	
21	3	33	41	47	53	57	61
21	4	38	46	54	62	68	76
21	5	41	53	61	71	77	89
21	6	46	58	68	80	88	102
21	7	51	65	75	87	97	111
21	8	56	70	82	96	104	122
21	9	59	75	89	103	113	133
21	10	64	80	94	110	122	142
21	11	67	85	101	117	129	151
21	12	70	90	106	124	136	162
21	13	75	95	113	131	145	171
21	14	78	100	118	138	152	180
21	15	83	105	123	145	159	189
21	16	86	110	130	152	168	198
21	17	89	115	135	159	175	207
21	18	94	118	140	166	182	216
21	19	97	123	147	173	189	225
21	20	100	128	152	178	196	232
21	21	105	133	157	185	205	241
22	1	20	22				
22	2	28	34	38	42	44	
22	3	34	42	48	54	58	64
22	4	38	48	56	66	70	80
22	5	44	54	64	74	82	94
22	6	48	60	72	84	90	106
22	7	52	66	78	92	100	116
22	8	58	72	86	100	108	128
22	9	62	78	92	108	118	138
22	10	66	84	98	114	126	148
22	11	70	88	104	122	134	158
22	12	74	94	110	130	142	168
22	13	78	98	116	136	150	178
22	14	82	104	122	144	158	186
22	15	86	108	128	150	166	196
22	16	90	114	134	158	174	206
22	17	92	118	140	164	182	214
22	18	96	124	146	172	188	224
22	19	100	128	152	178	196	232
22	20	104	132	158	186	204	242
22	21	108	138	162	192	212	250
22	22	112	142	168	198	218	260
23	1	21	23				
23	2	28	36	40	44	46	
23	3	35	43	51	57	61	67
23	4	40	50	58	68	74	84
23	5	45	57	67	77	85	97
23	6	50	64	74	86	94	110
23	7	55	69	81	95	103	121
23	8	60	76	88	104	114	132
23	9	63	81	95	111	121	143
23	10	68	86	102	120	130	154
23	11	73	91	107	127	139	163
23	12	76	96	114	134	148	174

Table G.20 *(continued)*

		One-tailed significance level, α					
		0.100	0.050	0.025	0.010	0.005	0.001
		Two-tailed significance level, α					
n_1	n_2	0.200	0.100	0.050	0.020	0.010	0.002
23	13	81	103	121	141	155	183
23	14	84	108	126	148	164	194
23	15	89	113	133	157	171	203
23	16	92	118	138	164	180	212
23	17	97	123	145	171	187	221
23	18	100	128	150	178	196	232
23	19	103	133	157	185	203	241
23	20	108	138	162	192	210	250
23	21	111	143	169	199	219	259
23	22	116	148	174	206	226	268
23	23	119	151	179	213	233	277
24	1	22	24				
24	2	30	36	42	46	48	
24	3	36	46	52	60	64	70
24	4	40	52	62	70	76	86
24	5	48	60	70	80	88	100
24	6	52	66	78	90	98	114
24	7	56	72	84	98	108	126
24	8	62	78	92	108	118	136
24	9	66	84	98	116	126	148
24	10	70	90	106	124	136	160
24	11	74	94	112	132	144	170
24	12	78	100	118	138	152	180
24	13	84	106	124	146	162	190
24	14	88	110	132	154	170	200
24	15	92	116	138	162	178	210
24	16	96	122	144	168	186	220
24	17	100	126	150	176	194	230
24	18	104	132	156	184	202	240
24	19	108	136	162	190	210	248
24	20	112	142	168	198	218	258
24	21	116	146	174	204	226	268
24	22	120	152	180	212	234	276
24	23	122	156	186	218	242	286
24	24	126	162	192	226	248	296
25	1	23	25				
25	2	32	38	44	48	50	
25	3	37	47	55	61	65	73
25	4	44	54	64	74	80	90
25	5	49	61	71	83	91	105
25	6	54	68	80	92	102	118
25	7	59	75	87	103	111	131
25	8	64	80	94	110	122	142
25	9	69	87	101	119	131	153
25	10	72	92	108	128	140	164
25	11	77	97	115	135	149	175
25	12	82	104	122	144	158	186
25	13	85	109	129	151	167	197
25	14	90	114	136	160	176	206
25	15	95	119	141	167	183	217
25	16	98	126	148	174	192	228
25	17	103	131	155	181	201	237
25	18	106	136	160	190	208	246
25	19	111	141	167	197	217	257
25	20	114	146	174	204	224	266
25	21	119	151	179	211	233	275
25	22	122	156	186	218	240	286
25	23	127	161	191	225	249	295
25	24	130	166	198	232	256	304
25	25	135	171	203	241	265	315

Source: L. R. Verdooren, "Extended critical values for Wilcoxon's test statistic," *Biometrika, 50*, 177-186 (1963).

Table G.21 Critical Values for the Spearman Rank-Order Correlation Coefficient

	Significance level for a one-tailed test at			
	.05	.025	.005	.001
	Significance level for a two-tailed test at			
N	.10	.05	.01	.002
5	.900	1.000		
6	.829	.886	1.000	
7	.715	.786	.929	1.000
8	.620	.715	.881	.953
9	.600	.700	.834	.917
10	.564	.649	.794	.879
11	.537	.619	.764	.855
12	.504	.588	.735	.826
13	.484	.561	.704	.797
14	.464	.539	.680	.772
15	.447	.522	.658	.750
16	.430	.503	.636	.730
17	.415	.488	.618	.711
18	.402	.474	.600	.693
19	.392	.460	.585	.676
20	.381	.447	.570	.661
21	.371	.437	.556	.647
22	.361	.426	.544	.633
23	.353	.417	.532	.620
24	.345	.407	.521	.608
25	.337	.399	.511	.597
26	.331	.391	.501	.587
27	.325	.383	.493	.577
28	.319	.376	.484	.567
29	.312	.369	.475	.558
30	.307	.363	.467	.549

If the observed value of ρ is greater than or equal to the tabled value for the appropriate level of significance, reject H_0. Note that the left-hand column is the number of pairs of scores, not the number of degrees of freedom.

Source: G. J. Glasser and R. F. Winter, "Critical values of the coefficient of rank correlation for testing the hypothesis of independence," *Biometrika, 48*, 444 (1961).

Table G.22 Probabilities Associated with Values as Large as Observed Value of χ_r^2 in the Friedman Two-Way Analysis of Variance by Ranks

k = 3

$N = 2$		$N = 3$		$N = 4$		$N = 5$	
χ_r^2	p	χ_r^2	p	χ_r^2	p	χ_r^2	p
0	1.000	.000	1.000	.0	1.000	.0	1.000
1	.833	.667	.944	.5	.931	.4	.954
3	.500	2.000	.528	1.5	.653	1.2	.691
4	.167	2.667	.361	2.0	.431	1.6	.522
		4.667	.194	3.5	.273	2.8	.367
		6.000	.028	4.5	.125	3.6	.182
				6.0	.069	4.8	.124
				6.5	.042	5.2	.093
				8.0	.0046	6.4	.039
						7.6	.024
						8.4	.0085
						10.0	.00077

$N = 6$		$N = 7$		$N = 8$		$N = 9$	
χ_r^2	p	χ_r^2	p	χ_r^2	p	χ_r^2	p
.00	1.000	.000	1.000	.00	1.000	.000	1.000
.33	.956	.286	.964	.25	.967	.222	.971
1.00	.740	.857	.768	.75	.794	.667	.814
1.33	.570	1.143	.620	1.00	.654	.889	.865
2.33	.430	2.000	.486	1.75	.531	1.556	.569
3.00	.252	2.571	.305	2.25	.355	2.000	.398
4.00	.184	3.429	.237	3.00	.285	2.667	.328
4.33	.142	3.714	.192	3.25	.236	2.889	.278
5.33	.072	4.571	.112	4.00	.149	3.556	.187
6.33	.052	5.429	.085	4.75	.120	4.222	.154
7.00	.029	6.000	.052	5.25	.079	4.667	.107
8.33	.012	7.143	.027	6.25	.047	5.556	.069
9.00	.0081	7.714	.021	6.75	.038	6.000	.057
9.33	.0055	8.000	.016	7.00	.030	6.222	.048
10.33	.0017	8.857	.0084	7.75	.018	6.889	.031
12.00	.00013	10.286	.0036	9.00	.0099	8.000	.019
		10.571	.0027	9.25	.0080	8.222	.016
		11.143	.0012	9.75	.0048	8.667	.010
		12.286	.00032	10.75	.0024	9.556	.0060
		14.000	.000021	12.00	.0011	10.667	.0035
				12.25	.00086	10.889	.0029
				13.00	.00026	11.556	.0013
				14.25	.000061	12.667	.00066
				16.00	.0000036	13.556	.00035
						14.000	.00020
						14.222	.000097
						14.889	.000054
						16.222	.000011
						18.000	.0000006

Source: Adapted from M. Friedman, "The use of ranks to avoid the assumption of normality implicit in the analysis of variance," *J. Amer. Statist. Ass.*, *32*, 688-689 (1973); with the kind permission of the author and the publisher.

For values of N and k larger than in this table, use Table G.7 with $df = k - 1$.

Table G.22 *(continued)*

b. k = 4

N = 2		N = 3		N = 4			
χ_r^2	p	χ_r^2	p	χ_r^2	p	χ_r^2	p
.0	1.000	.2	1.000	.0	1.000	5.7	.141
.6	.958	.6	.958	.3	.992	6.0	.105
1.2	.834	1.0	.910	.6	.928	6.3	.094
1.8	.792	1.8	.727	.9	.900	6.6	.077
2.4	.625	2.2	.608	1.2	.800	6.9	.068
3.0	.542	2.6	.524	1.5	.754	7.2	.054
3.6	.458	3.4	.446	1.8	.677	7.5	.052
4.2	.375	3.8	.342	2.1	.649	7.8	.036
4.8	.208	4.2	.300	2.4	.524	8.1	.033
5.4	.167	5.0	.207	2.7	.508	8.4	.019
6.0	.042	5.4	.175	3.0	.432	8.7	.014
		5.8	.148	3.3	.389	9.3	.012
		6.6	.075	3.6	.355	9.6	.0069
		7.0	.054	3.9	.324	9.9	.0062
		7.4	.033	4.5	.242	10.2	.0027
		8.2	.017	4.8	.200	10.8	.0016
		9.0	.0017	5.1	.190	11.1	.00094
				5.4	.158	12.0	.000072

For larger values of K or N, use Table G with df = k - 1.

(Footnote for Table G.23) If the rank sum for any treatment or sample is lower than the upper left value in the block or higher than the upper right value, then a significant difference among treatments has been established. Only after statistical significance has been established, equivalent to a significant *F* value in the analysis of variance, is it justified to consider the lower pair of the values in the block. If one or more treatment rank sums is found to be lower than the lower left-hand value, then these treatments may be considered as being significantly low. Similarly, any treatment rank sum that is higher than the lower right value in the block may be considered to be significantly high.

Table G.23 Selected Rank Totals for the Kramer Test

a. 5% Level

No. of judges.	Number of treatments, or samples ranked: 2	3	4	5	6	7	8	9	10	11	12	13	14	15	16	17	18	19	20
2	------- -------	------- -------	------- -------	------- 3-9	------- 3-11	------- 3-13	------- 4-14	------- 4-16	------- 4-18	------- 5-19	------- 5-21	------- 5-23	------- 5-25	------- 6-26	------- 6-28	------- 6-30	------- 7-31	------- 7-33	3-39 7-35
3	------- -------	------- 4-8	------- 4-11	4-14 5-13	4-17 6-15	4-20 6-18	4-23 7-20	5-25 8-22	5-28 8-25	5-31 9-27	5-34 10-29	5-37 10-32	5-40 11-34	6-42 12-36	6-45 12-39	6-48 13-41	6-51 14-43	6-54 14-46	7-56 15-48
4	------- -------	5-11 5-11	5-15 6-14	6-18 7-17	6-22 8-20	7-25 9-23	7-29 10-26	8-32 11-29	8-36 13-31	8-40 14-34	9-43 15-37	9-47 16-40	10-50 17-43	10-54 18-46	10-58 19-49	11-61 20-52	11-65 21-55	12-68 22-58	12-72 23-61
5	------- 6-9	6-14 7-13	7-18 8-17	8-22 10-20	9-26 11-24	9-31 13-27	10-35 14-31	11-39 15-35	12-43 17-38	12-48 18-42	13-52 20-45	14-56 21-49	14-61 23-52	15-65 24-56	16-69 25-60	16-74 27-63	17-78 28-67	18-82 30-70	18-87 31-74
6	7-11 7-11	8-16 9-15	9-21 11-19	10-26 12-24	11-31 14-28	12-36 16-32	13-41 18-36	14-46 20-40	15-51 21-45	17-55 23-49	18-60 25-53	19-65 27-57	19-71 29-61	20-76 31-65	21-81 32-70	22-86 34-74	23-91 36-78	24-96 38-82	25-101 40-86
7	8-13 8-13	10-18 10-18	11-24 13-22	12-30 15-27	14-35 17-32	15-41 19-37	17-46 22-41	18-52 24-46	19-58 26-51	21-63 28-56	22-69 30-61	23-75 33-65	25-80 35-70	26-86 37-75	27-92 39-80	29-97 42-84	30-103 44-89	31-109 46-94	32-115 48-99
8	9-15 10-14	11-21 12-20	13-27 15-25	15-33 17-31	17-39 20-36	18-46 23-41	20-52 25-47	22-58 28-52	24-64 31-57	25-71 33-63	27-77 36-68	29-83 39-73	30-90 41-79	32-96 44-84	33-103 47-89	35-109 49-95	37-115 52-100	38-122 54-106	40-128 57-111
9	11-16 11-16	13-23 14-22	15-30 17-28	17-37 20-34	19-44 23-40	22-50 26-46	24-57 29-52	26-64 32-58	28-71 35-64	30-78 38-70	32-85 41-76	34-92 45-81	36-99 48-87	38-106 51-93	40-113 54-99	42-120 57-105	44-127 60-111	45-135 63-117	47-142 66-123
10	12-18 12-18	15-25 16-24	17-33 19-31	20-40 23-37	22-48 26-44	25-55 30-50	27-63 33-57	30-70 37-63	32-78 40-70	34-86 44-76	37-93 47-83	39-101 51-89	41-109 54-96	44-116 57-103	46-124 61-109	48-132 64-116	51-139 68-122	53-147 71-129	55-155 75-135
11	13-20 14-19	16-28 18-26	19-36 21-34	22-44 25-41	25-52 29-48	28-60 33-55	31-68 37-62	34-76 41-69	36-85 45-76	39-93 49-83	42-101 53-90	45-109 57-97	47-118 60-105	50-126 64-112	53-134 68-119	55-143 72-126	58-151 76-133	60-160 80-140	63-168 84-147
12	15-21 15-21	18-30 19-29	21-39 24-36	25-47 28-44	28-56 32-52	31-65 37-59	34-74 41-67	38-82 45-75	41-91 50-82	44-100 54-90	47-109 58-98	50-118 63-105	53-127 67-113	56-136 71-121	59-145 76-128	62-154 80-136	65-163 84-144	68-172 89-151	71-181 93-159
13	16-23 17-22	20-32 21-31	24-41 26-39	27-51 31-47	31-60 35-56	35-69 40-64	38-79 45-72	42-88 50-80	45-98 54-89	49-107 59-97	52-117 64-105	56-126 69-113	59-136 74-121	62-146 78-130	66-155 83-138	69-165 88-146	73-174 93-154	76-184 97-163	79-194 102-171
14	17-25 18-24	22-34 23-33	26-44 28-42	30-54 33-51	34-64 38-60	38-74 44-68	42-84 49-77	46-94 54-86	50-104 59-95	54-114 65-103	57-125 70-112	61-135 75-121	65-145 80-130	69-155 85-139	73-165 91-147	76-176 96-156	80-186 101-165	84-196 106-174	88-206 111-183
15	19-26 19-26	23-37 25-35	28-47 30-45	32-58 36-54	37-68 42-63	41-79 47-73	46-89 53-82	50-100 59-91	54-111 64-101	58-122 70-110	63-132 75-120	67-143 81-129	71-154 87-138	75-165 92-148	79-176 98-157	84-186 104-166	88-197 109-176	92-208 115-185	96-219 121-194
16	20-28 21-27	25-39 27-37	30-50 33-47	35-61 39-57	40-72 45-67	45-83 51-77	49-95 57-87	54-106 63-97	59-117 69-107	63-129 75-117	68-140 81-127	73-151 87-137	77-163 93-147	82-174 100-156	86-186 106-166	91-197 112-176	95-209 118-186	100-220 124-196	104-232 130-206
17	22-29 22-29	27-41 28-40	32-53 35-50	38-64 41-61	43-76 48-71	48-88 54-82	53-100 61-92	58-112 67-103	63-124 74-113	68-136 81-123	73-148 87-134	78-160 94-144	83-172 100-155	88-184 107-165	93-196 113-176	98-208 120-186	103-220 126-197	108-232 133-207	113-244 139-218
18	23-31 24-30	29-43 30-42	34-56 37-53	40-68 44-64	46-80 51-75	51-93 58-86	57-105 65-97	62-118, 72-108	68-130 79-119	73-143 86-130	79-155 93-141	84-168 100-152	90-180 107-163	95-193 114-174	100-206 121-185	106-218 128-196	111-231 135-207	116-244 142-218	121-257 149-229
19	24-33 25-32	30-46 32-44	37-58 39-56	43-71 47-67	49-84 54-79	55-97 62-90	61-110 69-102	67-123 76-114	73-136 84-125	78-150 91-137	84-163 99-148	90-176 106-160	96-189 114-171	102-202 121-183	107-216 128-195	113-229 136-206	119-242 143-218	124-256 151-229	130-269 158-241
20	26-34 26-34	32-48 34-46	39-61 42-58	45-75 50-70	52-88 57-83	58-102 65-95	65-115 73-107	71-129 81-119	77-143 89-131	83-157 97-143	90-170 105-155	96-184 112-168	102-198 120-180	108-212 128-192	114-226 136-204	120-240 144-216	126-254 152-228	132-268 160-240	139-281 168-252
21	27-36 28-35	34-50 36-48	41-64 44-61	48-78 52-74	55-92 61-86	62-106 69-99	68-121 77-112	75-135 86-124	82-149 94-137	89-163 102-150	95-178 110-163	102-192 119-175	108-207 127-188	115-221 135-201	121-236 144-213	128-250 152-226	134-265 160-239	141-279 169-251	147-294 177-264
22	28-38 29-37	36-52 38-50	43-67 46-64	51-81 55-77	58-96 64-90	65-111 73-103	72-126 81-117	80-140 90-130	87-155 99-143	94-170 108-156	101-185 116-170	108-200 125-183	115-215 134-196	122-230 143-209	129-245 151-223	135-261 160-236	142-276 169-249	149-291 178-262	156-306 186-276
23	30-39 31-38	38-54 40-52	46-69 49-66	53-85 58-80	61-100 67-94	69-115 76-108	76-131 85-122	84-146 95-135	91-162 104-149	99-177 113-163	106-193 122-177	114-208 131-191	121-224 141-204	128-240 150-218	136-255 159-232	143-271 168-246	150-287 178-259	157-303 187-273	165-318 196-287
24	31-41 32-40	40-56 41-55	48-72 51-69	56-88 61-83	64-104 70-96	72-120 80-112	80-136 90-126	88-152 99-141	96-168 109-155	104-184 119-169	112-200 128-184	120-216 138-198	127-233 147-213	135-249 157-227	143-265 167-241	151-281 177-255	158-298 186-270	166-314 196-284	174-330 206-298
25	33-42 33-42	41-59 43-57	50-75 53-72	59-91 63-87	67-108 73-102	76-124 84-116	84-141 94-131	92-158 104-146	101-174 114-151	109-191 124-176	117-208 134-191	126-224 144-206	134-241 154-221	142-258 164-236	150-275 175-250	158-292 185-265	166-309 195-280	174-326 205-295	182-343 215-310

Table G.23 *(continued)*

No. of judges.	Number of treatments, or samples ranked: 2	3	4	5	6	7	8	9	10	11	12	13	14	15	16	17	18	19	20
26	34-44 35-43	43-61 45-59	52-78 56-74	61-95 66-90	70-112 77-105	79-129 87-121	88-146 98-136	97-163 108-152	106-180 119-167	114-198 129-183	123-215 140-198	132-232 151-213	140-250 161-229	149-267 172-244	157-285 182-260	166-302 193-275	174-320 203-291	183-337 214-306	191-355 225-321
27	35-46 36-45	45-63 47-61	55-80 58-77	64-98 69-93	73-116 80-109	83-133 91-125	92-151 102-141	101-169 113-157	110-187 124-173	119-205 135-189	129-222 146-205	138-240 157-221	147-258 168-237	156-276 179-253	165-294 190-269	174-312 201-285	182-331 212-301	191-349 223-317	200-367 234-333
28	37-47 38-46	47-65 49-63	57-83 60-80	67-101 72-96	76-120 83-113	86-138 95-129	96-156 106-146	106-174 118-162	115-193 129-179	125-211 140-196	134-230 152-212	144-248 163-229	153-267 175-245	162-286 186-262	172-304 198-278	181-323 209-295	191-341 221-311	200-360 232-328	209-379 244-344
29	38-49 39-48	49-67 51-65	59-86 63-82	69-105 74-100	80-123 86-117	90-142 98-134	100-161 110-151	110-180 122-168	120-199 134-185	130-218 146-202	140-237 158-219	150-256 170-236	160-275 182-253	169-295 194-270	179-314 206-287	189-333 218-304	199-352 230-321	208-372 241-339	218-391 253-356
30	40-50 41-49	51-69 53-67	61-89 65-85	72-108 77-103	83-127 90-120	93-147 102-138	104-166 114-156	114-186 127-173	125-205 139-191	135-225 151-209	145-245 164-226	156-264 176-244	166-284 189-261	176-304 201-279	186-324 213-297	197-343 226-314	207-363 238-332	217-383 251-349	227-403 263-367
31	41-52 42-51	52-72 55-69	64-91 67-88	75-111 80-106	86-131 93-124	97-151 106-142	108-171 119-160	119-191 131-179	130-211 144-197	140-232 157-215	151-252 170-233	162-272 183-251	173-292 196-269	183-313 208-288	194-333 221-306	204-354 234-324	215-374 247-342	226-394 260-360	236-415 273-378
32	42-54 43-53	54-74 56-72	66-94 70-90	77-115 83-109	89-135 96-128	100-156 109-147	112-176 123-165	123-197 136-184	134-218 149-203	146-238 163-221	157-259 176-240	168-280 189-259	179-301 202-278	190-322 216-296	201-343 229-315	212-364 242-334	223-385 256-352	234-406 269-371	245-427 282-390
33	44-55 45-54	56-76 58-74	68-97 72-93	80-118 86-112	92-139 99-132	104-160 113-151	116-181 127-170	128-202 141-189	139-224 154-209	151-245 168-228	163-266 182-247	174-288 196-266	186-309 209-286	197-331 223-305	209-352 237-324	220-374 251-343	231-396 264-363	243-417 278-382	254-439 292-401
34	45-57 46-56	58-78 60-76	70-100 74-96	83-121 88-116	95-143 103-135	108-164 117-155	120-186 131-175	132-208 145-195	144-230 159-215	156-252 174-234	168-274 188-254	180-296 202-274	192-318 216-294	204-340 231-313	216-362 245-333	228-384 259-353	240-406 273-373	252-428 287-393	263-451 302-412
35	47-58 48-57	60-80 62-78	73-102 77-98	86-124 91-119	98-147 106-139	111-169 121-159	124-191 135-180	136-214 150-200	149-236 165-220	161-259 179-241	174-281 194-261	186-304 209-281	199-326 223-302	211-349 238-322	223-372 253-342	236-394 267-363	248-417 282-383	260-440 297-403	272-463 311-424
36	48-60 49-59	62-82 64-80	75-105 79-101	88-128 94-122	102-150 109-143	115-173 124-164	128-196 139-185	141-219 155-205	154-242 170-226	167-265 185-247	180-288 200-268	193-311 215-289	205-335 230-310	218-358 245-331	231-381 260-352	244-404 276-372	256-428 291-393	269-451 306-414	282-474 321-435
37	50-61 51-60	63-85 66-82	77-108 81-104	91-131 97-125	105-154 112-147	118-178 128-168	132-201 144-189	145-225 159-211	159-248 175-232	172-272 190-254	185-296 206-275	199-319 222-296	212-343 237-318	225-367 253-339	238-391 268-361	251-415 284-382	265-438 300-403	278-462 315-425	291-486 331-446
38	51-63 52-62	65-87 68-84	80-110 84-106	94-134 100-128	108-158 116-150	122-182 132-172	136-206 148-194	150-230 164-216	164-254 180-238	177-279 196-260	191-303 212-282	205-327 228-304	219-351 244-326	232-376 260-348	246-400 276-370	259-425 292-392	273-449 308-414	286-474 324-436	300-498 341-457
39	52-65 53-64	67-89 70-86	82-113 86-109	97-137 102-132	111-162 119-154	126-186 135-177	140-211 152-199	154-236 168-222	169-260 185-244	183-285 202-266	197-310 218-289	211-335 235-311	225-360 251-334	239-385 268-356	253-410 284-379	267-435 301-401	281-460 317-424	295-485 334-446	309-510 350-469
40	54-66 55-65	69-91 72-88	84-116 88-112	99-141 105-135	114-166 122-158	129-191 139-181	144-216 156-204	159-241 173-227	173-267 190-250	188-292 207-273	203-317 224-296	217-343 241-319	232-368 258-342	246-394 275-365	261-419 292-388	275-445 309-411	290-470 326-434	304-496 343-457	318-522 360-480
41	55-68 56-67	71-93 73-91	87-118 91-114	102-144 108-138	117-170 126-161	133-195 143-185	148-221 160-209	163-247 178-232	178-273 195-256	193-299 213-279	208-325 230-303	223-351 248-326	238-377 265-350	253-403 282-374	268-429 300-397	283-455 317-421	298-481 335-444	313-507 352-468	327-534 370-491
42	57-69 58-68	73-95 75-93	89-121 93-117	105-147 111-141	121-173 129-165	136-200 147-189	152-226 165-213	168-252 182-238	183-279 200-262	199-305 218-286	214-332 236-310	230-358 254-334	245-385 272-358	260-412 290-382	276-438 308-406	291-465 326-430	306-492 344-454	322-518 362-476	337-545 380-502
43	58-71 59-70	75-97 77-95	91-124 95-120	108-150 114-144	124-177 132-169	140-204 150-194	156-231 169-218	172-258 187-243	188-285 206-267	204-312 224-292	220-339 242-317	236-366 261-341	252-393 279-366	268-420 297-391	283-448 316-415	299-475 334-440	315-502 353-464	330-530 371-489	346-557 389-514
44	60-72 61-71	77-99 79-97	93-127 98-122	110-154 117-147	127-181 135-173	144-208 154-198	160-236 173-223	177-263 192-248	193-291 211-273	210-318 229-299	226-346 248-324	242-374 267-349	258-402 286-374	275-429 305-399	291-457 324-424	307-485 343-449	323-513 361-475	339-541 380-500	355-569 399-525
45	61-74 62-73	78-102 81-99	96-129 100-125	113-157 119-151	130-185 139-176	147-213 158-202	164-241 177-228	181-269 197-253	198-297 216-279	215-325 235-305	232-353 254-331	248-382 274-356	265-410 293-382	282-438 312-408	298-467 332-433	315-495 351-459	332-523 370-485	348-552 390-510	365-580 409-536
46	62-76 63-75	80-104 83-101	98-132 103-127	116-160 122-154	133-189 142-180	151-217 162-206	168-246 181-233	186-274 201-259	203-303 221-285	220-332 241-311	238-360 260-338	255-389 280-364	272-418 300-390	289-447 320-416	306-476 340-442	323-505 359-469	340-534 379-495	357-563 399-521	374-592 419-547
47	64-77 65-76	82-106 85-103	100-135 105-130	119-163 125-157	137-192 145-184	155-221 165-211	172-251 186-237	190-280 206-264	208-309 226-291	226-338 246-318	243-368 267-344	261-397 287-371	279-426 307-398	296-456 327-425	314-485 348-451	331-515 368-478	348-545 388-505	366-574 408-532	383-604 428-559
48	65-79 66-78	84-108 87-105	103-137 107-133	121-167 128-160	140-196 149-187	158-226 169-215	176-256 190-242	195-285 211-269	213-315 231-297	231-345 252-324	249-375 273-351	267-405 293-379	285-435 314-406	303-465 335-433	321-495 355-461	339-525 376-488	357-555 397-515	375-585 418-542	392-616 438-570
49	67-80 68-79	86-110 89-107	105-140 110-135	124-170 131-163	143-200 152-191	162-230 173-219	181-260 194-247	199-291 215-275	218-321 236-303	236-352 258-330	255-382 279-358	273-413 300-386	292-443 321-414	310-474 342-442	329-504 363-470	347-535 385-497	365-566 406-525	384-596 427-553	402-627 448-581
50	68-82 69-81	88-112 91-109	107-143 112-138	127-173 134-166	146-204 155-195	165-235 177-223	185-265 198-252	204-296 220-280	223-327 242-308	242-358 263-337	261-389 285-365	280-420 306-394	299-451 328-422	317-483 350-450	336-514 371-479	355-545 393-507	374-576 415-535	392-608 436-564	411-639 458-592

Table G.23 *(continued)*

No. of judges.	2	3	4	5	6	7	8	9	10	11	12	13	14	15	16	17	18	19	20
51	70-83 71-82	90-114 92-112	110-145 114-141	130-176 136-170	149-208 158-199	169-239 181-227	189-270 203-256	208-302 225-285	228-333 247-314	247-365 269-343	267-396 291-372	286-428 313-401	305-460 335-430	325-491 357-459	344-523 379-488	363-555 401-517	382-587 424-545	401-619 446-574	420-651 468-603
52	71-85 72-84	92-116 94-114	112-148 117-143	132-180 139-173	153-211 162-202	173-243 184-232	193-275 207-261	213-307 229-291	233-339 252-320	253-371 274-350	273-403 297-379	292-436 320-408	312-468 342-438	332-500 365-467	352-532 387-497	371-565 410-526	391-597 432-556	410-630 455-585	430-662 478-614
53	72-87 74-85	93-119 96-116	114-151 119-146	135-183 142-176	156-215 165-206	176-248 188-236	197-280 211-266	217-313 234-296	238-345 257-326	258-378 280-356	278-411 303-386	299-443 326-416	319-476 349-446	339-509 372-476	359-542 395-506	379-575 418-536	399-608 441-566	419-641 464-596	439-674 487-626
54	74-88 75-87	95-121 98-118	117-153 121-149	138-186 145-179	159-219 168-210	180-252 192-240	201-285 215-271	222-318 239-301	243-351 262-332	264-384 286-362	284-418 309-393	305-451 333-423	326-484 356-454	346-518 380-484	367-551 403-515	387-585 427-545	408-618 450-576	428-652 474-606	449-685 497-637
55	75-90 76-89	97-123 100-120	119-156 124-151	141-189 148-182	162-223 172-213	184-256 196-244	205-290 220-275	227-323 243-307	248-357 267-338	269-391 291-369	290-425 315-400	311-459 339-431	332-493 363-462	353-527 387-493	374-561 411-524	395-595 435-555	416-629 459-586	437-663 483-617	458-697 507-648
56	77-91 78-90	99-125 102-122	121-189 126-154	143-193 151-185	165-227 175-217	187-261 199-249	209-295 224-280	231-329 248-312	253-363 273-343	274-398 297-375	296-432 322-406	318-466 346-438	339-501 370-470	361-535 395-501	382-570 419-533	403-605 444-564	425-639 468-596	446-674 493-627	467-709 517-659
57	78-93 79-92	101-127 104-124	124-161 129-156	146-196 153-189	169-230 178-221	191-265 203-253	213-300 228-285	236-334 253-317	258-369 278-349	280-404 303-381	302-439 328-413	324-474 353-445	346-509 377-478	368-544 402-510	390-579 427-542	412-614 452-574	433-650 477-606	455-685 502-638	477-720 527-670
58	80-94 81-93	103-129 106-126	126-164 131-159	149-199 156-192	172-234 182-224	195-269 207-257	218-304 232-290	240-340 258-322	263-375 283-355	285-411 308-388	308-446 334-420	330-482 359-453	353-517 384-486	375-553 410-518	397-589 435-551	420-624 461-583	442-660 486-616	464-696 511-649	486-732 537-681
59	81-96 82-95	105-131 108-128	128-167 133-162	152-202 159-195	175-238 185-228	198-274 211-261	222-309 237-294	245-345 262-328	268-381 288-361	291-417 314-394	314-453 340-427	337-489 366-460	360-525 392-493	382-562 417-527	405-598 443-560	428-634 469-593	450-671 495-626	473-707 521-659	496-743 547-692
60	82-98 84-96	107-133 110-130	131-169 136-164	155-205 162-198	178-242 188-232	202-278 215-265	226-314 241-299	249-351 267-333	273-387 293-367	296-424 320-400	320-460 346-434	343-497 372-468	366-534 399-501	390-570 425-535	413-607 451-569	436-644 478-602	459-681 504-636	482-718 530-670	505-755 557-703
61	84-99 85-98	108-136 112-132	133-172 138-167	157-209 165-201	182-245 192-235	206-282 218-270	230-319 245-304	254-356 272-338	278-393 299-372	302-430 325-407	326-467 352-441	349-505 379-475	373-542 406-509	397-579 432-544	420-617 459-578	444-654 486-612	468-691 513-646	491-729 540-680	515-766 566-715
62	85-101 87-99	110-138 113-135	135-175 141-169	160-212 168-204	185-249 195-239	210-286 222-274	234-324 249-309	259-361 277-343	283-399 304-378	307-437 331-413	332-474 358-448	356-512 386-482	380-550 413-517	404-588 440-552	428-626 467-587	452-664 495-621	476-702 522-656	500-740 549-691	524-778 576-726
63	87-102 88-101	112-140 115-137	138-177 143-172	163-215 171-207	188-253 198-243	213-291 226-278	238-329 254-313	263-367 281-349	288-405 309-384	313-443 337-419	337-482 364-455	362-520 392-490	387-558 420-525	411-597 448-560	436-635 475-596	460-674 503-631	485-712 531-666	509-751 558-702	534-7[illegible] 586-737
64	88-104 89-103	114-142 117-139	140-180 145-175	166-218 173-211	191-257 202-246	217-295 230-282	242-334 258-318	268-372 286-354	293-411 314-390	318-450 342-426	343-489 371-461	368-528 399-497	394-566 427-533	419-605 455-569	444-644 483-605	468-684 512-640	493-723 540-676	518-762 568-712	543-801 596-748
65	90-105 91-104	116-144 119-141	142-183 148-177	169-221 176-214	195-260 205-250	221-299 233-287	246-339 262-323	272-378 291-359	298-417 319-396	324-456 348-432	349-496 377-468	375-535 405-505	400-575 434-541	426-614 463-577	451-654 491-614	477-693 520-650	502-733 549-686	527-773 577-723	553-812 606-759
66	91-107 92-106	118-146 121-143	145-185 150-180	171-225 179-217	198-264 208-254	224-304 237-291	251-343 266-328	277-383 295-365	303-423 325-401	329-463 354-438	355-503 383-475	381-543 412-512	407-583 441-549	433-623 470-586	459-663 499-623	485-703 529-659	511-743 558-696	536-784 587-733	562-824 616-770
67	93-108 94-107	120-148 123-145	147-188 152-183	174-228 182-220	201-268 212-257	228-308 241-295	255-348 271-332	281-389 300-370	308-429 330-407	335-469 359-445	361-510 389-482	388-550 419-519	414-591 448-557	440-632 478-594	467-672 507-632	493-713 537-669	519-754 567-706	545-795 596-744	572-835 626-781
68	94-110 95-109	122-150 125-147	149-191 155-185	177-231 185-223	204-272 215-261	232-312 245-299	259-353 275-337	286-394 305-375	313-435 335-413	340-476 365-451	367-517 395-489	394-558 425-527	421-599 455-565	448-640 485-603	474-682 515-641	501-723 546-678	528-764 576-716	554-806 606-754	581-847 636-792
69	95-112 97-110	124-152 127-149	152-193 157-188	180-234 188-226	208-275 218-265	235-317 249-303	263-358 279-342	291-399 310-380	318-441 340-419	346-482 371-457	373-524 401-496	400-566 432-534	428-607 462-573	455-649 493-611	482-691 524-649	509-733 554-688	536-775 585-726	563-817 615-765	591-858 646-803
70	97-113 98-112	125-155 129-151	154-196 160-190	183-237 191-229	211-279 221-269	239-321 252-308	267-363 283-347	295-405 314-386	323-447 345-425	351-489 376-464	379-531 407-503	407-573 438-542	435-615 470-580	462-658 501-619	490-700 532-658	517-743 563-697	545-785 594-736	573-827 625-775	600-870 656-814
71	98-115 100-113	127-157 131-153	156-199 162-193	185-241 193-233	214-283 225-272	243-325 256-312	271-368 288-351	300-410 319-391	328-453 351-430	357-495 382-470	385-538 414-509	413-581 445-549	441-624 477-588	470-666 508-628	498-709 540-667	526-752 571-707	554-795 603-746	582-838 634-786	610-881 666-825
72	100-116 101-115	129-159 133-155	159-201 164-196	188-244 196-236	217-287 228-276	247-329 260-316	276-372 292-356	305-415 324-396	333-459 356-436	362-502 388-476	391-545 420-516	420-588 452-556	448-632 484-596	477-675 516-636	505-719 548-676	534-762 580-716	562-806 612-756	591-849 644-796	619-893 676-836
73	101-118 102-117	131-161 135-157	161-204 167-198	191-247 199-239	221-290 231-280	250-334 264-320	280-377 296-361	309-421 329-401	338-465 361-442	368-508 394-482	397-552 426-523	426-596 458-564	455-640 491-604	484-684 523-645	513-728 556-685	542-772 588-726	571-816 621-766	600-860 653-807	629-904 685-848
74	103-119 104-118	133-163 136-160	163-207 169-201	194-250 202-242	224-294 235-283	254-338 268-324	284-382 301-365	314-426 333-407	344-470 366-448	373-515 399-489	403-559 432-530	432-604 465-571	462-648 498-612	491-693 531-653	521-737 564-694	550-782 597-735	580-826 630-776	609-871 662-818	638-916 695-859
75	104-121 105-120	135-165 138-162	166-209 172-203	197-253 205-245	227-298 238-287	258-342 272-328	288-387 305-370	318-432 338-412	349-476 372-453	379-521 405-495	409-566 438-537	439-611 472-578	469-656 505-620	499-701 538-662	529-746 572-70[illegible]	559-791 605-745	588-837 639-786	618-882 672-828	648-927 705-870

Table G.23 *(continued)*

b. 1% Level

No. of judges	Number of treatments, or samples ranked: 2	3	4	5	6	7	8	9	10	11	12	13	14	15	16	17	18	19	20
2	------- -------	------- -------	------- -------	------- -------	------- -------	------- -------	------- -----	------- -------	------- 3-19	------- 3-21	------- 3-23	------- 3-25	------- 3-27	------- 3-29	------- 3-31	------- 3-33	------- 4-34	------- 4-36	------- 4-38
3	------- -------	------- -------	------- -------	------- 4-14	------- 4-17	------- 4-20	------- 5-22	------- 5-25	4-29 6-27	4-32 6-30	4-35 6-33	4-38 7-35	4-41 7-38	4-44 7-41	4-47 8-43	4-50 8-46	4-53 9-48	4-56 9-51	5-58 9-54
4	------- -------	------- -------	------- 5-15	5-19 6-18	5-23 6-22	5-27 7-25	6-30 8-28	6-34 8-32	6-38 9-35	6-42 10-38	7-45 10-42	7-49 11-45	7-53 12-48	7-57 13-51	8-60 13-55	8-64 14-58	8-68 15-61	8-72 15-65	9-75 16-68
5	------- -------	------- 6-14	6-19 7-18	7-23 8-22	7-28 9-26	8-32 10-30	8-37 11-34	9-41 12-38	9-46 13-42	10-50 14-46	10-55 15-50	11-59 16-54	11-64 17-58	12-68 18-62	12-73 19-66	13-77 20-70	13-82 21-74	14-86 22-78	14-91 23-82
6	------- -------	7-17 8-16	8-22 9-21	9-27 10-26	9-33 12-30	10-38 13-35	11-43 14-40	12-48 16-44	13-53 17-49	13-59 18-54	14-64 20-58	15-69 21-63	16-74 23-67	16-80 24-72	17-85 25-77	18-90 27-81	19-95 28-86	19-101 29-91	20-106 31-95
7	------- 8-13	8-20 9-19	10-25 11-24	11-31 12-30	12-37 14-35	13-43 16-40	14-49 18-45	15-55 19-51	16-61 21-56	17-67 23-61	18-73 25-66	19-79 26-72	20-85 28-77	21-91 30-82	22-97 32-87	23-103 33-93	24-109 35-98	25-115 37-103	26-121 39-108
8	9-15 9-15	10-22 11-21	11-29 13-27	13-35 15-33	14-42 17-39	16-48 19-45	17-55 21-51	19-61 23-57	20-68 25-63	21-75 28-68	23-81 30-74	24-88 32-80	25-95 34-86	27-101 36-92	28-108 38-98	29-115 40-104	31-121 42-110	32-128 45-115	33-135 47-121
9	10-17 10-17	12-24 12-24	13-32 15-30	15-39 17-37	17-46 20-43	19-53 22-50	21-60 25-56	22-68 27-63	24-75 30-69	26-82 32-76	27-90 35-82	29-97 37-89	31-104 40-95	32-112 42-102	34-119 45-108	35-127 47-115	37-134 50-121	39-141 52-128	40-149 55-134
10	11-19 11-19	13-27 14-26	15-35 17-33	18-42 20-40	20-50 23-47	22-58 25-55	24-66 28-62	26-74 31-69	28-82 34-76	30-90 37-83	32-98 40-90	34-106 43-97	36-114 46-104	38-122 49-111	40-130 52-118	42-138 54-126	44-146 57-133	46-154 60-140	48-162 63-147
11	12-21 13-20	15-29 16-28	17-38 19-36	20-46 22-44	22-55 25-52	25-63 29-59	27-72 32-67	30-80 35-75	32-89 35-82	34-98 42-90	37-106 45-98	39-115 48-106	41-124 52-113	44-132 55-121	46-141 58-129	48-150 62-136	50-159 65-144	53-167 68-152	55-176 72-159
12	14-22 14-22	17-31 18-30	19-41 21-39	22-50 25-47	25-54 28-56	28-68 32-64	31-77 36-72	33-87 39-81	36-96 43-89	39-105 47-97	42-114 50-106	44-124 54-114	47-133 58-122	50-142 62-130	52-152 65-139	55-161 69-147	57-171 73-155	60-180 76-164	63-189 80-172
13	15-24 15-24	18-34 19-33	21-44 23-42	25-53 27-51	28-63 31-60	31-73 35-69	34-83 39-78	37-93 44-86	40-103 48-95	43-113 52-104	46-123 56-113	50-132 60-122	53-142 64-131	56-152 68-140	58-163 72-149	61-173 76-158	64-183 80-167	67-193 85-175	70-203 89-184
14	16-26 17-25	20-36 21-35	24-46 25-45	27-57 30-54	31-67 34-64	34-78 39-73	38-88 43-83	41-99 48-92	45-109 52-102	48-120 57-111	51-131 61-121	55-141 66-130	58-152 70-140	62-162 75-149	65-173 79-159	68-184 84-168	72-194 88-178	75-205 93-187	78-216 97-197
15	18-27 18-27	22-38 23-37	26-49 28-47	30-60 32-58	34-71 37-68	37-83 42-78	41-94 47-88	45-105 52-98	49-116 57-108	53-127 62-118	56-139 67-128	60-150 72-138	64-161 76-149	68-172 81-159	71-184 86-169	75-195 91-179	79-206 96-189	82-218 101-199	86-229 106-209
16	19-29 19-29	23-41 25-39	28-52 30-50	32-64 35-61	36-76 40-72	41-87 46-82	45-99 51-93	49-111 56-104	53-123 61-115	57-135 67-125	62-146 72-136	66-158 77-147	70-170 83-157	74-182 88-168	78-194 93-179	82-206 99-189	86-218 104-200	90-230 109-211	94-242 115-221
17	20-31 21-30	25-43 26-42	30-55 32-53	35-67 38-64	39-80 43-76	44-92 49-87	49-104 55-98	53-117 60-110	58-129 66-121	62-142 72-132	67-154 78-143	71-167 83-155	76-179 89-166	80-192 95-177	85-204 101-188	89-217 106-200	93-230 112-211	98-242 118-222	102-255 124-233
18	22-32 22-32	27-45 28-44	32-58 34-56	37-71 40-68	42-84 46-80	47-97 52-92	52-110 59-103	57-123 65-115	62-136 71-127	67-149 77-139	72-162 83-151	77-175 89-163	82-188 95-175	86-202 102-186	91-215 108-198	96-228 114-210	101-241 120-222	106-254 126-234	110-268 132-246
19	23-34 24-33	29-47 30-46	34-61 36-59	40-74 43-71	45-88 49-84	50-102 56-96	56-115 62-109	61-129 69-121	67-142 76-133	72-156 82-146	77-170 89-158	82-184 95-171	88-197 102-183	93-211 108-196	98-225 115-208	103-239 122-220	108-253 128-233	113-267 135-245	119-280 141-258
20	24-36 25-35	30-50 32-48	36-64 38-62	42-78 45-75	48-92 52-88	54-106 59-101	60-120 66-114	65-135 73-127	71-149 80-140	77-163 87-153	82-178 94-166	88-192 101-179	94-206 108-192	99-221 115-205	105-235 122-218	110-250 129-231	116-264 136-244	121-279 143-257	127-293 150-270
21	26-37 26-37	32-52 33-51	38-67 41-64	45-81 48-78	51-96 55-92	57-111 63-105	63-126 70-119	69-141 78-132	75-156 85-146	82-170 92-160	88-185 100-173	94-200 107-187	100-215 115-200	106-230 122-214	112-245 130-227	118-260 137-241	123-276 145-254	129-291 152-268	135-306 159-282
22	27-39 28-38	34-54 35-53	40-70 43-67	47-85 51-81	54-100 58-96	60-116 66-110	67-131 74-124	74-146 82-138	80-162 90-152	86-178 98-166	93-193 106-180	99-209 113-195	106-224 121-209	112-240 129-223	118-256 137-237	125-271 145-251	131-287 153-265	137-303 161-279	144-318 168-294
23	28-41 29-40	36-56 37-55	43-72 45-70	50-88 53-85	57-104 62-99	64-120 70-114	71-136 78-129	78-152 86-144	85-168 95-158	91-185 103-173	98-201 111-188	105-217 119-203	112-233 128-217	119-249 136-232	125-266 144-247	132-282 153-261	139-298 161-276	145-315 169-291	152-331 178-305
24	30-42 30-42	37-59 39-57	45-75 47-73	52-92 56-88	60-108 65-103	67-125 73-119	75-141 82-134	82-158 91-149	89-175 99-165	96-192 108-180	104-208 117-195	111-225 126-210	118-242 134-226	125-259 143-241	132-276 152-256	139-293 160-272	146-310 169-287	154-326 178-302	161-343 187-317
25	31-44 32-43	39-61 41-59	47-78 50-75	55-95 59-91	63-112 68-107	71-129 77-123	78-147 86-139	86-164 95-155	94-181 104-171	101-199 113-187	109-216 123-202	117-233 132-218	124-251 141-234	132-268 150-250	139-286 159-266	147-303 168-282	154-321 177-298	162-338 187-313	169-356 196-329

Table G.23 *(continued)*

No. of judges.	Number of treatments, or samples ranked 2	3	4	5	6	7	8	9	10	11	12	13	14	15	16	17	18	19	20
26	33-45 33-45	41-63 42-62	49-81 52-78	57-99 61-95	66-116 71-111	74-134 80-128	82-152 90-144	90-170 100-160	98-188 109-177	106-206 119-193	114-224 128-210	122-242 138-226	130-260 147-243	138-278 157-259	146-296 167-275	154-314 176-292	162-332 186-308	170-350 195-325	178-368 205-341
27	34-47 35-46	43-65 44-64	51-84 54-81	60-102 64-98	69-120 74-115	77-139 84-132	86-157 94-149	94-176 104-166	103-194 114-183	111-213 124-200	120-231 134-217	128-250 144-234	137-268 154-251	145-287 164-268	153-306 174-285	162-324 184-302	170-343 194-319	178-362 204-336	186-381 214-353
28	35-49 36-48	44-68 46-66	54-86 56-84	63-105 67-101	72-124 77-119	81-143 88-136	90-162 98-154	99-181 108-172	108-200 119-189	116-220 129-207	125-239 140-224	134-258 150-242	143-277 161-259	152-296 171-277	160-316 182-294	169-335 192-312	178-354 202-330	186-374 213-347	195-393 223-365
29	37-50 37-50	46-70 48-68	56-89 59-86	65-109 69-105	75-128 80-123	84-148 91-141	94-167 102-159	103-187 113-177	112-207 124-195	122-226 135-213	131-246 145-232	140-266 156-250	149-286 167-268	158-306 178-286	167-326 189-304	177-345 200-322	186-365 211-340	195-385 222-358	204-405 233-376
30	38-52 39-51	48-72 50-70	58-92 61-89	68-112 72-108	78-132 83-127	88-152 95-145	97-173 106-164	107-193 117-183	117-213 129-201	127-233 140-220	136-254 151-239	146-274 163-257	155-295 174-276	165-315 185-295	175-335 197-313	184-356 208-332	194-376 219-351	203-397 230-370	212-418 242-388
31	39-54 40-53	50-74 51-73	60-95 63-92	71-115 75-111	81-136 86-131	91-157 98-150	101-178 110-169	112-198 122-188	122-219 133-206	132-240 145-227	142-261 157-246	152-282 169-265	162-303 180-285	172-324 192-304	182-345 204-323	192-366 216-342	202-387 228-361	211-409 239-381	221-430 251-400
32	41-55 41-55	52-76 53-75	62-98 65-95	73-119 77-115	84-140 90-134	95-161 102-154	105-183 114-174	116-204 126-194	126-226 138-214	137-247 151-233	147-269 163-253	158-290 175-273	168-312 187-293	179-333 199-313	189-355 212-332	199-377 224-352	209-399 236-372	220-420 248-392	230-442 260-412
33	42-57 43-56	53-79 55-77	65-100 68-97	76-122 80-118	87-144 93-138	98-166 105-159	109-188 118-179	120-210 131-199	131-232 143-220	142-254 156-240	153-276 169-260	164-298 181-281	174-321 194-301	185-343 206-322	196-365 219-342	207-387 232-362	217-410 244-383	228-432 257-403	239-454 270-423
34	44-58 44-58	55-81 57-79	67-103 70-100	78-126 83-121	90-148 96-142	102-170 109-163	113-193 122-184	124-216 135-205	136-238 148-226	147-261 161-247	158-284 174-268	170-306 187-289	181-329 201-309	192-352 214-330	203-375 227-351	214-398 240-372	225-421 253-393	237-443 266-414	248-466 279-435
35	45-60 46-59	57-83 59-81	69-106 72-103	81-129 86-124	93-152 99-146	105-175 113-167	117-198 126-189	129-221 140-210	141-244 153-232	152-268 167-253	164-291 180-275	176-314 194-296	187-338 207-318	199-361 221-339	210-385 234-361	222-408 248-382	234-431 261-404	245-455 275-425	257-478 288-447
36	46-62 47-61	59-85 61-83	71-109 74-106	84-132 88-128	96-156 102-150	109-179 116-172	121-203 130-194	133-227 144-216	145-251 158-238	157-275 172-260	170-298 186-282	182-322 200-304	194-346 214-326	206-370 228-348	218-394 242-370	230-418 256-392	242-442 270-414	254-466 284-436	265-491 298-458
37	48-63 48-63	61-87 63-85	74-111 77-108	86-136 91-131	99-160 105-154	112-134 120-176	125-208 134-199	137-233 149-221	150-257 163-244	163-281 177-267	175-306 192-289	188-330 206-312	200-355 221-334	213-379 235-357	225-404 249-380	237-429 264-402	250-453 278-425	262-478 293-447	274-503 307-470
38	49-65 50-64	62-90 64-88	76-114 79-111	89-139 94-134	102-164 109-157	116-188 123-181	129-213 138-204	142-238 153-227	155-263 168-250	168-288 183-273	181-313 198-296	194-338 213-319	207-363 227-343	219-389 242-366	232-414 257-389	245-439 272-412	258-464 287-435	271-489 302-458	283-515 317-481
39	51-66 51-66	64-92 66-90	78-117 84-114	92-142 97-137	105-168 112-161	119-193 127-185	133-218 142-209	146-244 158-232	160-269 173-256	173-295 188-280	186-321 204-303	200-346 219-327	213-372 234-351	226-398 249-375	240-423 265-398	253-449 280-422	266-475 295-446	279-501 311-469	292-527 326-493
40	52-68 53-67	66-94 68-92	80-120 84-116	94-146 99-141	109-171 115-165	123-197 131-189	137-223 146-214	150-250 162-238	164-276 178-262	178-302 194-286	192-328 209-311	206-354 225-335	219-381 241-359	233-407 257-383	247-433 272-408	260-460 288-432	274-486 304-456	288-512 320-480	301-539 335-505
41	53-70 54-69	68-96 70-94	83-122 86-119	97-149 102-144	112-175 118-169	126-202 134-194	140-229 150-219	155-255 167-243	169-282 183-268	183-309 199-293	198-335 215-318	212-362 231-343	226-389 248-367	240-416 264-392	254-443 280-417	268-470 296-442	282-497 312-467	296-524 329-491	310-551 345-516
42	55-71 56-70	70-98 72-96	85-125 88-122	100-152 105-147	115-179 121-173	130-206 138-198	144-234 155-223	159-261 171-249	174-288 188-274	189-315 204-300	203-343 221-325	218-370 238-350	232-398 254-376	247-425 271-401	261-453 288-426	276-480 304-452	290-508 321-477	305-535 338-502	319-563 354-528
43	56-73 57-72	72-100 74-98	87-128 91-124	103-155 108-150	118-183 125-176	133-211 142-202	148-239 159-228	164-266 176-254	179-294 193-280	194-322 210-306	209-350 227-332	224-378 244-358	239-406 261-384	254-434 278-410	269-462 295-436	284-490 312-462	299-518 330-487	313-547 347-513	328-575 364-539
44	58-74 58-74	73-103 75-101	89-131 93-127	105-159 110-154	121-187 128-180	137-215 145-207	152-244 163-233	168-272 180-260	184-300 198-286	199-329 215-313	215-357 233-339	230-386 250-366	245-415 268-392	261-443 286-418	276-472 303-445	291-501 321-471	307-529 338-498	322-558 356-524	337-587 373-551
45	59-76 60-75	75-105 77-103	92-133 95-130	108-162 113-157	124-191 131-184	140-220 149-211	156-249 167-238	172-278 185-265	188-307 203-292	204-336 221-319	220-365 239-346	236-394 257-373	252-423 275-400	268-452 293-427	284-481 311-454	299-511 329-481	315-540 347-508	331-569 365-535	346-599 383-562
46	60-78 61-77	77-107 79-105	94-136 97-133	111-165 116-160	127-195 134-188	144-224 153-215	160-254 171-243	177-283 189-271	193-313 208-298	210-342 226-326	226-372 245-353	242-402 263-381	259-431 282-408	275-461 300-436	291-491 318-464	307-521 337-491	323-551 355-519	339-581 374-546	355-611 392-574
47	62-79 63-78	79-109 81-107	96-139 100-135	113-169 119-163	130-199 137-192	147-229 156-220	164-259 175-248	181-289 194-276	198-319 213-304	215-349 232-332	232-379 251-360	248-410 269-389	265-440 288-417	282-470 307-445	298-501 326-473	315-531 345-501	331-562 364-529	348-592 383-557	365-622 402-585
48	63-81 64-80	81-111 83-109	98-142 102-138	116-172 121-167	133-203 141-195	151-233 160-224	168-264 179-253	186-294 198-282	203-325 218-310	220-356 237-339	237-387 257-367	255-417 276-396	272-448 295-425	289-479 315-453	306-510 334-482	323-541 353-511	340-572 373-539	357-603 392-568	374-634 411-597
49	65-82 65-82	83-113 85-111	101-144 104-141	119-175 124-170	137-206 144-199	155-237 164-228	172-269 183-258	190-300 203-287	208-331 223-316	225-363 243-345	243-394 262-375	261-425 282-404	278-457 302-433	296-488 322-462	313-520 342-491	331-551 361-521	348-583 381-550	365-615 401-579	383-646 421-608
50	66-84 67-83	84-116 87-113	103-147 107-143	121-179 127-173	140-210 147-203	158-242 167-233	176-274 187-263	195-305 208-292	213-337 228-322	231-369 248-352	249-401 268-382	267-433 289-411	285-465 309-441	303-497 329-471	321-529 349-501	338-562 370-530	356-594 390-560	374-626 410-590	392-658 430-620

Table G.23 *(continued)*

No. of judges.	2	3	4	5	6	7	8	9	10	11	12	13	14	15	16	17	18	19	20
51	67-86 68-85	86-118 88-116	105-150 109-146	124-182 130-176	143-214 150-207	162-246 171-237	180-279 192-267	199-311 212-298	218-343 233-328	236-376 254-358	255-408 274-389	273-441 295-419	291-474 316-449	310-506 336-480	328-539 357-510	346-572 376-540	365-604 399-570	383-637 419-601	401-670 440-631
52	69-87 70-86	88-120 90-118	108-152 111-149	127-185 132-180	146-218 153-211	165-251 175-241	184-284 196-272	203-317 217-303	222-350 238-334	241-383 259-365	260-416 280-396	279-449 301-427	298-482 323-457	317-515 344-488	336-548 365-519	354-582 386-550	373-615 407-581	392-648 428-612	410-682 449-643
53	70-89 71-88	90-122 92-120	110-155 114-151	130-188 135-183	149-222 157-214	169-255 178-246	188-289 200-277	208-322 221-309	227-356 243-340	247-389 265-371	266-423 286-403	285-457 308-434	305-490 329-466	324-524 351-497	343-558 373-528	362-592 394-560	381-626 416-591	400-660 437-623	419-694 459-654
54	72-90 73-89	92-124 94-122	112-158 116-154	132-192 138-186	152-226 160-218	172-260 182-250	192-294 204-282	212-328 226-314	232-362 248-346	252-396 270-378	272-430 292-410	291-465 314-442	311-499 336-474	331-533 358-506	350-568 380-538	370-602 402-570	390-636 425-601	409-671 447-633	429-705 469-665
55	73-92 74-91	94-126 96-124	114-161 118-157	135-195 141-189	156-229 163-222	176-264 186-254	196-299 208-287	217-333 231-319	237-368 253-352	257-403 276-384	278-437 298-417	298-472 321-449	318-507 343-482	338-542 366-514	358-577 388-547	378-612 411-579	398-647 433-612	418-682 456-644	438-717 478-677
56	74-94 75-93	96-128 98-126	117-163 121-159	138-198 143-193	159-233 166-226	180-268 189-259	200-304 212-292	221-339 235-325	242-374 258-358	263-409 281-391	283-445 304-424	304-480 327-457	324-516 350-490	345-551 373-523	365-587 396-556	386-622 419-589	406-658 442-622	427-693 465-655	447-729 488-688
57	76-95 77-94	97-131 100-128	119-166 123-162	140-202 146-196	162-237 170-229	183-273 193-263	205-308 216-297	226-344 240-330	247-380 263-364	268-416 287-397	289-452 310-431	310-488 333-465	331-524 357-498	352-560 380-532	373-596 404-565	394-632 427-599	415-668 451-632	435-705 474-666	456-741 497-700
58	77-97 78-96	99-133 102-130	121-169 125-165	143-205 149-199	165-241 173-233	187-277 197-267	209-313 220-302	230-350 244-336	252-386 268-370	273-423 292-404	295-459 316-438	316-496 340-472	338-532 364-506	359-569 388-540	380-606 412-574	402-642 435-609	423-679 459-643	444-716 483-677	465-753 507-711
59	79-98 80-97	101-135 103-133	124-171 128-167	146-208 152-202	168-245 176-237	190-282 200-272	213-318 225-306	235-355 249-341	257-392 273-376	279-429 298-410	301-466 322-445	323-503 346-480	344-541 371-514	366-578 395-549	388-615 419-584	410-652 444-618	431-690 468-653	453-727 492-688	475-764 517-722
60	80-100 81-99	103-137 105-135	126-174 130-170	149-211 155-205	171-249 179-241	194-286 204-276	217-323 229-311	239-361 254-346	262-398 278-382	284-436 303-417	306-474 328-452	329-511 353-487	351-549 378-522	373-587 402-558	395-625 427-593	418-662 452-628	440-700 477-663	462-738 502-698	484-776 526-7[illegible]
61	82-101 82-101	105-139 107-137	128-177 132-173	151-215 157-209	175-252 183-244	198-290 208-280	221-328 233-316	244-366 258-352	267-404 283-388	289-443 309-423	312-481 334-459	335-519 359-495	358-557 384-531	380-596 410-566	403-634 435-602	426-672 460-638	448-711 485-674	471-749 511-709	493-788 536-745
62	83-103 84-102	107-141 109-139	130-180 135-175	154-218 160-212	178-256 186-248	201-295 211-285	225-333 237-321	248-372 263-357	271-411 288-394	295-449 314-430	318-488 340-466	341-527 366-502	364-566 391-539	387-605 417-575	411-643 443-611	434-682 468-648	457-721 494-684	480-760 520-720	503-799 546-756
63	84-105 85-104	109-143 111-141	133-182 137-178	157-221 163-215	181-260 189-252	205-299 215-289	229-338 241-326	253-377 267-363	276-417 294-399	300-456 320-436	324-495 346-473	347-535 372-510	371-574 398-547	395-613 424-584	418-653 451-620	442-692 477-657	465-732 503-694	488-772 529-731	512-811 555-768
64	86-106 87-105	110-146 113-143	135-185 139-181	160-224 166-218	184-264 192-256	209-303 219-293	233-343 245-331	257-383 272-368	281-423 299-405	306-462 325-443	330-502 352-480	354-542 379-517	378-582 405-555	402-622 432-592	426-662 458-630	450-702 485-667	473-743 512-704	497-783 538-742	521-823 565-779
65	87-108 88-107	112-148 115-145	137-188 142-183	162-228 169-221	187-268 196-259	212-308 223-297	237-348 250-335	262-388 277-373	286-429 304-411	311-469 331-449	335-510 358-487	360-550 385-525	384-591 412-563	409-631 439-601	433-678 466-639	458-712 493-677	482-753 520-715	506-794 547-753	530-835 575-790
66	89-109 90-108	114-150 117-147	140-190 144-186	165-231 171-225	190-272 199-263	216-312 226-302	241-353 254-340	266-394 281-379	291-435 309-417	316-476 336-456	341-517 364-494	366-558 391-533	391-599 419-571	416-640 446-610	441-681 474-648	466-722 502-686	490-764 529-725	515-805 557-763	540-846 584-802
67	90-111 91-110	116-152 119-149	142-193 146-189	168-234 174-228	194-275 202-267	219-317 230-306	245-358 258-345	271-399 286-384	296-441 314-423	322-482 342-462	347-524 370-501	373-565 398-540	398-607 426-579	423-649 454-618	448-691 482-657	474-732 510-696	499-774 538-735	524-816 566-774	549-858 594-813
68	91-113 92-112	118-154 120-152	144-196 149-191	171-237 177-231	197-279 205-271	223-321 234-310	249-363 262-350	275-405 291-389	301-447 319-429	327-489 347-469	353-531 376-508	379-573 404-548	405-615 433-587	430-658 461-627	456-700 490-666	482-742 518-706	507-785 547-745	533-827 575-785	558-870 604-824
69	93-114 94-113	120-156 122-154	147-198 151-194	173-241 180-234	200-283 209-274	227-325 237-315	253-368 266-355	280-410 295-395	306-453 324-435	332-496 353-475	359-538 382-515	385-581 411-555	411-624 440-595	437-667 469-635	464-709 498-675	490-752 527-715	516-795 555-756	542-838 584-796	568-881 613-836
70	94-116 95-115	122-158 124-156	149-201 153-197	176-244 183-237	203-287 212-278	230-330 241-319	257-373 270-360	284-416 300-400	311-459 329-441	338-502 359-481	365-545 388-522	391-589 417-563	418-632 447-603	445-675 476-644	471-719 505-685	498-762 535-725	524-806 564-766	551-849 594-806	577-893 623-847
71	96-117 97-116	123-161 126-158	151-204 156-199	179-247 185-241	206-291 215-282	234-334 245-323	261-378 275-364	289-421 304-406	316-465 334-447	343-509 364-488	371-552 394-529	398-596 424-570	425-640 454-611	452-684 483-653	479-728 513-694	506-772 543-735	533-816 573-776	560-860 603-817	587-904 633-858
72	97-119 98-118	125-163 128-160	153-207 158-202	182-250 188-244	210-294 218-286	238-338 249-327	265-383 279-369	293-427 309-411	321-471 339-453	349-515 370-494	376-560 400-536	404-604 430-578	431-649 461-619	459-693 491-661	486-738 521-703	514-782 551-745	541-827 582-786	569-871 612-828	596-916 642-870
73	99-120 100-119	127-165 130-162	156-209 160-205	184-254 191-247	213-298 222-289	241-343 252-332	270-387 283-374	298-432 314-416	326-477 345-458	354-522 375-501	382-567 406-543	410-612 437-585	438-657 468-627	466-702 498-670	494-747 529-712	522-792 560-754	550-837 591-796	578-882 621-839	605-928 652-881
74	100-122 101-121	129-167 132-164	158-212 163-207	187-257 194-250	216-302 225-293	245-347 256-336	274-392 287-379	302-438 318-422	331-483 350-464	360-528 381-507	388-574 412-550	417-619 443-593	445-665 474-636	473-711 506-678	502-756 537-721	530-802 568-764	558-848 599-807	586-894 631-849	615-939 662-892
75	101-124 102-123	131-169 134-166	160-215 165-210	190-260 197-253	219-306 228-297	249-351 260-340	278-397 291-384	307-443 323-427	336-489 355-470	365-535 386-514	394-581 418-557	423-627 450-600	452-673 481-644	481-719 513-687	509-766 545-730	538-812 576-774	567-858 608-817	595-905 640-860	624-951 672-903

Source: A. Kramer et al., *Chem. Senses & Flavor 1*, 121-133 (1974).

Table G.24 Probabilities Associated with Values as Large as Observed Values of H in the Kruskal-Wallis One-Way Analysis of Variance by Ranks

Sample sizes n_1	n_2	n_3	H	p
2	1	1	2.7000	.500
2	2	1	3.6000	.200
2	2	2	4.5714	.067
			3.7143	.200
3	1	1	3.2000	.300
3	2	1	4.2857	.100
			3.8571	.133
3	2	2	5.3572	.029
			4.7143	.048
			4.5000	.067
			4.4643	.105
3	3	1	5.1429	.043
			4.5714	.100
			4.0000	.129
3	3	2	6.2500	.011
			5.3611	.032
			5.1389	.061
			4.5556	.100
			4.2500	.121
3	3	3	7.2000	.004
			6.4889	.011
			5.6889	.029
			5.6000	.050
			5.0667	.086
			4.6222	.100
4	1	1	3.5714	.200
4	2	1	4.8214	.057
			4.5000	.076
			4.0179	.114
4	2	2	6.0000	.014
			5.3333	.033
			5.1250	.052
			4.4583	.100
			4.1667	.105
4	3	1	5.8333	.021
			5.2083	.050
			5.0000	.057
			4.0556	.093
			3.8889	.129
4	3	2	6.4444	.008
			6.3000	.011
			5.4444	.046
			5.4000	.051
			4.5111	.098
			4.4444	.102
4	3	3	6.7455	.010
			6.7091	.013
			5.7909	.046
			5.7273	.050
			4.7091	.092
			4.7000	.101
4	4	1	6.6667	.010
			6.1667	.022
			4.9667	.048
			4.8667	.054
			4.1667	.082
			4.0667	.102
4	4	2	7.0364	.006
			6.8727	.011
			5.4545	.046
			5.2364	.052
			4.5545	.098
			4.4455	.103
4	4	3	7.1439	.010
			7.1364	.011
			5.5985	.049
			5.5758	.051
			4.5455	.099
			4.4773	.102
4	4	4	7.6538	.008
			7.5385	.011
			5.6923	.049
			5.6538	.054
			4.6539	.097
			4.5001	.104
5	1	1	3.8571	.143
5	2	1	5.2500	.036
			5.0000	.048
			4.4500	.071
			4.2000	.095
			4.0500	.119

Table G.24 *(continued)*

Sample sizes n_1	n_2	n_3	H	p
5	2	2	6.5333	.008
			6.1333	.013
			5.1600	.034
			5.0400	.056
			4.3733	.090
			4.2933	.122
5	3	1	6.4000	.012
			4.9600	.048
			4.8711	.052
			4.0178	.095
			3.8400	.123
5	3	2	6.9091	.009
			6.8218	.010
			5.2509	.049
			5.1055	.052
			4.6509	.091
			4.4945	.101
5	3	3	7.0788	.009
			6.9818	.011
			5.6485	.049
			5.5152	.051
			4.5333	.097
			4.4121	.109
5		1	6.9545	.008
			6.8400	.011
			4.9855	.044
			4.8600	.056
			3.9873	.098
			3.9600	.102
5	4	2	7.2045	.009
			7.1182	.010
			5.2727	.049
			5.2682	.050
			4.5409	.098
			4.5182	.101
5	4	3	7.4449	.010
			7.3949	.011
			5.6564	.049
			5.6308	.050
			4.5487	.099
			4.5231	.103
5	4	4	7.7604	.009
			7.7440	.011
			5.6571	.049
			5.6176	.050
			4.6187	.100
			4.5527	.102
5	5	1	7.3091	.009
			6.8364	.011
			5.1273	.046
			4.9091	.053
			4.1091	.086
			4.0364	.105
5	5	2	7.3385	.010
			7.2692	.010
			5.3385	.047
			5.2462	.051
			4.6231	.097
			4.5077	.100
5	5	3	7.5780	.010
			7.5429	.010
			5.7055	.046
			5.6264	.051
			4.5451	.100
			4.5363	.102
5	5	4	7.8229	.010
			7.7914	.010
			5.6657	.049
			5.6429	.050
			4.5229	.099
			4.5200	.101
5	5	5	8.0000	.009
			7.9800	.010
			5.7800	.049
			5.6600	.051
			4.5600	.100
			4.5000	.102

Source: Adapted and abridged from W. H. Kruskal and W. A. Wallis, "Use of ranks in one-criterion variance analysis," *J. Amer. Statist. Ass.*, *47*, 614-617 (1952); with the kind permission of the authors and the publisher. (The corrections to this table given by the authors in Errata, *J. Amer. Statist. Ass.*, *48*, 910, have been incorporated.)

Table G.25 Critical Values of Page's L[a]

Number of subjects	3	4	5	6	7	8	9	10
	—	—	109	178	260	388	544	726
2	—	60	106	173	201	376	520	696
	28	58	103	166	252	362	500	670
	—	89	160	260	394	567	790	1056
3	42	87	155	252	382	549	761	1019
	41	84	150	244	370	532	736	987
	56	117	210	341	516	743	1032	1382
4	55	114	204	331	501	722	999	1339
	54	111	197	321	487	701	971	1301
	70	145	259	420	637	917	1273	1704
5	68	141	251	409	620	893	1236	1656
	66	137	244	397	603	869	1204	1614
	83	172	307	499	757	1000	1512	2025
6	81	167	299	486	737	1063	1472	1972
	79	163	291	474	719	1037	1436	1927
	96	108	355	577	876	1262	1750	2344
7	93	193	346	563	855	1232	1706	2288
	91	180	338	550	835	1204	1668	2238
	109	225	403	655	994	1433	1987	2662
8	106	220	393	640	972	1401	1940	2602
	104	214	384	625	950	1371	1900	2549
	121	252	451	733	1113	1603	2223	2980
9	119	246	441	717	1088	1569	2174	2915
	116	240	431	701	1065	1537	2131	2859
	134	278	499	811	1230	1773	2459	3296
10	131	272	487	793	1205	1736	2407	3228
	128	266	477	777	1180	1703	2361	3169
	147	305	546	888	1348	1943	2694	3612
11	144	298	534	869	1321	1905	2639	3541
	141	292	523	852	1295	1868	2592	3478
	160	331	593	965	1465	2112	2929	3927
12	156	324	581	946	1437	2072	2872	3852
	153	317	570	928	1410	2035	2822	3788
	172	358	642	1044	1585	2285	3163	4241
13	169	350	628	1022	1553	2240	3104	4164
	165	343	615	1003	1525	2201	3052	4097
	185	384	689	1121	1702	2453	3397	4556
14	181	376	674	1098	1668	2407	3335	4475
	178	368	661	1078	1639	2367	3281	4405
	197	410	736	1197	1818	2622	3631	4869
15	194	402	721	1174	1784	2574	3567	4786
	190	394	707	1153	1754	2532	3511	4714
	210	436	783	1274	1935	2790	3864	5183
16	206	427	767	1249	1899	2740	2740	3798
	202	420	754	1228	1868	2697	3741	5022
	223	463	830	1350	2051	2958	4098	5496
17	218	453	814	1325	2014	2907	4029	5407
	215	445	800	1303	1982	2862	3970	5330
	235	489	876	1427	2167	3126	4330	5808
18	231	479	860	1401	2130	3073	4260	5717
	227	471	846	1378	2097	3028	4199	5638
	248	515	923	1503	2283	3294	4563	6121
19	243	505	906	1476	2245	3240	4491	6027
	239	496	891	1453	2217	3193	4428	5946
	260	541	970	1579	2399	3461	4796	6433
20	256	531	953	1552	2360	3406	4722	6337
	251	522	937	1528	2325	3358	4657	6253
	273	567	1017	1656	2515	3629	5028	6745
21	268	556	999	1628	2475	3572	4952	6647
	263	547	983	1603	2430	3523	4886	6561

For the 0.1% level use upper number

For the 1% level use middle number

For the 5% level use lower number

The values within the lines are based on exact distributions; the rest of the table, for larger sample sizes and more treatments, are based on normal-deviate values.

[a]Should the calculated L value equal or exceed the value in the table, reject H_0 in favor of a ranked alternative hypothesis.

Source: Adapted from E. B. Page, "Test for linear ranks," *J. Amer. Stats. Assoc., 58*, 216-230 (1963).

Table G.25 *(continued)*

Number of subjects	Number of treatments: 3	4	5	6	7	8	9	10
	285	593	1063	1732	2631	3796	5260	7057
22	280	582	1045	1703	2589	3738	5182	6956
	275	573	1029	1678	2553	3687	5115	6868
	298	619	1110	1808	2747	3963	5492	7368
23	292	608	1091	1778	2704	3904	5413	7265
	288	598	1075	1753	2667	3852	5343	7176
	310	644	1157	1884	2863	4130	5724	7679
24	305	633	1138	1854	2819	4070	5643	7574
	300	624	1121	1828	2781	4017	5572	7483
	322	670	1203	1960	2978	4297	5955	7991
25	317	659	1184	1929	2934	4235	5873	7883
	312	649	1167	1903	2895	4181	5801	7790
	335	696	1250	2036	3094	4464	6187	8302
26	329	685	1230	2004	3048	4401	6103	8192
	324	675	1213	1977	3009	4346	6029	8097
	347	722	1296	2111	3209	4631	6418	8612
27	342	710	1276	2080	3163	4567	6332	8501
	337	700	1258	2052	3123	4511	6257	8404
	360	748	1343	2187	3325	4798	6650	8923
28	354	736	1322	2155	3277	4732	6562	8810
	349	726	1304	2127	3236	4675	6486	8711
	372	774	1389	2263	3440	4964	6881	9234
29	366	762	1368	2230	3392	4898	6792	9118
	361	751	1350	2202	3350	4840	6714	9017
	385	800	1436	2339	3555	5131	7112	9544
30	379	787	1414	2305	3506	5063	7021	9426
	373	777	1396	2276	3464	5004	6942	9324
	397	825	1482	2415	3670	5297	7343	9854
31	391	813	1460	2381	3621	5228	7251	9735
	385	802	1441	2351	3578	5168	7170	9631
	409	851	1528	2490	3786	5464	7574	10164
32	403	839	1506	2456	3735	5394	7480	10043
	398	827	1487	2426	3691	5333	7398	9937
	422	877	1575	2566	3901	5630	7804	10474
33	415	864	1552	2531	3849	5559	7709	10351
	410	853	1533	2500	3805	5497	7627	10244
	434	903	1621	2641	4016	5796	8035	10784
34	428	890	1598	2606	3964	5724	7939	10659
	422	878	1579	2575	3918	5661	7855	10550
	447	929	1667	2717	4131	5963	8266	11094
35	440	915	1644	2681	4078	5889	8168	10967
	434	904	1624	2649	4032	5825	8083	10856
	459	954	1714	2793	4246	6129	8496	11404
36	452	941	1690	2756	4192	6054	8397	11275
	447	929	1670	2724	4146	5990	8311	11163
	471	980	1760	2868	4361	6295	8727	11713
37	465	966	1736	2831	4307	6219	8626	11583
	459	954	1716	2799	4259	6154	8538	11469
	484	1006	1806	2944	4476	6461	8957	12023
38	477	992	1782	2906	4421	6385	8855	11890
	471	980	1761	2873	4373	6318	8766	11775
	496	1032	1853	3019	4591	6627	9188	12332
39	489	1018	1828	2981	4535	6549	9084	12198
	483	1005	1807	2948	4486	6482	8994	12081
	508	1057	1899	3094	4705	6793	9418	12641
40	501	1043	1874	3056	4649	6714	9313	12506
	495	1031	1853	3022	4600	6646	9222	12387

For the 0.1% level use upper number

For the 1% level use middle number

For the 5% level use lower number

Table G.25 *(continued)*

Number of subjects	Number of treatments: 3	4	5	6	7	8	9	10
	521	1083	1945	3170	4820	6959	9648	12951
41	514	1069	1920	3131	4763	6879	9542	12813
	507	1056	1898	3097	4713	6810	9450	12694
	533	1109	1991	3245	4935	7124	9878	13260
42	526	1094	1966	3206	4877	7044	9771	13121
	520	1081	1944	3171	4827	6974	9677	13000
	545	1134	2037	3321	5050	7290	10108	13569
43	538	1120	2012	3281	4991	7209	10000	13428
	532	1107	1990	3246	4940	7138	9905	13306
	558	1160	2084	3396	5164	7456	10338	13878
44	550	1145	2058	3355	5105	7374	10228	13736
	544	1132	2035	3320	5054	7302	10133	13611
	570	1186	2130	3471	5279	7622	10568	14187
45	563	1171	2104	3430	5219	7539	10457	14043
	556	1157	2081	3395	5167	7466	10360	13917
	582	1211	2176	3547	5394	7787	10798	14496
46	575	1196	2150	3505	5333	7703	10686	14350
	568	1183	2126	3469	5280	7630	10588	14223
	595	1237	2222	3622	5508	7953	11028	14804
47	587	1222	2195	3580	5147	7868	10914	14657
	580	1208	2172	3544	5394	7794	10815	14529
	607	1263	2268	3697	5623	8119	11258	15113
48	599	1247	2241	3655	5561	8033	11143	14964
	593	1234	2218	3618	5507	7958	11043	14835
	619	1288	2314	3772	5737	8284	11487	15422
49	612	1273	2287	3730	5675	8197	11371	15271
	605	1259	2263	3692	5621	8122	11271	15141
	632	1314	2360	3848	5852	8450	11717	15730
50	624	1298	2333	3804	5789	8362	11600	15579
	617	1284	2309	3767	5734	8286	11498	15446

For the 0.1% level use upper number

For the 1% level use middle number

For the 5% level use lower number

Table G.26 Critical Values of Jonckheere's J

			One-tailed significance level, α				
			0.100	0.050	0.025	0.010	0.005
			Two-tailed significance level, α				
n_1	n_2	n_3	0.200	0.100	0.050	0.020	0.010
2	1	1	5				
2	2	1	6	8			
2	2	2	8	10	12		
3	1	1	7	7			
3	2	1	9	9	11		
3	2	2	10	12	14	16	16
3	3	1	9	11	13	15	
3	3	2	11	15	17	19	21
3	3	3	13	17	19	23	23
4	1	1	7	9			
4	2	1	10	12	14	14	
4	2	2	12	14	16	18	20
4	3	1	11	13	15	17	19
4	3	2	14	16	20	22	24
4	3	3	15	19	23	25	27
4	4	1	14	16	18	22	22
4	4	2	16	20	22	26	28
4	4	3	18	22	26	30	32
4	4	4	20	24	28	32	36
5	1	1	9	11	11		
5	2	1	14	16	16	18	
5	2	2	14	16	18	22	22
5	3	1	13	15	19	21	23
5	3	2	15	19	21	25	27
5	3	3	17	21	25	29	31
5	4	1	15	19	21	25	27
5	4	2	18	22	24	28	30
5	4	3	19	25	29	33	35
5	4	4	22	28	32	36	40
5	5	1	17	21	25	27	31
5	5	2	19	25	27	33	35
5	5	3	21	27	31	37	39
5	5	4	25	31	35	41	45
5	5	5	27	33	39	45	49

n_1 is the number of scores in the largest sample, n_2 the number in the next largest sample, and n_3 the number in the smallest sample. Reject H_0 if calculated J value is equal to or larger than J in the table. If calculated J value is negative, ignore its negative sign when using this table. Adapted from C. Leach, *Introduction to Statistics, Biometrika, 41,* 133 (1961).

Source: C. Leach, Introduction to Statistics, John Wiley & Sons, 1979. A. R. Jonckheere, Biometrika, 1954, *41*, 133.

Table G.27 Random Numbers

22 17 68 65 84	68 95 23 92 35	87 02 22 57 51	61 09 43 95 06	58 24 82 63 47
19 36 27 59 46	13 79 93 37 55	39 77 32 77 09	85 52 05 30 62	47 83 51 62 74
16 77 23 02 77	09 61 87 25 21	28 06 24 25 93	16 71 13 59 78	23 05 47 47 25
78 43 76 71 61	20 44 90 32 64	97 67 63 99 61	46 38 03 93 22	69 81 21 99 21
03 28 28 26 08	73 37 32 04 05	69 30 16 09 05	88 69 58 28 99	35 07 44 75 47
93 22 53 64 39	07 10 63 76 35	87 03 04 79 88	08 13 13 85 51	55 34 57 72 69
78 76 58 54 74	92 38 70 96 92	52 06 79 79 45	82 63 18 27 44	69 66 92 19 09
23 68 35 26 00	99 53 93 61 28	52 70 05 48 34	56 65 05 61 86	90 92 10 70 80
15 39 25 70 99	93 86 52 77 65	15 33 59 05 28	22 87 26 07 47	86 96 98 29 06
58 71 96 30 24	18 46 23 34 27	85 13 99 24 44	49 18 09 79 49	74 16 32 23 02
57 35 27 33 72	24 53 63 94 09	41 10 76 47 91	44 04 95 49 66	39 60 04 59 81
48 50 86 54 48	22 06 34 72 52	82 21 15 65 20	33 29 94 71 11	15 91 29 12 03
61 96 48 95 03	07 16 39 33 66	98 56 10 56 79	77 21 30 27 12	90 49 22 23 62
36 93 89 41 26	29 70 83 63 51	99 74 20 52 36	87 09 41 15 09	98 60 16 03 03
18 87 00 42 31	57 90 12 02 07	23 47 37 17 31	54 08 01 88 63	39 41 88 92 10
88 56 53 27 59	33 35 72 67 47	77 34 55 45 70	08 18 27 38 90	16 95 86 70 75
09 72 95 84 29	49 41 31 06 70	42 38 06 45 18	64 84 73 31 65	52 53 37 97 15
12 96 88 17 31	65 19 69 02 83	60 75 86 90 68	24 64 19 35 51	56 61 87 39 12
85 94 57 24 16	92 09 84 38 76	22 00 27 69 85	29 81 94 78 70	21 94 47 90 12
38 64 43 59 98	98 77 87 68 07	91 51 67 62 44	40 98 05 93 78	23 32 65 41 18
53 44 09 42 72	00 41 86 79 79	68 47 22 00 20	35 55 31 51 51	00 83 63 22 55
40 76 66 26 84	57 99 99 90 37	36 63 32 08 58	37 40 13 68 97	87 64 81 07 83
02 17 79 18 05	12 59 52 57 02	22 07 90 47 03	28 14 11 30 79	20 69 22 40 98
95 17 82 06 53	31 51 10 96 46	92 06 88 07 77	56 11 50 81 69	40 23 72 51 39
35 76 22 42 92	96 11 83 44 80	34 68 35 48 77	33 42 40 90 60	73 96 53 97 86
26 29 13 56 41	85 47 04 66 08	34 72 57 59 13	82 43 80 46 15	38 26 61 70 04
77 80 20 75 82	72 82 32 99 90	63 95 73 76 63	89 73 44 99 05	48 67 26 43 18
46 40 66 44 52	91 36 74 43 53	30 82 13 54 00	78 45 63 98 35	55 03 36 67 68
37 56 08 18 09	77 53 84 46 47	31 91 18 95 58	24 16 74 11 53	44 10 13 85 57
61 65 61 68 66	37 27 47 39 19	84 83 70 07 48	53 21 40 06 71	95 06 79 88 54
93 43 69 64 07	34 18 04 52 35	56 27 09 24 86	61 85 53 83 45	19 90 70 99 00
21 96 60 12 99	11 20 99 45 18	48 13 93 55 34	18 37 79 49 90	65 97 38 20 46
95 20 47 97 97	27 37 83 28 71	00 06 41 41 74	45 89 09 39 84	51 67 11 52 49
97 86 21 78 73	10 65 81 92 59	58 76 17 14 97	04 76 62 16 17	17 95 70 45 80
69 92 06 34 13	59 71 74 17 32	27 55 10 24 19	23 71 82 13 74	63 52 52 01 41
04 31 17 21 56	33 73 99 19 87	26 72 39 27 67	53 77 57 68 93	60 61 97 22 61
61 06 98 03 91	87 14 77 43 96	43 00 65 98 50	45 60 33 01 07	98 99 46 50 47
85 93 85 86 88	72 87 08 62 40	16 06 10 89 20	23 21 34 74 97	76 38 03 29 63
21 74 32 47 45	73 96 07 94 52	09 65 90 77 47	25 76 16 19 33	53 05 70 53 30
15 69 53 82 80	79 96 23 53 10	65 39 07 16 29	45 33 02 43 70	02 87 40 41 45
02 89 08 04 49	20 21 14 68 86	87 63 93 95 17	11 29 01 95 80	35 14 97 35 33
87 18 15 89 79	85 43 01 72 73	08 61 74 51 69	89 74 39 82 15	94 51 33 41 67
98 83 71 94 22	59 97 50 99 52	08 52 85 08 40	87 80 61 65 31	91 51 80 32 44
10 08 58 21 66	72 68 49 29 31	89 85 84 46 06	59 73 19 85 23	65 09 29 75 63
47 90 56 10 08	88 02 84 27 83	42 29 72 23 19	66 56 45 65 79	20 71 53 20 25
22 85 61 68 90	49 64 92 85 44	16 40 12 89 88	50 14 49 81 06	01 82 77 45 12
67 80 43 79 33	12 83 11 41 16	25 58 19 68 70	77 02 54 00 52	53 43 37 15 26
27 62 50 96 72	79 44 61 40 15	14 53 40 65 39	27 31 58 50 28	11 39 03 34 25
33 78 80 87 15	38 30 06 38 21	14 47 47 07 26	54 96 87 53 32	40 36 40 96 76
13 13 92 66 99	47 24 49 57 74	32 25 43 62 17	10 97 11 69 84	99 63 22 32 98

Table G.27 *(continued)*

10 27 53 96 23	71 50 54 36 23	54 31 04 82 98	04 14 12 15 09	26 78 25 47 47
28 41 50 61 88	64 85 27 20 18	83 36 36 05 56	39 71 65 09 62	94 76 62 11 89
34 21 42 57 02	59 19 18 97 48	80 30 03 30 98	05 24 67 70 07	84 97 50 87 46
61 81 77 23 23	82 82 11 54 08	53 28 70 58 96	44 07 39 55 43	42 34 43 39 28
61 15 18 13 54	16 86 20 26 88	90 74 80 55 09	14 53 90 51 17	52 01 63 01 59
91 76 21 64 64	44 91 13 32 97	75 31 62 66 54	84 80 32 75 77	56 08 25 70 29
00 97 79 08 06	37 30 28 59 85	53 56 68 53 40	01 74 39 59 73	30 19 99 85 48
36 46 18 34 94	75 20 80 27 77	78 91 69 16 00	08 43 18 73 68	67 69 61 34 25
88 98 99 60 50	65 95 79 42 94	93 62 40 89 96	43 56 47 71 66	46 76 29 67 02
04 37 59 87 21	05 02 03 24 17	47 97 81 56 51	92 34 86 01 82	55 51 33 12 91
63 62 06 34 41	94 21 78 55 09	72 76 45 16 94	29 95 81 83 83	79 88 01 97 30
78 47 23 53 90	34 41 92 45 71	09 23 70 70 07	12 38 92 79 43	14 85 11 47 23
87 68 62 15 43	53 14 36 59 25	54 47 33 70 15	59 24 48 40 35	50 03 42 99 36
47 60 92 10 77	88 59 53 11 52	66 25 69 07 04	48 68 64 71 06	61 65 70 22 12
56 88 87 59 41	65 28 04 67 53	95 79 88 37 31	50 41 06 94 76	81 83 17 16 33
02 57 45 86 67	73 43 07 34 48	44 26 87 93 29	77 09 61 67 84	06 69 44 77 75
31 54 14 13 17	48 62 11 90 60	68 12 93 64 28	46 24 79 16 76	14 60 25 51 01
28 50 16 43 36	28 97 85 58 99	67 22 52 76 23	24 70 36 54 54	59 28 61 71 96
63 29 62 66 50	02 63 45 52 38	67 63 47 54 75	83 24 78 43 20	92 63 13 47 48
45 65 58 26 51	76 96 59 38 72	86 57 45 71 46	44 67 76 14 55	44 88 01 62 12
39 65 36 63 70	77 45 85 50 51	74 13 39 35 22	30 53 36 02 95	49 34 88 73 61
73 71 98 16 04	29 18 94 51 23	76 51 94 84 86	79 93 96 38 63	08 58 25 58 94
72 20 56 20 11	72 65 71 08 86	79 57 95 13 91	97 48 72 66 48	09 71 17 24 89
75 17 26 99 76	89 37 20 70 01	77 31 61 95 46	26 97 05 73 51	53 33 18 72 87
37 48 60 82 29	81 30 15 39 14	48 38 75 93 29	06 87 37 78 48	45 56 00 84 47
68 08 02 80 72	83 71 46 30 49	89 17 95 88 29	02 39 56 03 46	97 74 06 56 17
14 23 98 61 67	70 52 85 01 50	01 84 02 78 43	10 62 98 19 41	18 83 99 47 99
49 08 96 21 44	25 27 99 41 28	07 41 08 34 66	19 42 74 39 91	41 96 53 78 72
78 37 06 08 43	63 61 62 42 29	39 68 95 10 96	09 24 23 00 62	56 12 80 73 16
37 21 34 17 68	68 96 83 23 56	32 84 60 15 31	44 73 67 34 77	91 15 79 74 58
14 29 09 34 04	87 83 07 55 07	76 58 30 83 64	87 29 25 58 84	86 50 60 00 25
58 43 28 06 36	49 52 83 51 14	47 56 91 29 34	05 87 31 06 95	12 45 57 09 09
10 43 67 29 70	80 62 80 03 42	10 80 21 38 84	90 56 35 03 09	43 12 74 49 14
44 38 88 39 54	86 97 37 44 22	00 95 01 31 76	17 16 29 56 63	38 78 94 49 81
90 69 59 19 51	85 39 52 85 13	07 28 37 07 61	11 16 36 27 03	78 86 72 04 95
41 47 10 25 62	97 05 31 03 61	20 26 36 31 62	68 69 86 95 44	84 95 48 46 45
91 94 14 63 19	75 89 11 47 11	31 56 34 19 09	79 57 92 36 59	14 93 87 81 40
80 06 54 18 66	09 18 94 06 19	98 40 07 17 81	22 45 44 84 11	24 62 20 42 31
67 72 77 63 48	84 08 31 55 58	24 33 45 77 58	80 45 67 93 82	75 70 16 08 24
59 40 24 13 27	79 26 88 86 30	01 31 60 10 39	53 58 47 70 93	85 81 56 39 38
05 90 35 89 95	01 61 16 96 94	50 78 13 69 36	37 68 53 37 31	71 26 35 03 71
44 43 80 69 98	46 68 05 14 82	90 78 50 05 62	77 79 13 57 44	59 60 10 39 66
61 81 31 96 82	00 57 25 60 59	46 72 60 18 77	55 66 12 62 11	08 99 55 64 57
42 88 07 10 05	24 98 65 63 21	47 21 61 88 32	27 80 30 21 60	10 92 35 36 12
77 94 30 05 39	28 10 99 00 27	12 73 73 99 12	49 99 57 94 82	96 88 57 17 91
78 83 19 76 16	94 11 68 84 26	23 54 20 86 85	23 86 66 99 07	36 37 34 92 09
87 76 59 61 81	43 63 64 61 61	65 76 36 95 90	18 48 27 45 68	27 23 65 30 72
91 43 05 96 47	55 78 99 95 24	37 55 85 78 78	01 48 41 19 10	35 19 54 07 73
84 97 77 72 73	09 62 06 65 72	87 12 49 03 60	41 15 20 76 27	50 47 02 29 16
87 41 60 76 83	44 88 96 07 80	83 05 83 38 96	73 70 66 81 90	30 56 10 48 59

Source: Table XXXIII of Fisher and Yates, *Statistical Tables for Biological, Agricultural and Medical Research*, published by Longman Group Ltd., London (previously published by Oliver and Boyd Ltd., Edinburgh) and by permission of the authors and publishers.

Index

Addition law, 33-35 (*see also* Probability)
Analysis of variance (*see* ANOVA)
ANOVA (*see also* Friedman Two-Factor Ranked Analysis of Variance, Kramer Two-Factor Ranked Analysis of Variance, Kruskal-Wallis One-Factor Ranked Analysis of Variance, Page Test, Jonckheere Test)
  calculation of number of matrices, 226
  correction term, 141
  degrees of freedom, *df*, 141
  fixed-effects model, 247-257
    computation of, 248-250, 256-257
    definition of, 247-248
    expected mean square, 250-255
    summary of denominators, 256-257
    theory of, 250-255
[ANOVA]
  four-factor design, 222-226
  mean square, 141
  mean square estimate, 141
  mixed-effects model, 247-257
    computation of, 248-250, 256-257
    definition of, 247-248
    expected mean square, 250-255
    summary of denominators, 256-257
    theory of, 250-255
  MS, 141
  one-factor completely randomized design, 135-152
    assumptions for, 151-152
    computational formulas for, 143-146
    F and *t*, relationship between, 151, 179
    F ratio, computation of, 140-146
    F ratio, definition of, 139

[ANOVA]
logic of, 136-140
mathematical model, 152
multiple *t* tests and, 135-136
one- and two-tailed tests, 150-151
scaled data, analysis by, 152
worked examples, 147-150, 166-169
random-effects model, 247-257
computation of, 248-250, 256-257
definition of, 247-248
expected mean square, 250-255
summary of denominators, 256-257
theory of, 250-255
related versus unrelated samples design, 366-370
split-plot design, 259-278
between-subjects error degrees of freedom, 265
between-subjects error sum of squares, 264-265
cell total sum of squares, 263
computation of, 266-272
explanation of design, 260-261
logic of, 261-266
three-factor design, relation to, 260
two-factor design, relation to, 259
two-way matrix, 263
whole matrix, 262
worked example, 272-278
SS, 141
$SS_T = SS_B + SS_E$, 379-383
sum of squares, 17, 141
table of names of designs, 421-425
table of significance of *F*, 426-430
testing significance of *F*, 146

[ANOVA]
three-factor design, 211-222
computation of, 211-218
higher order interactions, inability to calculate, 221-222
relation to split-plot design, 260
three-way interactions, 211-213, 218-221
three-way matrix, 211-213
two-way interactions, 211-213, 218
two-way matrix, 211-213
whole matrix, 211-213
worked examples, 226-234, 237-243
two-factor design with interaction, 183-209
assumptions for, 200
cell total sum of squares, 198, 216
computation of, 193-195
interaction and consumer testing, 203-204
interaction, *F* values, 189-192
interaction, nature of, 183-192
interaction, two-way, 183-192
logic of, 193-195
need for more than one value per combination of conditions, 201-203
pooling insignificant interactions, 200-201
relation to one-factor design, 193
two-way interactions, inability to calculate, 201-203
two-way matrix, 197
whole matrix, 197
worked examples, 195-200, 204-209, 234-236

[ANOVA]
two-factor design without interaction, completely randomized, 179-180
two-factor design without interaction, repeated measures, 171-182
assumptions for, 178-179, 180-182
computation of, 173-176
logic of, 171-173, 193-194
mathematical model for, 178
randomized complete block design, 173
relation to split plot design, 259
worked example, 176-178
A.O.V. (*see* ANOVA)
A posteriori probability, 28
A priori probability, 27
Averages, 9-11

Between-subjects error term, 264-265
Between-subjects variance, 171
Between-treatments variance (*see also* ANOVA)
definition of, 136-140
relation to other variances, 142-143, 379-383
Bimodal, 11
Binomial coefficients, table of, 407
Binomial expansion, 373-378
Binomial probabilities, 55-61
tables of, 408-413
Binomial test, 57-90
abuse of, 84-85
assumptions for, 84-85
binomial expansion, 373-378
binomial expansion, application to probability calculations, 63-65
binomial probabilities, 55-61
chi-square, comparison with, 94-95

[Binomial test]
comparison of proportions, 78
nonparametric nature of, 70-71
normality, approximation to, 71-72
null hypothesis for, 61-63
one- and two-tailed tests, 68-70
one-sample nature, 73
Pascal's triangle, 375
relation to sign test, 79
samples too large for tables, 71-72
sensory difference tests, application to, 79-84
sign test, 73-78
significance, levels of, 66-68
tables for, 408-413
via z tests, 71-72
worked examples, 85-89
Bivariate normal distribution, 290
Block, 173

C (correction term), 141
Calculus
abscissa, 280, 283
for correlation, 280-283
for regression, 280-283
intercept, 282-283, 297
ordinate, 282, 283
slope, 280-283, 297
Carryover effects, 74
Category scales, 127-129, 366
Causality, relation to correlation, 293
Cell totals, 198, 263
Cell total sum of squares, 198
Central limit theorem, 52, 385
Chi-square, 91-110
application of, 99-101
assumptions for, 99-101
binomial test, comparison with, 94-95
Cochran Q test, 105-106
contingency coefficient, 107-108

[Chi-square]
correction for continuity, 100-104
definition of, 91
degrees of freedom, 93
distribution of, 104
expected frequencies, 93, 97, 99-100
free and fixed margins, 102-104
formula for, 91
McNemar test, 104-105
one-way classification, 92-96
single-sample test, 92-96
tables for, 415
two-way classification, 96-99, 101-104
2 × 2 classification, 101-104
Type I and Type II errors, 94
worked examples, 108-110
Yates' correction, 101
Classical probability, 27 (*see also* Probability)
Coefficient, 373
Coefficient of determination, 288-289
Cochran Q test, 105-106
Combinations, 41-44
explanation of, 41-42
worked examples, 42-44
Completely randomized design (*see* Independent samples design)
Confidence interval, 17, 385-387
Consumer research (*see* Consumer testing)
Consumer testing, 2-8, 203-204
ANOVA, use of, 203-204
definition of, 2-5
sampling for, 8
statistics, use of, 6
use of, 2-5
Contingency coefficient, 107-108
Continuity (*see* Correction for continuity)
Continuous variable, 22
Control groups, 8, 74
Correction factor (*see* Correction term)
Correction for continuity
for chi-square, 100-104
for Jonckheere test, 357
for Wilcoxon-Mann-Whitney rank sums test, 320, 324
Yates' correction, 101
Correction term, 17, 141 (*see also* ANOVA)
Correlation, Pearson's product moment correlation, 279-302 (*see also* Contingency coefficient, Spearman's ranked correlation)
assumptions for, 289-292
bivariate normal distribution, 290
calculus for, 280-283
coefficient of determination, r^2, 288-289
computation of, 286
Fisher's Z, 294-295
homoscedasticity, 292
interpretation of correlation coefficient, 283-285
mean of several coefficients, 294
relation to causality, 293
relation to ranked correlation, 327
relation to t, 293-294
significance of difference between coefficients, 294-295
significance, test of, 287-288, 293-294
tables of Fisher's Z, 443
tables of Pearson's product-moment correlation, 442
use of, 279-280
worked examples, 287, 300-302
Counterbalancing, 73-76

Degrees of freedom, *df*
ANOVA and, 17, 141
between-subjects error *df*, 265
chi-square and, 93, 99, 104
definition of, 93-94
t test and, 113
Dependent variable, 280
Descriptive statistics, 6
df (*see* Degrees of freedom)
Difference tests (*see* Sensory difference tests)
Discrete variable, 22
Duncan's test
computation of, 164-165
formula for, 157
tables for, 434-436
use of, 159-160
worked example, 165-166
Dunn test
formula for, 157
tables for, 440-441
use of, 158, 160
Dunnett test
formula for, 156-157
tables for, 437-439
use of, 158, 160
Duo-trio test, 80-85, 414
abuse of, 84-85
definition of, 80-84
relation to other tests, 414
statistics for, 82-84

Empirical probability, 28 (*see also* Probability)
EMS (*see* ANOVA, expected mean square)
Error variance
definition of, 136-140
relation to other variances, 142-143, 172, 379-383
Expected frequency
computation for chi-square, 93, 97
minimum value for chi-square, 99-100
Experimental psychology (*see* Psychophysics)
Exponent, 373

F ratios (*see* ANOVA)
F tests (*see* ANOVA)
Factorials
definition of, 40
tables of, 406
Fisher LSD test (*see* LSD test)
Fisher's *Z*
correlation and, 294-295
definition of, 294
tables of, 443
use of, 294-295
Fixed effects, 4, 6
Fixed margins for chi-square, 102-104
Free margins for chi-square, 102-104
Frequency distributions, 45
Friedman test (*see* Friedman Two-Factor Ranked Analysis of Variance)
Friedman Two-Factor Ranked Analysis of Variance, 332-337
assumptions for, 336
computation of, 332-334
logic of, 333-334
multiple comparisons for, 334-335
relation to Page test, 346
significance, test of, 334
tables for, 459-460
ties, 335-336
two-tailed nature of, 336
Type I error and, 334
use of, 332, 336
worked example, 336-337

Gaussian distribution (*see* Normal distribution
Geometric mean, 11

Graphic scales, use of t tests, 127-129

H_0 (*see* Null hypothesis)
Harmonic mean, 11
Hedonic scale, 12
Homoscedasticity
  correlation and, 292
  definition of, 122-123
  scaled data and, 127-129
  t tests and, 123-124, 127-129

Independent-samples design (*see also* Completely randomized design)
  ANOVA and, 366-370
  chi-square and, 99
  ranking tests and, 366-370
  t tests and, 111, 117-121, 123-124
  use of, 399
Independent variable, 280
Index, 373
Inferential statistics, 7
Interaction (*see* ANOVA)
Interquartile range, 13
Interval scale, definition, 21-22

Jonckheere test, 350-365
  assumptions for, 357-358
  computation of, 351-356
  correction for continuity, 357
  one- or two-tailed nature, 353-354, 357-358
  relation to Wilcoxon-Mann-Whitney sums test, 350-351
  relation to Page test, 347, 350-351
  samples too large for tables, 356-357
  significance, test of, 353-354, 359-362
[Jonckheere test]
  standard deviation for, 357, 359
  tables for, 472
  ties, 357, 358-362
  use of, 350-351, 358
  via z test, 356-357, 359-362
  worked example, 362-365
Judges, 3-4
  fixed and random effects, 4
  screening of, 3
  training of, 3

Kendall's ranked correlation coefficient, 331
Kendall's tau, 331
Kramer test (*see* Kramer two-factor ranked analysis of variance)
Kramer two-factor ranked analysis of variance, 337-341
  assumptions for, 339
  computation of, 338-339
  significance, test of, 338-339
  tables for, 461-466
  two-tailed nature of, 339
  Type I error and, 339
  use of, 337-338, 340
  worked example, 340-341
Kruskal-Wallis one-factor ranked analysis of variance, 341-346
  assumptions for, 344-345
  computation of, 341
  logic of, 342
  multiple comparisons for, 343-344
  relation to Wilcoxon-Mann-Whitney rank sums test, 341, 343
  significance, test of, 342-343
  tables for, 467-468
  ties, 344
  two-tailed nature of, 344
  use of, 341
  worked example, 345-346

Laplacean probability, 27 (*see also* Probability)
Levels of significance, 66-68
Linear regression (*see* Regression)
Logic of statistics, 57-70
LSD test
  computation of, 161-162
  formula for, 156-157
  split-plot ANOVA and, 278
  tables for, 416
  use of, 159-160
  worked example, 162-164

Magnitude estimation, 11, 127-129
Mann-Whitney test (*see* Wilcoxon-Mann-Whitney rank sums test, Mann-Whitney U test)
Mann-Whitney U test, 309-314
  alternatives to, 313
  assumptions for, 313
  computation of, 309-312
  correction for continuity, 320, 324
  logic of, 309
  one- or two-tailed test, 310
  relation to Wilcoxon-Mann-Whitney ranks sums test, 315
  samples too large for tables, 313
  significance, test of, 310, 313
  tables for, 446-453
  ties, 311-312
  use of, 309
  via *z* test, 313
  worked example, 314
Matched design (*see* Related samples design)
Matrices (*see* ANOVA)
Matrix (*see* ANOVA)
McNemar test, 104-105
Mean
  arithmetic, 9-10
  geometric, 11
  harmonic, 11
Mean deviation, 13-14
Median, 10
Mode, 10-11
Modulus, 13
MS (*see* ANOVA)
Multiple comparisons, 153-169
  Duncan's test, 154-160, 164-166
  Dunn test, 154-160
  Dunnett test, 154-160
  for Friedman two-factor ranked analysis of variance, 334-335
  for Kruskal-Wallis one-factor ranked analysis of variance, 343-344
  formulas for, 156-157
  logic of, 154-155
  LSD test, 154-164
  Newman-Keuls test, 154-160
  Sheffé test, 154-160
  tables for, 431-441
  Tukey HSD test, 154-160
  use of, 153, 158-160
  worked example of Duncan's test, 165-166
  worked example of LSD test, 162-164, 166-169
Multiplication law, 30-33 (*see also* Probability)

Newman-Keuls test
  formula for, 156-157
  tables for, 431-433
  use of, 159-160
Nominal scale, 20, 70
Nonparametric tests, 57-90, 91-110, 303-370
  binomial test and, 70-71
  definition of, 22-25
  normal distribution and, 22-25
  scaled data and, 303, 365-366
  use of, 22-25, 303-304
Normal distribution, 45-56
  areas under curve, 404-405
  central limit theorem, 52-55
  definition of, 45-46

[Normal distribution]
formula for, 45
parametric, nonparametric tests and, 22-25
sampling distributions and, 52-55, 124-127
shape of, 23, 46
significance, test of, 48-51
standard error, 52-55
statistical theory and, 52-55
occurrence in nature, 51-52
t tests and, 124-127
use of, 51-52
worked examples of *z* tests, 55-56
z scores, 47-51
z tests, 47-51
Null hypothesis
definition of, 61-63
for binomial tests, 61-63
for correlation, 287
for difference testing, *t* tests, 113-115, 118

One-sample test
binomial test as, 73
chi-square as, 96
t test as, 111-112
One-tailed tests
definition of, 68-69
correlation and, 288
use of, 69-70
Ordinal scale, 20

Page test, 346-350
assumptions for, 349
computation of, 347-349
logic of, 347-348
one-tailed nature of, 348, 357-358
relation to Friedman two-factor ranked analysis of variance, 346
relation to Jonckheere test, 347
significance, test of, 348
[Page test]
tables for, 469-471
use of, 346-347, 349
worked example, 349-350
Paired-comparison test
abuse of, 84-85
definition of, 80-84
relation to other tests, 414
statistics for, 82-84
Paired-samples design (*see* Related samples design)
Panel, 3
Parameter, definition of, 7
Parametric tests
categorization of, binomial tests, 70-71
definition of, 22-25
normal distribution and, 22-25
scaled data and, 365-366
use of, 22-25
Pascal's triangle
explanation of, 375
table of, 407
Pearson's product-moment correlation (*see* Correlation)
Permutations
explanation of, 40-41
worked examples, 42-44
Placebo, 8, 75
Placebo effect, 8
Plateau effects, 75
Poisson distribution, 72
Population, 3
Post hoc comparisons, post hoc tests (*see* Multiple comparisons)
Power, 94-95
comparison of binomial and chi-square tests, 94-95
definition of, 94
Preference tests (*see* Sensory preference tests)
Probability, 27-44
addition law, 33-35
a posteriori, 28
a priori, 27

[Probability]
classical, 27
computation of probabilities, 29-35
computation of total number of events, 29-30
definitions of, 27-29
empirical, 28
Laplacean, 27
multiplication law, 30-33
worked examples, 35-40
Proportions, binomial comparison of, 78
Psychophysics, 2-8
correct sampling for, 8
definition of, 2-5
statistics for, 6
use of, 2-5

Quality assurance
statistics and, 5-6, 399-401
Type I error and, 400
Type II error and, 400

r (*see* Correlation)
r^2 (*see* Coefficient of determination)
Random effects, 4, 6
Random numbers, table of, 473-474
Random sample, 3, 5, 7
Randomized complete-block ANOVA, 173
Range
definition and computation, 12-13
interquartile range, 13
Ranked data, 20
Rank sums test (*see* Wilcoxon-Mann-Whitney rank sums test)
Ratio scale, 21-22
Regression
assumptions for, 299-300
[Regression]
calculus for, 280-283, 295
computation for, 297
intercept, 280-283, 297
logic of, 295-297
slope, 280-283, 297
trend analysis, 300
use of, 279-280
worked example, 298-302
Related samples design, 73-76 (*see also* Repeated measures design)
ANOVA and, 366-370
carryover effects, 74
control group, 74
counterbalancing, 73-76
placebo, 75
plateau effects, 75
ranked analysis of variance and, 366-370
t test and, 111, 115-117
use of, 399
Repeated measures design (*see* Related samples design)
rho, ρ (*see* Spearman's ranked correlation)
R index, 389-397
analysis by Jonckheere test, 362
analysis by Wilcoxon-Mann-Whitney rank sums test, 270
by rating, 390-395
by ranking, 395-397
logic of, 390
statistical analysis, 393, 395
use of, 389-390

Sampling distributions
confidence intervals and, 385-387
definition, 52
normal distribution and, 52-55
Scale, types of, 19-22
Scaled data (*see* Scaling)

Scaling
ANOVA, use of, 152
category scales, 127-129, 366
end effects, 365-366
favored number effects, 366
graphic scales, 127-129
homoscedasticity, need for, 127-129
magnitude estimation, 127-129
nonparametric tests for analysis, 303, 365-366
normal distributions, need for, 127-129
parametric tests for analysis, 365-366
straight line (graphic) scales, 366
t tests, difficulties with analysis, 127-129
Screening (*see* Judges)
Sensory analysis (*see* Sensory Evaluation)
Sensory difference tests, 79-90
abuse of, 84-85
binomial statistics, application of, 79-84
duo-trio test, 80
paired comparison test, 80
philosophy, 89-90
relation between, 414
R index, 85-89, 389-397
triangle test, 80
use of, 89-90
Sensory evaluation, 2-8
ANOVA, application of, 203-204
definition of, 2-5
related- and unrelated-samples design, 366-370
sampling correctly, 8
statistics, use of, 6
use of, 2-5
Sensory preference tests
binomial statistics, 79-81, 83-84
comparison with sensory difference tests, 414
Sensory psychophysics (*see* Psychophysics)
Sheffé test
formula for, 156-157
tables for, 426-430
use of, 159-160
Sign test, 73-78
computation of, 76-78, 304-305
counterbalancing, 73
related-samples nature, 73-76
relation to binomial test, 73, 79
relation to Wilcoxon test, 305
as multiple comparisons for Friedman ranked analysis of variance, 334-335
Signal detection, 389-390
Skew, 11
Sorting, 40
Spearman's ranked correlation, 327-332 (*see also* Correlation)
assumptions for, 330
computation of, 327-329
logic of, 328, 331
relation to Pearson's product-moment correlation, 327
samples too large for tables, 329
significance, test of, 328-329
table for, 458
ties, 329-330
two-tailed nature, 328-329
Type I error and, 329
via *t* test, 329
worked example, 331-332
Spearman's ρ (*see* Spearman's ranked correlation)
Spread of scores, 12-17
SS (*see* ANOVA)
Standard deviation
computation of, 14-17
Jonckheere test and, 357, 359
normal distribution and, 46
Wilcoxon-Mann-Whitney rank sums test and, 324
worked example, 17-19
$\Sigma(X - \bar{X})^2 = \Sigma X^2 - \frac{\Sigma(X)^2}{N}$, 371-372

Standard error
confidence intervals and, 386-387
definition of, 53
t test and, 112
Statistic, definition of, 7
Statistical tables, 404-474
Statistics, use of, 6
Straight line scale (*see* Graphic scale)
Stratified sample, 5, 7
Studentized ranges, 431-433
Student's *t* test (*see t* test)
Sum of squares (*see* ANOVA)

t test
central limit theorem and, 385
confidence intervals and, 385-387
correlation, testing significance of, 293-294
definition of, 111
degrees of freedom, 113
F, relation to, 151, 179
independent-samples *t* test, unequal variance, 123-124
multiple *t* tests, misuse of, 135-136
normal distributions and, 124-127
null hypotheses and, 113, 114, 115
one sample *t* test, 111-115
relation to *z* test, 124-127
sampling distribution and, 124-127, 385-387
scaling data, analysis of, 127-129
Spearman's ranked correlation, testing significance of, 329
standard error and, 112, 386-387
Student, who was he?, 112
t distribution, 112
tables of significance, 416
[*t* test]
tables to assist computation of, 417-420
two-sample independent-samples test, 111, 117-121
two-sample related-samples test, 111, 115-117
worked examples, 114-117, 119-121, 129-134
Tables, 404-474
Tail (*see* One-tailed tests, Two-tailed tests)
Total variance (*see also* ANOVA)
definition of, 136-140
relation to other variances, 142-143, 172, 379-383
Training (*see* Judges)
Trend analysis, 300
Triangle test
abuse of, 84-85
definition of, 82-84
relation to other tests, 414
statistics for, 82-84
Trimodal, 11
Tukey HSD test
formula for, 157
tables for, 431-433
use of, 159-160
Two-tailed tests
ANOVA and, 150-151
correlation and, 288
definition of, 68-69
use of, 69-70
Type I error
definition of, 66-68
Friedman two-factor ranked analysis of variance and, 334
Kramer two-factor ranked analysis of variance and, 339
multiple comparisons and, 153
multiple *t* tests and, 135
power of tests and, 94-95
quality assurance and, 400
Spearman's ranked correlation coefficient, 329

Type II error
  definition of, 66-68
  power of tests and, 94-95
  quality assurance and, 400

Unrelated-samples design (*see* (Independent-samples design)

Variance (*see also* ANOVA)
  computation of, 14-17
  worked example, 17-19
  $\Sigma(X - \bar{X})^2 = \Sigma X^2 - \frac{\Sigma(X)^2}{N}$, 371-372

Wilcoxon-Mann-Whitney rank sums test, 315-327
  assumptions for, 320
  computation of, 316-319
  logic of, 315
  relation to Kruskal-Wallis one-factor analysis of variance, 341, 343-344
  relation to Mann-Whitney U test, 315
  samples too large for tables, 320
  significance, test of, 315, 320, 324-325
  standard deviation for, 324
  tables for, 454-457
  ties, 320, 321-325
  via *z* test, 320
  worked example, 325
Wilcoxon rank sums test (*see* Wilcoxon-Mann-Whitney rank sums test)
Wilcoxon signed rank test (*see* Wilcoxon test)
Wilcoxon test, 304-308
  assumptions for, 308
  computation of, 304-306
  logic of, 306
  one- or two-tailed test, 306
[Wilcoxon test]
  relation to sign test, 305
  sample too large for tables, 307
  significance, test of, 306-307
  tables for, 444-445
  ties, 306-307
  use of, 304
  via *z* test, 307
  worked example, 308-309
Within-treatments variance (*see also* ANOVA)
  definition of, 136-140
  relation to other variances, 142-143, 172, 379-383
Worked examples
  ANOVA
    one-factor completely randomized, 147-150, 166-169
    two-factor without interaction, repeated measures, 176-178, 180-182
    two-factor with interaction, 195-200, 204-209, 234-236
    three-factor, 226-234, 237-243
    five-factor, 243-246
    split-plot, 272-278
  binomial test, 85-89
  chi-square, 108-110
  combinations, 42-44
  correlation (Pearson's product moment), 287, 300-302
  Duncan's test, 165-166
  Friedman two-factor ranked analysis of variance, 336-337
  Jonckheere test, 362-365
  Kramer two-factor ranked analysis of variance, 340-341
  Kruskal-Wallis one-factor analysis of variance, 345-346
  LSD test, 162-164, 168-169, 208-209

[Worked examples]
Mann-Whitney U test, 314
Page test, 349-350
permutations, 42-44
probability, 35-40
regression, 298-299, 300-302
standard deviation, 17-19
t test
one-sample, 114-115, 129-131
two-sample, related-samples, 116-117, 131-132
two-sample, independent-samples, 119-121, 132-134
variance, 17-19
Wilcoxon-Mann-Whitney rank sums test, 325-327
Wilcoxon test, 308-309
z tests, 55-56

z scores, 47-51
computing probabilities for, 48-51
[*z* scores]
definition of, 47-48
significance, test of, 48-51
worked examples, 55-56
z test, 47-51
computing probabilities for, 48-51
definition of, 47-48
relation to *t* test, 124-127
significance, test of, 48-51
tables for, 404-405
testing significance for binomial test, 71-72
testing significance for Jonckheere test, 356-357, 359-362
testing significance for Mann-Whitney U test, 313
testing significance for Wilcoxon-Mann-Whitney rank sums test, 320
testing significance for Wilcoxon test, 307
worked examples, 55-56

Printed in the USA/Agawam, MA
August 20, 2014

595165.005